Frontiers in Earth Sciences

More information about this series at http://www.springer.com/series/7066

Sérgio Henrique Faria
Sepp Kipfstuhl · Anja Lambrecht

The EPICA-DML Deep Ice Core

A Visual Record

Springer

Sérgio Henrique Faria
Basque Centre for Climate Change (BC3)
Leioa
Spain

and

IKERBASQUE, the Basque Foundation for Science
Bilbao
Spain

Sepp Kipfstuhl
Alfred Wegener Institute (AWI)
Bremerhaven
Germany

Anja Lambrecht
Essen
Germany

ISSN 1863-4621 ISSN 1863-463X (electronic)
Frontiers in Earth Sciences
ISBN 978-3-662-55306-0 ISBN 978-3-662-55308-4 (eBook)
https://doi.org/10.1007/978-3-662-55308-4

Library of Congress Control Number: 2017943245

Printed on acid-free paper

This Springer imprint is published by Springer Nature
The registered company is Springer-Verlag GmbH Germany
The registered company address is: Heidelberger Platz 3, 14197 Berlin, Germany

Dedicated to . . .
Nobuhiko “Nobby” Azuma.
Thanks, Nobby, for being our inspiring Sensei.

Kolumban “Koli” Hutter.
Thanks, Koli, for believing on this book project.

Heinrich “Heinz” Miller.
Thanks, Heinz, for making EPICA-DML come true.

Preface

Compared with the hardships encountered by the pioneering polar explorers of the nineteenth and early twentieth centuries, contemporary expeditions to polar regions are considerably more comfortable and safe. Despite the increasing human acquaintance with the frozen worlds, deep-drilling campaigns in Antarctica and Greenland are still today far from being ordinary events. Such drilling campaigns have become indispensable for our understanding of the dynamics of polar ice sheets and the climate records stored within the ice. However, every drilling site has its peculiarities and demands specific solutions. Ice drilling technology is still in a maturing stage, and so are also the scientific methods applied to analyse ice cores on the field. Both activities (drilling and scientific research) repeatedly impose new technical and logistic challenges.

The purpose of this book is to offer an informal, descriptive introduction to the main product of deep ice-core drilling: the ice core itself. We accomplish this in a pragmatic manner by considering a particular deep-drilling project, the *European Project for Ice Coring in Antarctica* (EPICA) at *Dronning Maud Land* (DML), which retrieved the EPICA-DML (a.k.a. EDML) Deep Ice Core, a 2774 m deep ice core from an ice ridge in the Atlantic Sector of Antarctica. We review the history and main features of this project, and then embark on a visual analysis of the multiscale structure of the Antarctic Ice Sheet at the EDML site.

Seeing that there is still no general agreement about the terminology to describe the physical properties of polar ice, we introduce in Chap. 1 a glossary of useful terms. Chapters 2 and 3 of Part I summarize the history of EPICA and the main characteristics of the drilling site at Dronning Maud Land. Scientific instruments and techniques used to produce the images presented in this book are described in Chap. 4 together with a description of the physical properties of the EDML Deep Ice Core. Finally, Chap. 5 explains how to interpret the complete collection of visual stratigraphy images of the EDML Deep Ice Core, which are presented in Part II.

It is evident from its structure and style that this text does not emphasize technicalities. Rather, following the spur to multidisciplinarity (Faria 2009), this book aims at presenting a tour d'horizon to a larger audience, which includes engineers, geoscientists, physicists, mathematicians, and any other non-specialist in glaciology interested in deep ice cores. We hope that this broad general approach may help the reader to develop a basic intuition about some complex concepts of ice structure, ice sheet flow, paleoclimate records, and climate change, which shall eventually be integrated into a single holistic view.

There is excellent specialized literature dealing in depth with the quantitative analysis of the physical properties of polar ice cores, and bibliographic information about it is provided in suitable places through the text. Therefore, even though this book is primarily intended for non-specialists, we hope that it may be valuable also for glaciologists interested in more specialized texts and their relation to the "big picture".

Bilbao, Spain
Bremerhaven, Germany
Essen, Germany
November 2016

Sérgio Henrique Faria
Sepp Kipfstuhl
Anja Lambrecht

Acknowledgements

In this limited space, it is impossible to individually thank all the people that have made this work possible. We should, of course, start with those whom we dedicate this work to:

Heinz Miller was Chair of EPICA in the period 2001–2006. Until 2012, he was also Head of Glaciology at the Alfred Wegener Institute (AWI Bremerhaven, Germany), which was the institution in charge of the logistics at EDML. As such, Heinz was a major driving force for the materialization of the EPICA-DML project. A gifted leader and good friend, he has also revised an early draft of this book.

Koli Hutter was the editor of this book from its onset until early 2015. He advocated its importance to the publisher and coordinated the peer reviews of the manuscript, besides providing his own suggestions to the text. Through his distinctive editorial style, he used to exert a friendly pressure to finish this book, while simultaneously showing a paternal patience with the repeated delays caused by the hurdles faced by two young researchers moving about in pursuit of tenure and a senior researcher frequently busy with the preparations of countless expeditions to Greenland and Antarctica.

Nobby Azuma was officially not part of the EPICA community, but he has nevertheless played a decisive role in much of the research related to the physical properties of the EDML Deep Ice Core. Distinguished descendant from one of the noblest and most brilliant academic lineages of glaciology (the Nakaya–Higashi line), he never hesitated to share his profound knowledge of ice mechanics with others. Thanks to his input, the EDML Deep Ice Core has become a milestone in the physics of polar ice. In fact, many concepts, techniques and results presented in this book are directly or indirectly related to his views.

This work could not come true without the contributions of all members of the EPICA-DML project, including its administrative staff, technicians, drilling, logistics and science teams. We are deeply grateful to them all, and in particular to Cord Drücker, Hans Oerter, Daniel Steinhage, Anders Svensson, and Ilka Weikusat, who offered us not only illustrations, data, and other useful information presented in this book, but also friendship and encouraging words. We are also deeply grateful to an anonymous reviewer and to Thomas Stocker, who was so kind as to produce a thorough review of the whole manuscript. Evidently, any errors or misconceptions that remain are our own.

Finally, the editorial effort to produce a book like this one should not be underestimated. Here, we kindly acknowledge the excellent assistance of the whole Springer team, in particular Annett Büttner (Publishing Editor, Earth Sciences and Geography), Helen Rachner (Publishing Assistant), and Almas Schimmel (Production Editor), all based at Springer Heidelberg, Germany.

We issue here the official EPICA acknowledgement:

> This work is a contribution to the European Project for Ice Coring in Antarctica (EPICA), a joint European Science Foundation/European Commission scientific programme, funded by the EU and by national contributions from Belgium, Denmark, France, Germany, Italy, the Netherlands, Norway, Sweden, Switzerland, and the UK. The main logistic support was provided by IPEV and PNRA (at Dome C) and AWI (at Dronning Maud Land). This is EPICA publication no. 307.

At last, we close these preliminaries with the personal acknowledgements of each author:

Sérgio Henrique Faria

Besides all names, groups, and institutions already mentioned, I would like to thank Kumiko Goto-Azuma for inspiring discussions about ice-core climate records, Gonzalo Morcillo for sharing his scientific and artistic views of image analysis, ice-core line-scanning and visual stratigraphy, as well all participants of the Micro-DICE/ESF Exploratory Workshop "Towards a Microstructure-Based Flow Law of Ice" (Nuuk, Greenland 2011) for discussions about stratigraphy, folding, and ice-sheet flow. Special thanks go to the Basque Centre for Climate Change (BC3) and Ikerbasque, the Basque Foundation of Science, for providing assistance and a favorable environment for writing this book. Financial support from the Ramón y Cajal Grant RYC-2012-12167 of the Spanish Ministry of Economy, Industry and Competitiveness is kindly acknowledged. Travel support for meeting co-authors was partially granted by Micro-DICE (Micro-Dynamics of Ice), a Research Networking Programme of the European Science Foundation (ESF).

Anja Lambrecht

My thanks go to all people who gave me the opportunity to take part not only in the EPICA-DML project at Kohnen Station, but also previously in the 21st Antarctic Overwintering at Neumayer Station II, where my fascination for snow and ice began. I cannot mention all names here, because they are too many, but all of them were important. I had the big chance to hold in my hands one of the most fascinating materials in the world: polar ice. I could recognize different structures in the snow surface every morning while shovelling snow into the melter at Neumayer Station II, and I was always inspired by a new decoration of fantastic ice crystals in the cave of the seismic and magnetic observatory. Cold fingers were a constant companion in Antarctica, but I will not miss this time. The readers of this book hopefully can discover what a fascinating material polar ice is, without getting them. Just as the pictures remind us of the wonderful phenomena the Earth has reserved for us. There is still a lot to investigate, and even if you have no deep knowledge of the physics of ice, I hope you may appreciate the artistry of nature in these pictures. Lastly, my special thanks belong to my sister Astrid Lambrecht, whom I can rely on wherever I am.

Sepp Kipfstuhl

I guess we have already said everything that had to be said, so thank you all again, and let us start the next ice-core book!

Contributions Statement: The text was written by SHF, including images and figures provided by the sources explicitly cited. The line-scan images in Chap. 5 were originally produced by SK and AL, and have subsequently been edited, formatted, modified, and optimized for this book by SHF. All authors contributed ideas and suggestions to the manuscript.

Contents

1 Getting Started

Today, the appeal to the authority of experts is sometimes excused by the immensity of our specialized knowledge. [...] But in my view, the appeal to the authority of experts should be neither excused nor defended. It should, on the contrary, be recognized for what it is —an intellectual fashion— and it should be attacked by a frank acknowledgement of how little we know, and how much that little is due to people who have worked in many fields at the same time.

Sir Karl R. Popper (1994), p. x

1.1 Motivation

Nature does not have a preferred size scale to reveal its wonders. From the synchronized dynamics of the solar system to the symmetry of a crystal lattice, every scale discloses new manifestations of the natural order. In the past, scientists had already paid tribute to Nature's majesty on all these size scales. Scholars like Johannes Kepler (1571–1630) or René Descartes (1596–1650) could discourse with the same geniality and fascination on subjects as distinct as the motion of celestial bodies (Descartes 1644; Kepler 1619) and the crystalline structure of snow (Descartes 1637; Kepler 1611). Over the centuries, however, the human compulsion for categorization has replaced polymathy by scientific specialization, which greatly facilitates the engineering of knowledge into practical applications, but restricts our perception of the "whole", that is, of the multiscale character of natural processes.

Fortunately, some branches of science have to some extent resisted the rush of specialization, for distinct reasons. In the case of glaciology, which is the subject of this book, this resistance arises from the simple fact that glaciology is, by definition, a multidisciplinary field: it combines geosciences with continuum mechanics, thermodynamics, solid state physics, engineering and materials science, climatology, biology, environmental sciences, and many other specialities. As such, the glaciological community encompasses a great diversity of scientists, from mathematicians to ecologists, and other professionals sharing interest in natural ice. Paradoxically, it is precisely this multidisciplinary character of glaciology that constitutes its greatest strength, but also its main weakness: on the one hand it promotes collaborative research between diverse disciplines, but on the other hand it complicates the sharing of ideas and the precise definition of concepts, which are often differently interpreted within distinct specialities.

This book offers the opportunity to appreciate the multidisciplinary nature of polar glaciology in a very simple, visual manner. Its central topic, the local stratigraphy of the Antarctic ice sheet, is presented in a series of high-resolution images that serve as starting point for investigating the flow of polar ice, its chemical and physical structure, as well as the integrity of its paleoclimate records. Relevant aspects of the physics of polar ice are reviewed, including a glossary (Sect. 1.2), which aims at making clear and accurate the concepts discussed in this work, in a manner that both, trained and untrained readers, may equally comprehend and enjoy its contents.

1.2 Glossary

The choice of a consistent and unambiguous vocabulary for discussing science in a multidisciplinary environment always imposes a dilemma. A typical example is the disparity in the vocabulary of geologists and materials scientists when discussing the same microstructural features of polycrystals. In an attempt to be systematic without being too discrepant with the existing literature, we list below some important definitions and acronyms used in this book. This glossary is based on the terminology recently proposed by Faria et al. (2009, 2014b) and Cogley et al. (2011).

Air hydrate: See *clathrate hydrate*.
BHT zone: See *bubble–hydrate transition zone*.

S.H. Faria et al., *The EPICA-DML Deep Ice Core*, Frontiers in Earth Sciences,
https://doi.org/10.1007/978-3-662-55308-4_1

Brittle zone: Deep ice-core depth interval characterized by exceptionally brittle ice, which can be processed (viz. sawed) only after weeks or months of post-drilling relaxation. Experience shows that the brittle zone coincides roughly with the upper half of the *bubble–hydrate transition zone* (BHT zone, cf. id.), suggesting that the characteristic brittleness of this ice might be related to internal stresses generated by the enclosed pressure in air bubbles and their transformation into *clathrate hydrates* (cf. id.). The pressure/temperature shock during core retrieval induces the formation of post-drilling micro-cracks near the bubbles.

Bubble–hydrate transition (BHT) zone: Depth interval in an ice sheet, where air bubbles coexist with clathrate hydrates. It lies below the critical depth at which the overburden pressure of ice equals the clathrate-hydrate dissociation pressure (cf. *clathrate hydrates*).

Bubbly ice: Impermeable ice containing air bubbles. In glaciers and ice sheets it occurs in the depth zone between the firn–ice transition depth (cf. *firn*) and the bottom of the *bubble–hydrate transition zone* (cf. id.).

CFA: Continuous Flow Analysis.

Clathrate hydrate: Crystalline compound containing guest molecules enclosed in cage-like structures made up of hydrogen-bonded water molecules. Clathrate hydrates (CHs) are thermodynamically stable only at low temperatures and high pressures (viz. below a pressure-dependent *CH dissociation temperature* or above a temperature-dependent *CH dissociation pressure*). They are not chemical compounds, because the guest molecules are not bonded to the water molecules, but rather just physically trapped in the cages. When the guest molecules form gas under standard conditions, such compounds are also named *gas hydrates*. In particular, *air hydrates* are formed by atmospheric gases (mainly O_2 and N_2). In natural ice, air hydrates (AHs) form from air bubbles in the *bubble–hydrate transition zone* (BHT zone, cf. id.).

Cloudy band: Ice stratum with turbid appearance due to a high concentration of microinclusions. Experience shows a strong correlation between high impurity concentration and small grain sizes in cloudy-band ice.

Creep: Ductile plastic deformation regime of solids at relatively high temperatures (typically above ca. 30% of the melting point), which is strongly dependent on the *strain rate*. For polycrystals, the most common creep mechanisms are *diffusion creep* and *dislocation creep* (cf. id.).

Crystallite: See *grain*.

Crystallization: Crystal formation through freezing from a liquid, precipitation from a solution, or deposition from a gas. It should not be confounded with *recrystallization* (cf. id.). Crystallization consists of two major stages: *crystal nucleation* and *crystal growth* (cf. *nucleation*). These two processes should not be confused with their recrystallization counterparts: *grain nucleation* and *grain growth* (cf. id.).

Deep ice: Ice from deep depths (usually below several hundred metres) of a glacier or ice sheet (cf. *shallow ice*).

Deformation-related structure: Any structural feature in the ice that serves as evidence for its past or current deformation, e.g. dislocation walls, subgrain boundaries, slip bands, stratigraphic folds, etc.

DEP: Dielectric Profiling.

Diffusion creep: See *diffusional flow*.

Diffusional flow: Ductile deformation regime characterized by diffusional flux of matter through the body. In polycrystals, diffusional flow is often called *diffusion creep*, as it describes the creep by mass transport through or around the grains. The former is named *lattice diffusion creep* (or *Nabarro–Herring creep*), while the latter is called *grain-boundary diffusion creep* (or *Coble creep*).

Dislocation creep: Creep of (poly-)crystalline matter by the motion of dislocations through the crystalline lattice.

Dislocation wall: Lattice defect consisting of dislocations arranged in a two dimensional framework; the precursor of a subgrain boundary (cf. *subgrain*).

DML: Dronning Maud Land, Antarctica.

Drift ice: Floating ice that drifts with the wind or ocean currents, e.g. pack ice or icebergs (cf. *land ice* and *sea ice*).

Dynamic grain growth (DGG): Class of phenomenological recrystallization processes of grain coarsening in polycrystals *during deformation*. Several recovery and recrystallization processes may be simultaneously active during DGG, all competing for the minimization of both, the stored strain energy and the grain-boundary energy.
The essential feature of DGG (in comparison to other recrystallization processes) is the monotonic increase of the mean grain size with time. Owing to its dynamic nature, however, the diversified kinetics of DGG can generally not be compared with the simple kinetics predicted for *normal grain growth* (*NGG*, cf. id.).

Dynamic recrystallization: See *recrystallization*.

ECM: Electrical Conductivity Measurement.

EDC: EPICA Dome C (a deep-drilling site in Antarctica).

EDML: EPICA DML (a deep-drilling site in Antarctica).

Elementary structural process: A fundamental operation of structural change. In the context of recovery and recrystallization, it includes e.g. grain boundary migration and subgrain rotation. Several elementary processes may combine to produce different types of *phenomenological structural processes*[1] (cf. id.).

[1]Recovery and recrystallization are complex physical phenomena that are better understood if decomposed in a hierarchy of structural processes or mechanisms, here qualified as "elementary" and "phenomenological".

EPICA: European Project for Ice Coring in Antarctica.

Fabric: See *Lattice Preferred Orientation (LPO)*.

Firn: Sintered snow that has outlasted at least one ablation season (viz. the seasonal period of mass loss, typically during the summer). Firn is a porous material in a metamorphic stage intermediate between snow and *bubbly ice* (cf. id.), with a (partially) interconnected pore space that makes it permeable. Typical firn mass densities range from 350–400 kg/m^3 beneath the snow cover to 820–840 kg/m^3 near the firn–ice transition depth.

GBS: See *grain boundary sliding*.

GISP2: Greenland Ice Sheet Project 2 (a deep-drilling site in Greenland).

Glacier: Land-ice mass whose flow and shape are constrained by the bed topography (as distinguished from *ice caps* and *ice sheets*, cf. id.). Glaciers occur naturally in many forms. In particular, if it is nourished by an ice cap or ice sheet it is called an *outlet glacier*.

Grain: Connected volume in a polycrystalline solid composed of an uninterrupted (although possibly imperfect) crystalline lattice and bounded to other grains by *grain boundaries*. Also loosely called *crystallite*. It should be noticed the difference between grains of polycrystalline solids (e.g. ice) and the loose particles of crystalline granular media (e.g. snow).

Grain Boundary Sliding (GBS): Relative sliding of a pair of grains by shear movement at their common interface. The shear may be completely confined to the boundary, or occur within a zone immediately adjacent to it.

Grain growth: Class of phenomenological recrystallization processes of grain coarsening in polycrystals. If grain coarsening occurs *during deformation*, it is called *dynamic grain growth* (DGG; cf. id.). Otherwise, it is called *static grain growth*, which includes *normal grain growth* (NGG; cf. id.) as a special case. Grain growth should not be confused with crystal growth (cf. *crystallization*).

Grain nucleation: Class of phenomenological recrystallization processes involving the formation of new *nuclei* (viz. tiny new grains free of strain and deformation-related structures). Two types of grain nucleation mechanisms can generally be identified, here called "pseudo-" and "classical nucleation". During *classical nucleation*, a cluster of atoms/molecules spontaneously form a new embryo under the action of high internal stresses and thermally-activated fluctuations, which subsequently grows to become a new nucleus. Despite recurrent consideration of this mechanism in the glaciological literature (due, in part, to its analogy to crystal nucleation, cf. *crystallization*), it is currently acknowledged that classical nucleation is not relevant for polar ice.[2] During *pseudo-nucleation* a special combination of elementary recrystallization processes (e.g. SIBM, subgrain rotation and growth) takes place *within a small crystalline region* with high stored strain energy, giving rise to a little strain-free new grain called *pseudo-nucleus*.[3] Pseudo-nucleation most likely happens at grain boundaries and other zones of high stored strain energy, e.g. at air bubbles and solid inclusions.

Grain stereology: Spatial arrangement of grains in a polycrystal, including their sizes and shapes (cf. *orientation stereology* and *lattice preferred orientation*).

Grain subdivision: Phenomenological recovery process of formation of new *subgrain boundaries*. It involves the progressive rotation of certain portions of the grain, called *subgrains* (cf. id.), as well as the strengthening of dislocation walls through dislocation rearrangement and migration in regions with strong lattice curvature. If the misorientation across the new subgrain boundary increases with time, grain subdivision may give rise to *rotation recrystallization* (cf. id.).

GRIP: Greenland Ice-core Project (a deep-drilling site in Greenland).

Ice cap: Dome-shaped land-ice mass with predominant radial flow unconstrained by the underlying bed topography (as distinguished from a *glacier*, cf. id.). As such, ice caps show some similarities to *ice sheets* (cf. id.), but on a smaller scale that covers an area not greater than 50 000 km^2.

Ice sheet: Land-ice mass of continental size, which covers an area greater than 50 000 km^2 (cf. *ice cap*), and is sufficiently thick to have its flow unconstrained by the underlying bed topography (cf. *glacier*). Despite its thickness, its width-to-height aspect ratio is still generally larger than 10^2. A typical ice sheet consists of a central ice plateau, comprising one or several *domes* and *ridges*, and a marginal region characterized by *outlet glaciers*, *ice streams* and *ice shelves*. Ice sheets flow outward from the central

(Footnote 1 continued)
A somewhat similar hierarchical scheme for recrystallization has formerly been proposed by Drury and Urai (1990), but with the expressions "elementary/phenomenological process" replaced respectively by "basic process" and "mechanism". We favour here the qualifiers "elementary/phenomenological" (against the "process/mechanism" scheme) because these qualifiers facilitate the visualization of the hierarchy and leave us free to use the terms "process" and "mechanism" as synonyms.

[2]Calculations show (Cahn 1970; Humphreys and Hatherly 2004) that classical nucleation recrystallization is extremely unlikely to occur in single-phase polycrystals, owing to the high energies required for the creation and growth of classical nuclei, except if strong chemical driving forces are present, which is clearly not the case for polar ice.

[3]The prefix "pseudo-" is used here to emphasize that this nucleus is usually much larger than the nucleus formed by classical nucleation, but still small enough to be strain-free. It should be noticed that the distinction between pseudo-nucleation and a combination of SIBM-O with rotation recrystallization is basically a matter of scale: in the latter case the new crystallite is large enough to inherit a considerable amount of internal structures from the parent grain.

ice plateau with a small average surface slope. The margins usually slope more steeply, and most ice is discharged through ice streams and fast-flowing outlet glaciers, in some cases into the sea or into ice shelves. An ice sheet without ice shelves and other floating extensions is called an *inland ice sheet*[4] (cf. *inland ice*) and consists exclusively of meteoric ice. There are only two large ice sheets in the modern world, one in Greenland and another in Antarctica, the latter frequently subdivided into two nearly independent portions, called East and West Antarctic Ice Sheets, which are separated by the Transantarctic Mountains. Several other large ice sheets existed in the past, during glacial periods.

Ice shelf: A thick and extensive slab of meteoric and accreted marine ice, which is floating on the ocean while being attached to, and nourished by, an inland ice sheet. Nearly all ice shelves are found today in Antarctica, often filling embayments in the coastline.

Ice stream: A region in an ice sheet that flows much faster than the surrounding ice and shows no evidence of lateral rock boundaries. If such rock boundaries become evident, the ice stream is called a fast-flowing *outlet glacier* (cf. id.).

Iceberg: Loosely drifting chunk of ice calved from an ice shelf, tongue, or wall. Small icebergs are usually called *growlers* (characteristic length up to 10 m) or *bergy bits* (between 10 and 20 m in size). In Antarctica most icebergs are tabular in shape (length–height ratio > 5) with lengths in the range 500–2000 m and length–width ratios between 1 and 2. Arctic icebergs are generally much smaller than their Antarctic counterparts. Apart from their coexistence, icebergs and *sea ice* (cf. id.) are distinct things and should not be confounded with each other.

Inclusion: Localized deposit of undissolved chemical impurities observed in polar ice, like air bubbles, clathrate hydrates, or brine pockets. Inclusions no larger than a few micrometres are often called *microinclusions* (e.g. dust particles, microbubbles, etc.).

Inland ice: Ice underlain by land (cf. *land ice*).

Isotropic ice: In full *isotropic polycrystalline ice*. Ice with isotropic and homogeneous *orientation stereology* (cf. id.). In other words, homogeneous polycrystalline ice with no *LPO* (cf. id.).

Kink band: Crystalline zone limited by two nearly parallel *tilt boundaries* (cf. id.) of opposite sense. It is produced by the gradual lattice rotation of the interior zone with respect to the lattice of the two adjacent regions.

Land ice: Ice underlain by land or attached to the shore, thus including inland ice, ice shelves, ice tongues, landfast sea ice, etc., as distinguished from *drift ice* (cf. id.).

Lattice defect: Any imperfection of the crystalline lattice. Lattice defects are usually produced/eliminated during crystallization, deformation, recovery and recrystallization, and depending on their dimensions they are classified as *point defects* (vacancies, interstitials, etc.), *line defects* (dislocations, disclinations, etc.), *plane defects* (grain boundaries, stacking faults, phase interfaces, etc.), and *bulk defects* (fractures, pores, inclusions, etc.).

Lattice Preferred Orientation (LPO): Statistically preferred orientation of the crystalline lattices of a population of grains. In plural (LPOs): the directional pattern of lattice orientations in a polycrystalline region (cf. *orientation stereology*). In the glaciological literature, LPOs are often called *fabric* (Paterson 1994), while in materials science they are frequently termed *texture* (Humphreys and Hatherly 2004). In particular, a polycrystalline region with a random distribution of lattice orientations is said to have no LPO (viz. texture-free, random fabric).

Lattice structure: Arrangement of atoms in the unit cell of the crystalline lattice.

Logging: In full *ice-core logging*. Systematic preparation procedure for the future processing and archiving of ice-core pieces. It usually consists of a series of operations, including: cleaning; fitting the core and its immediate precursor together through the matching of core breaks; measuring the core length (which gives rise to the *logged depth*, cf. id.); and labelling (sometimes also cutting and packing) the core at regular length intervals called *bags*. Missing ice, though rare, must be handled consistently at this stage. Each of such bags eventually contains one or more ice-core pieces, which are ready for transportation, storage and processing, and are uniquely identified by an exclusive *bag number*.[5]

Logged depth: Also called *core-length depth*. Depth from the surface of a glacier or ice sheet, measured in terms of the length of the logged ice core (cf. *logging*). The logged depth is identical to neither the real (vertical) depth nor the borehole depth, because the borehole is never perfectly straight and vertical, and its profile deforms over time with the ice flow.[6]

LPO: See *lattice preferred orientation*.

Marine ice: Ice formed by the freezing of seawater at the base of an ice shelf (cf. *meteoric ice*).

[4] According to this definition, which derives from Kirchner and Faria (2009), the term "ice sheet" generally refers to "inland ice sheet *and* related ice shelves".

[5] A detailed description of a logging procedure, in the case of the North-GRIP Deep Ice Core, is presented by Hvidberg et al. (2002).

[6] More information on borehole deformation is available in several studies (e.g. Dahl-Jensen and Gundestrup 1987; Etheridge 1989; Garfield and Ueda 1976; Gundestrup and Hansen 1984; Ryser et al. 2014). For the particular case of the EDML borehole, see Weikusat et al. (2017).

Megadune: Very large, antidunal formation that is part of a wavy structure with a few kilometres in wavelength and several metres in amplitude. Such vast structures are visible from air- and space-borne platforms as alternating bands formed transversally to persistent katabatic winds (cf. *sastruga*).

Meteoric ice: Freshwater ice formed by the sintering of ancient snowfalls (cf. *marine ice*).

Microbubble: Air bubble not larger than a critical diameter (e.g. 100 μm in shallow ice). The critical microbubble diameter is usually defined by the typical bimodal size distribution of air bubbles in natural ice, and it reduces with depth owing to the increasing overburden pressure and the transfer of gas to large bubbles via Ostwald ripening. *Primary microbubbles* are those formed together with normal bubbles in the snow and firn layers. In contrast, microbubbles created by post-drilling relaxation of the ice core are called *secondary microbubbles*. See also *inclusion*.

Microinclusion: See *inclusion*.

Microshear: Strong, localized shear across a grain that experiences a highly inhomogeneous shear deformation. It culminates with the formation of a new, flat subgrain boundary parallel to the shear plane, called *microshear boundary* (cf. *slip bands*). Microshear is generally related to *grain boundary sliding* (cf. id.).

Microstructure: Collection of all microscopic structural features in a polycrystal, including its lattice defects, inclusions, and *orientation stereology* (cf. id.).

Migration recrystallization: In full *strain-induced migration recrystallization*. Class of phenomenological recrystallization processes based on the elementary *SIBM* mechanism (cf. id.). If *nucleation* (cf. id.) is involved in the process, we may call it *nucleated migration recrystallization* (SIBM-N), where the suffix "-N" stands for "new grain". Otherwise, i.e. if the migration of boundaries occurs without formation of new grains, we may call it *ordinary migration recrystallization* (SIBM-O), where the suffix "-O" stands for "old grain".[7]

Multiscale structure: The collection of all structural features in a material observed on multiple size scales. In ice sheets such structural features occur on the nano- and microscale (e.g. dislocations, air hydrates), on the mesoscale (e.g. cloudy bands) and on large scales (megadunes, subglacial lakes, etc.). Some of them may appear also on a range of scales (e.g. stratigraphic folds). Most of them are evolving structures that interact with each other as well as with the ice-sheet flow and the environment via the *SFEI* (cf. id.).

NBSAE: Norwegian–British–Swedish Antarctic Expedition.

Normal grain growth (NGG): Phenomenological recrystallization process of grain coarsening in polycrystals, resulting from "the interaction between the topological requirements of space-filling and the geometrical needs of (grain-boundary) surface-tension equilibrium" (quoted from Smith, 1952). By definition, grain coarsening during NGG is *statistically uniform* and *self-similar*, grain-boundary migration is *exclusively* driven by minimization of the grain-boundary area (and associated free energy), and the grain stereology is close to a configuration of "surface-tension equilibrium" (so-called "foam-like structure"). Owing to these essential features, NGG is generally regarded as a static recrystallization process (cf. *recrystallization*) taking place before/after deformation (cf. *dynamic grain growth*). Mathematical and physical arguments strongly suggest that the kinetics of NGG is parabolic with respect to the mean grain radius.[8]

NorthGRIP: North-Greenland Ice-Core Project, also abbreviated as *NGRIP* (a deep-drilling site in Greenland).

Nucleation: Formation of minute bits of crystalline matter called *nuclei*, which occur as tiny new crystals in the case of *crystallization* (cf. id.), or tiny new grains in the case of *recrystallization* (cf. id.). In this work, crystal nucleation is not considered, so that the term "nucleation" is generally used for brevity as a contraction of *grain nucleation* (cf. id.), unless explicitly mentioned otherwise.

Orientation stereology: Spatial arrangement of lattice orientations in a polycrystal, i.e. the combination of *grain stereology* and *LPO*.

Outlet glacier: See *glacier*.

Pack ice: See *sea ice*.

Phenomenological structural process: Any combination of elementary structural processes that gives rise to general changes in the structure of the polycrystal (cf. *elementary structural process*). Examples of phenomenological processes are nucleation and grain subdivision.

[7] The definition adopted here is based on the concept of "grain-boundary migration recrystallization" originally described by Beck and Sperry (1950). Notice that this definition is identical neither to that used by Poirier (1985) nor by Humphreys and Hatherly (2004), and it is also quite distinct from some loose connotations invoked in the glaciological literature. The terms SIBM-N and SIBM-O are not standard in the literature, but they are nevertheless adopted here because they describe quite precisely the kind of information obtained from microscopic analyses of ice-core sections. There is unfortunately no one-to-one relation between SIBM-N/SIBM-O and the expressions "multiple/single subgrain SIBM" used e.g. by Humphreys and Hatherly (2004).

[8] As explained by Smith (1952), the interest in NGG comes from the fact that its kinetics depends solely on the properties of the migrating boundaries and is otherwise independent of the medium or its deformation history. This means that the theory underlying the NGG kinetics is not restricted to polycrystals: similar coarsening phenomena are also observed in foams, some tissues, and many other cellular media.

Plate-like inclusion (PLI): Microscopic, flat cavity with hexagonal symmetry lying on the basal plane of its hosting ice crystallite, i.e. a thin negative crystal. With basal sizes up to a few hundred micrometres and thicknesses not larger than several micrometres (typical aspect ratio > 20), PLIs are usually filled with air, occur in a wide range of temperatures, and are supposed to be produced by stress relaxation within the ice. Owing to their thinness, PLIs often appear translucent in transmitted light microscopy (resembling air hydrates), although some may appear dark depending on their thickness and inclination. PLIs should not be confused with *Tyndall figures* (cf. id.).

Polygonization: Special type of recovery mechanism for the formation of *tilt boundaries*. It is a particular case of *grain subdivision* (cf. id.), by restricting it to tilting (bending) of crystallographic planes. In ice, polygonization is often used in reference to the bending of basal planes.

Proxy: Chemical or physical property that may not in itself be particularly relevant, but that serves as a surrogate for a relevant but unobservable or immeasurable property or condition. In particular, *climate proxies* (e.g. electric conductivity, air content, stable isotope concentration, etc.) are used as indirect records of past climate conditions experienced by the ice body (e.g. atmospheric composition, insolation, past temperature, etc.).

Pseudo-nucleus: See *grain nucleation*.

Recovery: Release of the stored strain energy by any thermomechanical process of microstructural change other than recrystallization. The qualifiers *dynamic* and *static* denote recovery phenomena occurring *during* and *prior/after* deformation, respectively. Frequently (especially under dynamic conditions), recovery and recrystallization coexist and may even be complementary (e.g. during rotation recrystallization), so that the distinction between them is sometimes very difficult.

Recrystallization: Any reorientation (and resulting rearrangement) of the crystalline lattice caused by grain boundary migration and/or formation of new grain boundaries, therefore including SIBM, RRX, DGG and NGG[9] (cf. *recovery*). The qualifiers *dynamic* and *static* denote recrystallization phenomena occurring *during* and *prior/after* deformation, respectively. Further classification schemes often invoked in the literature include the qualifiers *continuous/discontinuous* and *continual/discontinual*, used to specify the spatial homogeneity and temporal continuity of the recrystallization process, respectively. These classifications are, however, not always unique and are therefore of limited use.

Rotation recrystallization (RRX): Phenomenological recrystallization process responsible for the formation of new *grain boundaries*. It proceeds from the mechanism of *grain subdivision*, and as such it involves the progressive rotation of subgrains as well as the migration of subgrain boundaries through regions with lattice curvature. Notice that this recrystallization process does not require significant migration of pre-existing grain boundaries, in contrast to migration recrystallization.

Sastruga: Any of a series of low, irregular ridges of hard snow alternating with wind-eroded furrows parallel to the wind direction. Often mentioned in plural (*sastrugi*), they have typical lengths of metres and heights less than a metre (cf. *megadune*). The alternative spelling zastruga/zastrugi is also commonly used, alluding to the Russian origin of the word.

Sea ice: Any form of ice originated from the freezing of the sea surface. *Ice floes* are pieces of sea ice, which can be driven together into a nearly compact drifting mass called *pack ice*. If sea ice forms a sheet firmly attached to the shore, it is named *land-fast sea ice*. Apart from their coexistence, sea ice and *icebergs* (cf. id.) are distinct things and should not be confounded with each other.

SFEI: Structure–form–environment interplay. Generally, the environment influences the form; changes in the form affect the environment; environmental changes act on the structure; the structure modulates the form evolution; the evolving form alters the structure. In the case of glaciers and ice sheets it denotes the multiscale interactions between the deforming ice, its structure, and the environment.

Shallow ice: Ice from shallow depths (usually the upper few hundred metres) of a glacier or ice sheet (cf. *deep ice*).

SIBM: See *strain-induced boundary migration*.

SIBM-N/SIBM-O: See *migration recrystallization*.

Slip bands: Series of parallel layers with intense slip activity and high amount of intracrystalline lattice defects (especially dislocations). Slip bands in ice always appear in groups parallel to the basal planes and are indicative of a nearly homogeneous shear deformation of the respective grain via basal slip (cf. *microshear*).

Static recrystallization: See *recrystallization*.

Stored strain energy: Fraction of the mechanical energy expended during deformation that is stored in the material in diverse types of intracrystalline lattice defects, e.g. dislocations, stacking faults, subgrain boundaries, etc.

Strain-induced boundary migration (SIBM): Elementary recrystallization process of grain boundary motion driven by minimization of the stored strain energy. It involves the migration of a grain boundary towards a region of high stored strain energy. The migrating

[9] In contrast to the definition adopted here, some authors reserve the term "recrystallization" solely for those processes driven by the stored strain energy, therefore excluding NGG (see *normal grain growth*) from its definition. Other authors (especially in the older literature) loosely use "recrystallization" as a synonym for SIBM-N (cf. *migration recrystallization*).

boundary heals the highly energetic lattice defects in that region, therefore promoting a net reduction in the total stored strain energy of the polycrystal. See also *migration recrystallization*.

Subglacial structure: Any structural feature underneath the glacier or ice sheet, ranging from till and rocks to channels and lakes.

Subgrain: Sub-domain of a grain, delimited by a *subgrain boundary* and characterized by a lattice orientation that is similar, but not identical, to that of the rest of the grain. In ice, the lattice misorientation across a subgrain boundary is limited to a few degrees.

Texture: See *Lattice Preferred Orientation (LPO)*.

Tilt boundary: Special type of subgrain boundary in which the misorientation axis is tangential to the boundary interface.

Twist boundary: Special type of subgrain boundary in which the misorientation axis is orthogonal to the boundary interface.

Tyndall figure: Negative crystals in ice filled with liquid water and vapour, which are produced by internal melting (e.g. through exposition to radiation). Tyndall figures should not be confused with *plate-like inclusions* (cf. id.).

Wind crust: Thin layer of hard snow with high mass density, produced on the glacier or ice-sheet surface by strong, persistent winds combined with appropriate conditions of humidity, temperature, insolation, etc.

2 Antarctica and EPICA

The first day or so, we all pointed to our countries. The third or fourth day, we were pointing to our continents. By the fifth day, we were aware of only one Earth.
Prince Sultan bin Salman bin Abdulaziz al Saud, Saudi Arabian astronaut (Sagan 1997, p. 139)

2.1 The Antarctic Ice Sheet

In German, an ice sheet of the type covering Antarctica is usually called an *Eisschild* or *Eispanzer*, which can loosely be translated as an "ice shield" or "ice armour", respectively. While the English term "sheet" better describes the thin geometry of an ice sheet,[1] there is no doubt that the German names best convey the idea of a massive bulk of ice over Antarctica: with a volume of circa $25 \times 10^6\,\text{km}^3$ (including ice shelves) and an average thickness close to 2 km spread over an area larger than $13 \times 10^6\,\text{km}^2$, the Antarctic Ice Sheet contains about 23 million gigatonnes of ice, which represent roughly 67% of world's freshwater and a potential contribution to global sea-level rise of 58 m (Lemke et al. 2007; Vaughan et al. 2013). As if these figures were not impressive enough, the Antarctic ice plateau is also the coldest desert in the world, with annual mean temperatures far below freezing and precipitation values comparable with the driest deserts on the globe.

As mighty as it may seem, the Antarctic Ice Sheet is not insensitive to external agents. Similar to other polycrystalline solids, ice yields in a ductile manner at sufficiently high temperatures, slowly flowing under the burden of its own weight in a highly viscous fluid-like regime of plastic deformation known as *creep*. As reference, metals and ceramics usually start to creep at temperatures above circa 30 and 50% of their melting points, respectively (Frost and Ashby 1982; Phillips 2001). Thus, recalling that ice as cold as $-40\,°\text{C}$ (which is a common temperature for Antarctic ice) is already at 85% of its melting point, it should come as no surprise that glaciers and ice sheets creep: this is an expected and unavoidable phenomenon, simply caused by the fact that the Earth's surface is always considerably hot for ice.

This creep regime allows the ice in Antarctica to continually flow away over the centuries and millennia under the action of its own colossal weight, creeping towards the ocean at a pace from several to many metres per year (Bamber et al. 2000), until it eventually melts beneath ice shelves or calves into icebergs (Kirchner and Faria 2009; Kristensen 1983; Rignot et al. 2004). This ceaseless creep is mainly controlled by the ice temperature, and nourished by snowfall on its surface. As a consequence, climatic changes have a profound impact on the *mass balance* of the ice sheet, which is basically the difference between the gain through precipitation from the atmosphere (*snow accumulation*) and the loss through melting and calving to the ocean (*ice ablation*). If the balance is positive, the ice sheet grows, otherwise it dwindles.

The resulting climate-driven waxing and waning of the Antarctic Ice Sheet reflects its fundamental role in the Earth's climate system, which is characterized by the ability of the ice sheet to interact with the atmosphere and the ocean on a global scale (Bamber et al. 2007; Hemming 2004; Oerlemans 2001; Siegert 2001). Evidences of these interactions can be easily found, e.g. in paleoclimate records showing the differences in the global climate before and after the emergence of large ice sheets in the early Oligocene (ca. 34 million years ago; Zachos et al. 2001), or in sea-floor sediments and paleoglacial landforms revealing massive ice discharges from ancient ice sheets (Bond et al. 1992; De Angelis and Kleman 2007).

Crucial is that such ice–environment interactions remain recorded also in the ice sheet itself. Within and together with the snow, further substances (aerosols, isotopes, etc.) are also deposited from the atmosphere onto the ice sheet surface. These deposits contain information about the atmospheric

[1] The Antarctic Ice Sheet has a width-to-height aspect ratio of $\approx 1.8 \times 10^3$, which is of the same order of magnitude as the aspect ratio of a typical sheet of A4 office paper (cf. definition of *ice sheet* in Sect. 1.2).

S.H. Faria et al., *The EPICA-DML Deep Ice Core*, Frontiers in Earth Sciences,
https://doi.org/10.1007/978-3-662-55308-4_2

composition in the period between the formation of snow crystals in the atmosphere, their precipitation and metamorphism into porous firn, and the subsequent transformation of firn into impermeable bubbly ice. The outcome of this metamorphic process is the chemical and structural stratification of polar ice, which can be observed in all ice cores retrieved from Antarctica (and Greenland). Many of these stratified impurities have already been identified as climate *proxies* (cf. Glossary in Sect. 1.2), which render the Antarctic ice sheet a unique archive of Earth's climate in the past million years (EPICA Community Members 2004; EPICA community members 2006; Petit et al. 1999; Watanabe et al. 2003).

At this point it is important to emphasize that the *active* interaction of the ice sheet with its environment (through its flow, albedo, melting, etc.) and its *passive* recording of the past climate are intimately coupled. For instance, experience shows that climate proxies in form of microscopic inclusions and dissolved impurities in polar ice can have a marked impact on the ice *microstructure* (see Glossary in Sect. 1.2), as exhibited e.g. in the conspicuous correlation between mean grain size of ice and impurity concentration within *cloudy bands* identified in polar ice cores (Faria et al. 2010, 2014a; Gow and Meese 2007a; Gow and Williamson 1976). On the other hand, as in any other polycrystalline material, the microstructure plays an essential role in defining the mechanical properties of ice, including its flow (also called *rheology*). From these facts we conclude that climate proxies not only build up the paleoclimate records of the ice sheet, but they also have a significant effect on the rheology of polar ice (Faria et al. 2009, 2014b; Gow and Meese 2007b; Lhomme et al. 2005; Paterson 1991, 1994).

It happens, however, that paleoclimate records are not always well preserved in polar ice. The main threat to record integrity is the ice flow, which generates considerable shearing in the lower part of the ice sheet. This shearing often causes disturbances and alterations in the ice strata (e.g. folds) that eventually lead to the disruption of the ice stratigraphy (Alley et al. 1997; Faria et al. 2010; NEEM Community Members 2013). Such disruptions render the interpretation of paleoclimate records in the deepest portion of ice cores (which is generally also the oldest and most valuable) a formidable task.

In summary, inclusions and dissolved impurities transferred from the atmosphere into the ice make up climate records that can have significant effects on the microstructure and rheology of polar ice. In turn, climate records stored within the ice are generally not static: the same ice that preserves paleoclimate records can also operate on them. The most serious operation is the disruption of the local stratigraphy by the ice flow. Fortunately, many of these operations leave temporary imprints on the ice structure, which can be identified and analysed through appropriate techniques. Some of these techniques are described in the next chapters. The identification of such temporary imprints allows glaciologists to devise new methods for reconstructing disrupted climate records.

2.2 Understanding the SFEI

The interactions between impurities, microstructure, and mechanical properties of polar ice, outlined in the last section, are particular examples of a comprehensive multiscale interpretation of the dynamics of natural systems called *Structure–Form–Environment Interplay* (*SFEI*; cf. Fig. 2.1 and Glossary in Sect. 1.2). Generally, the term "structure" encompasses all constitutive features of the system that determine its symmetry and response on diverse scales. The notion of "form" is defined by the interior and boundaries of the system, so that changes in form include all morphological processes of deformation, flow, growth, decay, etc., undergone by the system. Finally, "environment" stands for all exterior circumstances and boundary conditions necessary to fully characterize the system in relation to its surroundings.

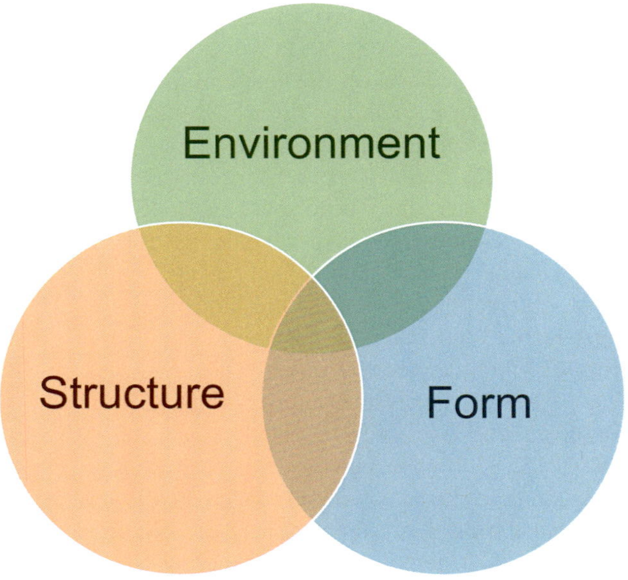

Fig. 2.1 Schematic representation of the Structure–Form–Environment Interplay (SFEI). The environment acts on form and structure; changes in form affect structure and environment; the structure influences form and environment

It must be emphasized that the general characterization of structure, form, and environment is far from resulting in a sharp trichotomy. In fact, these three concepts generally overlap considerably, so that their precise distinction depends largely on the context and the scale under consideration. For instance, in a large-scale ice-sheet model, the pore space of firn is part of the ice microstructure, while the atmosphere on the ice sheet belongs to the environment. In contrast, for a glaciologist interested in the microscopic genesis of climate records, the pore space of firn defines the form, while the air inside the pores constitutes the environment.

For the sake of illustration, we may consider the notion of SFEI with regard to the problem of a large armada of *icebergs* drifting under the action of winds and ocean currents, as it occurs e.g. in the North Atlantic during *Heinrich events* or in the Southern Ocean during the collapse of Antarctic ice shelves:

- If the modeller is interested in the fate of a single iceberg, then the concept of *form* comprises the size and shape of that particular iceberg, while the ice crystallites, inclusions and impurities within the iceberg define its (micro- and macro-) *structure*. Accordingly, melting and fragmentation of the iceberg represent interactions between *form* and *environment*.
- A very different picture of SFEI emerges if the modeller is interested in the dispersion of the whole armada. In this case, the *form* consists of the spatial distribution (i.e. instantaneous positions) of all icebergs, while the sizes and shapes of individual icebergs define the ("micro-")*structure* of the armada. Melting and fragmentation of icebergs represent now interactions between *structure* and *environment*.

The contextual flexibility in the definitions of structure, form, and environment within the SFEI are typical of complex systems with *nested* or *multiscale hierarchy*: the form of a small-scale system Σ becomes part of the structure of a larger system Λ, which is made up of many copies of the former (i.e. $\Sigma \subset \Lambda$). During this *upscaling* process, the definition of environment also changes, since form and environment are complementary concepts (the environment is exterior to the form).

To further explore the concept of SFEI within the context of Antarctica, we briefly review in the next section some of the typical structures observed in Antarctic ice, how they emerge, and how they interact with climate proxies and ice mechanical properties. Further details are discussed also in Chap. 4.

2.3 The Multiscale Structure of Antarctica

In the previous sections we introduced the notion of Antarctica as a *complex system*, with structural features on diverse scales, which evolves according to the framework of SFEI. Here we analyse this view in more detail, by investigating the formation and evolution of some key structural features of Antarctic ice: air inclusions.

Environmental conditions initially determine not only the size and shape of snow crystals (Fig. 2.2), but also the concentration of aerosols (mineral dust, sea salt, and other trace species) captured by the falling snow that accumulates on the ice-sheet surface (McConnell et al. 2014; Fischer et al. 1998, 2007b). Stable isotope ratios of precipitation ($HD^{16}O$ and $H_2^{18}O$ relative to $H_2^{16}O$) are commonly used as proxies for the temperature at the time of snow formation, even though seasonal variations are usually lost by diffusive mixing (Cuffey and Steig 1998). Some types of snow crystals may be transported over long distances by strong winds, eventually accumulating selectively in a variety of surface patterns on multiple size scales, ranging from millimetre-thick wind crusts, to snowdrifts and sastrugi, up to vast megadunes (Birnbaum et al. 2010; Frezzotti et al. 2002; cf. Glossary in Sect. 1.2). These multiscale surface structures are then further modified by wind erosion, sublimation, and the rapid metamorphism of snow, triggered by direct exposition to solar radiation, wind, moisture, and temperature gradients (Colbeck 1983; Davis et al. 1996).

The surface of the Antarctic Ice Sheet essentially consists of such patches of multiscale surface structures made up of different types of compacted and eroded snow, which have been discontinuously formed during snowfall and wind events over several years (Fig. 2.3). The diverse granular compositions found in these multiscale surface structures are characterized by contrasts in grain size and shape, lattice preferred orientations, porosity, impurity and moisture content, which have their origins in intricate chemical and metamorphic processes in the snowpack. Air–snow exchanges of trace compounds affect the chemistry of the snow cover and the lower atmosphere, therefore interfering with the snow metamorphism (Dominé and Shepson 2002; Grannas et al. 2007). Besides the evident consequences for the formation of climate proxies, changes in snow chemistry and metamorphism can alter the surface albedo and permeability of the snow cover, with implications for the validation of radar backscatter signals from altimetry/interferometry surveys (RADARSAT, CryoSat, etc.; West et al. 1996).

Subsequent compaction and sintering of the old snow, driven by the increasing overburden of new snowfalls, leads to the formation with depth of the porous material known as *firn* (Figs. 2.4 and 2.5). In dry polar regions, the firn layer extends from the near surface[2] down to 50–100 m depth, with a gradual mass density increase (called *firn metamorphism/densification*, or *firnification*) from about 300–350 kg/m^3 to 840 kg/m^3, the latter defining the transition from

[2]It must be warned that the definition of firn presented in the Glossary (Sect. 1.2) requires some care, when applied to the dry, cold, and windy Antarctic plateau. As illustrated in Fig. 2.3, snow accumulation on this region is so low that its snow cover is discontinuous. That is, the plateau's surface is generally characterized by a patchy coexistence of shallow snowdrifts and sastrugi with firn-exposed regions that may experience mass loss (ablation) through wind erosion and sublimation for several years. Consequently, firn with distinct physical properties (e.g. mass density, crystalline and pore structures, mechanical strength, etc.) can be found in neighbouring patches near the surface, making the distinction among old snow and various types of firn not so straightforward.

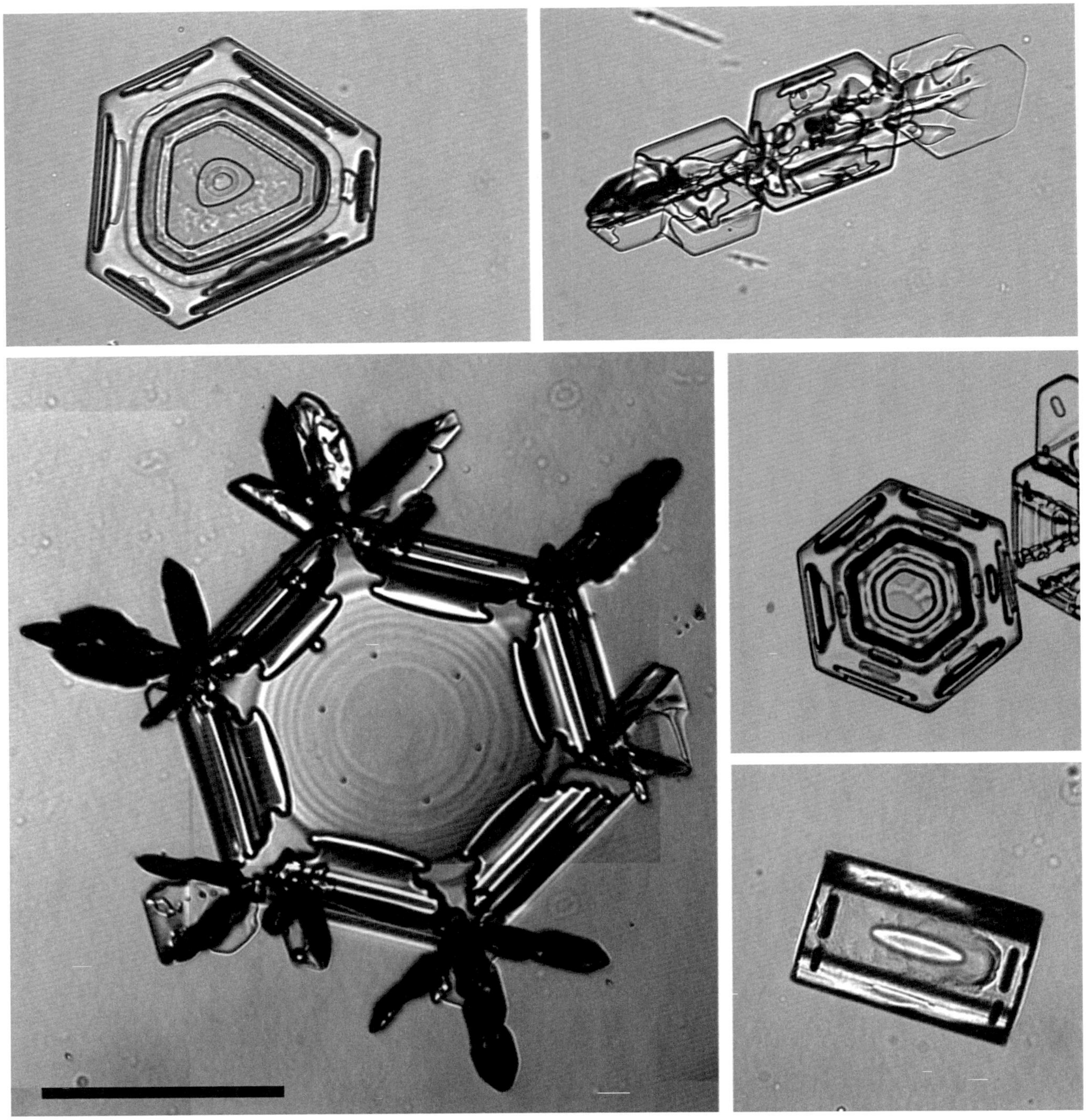

Fig. 2.2 Typical snow crystals at Kohnen Station, Dronning Maud Land, Antarctica. Fresh snow that falls on the cold and dry Antarctic plateau is usually monocrystalline (therefore the term *snow crystal*). In contrast, fresh snowflakes (viz. feathery aggregations of snow crystals typical of snowfalls in mid-latitudes) are rare. Nevertheless, snow crystals already deposited on the surface may sometimes aggregate into flakes, being eventually blown by strong winds up to considerable heights and distances. Scale bar: 1 mm

firn to *bubbly ice*. In the first 10–20 m below surface, seasonal temperature variations can penetrate into the firn and air exchange with the outer atmosphere occurs mainly through forced convection, in response to pressure gradients at the surface (Alley et al. 1982; Bender et al. 1994; Colbeck 1989). Deeper down, gas transport occurs predominantly via diffusion driven by water-vapour density gradients (Colbeck 1983).

Evidently, the air exchange and transport within the firn is strongly constrained by the geometrical and topological

Fig. 2.3 Typical "patchwork" of multiscale surface structures on the Antarctic Ice Sheet. It consists of various types of snow covers with height variations not larger than many centimetres: flat-fissured (*bottom-right*), crinkly (*bottom and centre-left*), loosely rough (*centre*), and wavy-stepped (*centre-right*). Each of these covers is characterized by certain grain and pore microstructures, and has been produced by a particular combination of snow deposition, drift, sublimation, and erosion processes. The predominant wind direction in this site is from the *bottom-right* to the *top-left*. Image width corresponds approximately to 3 m at the bottom of the photograph

characteristics of the pore space, which in turn depend on the original structure of the snow cover and the whole history of firn metamorphism. Owing to this, a great variety of firn densification models have been proposed in the literature (e.g. Alley 1987; Herron and Langway 1980; Maeno and Ebinuma 1983; Barnola et al. 1991; Salamatin et al. 2009; Arnaud et al. 2000; Arthern et al. 2010; Freitag et al. 2013; Goujon et al. 2003). Common to all these models is the observance of two sintering stages, described by Anderson and Benson (1963) in terms of the mass density of firn ϱ_f as follows (the critical densities mentioned below are just approximate, since they actually depend on several factors, like temperature or impurity content of the firn layer, as discussed by Arnaud et al. 2000):

1. *Snow and shallow-firn metamorphism* ($\varrho_f < 550\,kg/m^3$). Densification is controlled by the structural rearrangement of unbounded snow and bounded firn particles by mechanical packing and grain-boundary sliding. The first critical density ($\approx 550\,kg/m^3$) corresponds to the limit beyond which grain-boundary sliding is no longer effective as the controlling densification process, giving place to intracrystalline plastic deformation (creep).
2. *Deep firn metamorphism* ($550\,kg/m^3 \leq \varrho_f <$ 820–840 kg/m^3). Firnification is controlled by intracrystalline power-law creep and recrystallization. The last critical density ($\approx$820–840 kg/m^3) defines the *pore close-off depth interval*, also called *firn–ice transition zone*, in which intercommunicating pores are pinched off and the air becomes entrapped in independent bubbles within the newly formed bubbly ice.

In addition to these two stages, Maeno and Ebinuma (1983) identified another critical density ($\approx 730\,kg/m^3$), which divides the second stage into two sub-stages. Although this critical density could be identified in the firnification curves of several firn cores, its physical grounds (proposed by Maeno and Ebinuma 1983, in terms of topological changes in the pore space) remained obscure for long time. This issue has been recently revisited by Kipfstuhl et al. (2009) and Faria et al. (2014b), who proved through observations and theory that the Maeno–Ebinuma critical density is actually

Fig. 2.4 The birth of Antarctic firn. *Left*: In a sense, this tablet may be regarded as the "Rosetta stone" of polar glaciology. It displays the uppermost 50 cm of firn (corresponding to several years of accumulation at DML), and reveals the genesis of the polar ice microstructure. Over the years, recurrent burial of diverse types of surface structures gives rise to a vertical sequence of fine- and coarse-pored strata, which are easily identified on the tablet when it is observed against sunlight. Each stratum has its own characteristic microstructure, impurity content, etc. Notice how the irregular structures at the top become gradually more compacted and regular with depth. *Right*: Set-up used for observing the tablet shown on the left. Firn blocks extracted from a nearby snow pit are piled together and used as a support for the tablet. Erosion and sublimation (intensified by exposing of the tablet to wind and sunlight) enhance the microstructural contrasts

caused not only by topological changes in the pore space, but mainly by changes in the crystalline structure and mechanical properties of the firn skeleton:

2.a. *Upper (deep) firn metamorphism* ($550\,kg/m^3 \leq \varrho_f < 730\,kg/m^3$). Densification is controlled by intracrystalline power-law creep with strain hardening, characterised by the development of many regions with high internal stresses. The topology of the pore space reaches a state where its contribution to strain accommodation becomes negligible.

2.b. *Lower (deep) firn metamorphism* ($730\,kg/m^3 \leq \varrho_f < 820$–$840\,kg/m^3$). Internal stress concentrations become so high that the stored strain energy in many regions reaches the critical level to trigger dynamic recrystallization (SIBM-O and SIBM-N, cf. Sect. 1.2), causing strain softening and redistribution of internal stresses.

At the critical depth for pore close-off (named *firn–ice transition zone*), firn converts into bubbly ice, and the formerly connected pore space splits into many separated air bubbles, which take up approximately 10% of the total volume of bubbly ice (Stauffer et al. 1985). These air bubbles represent a *unique climate archive* for the reconstruction of past changes in atmospheric composition. Owing to the above-mentioned exchange and transport of air within the firn, however, the precise dating of these air bubbles (also known as the *ice–air age difference*) turns out to be a challenging task: gas inclusions

Fig. 2.5 Multiscale structure of an Antarctic firn sample from some metres depth (EDML B36 firn core). *Right*: Millimetre-thick wind crust (near the 68 mark) sandwiched between coarse-pored material, which is in turn surrounded by fine-pored firn. *Left*: Photomicrographs of horizontal sections showing the microstructures of the coarse- and fine-pored strata. The pore space appears in *black* and the ice skeleton in *light grey*. Some grain boundaries (faint dark lines inside the ice skeleton) can also be identified. The diameter of each micrograph is ca. 35 mm. In spite of the profound microstructural changes caused in firn by recrystallization and metamorphism, essential microstructural contrasts remain recorded in deeper ice (Bendel et al. 2013; Ueltzhöffer et al. 2010) as variations in the mean grain size, as well as in the number, average size, and spatial distribution of air bubbles

are always younger than the surrounding ice matrix. Air in firn pores down to 50–100 m depth still is in exchange with the atmosphere through connected pathways in the pore network. As a consequence, ice underneath the firn–ice transition depth can be a few thousand years older than the air entrapped in its bubbles (Barnola et al. 1987; Bender 2002; Raynaud et al. 2007; Schwander and Stauffer 1984).

The main source of complication comes from the fact that the firn mass density shows strong stratification, so that the resulting depth-variations in the pore structure entail critical differences in the exchange and transport of air between adjacent layers. High-resolution density studies of Arctic and Antarctic firn (Gerland et al. 1999; Hori et al. 1999; Freitag et al. 2004; Hörhold et al. 2011, 2012) revealed that the initially strong density stratification (viz. *density variability*) caused by irregular snow deposition (Figs. 2.4 and 2.5) rapidly decreases within the shallow firn. Notwithstanding, the density variability increases again below 10–30 m depth, where the firn reaches a density of ca. 600–650 kg/m^3, and it continues to increase, correlating with the impurity content, down to the firn–ice transition. Finally, in the bubbly ice zone, the density variability slowly diminishes again. These results suggest that density variations at the firn–ice transition, which are crucial for the dating of air bubbles, may be mainly caused by variations in impurity content (Hörhold et al. 2012; Freitag et al. 2013).

Besides, recent studies by Ueltzhöffer et al. (2010) and Bendel et al. (2013) have shown that the *collective bubble structure* of bubbly ice (i.e. statistical data on bubble sizes, shapes, and spatial distribution) can be used to reconstruct the primordial pore structure of firn (viz. the *firn paleo-pore structure*). Such reconstructions tell us that layers with high impurity concentration, which are most frequent in glacial periods, reach the critical density for pore close-off first, therefore diminishing the ice–air age difference in ice from glacial periods.

In summary, a correct interpretation of climate records in the upper part of polar ice cores depends on identifying the links between atmospheric composition and the climate signal stored in the ice, together with a reliable dating of the ice matrix and its inclusions (Legrand and Mayewski 1997). Both issues are strongly connected to the structural properties of snow and firn, as well as their accumulation rate and metamorphism. As if the above arguments were not enough to underline the importance of structural studies of snow and firn, it has been recently recognized (Cuffey 2008; Helsen et al. 2008) that variations in the rate of firn metamorphism and snow accumulation over years and decades complicate assessments of Antarctic mass changes based on altimetry surveys, to such an extent that they represent today the main source of uncertainty in predictions of future seal level rise.

Beneath the firn–ice transition, air bubbles gradually decrease in size and the enclosed gas pressure increases with depth, in order to counterbalance the overburden pressure of ice (Bendel et al. 2013; Lipenkov et al. 1997; Salamatin et al. 1997, 2009; Ueltzhöffer et al. 2010). This increase of gas pressure with depth does not proceed indefinitely, though. Below a certain critical depth (which depends mainly on the pressure and temperature of the ice matrix), the ice–bubble composite structure is no longer thermodynamically stable, so that the gas and water molecules rearrange themselves into crystalline compounds called *clathrate hydrates* (cf. Glossary in Sect. 1.2; Shoji and Langway, Jr. 1982; Kipfstuhl et al. 2001; Salamatin et al. 2001). Clathrate hydrates in polar ice are composed of air, therefore being frequently called *air hydrates*.

The bubble–hydrate conversion in ice sheets is an intriguing microstructural phenomenon, which remains poorly understood. Air hydrates in polar ice seem to refuse complying with the simple thermodynamic theory that successfully explains hydrate formation in other environments, like pipelines or the sea floor (Miller 1969; Sloan 1998): not all bubbles in polar ice transform into air hydrates at the same critical depth predicted by the simple theory (which corresponds to the depth at which the overburden pressure of ice equals the theoretical clathrate-hydrate dissociation pressure, cf. Glossary in Sect. 1.2). Rather, there exists a large depth interval that comprises the *bubble–hydrate transition zone*, which generally spans circa 500 m (corresponding to thousands of years), where bubbles and hydrates coexist (Fig. 2.6; Faria et al. 2010, 2009; Kipfstuhl et al. 2001; Narita et al. 1999; Pauer et al. 1996; Shoji and Langway, Jr. 1982, 1987; Uchida et al. 1994a, 2014).

Within this zone, which in the case of EDML covers the 800–1200 m depth interval (Faria et al. 2010, 2009), some interesting physical and chemical phenomena take place. In particular, there occurs a depth-dependent gas fractionation, with N_2-enriched bubbles co-existing with O_2-enriched hydrates (Ikeda et al. 1999; Nakahara et al. 1988). This fractionation is significant, with N_2/O_2 concentration ratios within clathrate hydrates reaching values around 2 or less at the upper part of the bubble–hydrate transition zone. Deeper down, this ratio gradually increases with depth again, moving back to the standard average atmospheric ratio of 3.7, which is achieved underneath the transition zone and coincides with the N_2/O_2 ratio originally found in the air bubbles above this zone. A similar fractionation phenomenon is observed also in the case of carbon dioxide (Stauffer and Tschumi 2000), with air hydrates at the upper part of the bubble–hydrate transition zone being depleted of CO_2.

The current explanation for this local air fractionation phenomenon lies in a combination of the lower hydrate dissociation pressure of O_2 and its ability to diffuse through ice faster than other air components, like N_2 or CO_2 (Ikeda-Fukazawa et al. 2005, 2001; Nedelcu et al. 2009; Severinghaus and Battle 2006; Stauffer and Tschumi 2000). Thus, the air diffusion from bubbles to hydrates through the ice matrix, caused by the higher pressure in the bubbles compared to the hydrate dissociation pressure, turns out to be more effective for O_2 than for N_2 or CO_2.

From the viewpoint of ice-core paleoclimate records, the phenomenon of bubble–hydrate air fractionation may cause serious complications for small-scale, high-resolution studies. Such studies are crucial for identifying abrupt variations in the past composition of Earth's atmosphere, and are especially important for the study of deep ice records, where the stratigraphy has already thinned considerably due to the ice-sheet flow. It has also been recognized that small changes in the N_2/O_2 ratio (in the per-mil range) correlate well with the local summertime insolation and could, in theory, be used as an absolute dating tool for ice cores (Bender 2002; Fujita et al. 2009; Suwa and Bender 2008). In both cases, the gas diffusion associated with air fractionation in the bubble–hydrate transition zone has the potential to spoil the integrity of high-resolution paleoclimate records.

The complications caused by air fractionation are however not limited to the bubble–hydrate transition zone: air hydrates in deep ice cores extracted and stored at atmospheric pressure are no longer in thermodynamic equilibrium, and therefore they tend to slowly dissociate air back into bubbles

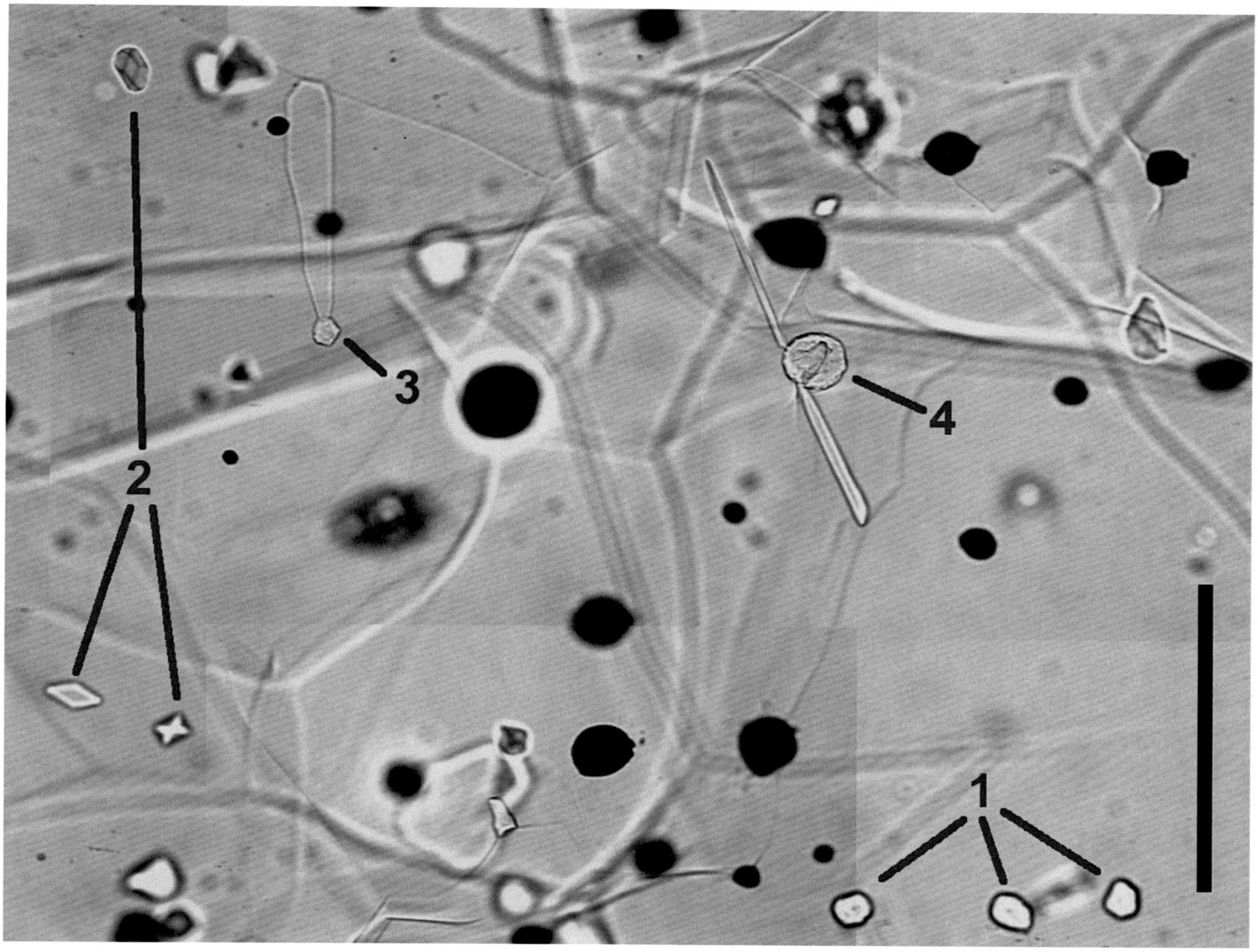

Fig. 2.6 Mosaic micrograph of the interior of a fresh EDML deep ice-core sample from 1055 m depth. It shows air bubbles and hydrates coexisting within the bubble–hydrate transition zone. Air bubbles appear *dark* and round, while air hydrates seem translucent. Defocused lines crossing the image in various directions are by-products of sample preparation, viz. grain-boundary marks (sublimation grooves) etched on both surfaces of the ice sample. Four types of air hydrates are explicitly identified: (**1**) rugged spheroids; (**2**) polyhedral shapes; (**3**) rugged spheroid with a large multifaceted protrusion; (**4**) rugged spheroid with two slender prolongations. The sample is "fresh" in the sense that it was prepared and analysed shortly after drilling, in order to minimize the effects of post-drilling relaxation (plate-like inclusions, secondary microbubbles, decomposing hydrates, etc.). Vertical scale bar: 1 mm

(Gow and Williamson 1975; Stauffer and Tschumi 2000). The rate of hydrate dissociation is mainly determined by the storage temperature (under atmospheric pressure) and may take between few and many years. This means that the coexistence of bubbles with hydrates in stored ice cores is inevitable, and consequently so are also the local diffusion and fractionation of air components, which can be minimized, but not fully avoided, by low temperature storage (Miyamoto et al. 2009; Gow 1971; Uchida et al. 1994b). For these and other reasons, a clear understanding of the interplay between ice microstructure and climate proxies is essential for the correct interpretation of high-resolution paleoclimate records.

Below the bubble–hydrate transition zone, air hydrates continue to evolve over millennia in curious ways (Fig. 2.6): from rugged polycrystalline spheroids to multifaceted, polyhedral, and slender shapes, as well as smooth single-crystalline globules (Kipfstuhl et al. 2001; Pauer et al. 1999; Shoji and Langway, Jr. 1982; Uchida et al. 1994a). The causes of these varied recrystallization and metamorphic processes remain unclear.

Studies of climate records in the lowest tens of metres of deep ice cores are usually impaired by stratigraphic disturbances (e.g. folds) and interactions with subglacial encroachments (e.g. soil particles, rock fragments, melt water, etc.). Additionally, regardless of careful storage at low temperatures (usually between −20 and −50 °C), post-drilling relaxation structures emerge quickly and unavoidably in deep ice cores. They consist mostly of the previously mentioned

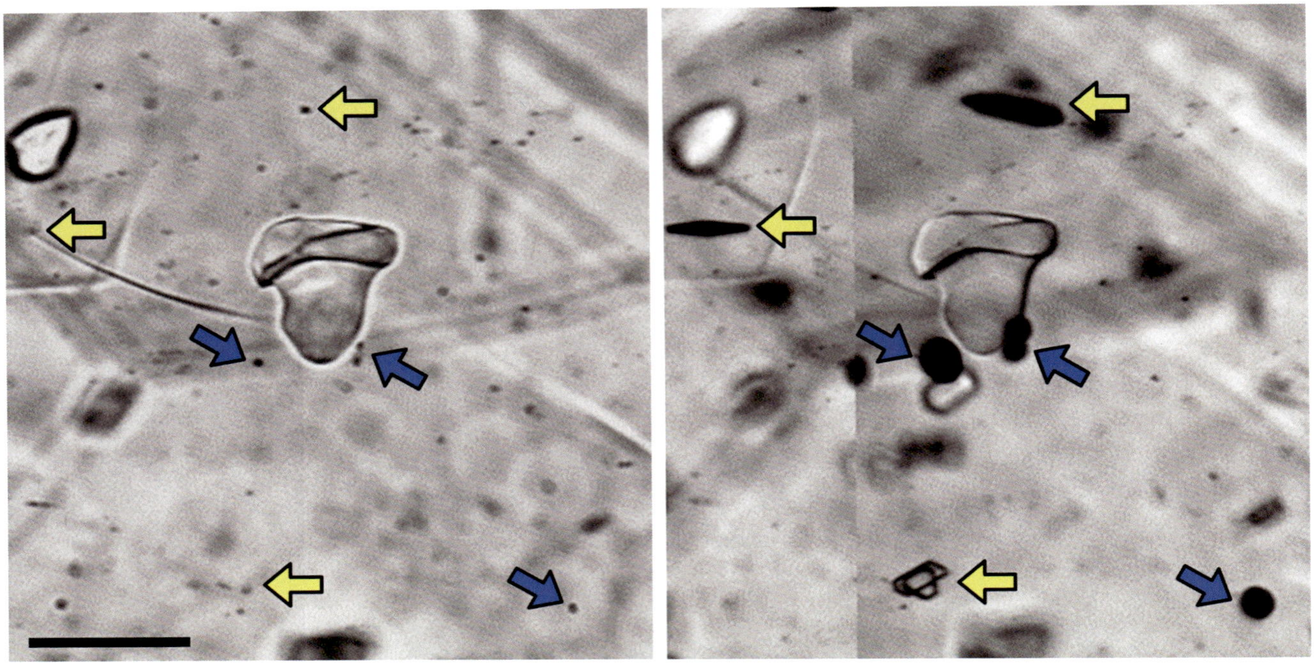

Fig. 2.7 Examples of post-drilling relaxation microstructures. *Left*: Fresh EDML ice core sample from 1095 m depth. Noticeable are a large hydrate in the centre and another smaller one at the *top-left* part of the image. The many *black dots* are microinclusions characteristic of cloudy-band ice (see Glossary in Sect. 1.2). *Right*: The same sample after three years of relaxation under standard storage conditions (−30 °C and atmospheric pressure). At the locations of some microinclusions, secondary microbubbles (inclined *dark-blue arrows*) or plate-like inclusions (*horizontal light-yellow arrows*) have appeared. Interestingly, other microinclusions seem unaltered and show no signs of relaxation, possibly due to distinct physico-chemical properties. Scale bar: 100 μm. After Weikusat et al. (2012)

decomposing hydrates, microscopic relaxation bubbles called *secondary microbubbles*, and flat hexagonal cavities (i.e. negative crystals) known as *plate-like inclusions* (PLIs); cf. Fig. 2.7 and Glossary in Sect. 1.2 (Gow 1971; Nedelcu et al. 2009; Weikusat et al. 2012). Secondary microbubbles and PLIs are generally filled with fractionated (oxygen-enriched) air that was originally dissolved in the ice matrix or stored in air bubbles and hydrates. The characteristic O_2 enrichment of these relaxation structures suggests that the gas filling process occurs via molecular diffusion through the ice matrix, in a similar fashion as the filling of new air hydrates in the bubble–hydrate transition zone. Consequently, studies of PLIs and secondary microbubbles may help to clarify the mechanisms of gas diffusion in polar ice, in particular those related to air fractionation and metamorphism of air hydrates.

2.4 The Birth of EPICA

In a certain sense, one may say that the origins of the European Project for Ice Coring in Antarctica (EPICA) can be traced back to the end of 19th century, when Carl Weyprecht (1838–1881) developed the concept of "International Polar Year" (IPY). The basic purpose of IPY was to convince scientists and policy makers that a coordinated international effort was needed to investigate the geophysics of polar regions. Weyprecht did not live to see his idea materialized: the *First International Polar Year* occurred in the period 1882–1883, shortly after his death. It became a milestone of polar glaciology, with twelve participating nations coordinated by an International Polar Commission, chaired by Georg von Neumayer (1826–1909). Besides many scientific advances and the completion of 15 polar expeditions (including two to Antarctica), the principal legacy of the First IPY was to set a precedent for international cooperation in science.

The example of international cooperation given by the First IPY inspired many subsequent scientific events and activities, including the *Second International Polar Year*, half century later (1932–1933). World War II imposed a hiatus in non-military polar research, but international research activities resumed shortly afterwards, following the spirit established by the first two IPYs. In particular, close scientific collaboration between Norway, Sweden, and Great Britain, initiated by Hans Ahlmann (1889–1974), culminated with the 1949–1952 *Norwegian–British–Swedish Antarctic Expedition* (NBSAE) led by John Giæver, Valter Schytt, and Gordon Robin (Schytt 1974), which recovered a nearly 100 m deep ice core from the *Maudheim* site on Quar Ice Shelf, Dronning Maud Land (DML), Antarctica (Faria et al. 2014a; Schytt

1958). NBSAE became a landmark in the history of geosciences not only because it provided *the very first Antarctic ice core* and launched the investigations of Antarctica's multiscale structure, but also because it established many aspects of the paradigm of scientific research to be pursued during the succeeding International Polar Year, which was expanded and renamed the *International Geophysical Year* (IGY, 1957–1958).

One of the main goals of IGY was to redirect the technology developed during World War II towards applications to geophysical research. Among the many scientific advances associated with this event, one could mention the verification of Alfred Wegener's theory of continental drift, the launching of the first artificial satellites (Sputnik series, Explorer series, Vanguard), and the discovery of the Van Allen radiation belt. Additionally, it established an international research cooperation programme in Antarctic glaciology of unprecedented proportions: 12 nations, 48 Antarctic stations, and diverse traverses over the Antarctic ice sheet, which provided not only the first informed estimates of Antarctica's total ice volume, but also two intermediate-depth ice cores extracted from *Byrd Station* (1957–1958, 307 m deep) and *Little America V* (1958–1959, 264 m deep; Langway 2008). Finally, another historical outcome from this period was the preparation of the *Main Antarctic Treaty*, ratified in 1961 by the twelve countries active in Antarctica during the IGY. Today, this Treaty still plays a fundamental role in preserving the environment and regulating human activity in Antarctica.

After the IGY in the late 1950s, a new era of polar glaciology was established, marked by the quest for ever deeper and older polar ice cores. In this vein, two prominent U.S. Army scientists, Henri Bader (a key person behind the IGY ice-core drilling campaigns) and his colleague B. Lyle Hansen (a world-renowned ice-core driller) started devising a plan to reach bedrock depths in Greenland and Antarctica (Bader 1962). In 1966, after a strenuous six-year field effort, the *first deep ice core reaching bedrock* (1387 m depth) was extracted from *Camp Century*, Northwest Greenland, a subterranean nuclear-powered overwintering facility for 200 persons operated by the U.S. Army (Ueda and Garfield 1968).

Surprisingly, in a genuine tour de force the same team and equipment continued their lucky streak by reaching bedrock in Antarctica less than two field seasons later, with the extraction of a 2164 m deep ice core from *Byrd Station* in 1968 (Ueda and Garfield 1969).

Whereas U.S. scientists primarily performed the aforementioned drilling operations, the ice-core laboratory analyses demanded an internationally coordinated action, mainly involving Denmark, Switzerland, and the United States. This collaboration culminated with the creation of the American–Danish–Swiss drilling initiative GISP (*Greenland Ice Sheet Project*), headed by Willi Dansgaard (University of Copenhagen), Chester Langway, Jr. (U.S. Army Polar Research Unit), and Hans Oeschger (University of Bern). Three bedrock core drillings were originally planned for GISP, the major one at Summit, the highest point of the Greenland ice sheet. However, financial restrictions from U.S. side forced the selection of only one location, the logistically more convenient *Dye-3* site, a U.S. radar station near Greenland's southern coast. By 1981 bedrock was reached at Dye-3, yielding a 2037 m deep ice core.

The experience acquired with GISP, combined with the prevalent trend towards a unified European community (including the establishment of the *European Science Foundation*, ESF, in 1974) motivated European scientists to pursue joint inter-European deep drilling initiatives. The first of them, called GRIP (*Greenland Ice Core Project*), was coordinated by a steering committee chaired by Bernhard Stauffer (University of Bern) and involved eight European countries (Denmark, Switzerland, France, Germany, United Kingdom, Italy, Iceland and Belgium). Its aim was to drill an ice core down to Greenland's bedrock at Summit (the same drilling site originally planned for GISP). Simultaneously, a companion U.S. drilling project called GISP2 (*Greenland Ice Sheet Project II*) was aiming at similar drilling objectives just circa 30 km west of Summit. Collaboration on the field between the two drilling parties was very good and helped to save costs for both projects. In 1992 the GRIP ice-coring team reached bedrock at 3029 m depth. Approximately one year later a 3053 m deep ice core was recovered to bedrock depth at the GISP2 site.

These two sibling cores provided a wealth of information about the climate history of the North-Atlantic region in the last 115 ka. In particular, they revealed a remarkably turbulent climate during the last glacial period (14–114 ka BP), characterized by what became known as *Dansgaard–Oeschger events*: rapid climate fluctuations at intervals ranging from few hundred to few thousand years (Dansgaard et al. 1993). Such abrupt changes could be identified not only in other Greenland cores (e.g. Camp Century, Dye-3) but also in marine sediment records of the North Atlantic, which indicate large-scale shifts in ocean currents (Bond et al. 1993; Johnsen et al. 1992). In contrast, the identification of similar events in Antarctica was hampered by the generally lower snow accumulation, which increases the chance of finding older climate records, but reduces their chronological resolution.

Actually, the situation was such that, by the end of the GRIP–GISP2 drilling projects, there was only one Antarctic ice core extending through the last glacial period and beyond: the *Vostok* Deep Ice Core, extracted at the homonymous Soviet research station, which was established in 1957 during the IGY (Fig. 2.8). The Vostok core revealed a close connection between climate and the atmospheric concentration of greenhouse gases back to 420 ka BP (Jouzel et al. 1993; Petit et al. 1999), but it showed no clear evidence of an Antarctic counterpart to Dansgaard–Oeschger events (Paterson 1994).

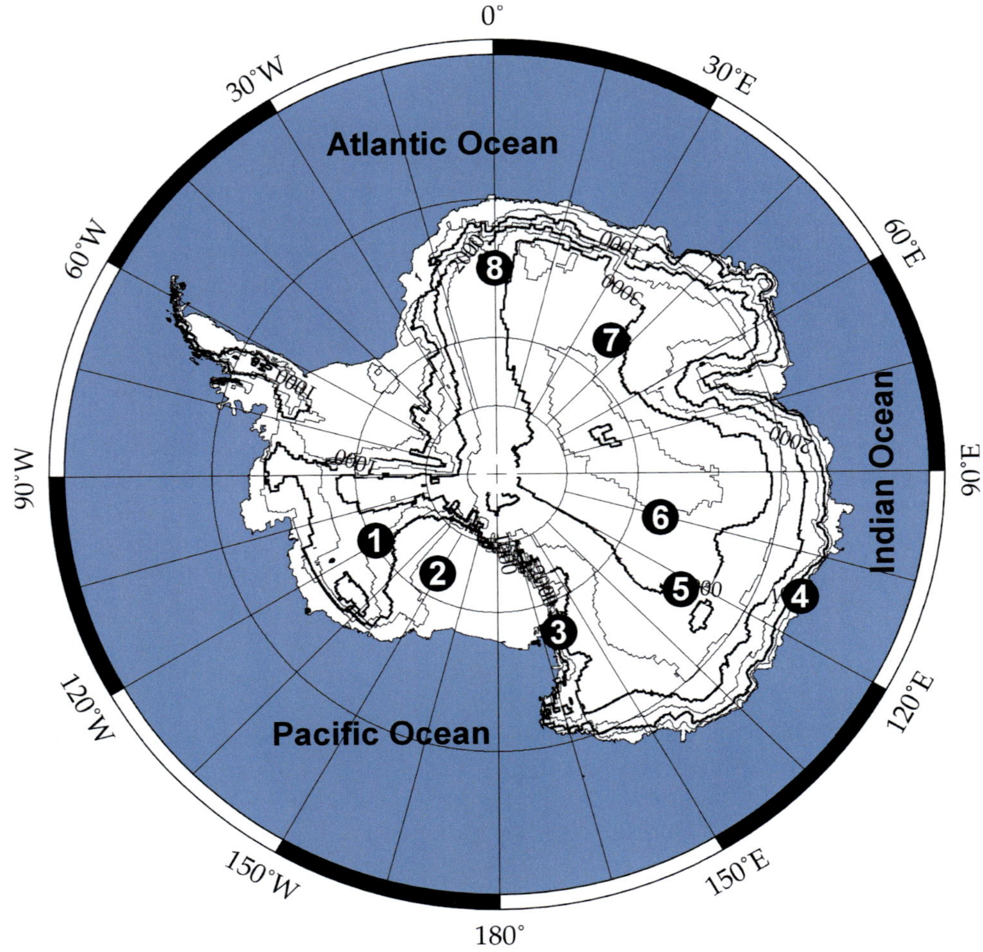

Fig. 2.8 Location of some deep ice-coring sites in Antarctica. ❶ Byrd, ❷ Siple Dome, ❸ Taylor Dome, ❹ Law Dome, ❺ EPICA Dome C, ❻ Vostok, ❼ Dome Fuji, ❽ EPICA-DML. After Steinhage (2001). Courtesy of D. Steinhage (AWI)

These results made patent that high-resolution records from Antarctica's South-Atlantic sector were necessary to complement the records obtained in central Greenland. On the other hand, there was also the lasting desire to go further back in time with ever older ice cores. It was in this spirit that the Antarctic counterpart to GRIP, called the *European Project for Ice Coring in Antarctica* (EPICA) came out in the early 1990s. EPICA was conceived as a multinational inter-European consortium with the objective of drilling two ice cores to bedrock on the East-Antarctic Plateau within a decade. At *Dome C*, in the Indian Sector, the target was old ice, with the largest possible amount of glacial–interglacial cycles in one record. At *Dronning Maud Land* (DML), in the Atlantic Sector, the aim was to obtain a South-Atlantic counterpart to the high-resolution Greenland records, covering at least one complete climate cycle (Fig. 2.8).

2.5 EPICA Facts and Figures

The European Project for Ice Coring in Antarctica (EPICA) was organized in a threefold funding scheme:

1. as two successive programmes of the European Science Foundation (ESF);
2. as 3 + 1 projects within successive Framework Programmes of the European Commission (EC);
3. as a number of national contributions from the ten European nations formally involved in EPICA: Belgium, Denmark, France, Germany, Italy, the Netherlands, Norway, Sweden, Switzerland, and the United Kingdom.

EPICA started as a five-year ESF Research Programme in January 1996, which was then extended for further five years in 2001, and eventually continued for one extra year until the end of 2006. During this eleven-year period ESF contributed a total of EUR 490,000 to the Programme (Oerter et al. 2009), which were made available mainly for coordination, meetings, and publications. The EPICA Programme was managed by a Scientific Steering Committee chaired from 1996 to 2001 by Jean Jouzel (Climate and Environment Sciences Laboratory, LSCE, Saclay), and subsequently by Heinz Miller (Alfred Wegener Institute, AWI, Bremerhaven) until 2006.

Table 2.1 The EPICA–EC grants (after Oerter et al. 2009)

Project	Reference[†]	Duration	Grant[‡]
EPICA 1	ENV4-CT95-0074	Feb 1996 – Jan 1999	5.0
EPICA 2	ENV4-CT98-0702	Feb 1999 – Apr 2001	2.9
EPICA 3	EVK2-CT-2000-00077	May 2001 – Apr 2004	2.4
EPICA-MIS	STREP-003868	Dec 2004 – May 2008	2.5

[†]Sometimes the dashes and/or the letters "CT" are omitted
[‡]Million EUR

Table 2.2 Summary data on the two EPICA drilling sites

	Dome C[†]	Dronning maud land[‡]
Ice-core notation	EDC96 and EDC99	EDML
Station name	Concordia	Kohnen
Location	75°06'04"S, 123°20'52"E	75°00'06"S, 00°04'04"E
Elevation (a.s.l.)	3233 m	2892 m
Ice thickness	3309±22 m	2782±10 m
Ice-core length	3259.72 m	2774.15 m
Ice-core age	ca. 800,000 years	ca. 150,000 years (at 2416 m) possibly 250,000 years (near bedrock)
Accumulation rate (annual mean)	25.0 kg $m^{-2}a^{-1}$ (present time scale) 25.6 kg $m^{-2}a^{-1}$ (using Tambora) 25.4 kg $m^{-2}a^{-1}$ (last 1000 years)	64.0±0.5 kg $m^{-2}a^{-1}$(last 1000 and 4000 years) 65 kg $m^{-2}a^{-1}$ (from radar)

[†] http://www.esf.org (retrieved on 30 Nov 2015, search word: EPICA)
[‡] Ibid. and Oerter et al. (2009)

Field work of EPICA was co-funded by three subsequent EC grants within the 4th–6th European Framework Programmes (Table 2.1), as well as one Specific Targeted Research Project, called EPICA-MIS (Enhanced Paleoreconstruction and Integrated Climate Analysis through Marine and Ice-core Studies). According to Wilhelms et al. (2014), the EC grants amounted to circa 25% of the total budget, while multiple national contributions (which are difficult to estimate precisely) covered the remaining 75%.

The main logistic support for Dome C was provided by Italy and France through PNRA (National Research programme in Antarctica) and IPEV (Institute Paul Emile Victor), respectively. On the other hand, Germany (AWI) was in charge of the logistics for Dronning Maud Land.

Scientific activities were planned and coordinated by a Scientific Committee led by Bernhard Stauffer (University of Bern) in the period 1996–2002, and then by Eric Wolff (British Antarctic Survey, BAS, Cambridge) until 2006. Research was arranged in five basic consortia: isotopes, gases, chemistry, dust, and ice physical properties. Smaller teams dealt with ice sheet modelling and meteorology, while two special groups were formed during the course of the project for dating and biological studies.

The EPICA science plan, formulated in 1994, addressed many of the major paleoclimate issues of the time, especially that of investigating past climate cycles (with an expected time coverage of 500,000 years, much less than was eventually achieved), and of understanding the geographical and temporal scope of rapid climate changes during glacial periods. To achieve these goals EPICA used drill rigs heavily based on designs previously developed by Danish scientists and colleagues for Greenland deep-drilling programmes (cf. Sects. 3.2 and 3.3).

At *Dome C* (75°06'S, 123°21'E, also known as EDC; cf. Table 2.2), next to the all-year French–Italian Concordia Station, drilling started in 1996, but the drill got stuck at 788 m depth and the borehole was abandoned in 1999 (EPICA Community Members 2004). This first core became known as EDC96. Drilling of the second core (EDC99, sometimes also called EDC2) started in 1999, circa 10 m apart from the EDC96 borehole, and stopped at a logged depth[3] of 3259.72 m in December 2004, nearly 15 m above bedrock (Jouzel et al. 2007), after seismic soundings suggested the presence of melt water just below. Ice at the bottom of the EDC99 core is estimated to be older than 800 kaBP (Jouzel et al. 2007; Parrenin et al. 2007).

In *Dronning Maud Land* (DML), an extensive pre-site survey (Steinhage 2001) led to a drilling site at 75°00'S, 00°04'E, now known as Kohnen Station (Table 2.2). Deep drilling started in January 2002 and finished in January 2006 at a logged depth of 2774.15 m, nearly 10 m above bedrock, after subglacial water poured into the borehole (Oerter et al. 2009). Dating of the core could be performed unambiguously down to 2416 m depth, corresponding to circa 150 kaBP (Ruth

[3]For a definition of *logged depth*, see the Glossary in Sect. 1.2.

et al. 2007). For the remaining 358 m of the core (approximately 13% of the core length, possibly corresponding to further 100,000 years) the core could not be dated, because of flow disturbances that corrupted the stratigraphy (Faria et al. 2010, 2014b; Ruth et al. 2007, see Chap. 4). Thus, the paleoclimate records of the EPICA-DML Deep Ice Core (usually called EDML) cover a complete glacial cycle in relatively high resolution, and serve as a direct southern-hemisphere counterpart to the Greenland records (e.g. GRIP, GISP2, and the more recent NorthGRIP and NEEM ice-core records; EPICA community members 2006).

These findings and many others have been highlighted in more than 300 scientific publications to date, including two EPICA community papers (EPICA Community Members 2004; EPICA community members 2006) and a special EPICA issue (Barbante et al. 2010). EPICA data have featured heavily in authoritative assessment reports, such as those of SCAR (Scientific Committee on Antarctic Research), or the 4th and 5th Assessment Reports of IPCC (Intergovernmental Panel on Climate Change). The achievements of EPICA were also recognized by the Descartes Prize for Excellence in Scientific Collaborative Research 2007, awarded to EPICA by the European Commission. Likewise, several EPICA scientists have been honoured with awards from respected bodies like the European Geophysical Union (EGU) and others.

3 The EDML Drilling Site

Thou, which hast shewed me great and sore troubles, shalt quicken me again, and shalt bring me up again from the depths of the earth.

Psalm 71: 20,
The Holy Bible (2003) Authorized Version

3.1 Dronning Maud Land

To become a polar explorer in the early 20th century was an exciting call for many, but vocation for just a few properly gifted. The occupation required not only intrepidity and exceptional physical condition, but also talent to convince potential sponsors of the predictability of unpredictable expeditions and nerves to withstand the perpetual fight against time. Needless to say that all these virtues have been dramatically epitomized in the race for the South Pole undertaken by Robert Falcon Scott (1868–1912) and Roald Amundsen (1872–1928) in the austral summer 1911–1912. However, the polar explorer's misery would not end with the conquest of the poles. Rather, the conquering just paved the way for a new race, this time for charting, exploration of natural resources, and territory claims.

It was in this spirit that during the 1920s, after the World War I hiatus, a new rush of explorers burst through Antarctic waters and—most extraordinarily—its skies (from this time on aircraft would prove themselves invaluable for Antarctic research; Baschin 1929; Byrd 1930). No piece of land should be spared a claim, no geographical location should be spared the name of a royalty or other generous sponsor. Owing to their long experience in Antarctica since the conquest of the South Pole, Norwegians and British were especially prolific in this respect. Notwithstanding, whereas the Norwegians proved to be faster in the race for the South Pole, it transpired that the British were more effective in territorial claims: by the end of the decade, most of East Antarctica would be claimed by the British Commonwealth, leaving untouched only a largely unexplored territory bordered by Coats Land on the west (20°W) and Enderby Land on the east (45°E; Baschin 1931). Although different parts of that unexplored region were back then loosely referred to as *Dronning Maud Land*—after Queen Maud of Norway (1869–1938), consort of King Haakon VII—it was only in 1939, after Queen Maud's decease, that Norway finally instituted a formal claim for the Antarctic region between 20°W and 45°E (Fig. 3.1).

While the coast of Dronning Maud Land (DML) became reasonably well charted and examined in the course of the 20th century, its inland remained poorly explored until the establishment of EPICA. Consequently, in order to determine a suitable location for the EPICA-DML (EDML) deep ice drilling, extensive pre-site surveys turned out to be mandatory. They lasted several years (from 1994 to 1999 and beyond) and extended over an area of circa 10^6 km^2 in the central-north-western part of DML (Fig. 3.1). The adopted survey methods included a great number of airborne radio-echo sounding measurements of ice thickness and bedrock topography, ground-based measurements of accumulation and ice movement, and a mapping project.

Surface topography was determined from ERS-1 satellite altimetry data (Bamber and Bindschadler 1997; Huybrechts et al. 2000), while extensive airborne radio-echo sounding (RES) surveys, most of them carried out by AWI in the period 1994–1999 (5 campaigns, 89 flights, 91,500 km of radar profiles), provided a wealth of data about ice thickness, stratigraphy, and bedrock topography (Steinhage 2001; Steinhage et al. 1999, 2001). Results of some of these surveys are shown in Figs. 3.2, 3.3, 3.4, 3.5 and 3.6. Particularly interesting is the

S.H. Faria et al., *The EPICA-DML Deep Ice Core*, Frontiers in Earth Sciences,
https://doi.org/10.1007/978-3-662-55308-4_3

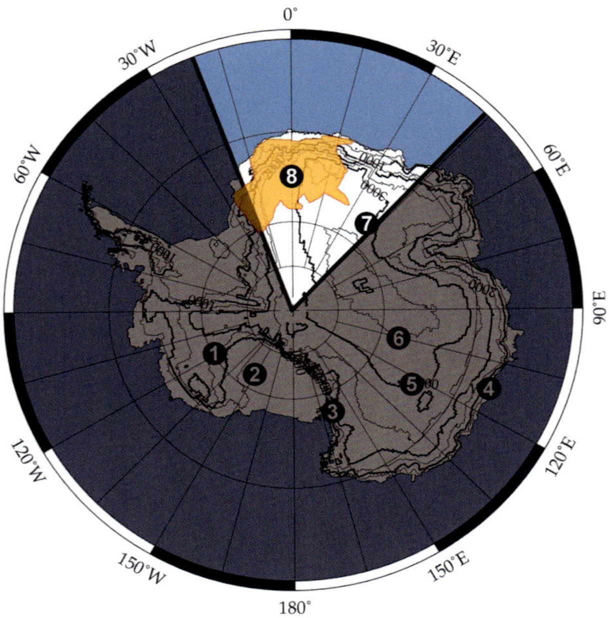

Fig. 3.1 Similar to Fig. 2.8, but with a superposed screen that highlights the Dronning Maud Land sector. The region coloured in *gold* defines the EPICA survey area, in which the optimal drill site for the EPICA-DML core (marked with ❽) was eventually chosen. After Oerter et al. (2009) and Steinhage (2001)

long RES composition (Profile 993136) along the flow-line connecting Dome F to the EDML site, shown in Figs. 3.2 and 3.3: the EDML site lies on an ice ridge, which stems from Dome F, circa 1200 km upstream. Consequently, the radar profile along the ridge (which, by definition, coincides with a flow-line, cf. Fig. 3.7) gives rise to a two-dimensional stratigraphic map, on which one can trace the complete trajectory of any EDML ice particle (e.g. the dashed orange line in Fig. 3.3).

Snow accumulation and ice-sheet surface velocities have been repeatedly measured since 1995 by means of snow pit sampling, firn-core drilling, temperature measurements at 10 m depth, ice-penetrating radar, and static GPS measurements (Oerter et al., 2009). Based on these data, the velocity field around the EDML site could be modelled (Fig. 3.7; Calov et al. 1998; Huybrechts et al. 2000; Steinhage 2001).

Out of 24 potential drilling positions around the EDML site (respectively named DML01–DML24), the EPICA Steering Committee eventually selected in 1999 a location 1.6 km east of DML05 as the optimal site for drilling the EDML Deep Ice Core. On 10 January 2001, after two seasons building Kohnen Station and the respective drill trench, the borehole was located at 75°00'09"S, 00°04'06"E at an elevation of

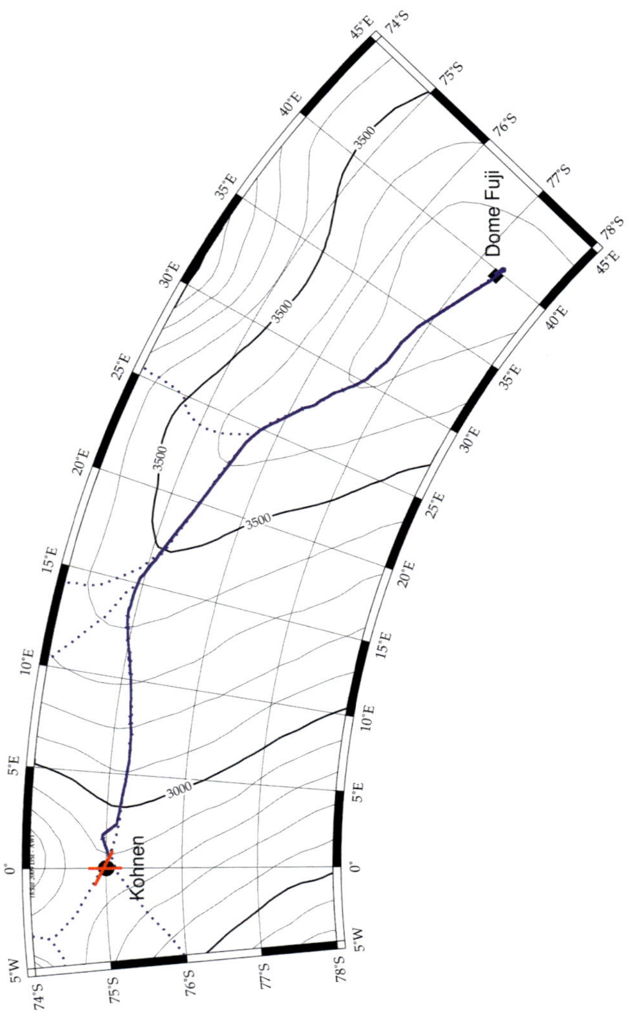

Fig. 3.2 Map of the region around the ice ridge connecting the summit of East DML (Dome F) to the EDML site. Three radio-echo sounding (RES) paths are shown. The first (*blue solid line*) is a long stacked profile along the flow-line between the stations Dome Fuji (on Dome F) and Kohnen (in the EDML site), which is shown in Fig. 3.3 (Profiles 993113, 993132, 993136, 023150, 033141). The other two (*red solid lines*) are high-resolution short profiles crossing the EDML site north- and eastwards, and are shown in Figs. 3.4 and 3.5 (Profiles 983101 and 023150, respectively). *Blue dotted lines* mark the ice divide (cf. Fig. 3.7). After Steinhage (2001). Courtesy of D. Steinhage (AWI)

2892 m (cf. Fig. 3.6 and Table 2.2). The surface flow velocity of the ice at the drill site is 75.6 cm/a with an azimuth of 273.4° (Fig. 3.7; Wesche et al. 2007). The mean annual temperature derived from measurements at 10 m depth is −44.6 °C, with a recent temperature range (registered by the automatic weather station at Kohnen) between −70 and −17 °C (Oerter et al. 2009).

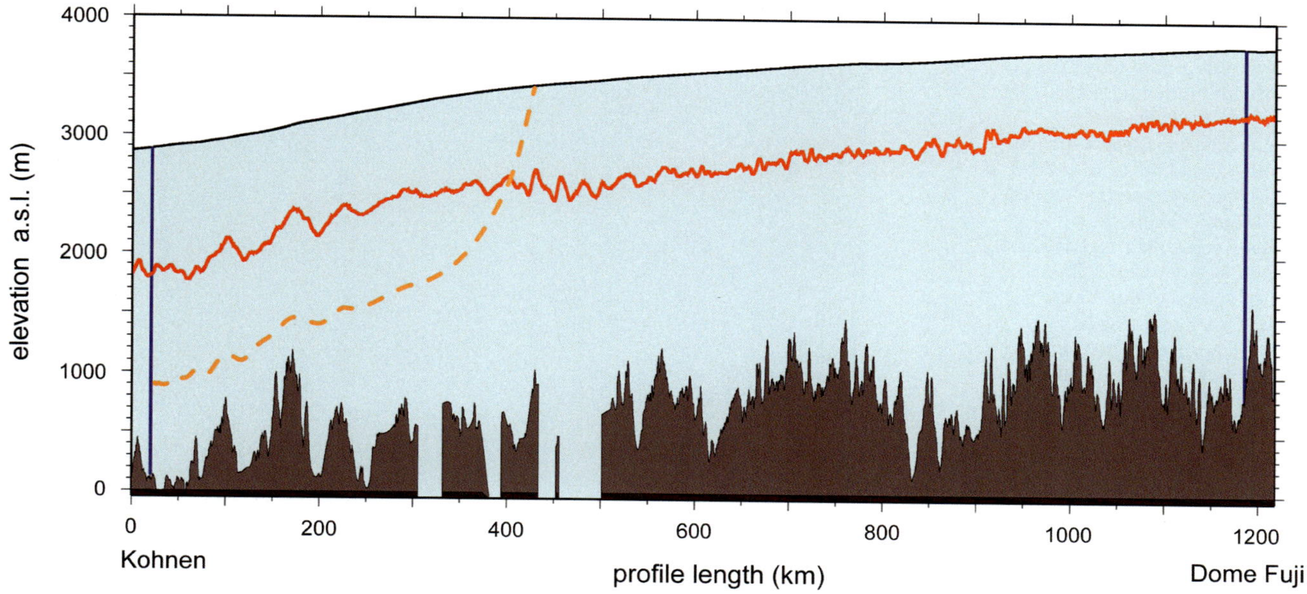

Fig. 3.3 Selective section representation of the composition of radio-echo soundings (RES) along the flow line between the stations Dome Fuji and Kohnen, illustrated in Fig. 3.2 (Profiles 993113, 993132, 993136, 023150, 033141). Selected are three topographic profiles: the ice-sheet surface (*black line*, separating the ice in *light blue* and the atmosphere in *white*), the bedrock (*brown*), and one internal horizon representing the ice stratigraphy (*red line*). The *two blue vertical lines* indicate the locations of the Dome Fuji and EDML boreholes, respectively. Four portions of the bedrock could not be resolved from the RES data (around 320, 390, 440 and 470 km). The *red horizon* describes the I5 isochron layer identified by RES at 1062 m depth, with an age of ca. 25 kaBP (cf. Figs. 3.4 and 3.5). The *dashed orange line* represents a rough sketch of the path followed by an ice particle from the ice-sheet surface to its present position in the EDML core. It illustrates the fact that EDML ice at a depth around 2 km has originated from a surface location circa 400 km upstream along the flow-line. After Steinhage (2001). Courtesy of D. Steinhage (AWI)

3.2 Kohnen Station

The construction of a polar research base is always subjected to natural and technical constraints on design and procedures. In the case of the EDML drilling camp, the Alfred Wegener Institute (AWI), Bremerhaven, performed all the design, construction, and related logistics, according to the following guidelines (Drücker et al. 2002; Oerter et al. 2009):

- The base has to operate as a summer camp, which has to give full support to the ice-core drilling operation.
- Fuel and heavy cargo for building and maintaining the camp must be shipped by the Research Vessel Polarstern (or other supply vessels) to Neumayer Station II at Atka Bay, DML, and transported from there by sledge traverses to the EDML site, circa 760 km inland. For lighter cargo, additional aircraft support is also provided.
- The camp has to accommodate 20 persons during the drilling season, with additional short-term accommodation for up to 7 persons.
- Sufficient energy for operation of the camp, as well as for drilling and scientific activities, must be generated on site.
- Mass and volume of all building material has to be as low as possible for transportation.
- Environmental impact due to transport, construction, and operation of the camp has to be as low as possible.
- Only a short time window is available for the construction of the camp, which has to be simple and reliable to withstand the climatic conditions at the construction site.
- Deep drilling should start in the austral summer season 2001/2002 and should take between three and four summer seasons.
- The camp should be operational for approximately 20 years.

Based on these requirements, the EDML central building was designed as a steel platform on 16 pillars with eleven 20-feet containers on top of it (Figs. 3.8-top and 3.9). The first seven containers (radio room, mess room, kitchen, bathroom, and sleeping rooms) had actually been part of a former Antarctic station: AWI's *Filchner Station*, which was salvaged in February 1999 from the giant tabular iceberg A-38B, while it was drifting in the Weddell Sea[1] (Ams et al. 2001). Four

[1] The giant tabular iceberg A-38 calved off the Filchner–Ronne Ice Shelf in October 1998. It had an approximate size of 150 × 50 km and split into two halves (A-38A and A-38B) shortly after calving. The two parts drifted along the Weddell Gyre in direction of the Antarctic Peninsula and then northwards, eventually grounding north-east of South Georgia Island in January 2004, where they disintegrated within months (Jansen et al. 2005a, b).

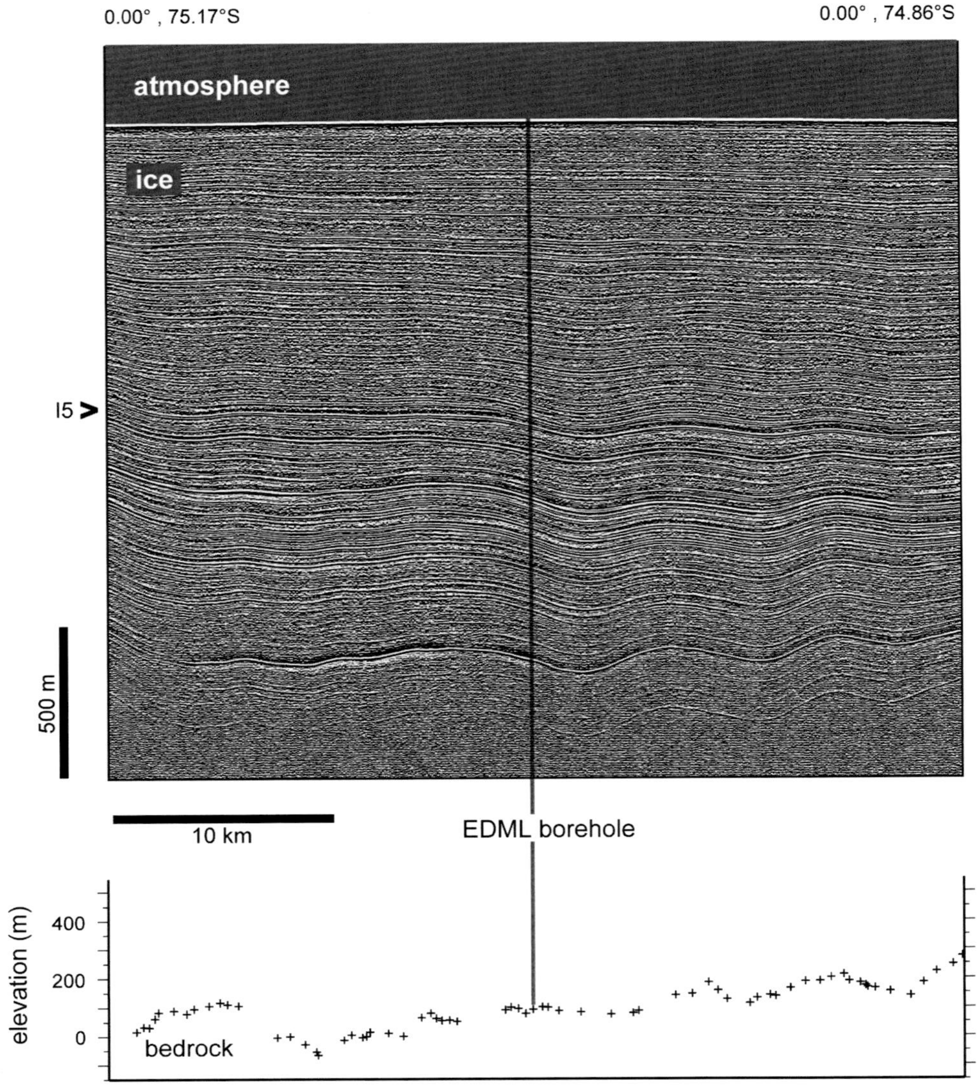

Fig. 3.4 Radio-echo sounding (RES) section (Profile 983101) of a northwards profile across the EDML drilling site (cf. Fig. 3.2). For clarity, the bedrock topography data is shown in the *bottom diagram*. As usual in RES studies, the lowest portion of the record shows no reflection horizons (echo-free zone) and has therefore been excluded. The precise location of the EDML borehole and the internal reflection horizon I5 are marked (cf. Figs. 3.3, 3.5 and 3.6). After Steinhage (2001). Courtesy of D. Steinhage (AWI)

additional containers on the platform (snow melter, power supply, store, and workshop) completed the design of the EDML central building. Besides the eight beds offered in the central building, additional housing was included as mobile units according to the needs during the drilling operation (Fig. 3.8-top).

In contrast to the lifted configuration of the central building, the venue for ice-core drilling and preliminary physical analyses was conceived as a 66 m long, 4.8 m wide, and 6 m deep trench in the snow and firn, with wooden floor and roof (Figs. 3.8-bottom, 3.10 and 3.11). This design, which is an improved version of former drill trenches built in Greenland, has proven to be very efficient in the low-accumulation, slow-flowing EDML site. The underground structure offers ample space, without exposing itself to the elements. Furthermore, the snow and firn walls act as natural thermal insulators, maintaining the temperature in the trench low and nearly constant during the whole drilling season ($\approx -25\,°C$).

The EDML trench was divided in three sections (Figs. 3.10, 3.11, 3.12, and 3.13): one dedicated for drilling ("drill trench"), one for ice-core logging and preliminary analyses ("science trench"), and another one for temporary ice-core storage ("core buffer"). Access for persons and material was given by a ramp at the front, a lift in the

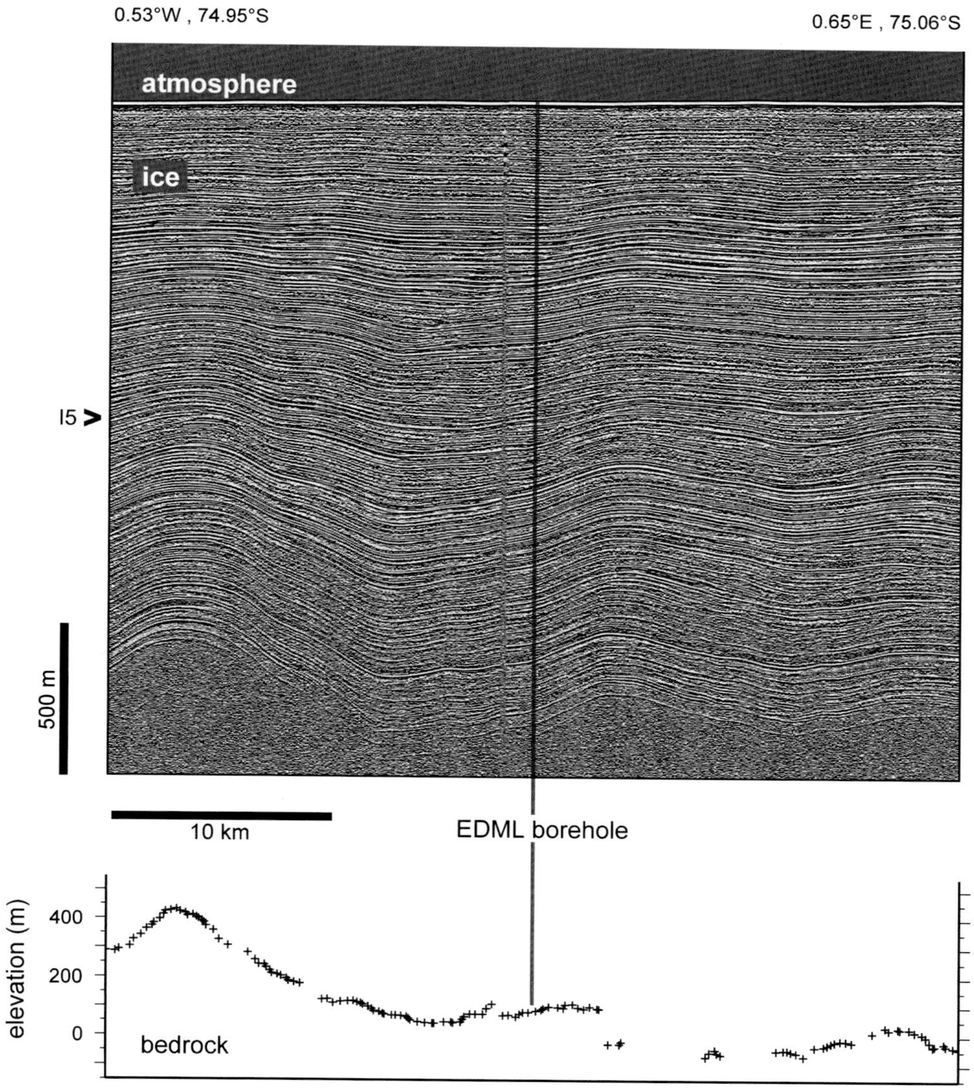

Fig. 3.5 Radio-echo sounding (RES) section (Profile 023150) of an eastwards profile across the EDML drilling site (cf. Fig. 3.2). For clarity, the bedrock topography data is shown in the *bottom diagram*. As usual in RES studies, the lowest portion of the record shows no reflection horizons (echo-free zone) and has therefore been excluded. The precise location of the EDML borehole and the internal reflection horizon I5 are marked (cf. Figs. 3.3, 3.4 and 3.6). After Steinhage (2001). Courtesy of D. Steinhage (AWI)

middle, and an emergency exit ladder at the end (in the core buffer). In the first drilling season (2001/2002), an additional cave was hand-excavated at the end of the core buffer for hosting a fully equipped ice microscopy laboratory for the microstructure mapping (μSM) of fresh ice-core samples (red dashed outline labelled μSM in Fig. 3.10). In this laboratory, delicate instruments (computers, video recorders, etc.) were kept in a small heated cabin built in a portion of the cave, while microscopes and other robust equipment were operated outside the cabin, in the cave's naturally cold environment.

The drill apparatus used at EDML was a 12-metre version of the Hans Tausen drill (Johnsen et al. 2007), which is essentially the same drill model used for the deep drillings at NorthGRIP (Greenland) and EPICA Dome C (Antarctica). It consists primarily of a winch, a tiltable tower, and a down-hole unit (containing all the electronics, mechanical parts, core barrel, and drill head) mechanically suspended and electrically alimented by a 3.5 km long stainless steel wire armoured cable (Figs. 3.10, 3.11, 3.14, and 3.15). Wooden beams mounted in the floor provided foundation for the drill tower and the winch. The tiltable tower rotates about its half-length to a

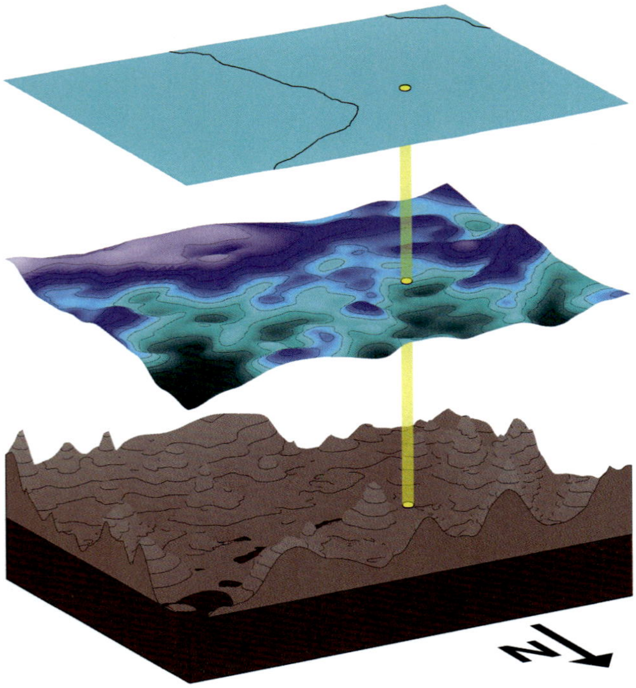

Fig. 3.6 Exploded perspective (viewed from 30°E and 20° elevation) of the surroundings of the EDML drilling site, illustrating the remote-sensed topography of the ice-sheet surface, the bedrock, and an internal isochron layer (all with 25× vertical magnification). The image covers a region extending 70 km eastwards and 50 km northwards (cf. *arrow* towards north). Depicted in the centre is the I5 isochron layer identified in Figs. 3.3, 3.4 and 3.5. Bedrock is represented in *brown*, with *dark brown* denoting the region below sea level. The topography of ice layers is indicated by hues ranging from *dark cyan* to *blue*, up to *light violet*. Isolines mark height intervals of 50 m on the ice surfaces, and 100 m on the bedrock surface. The position of the EDML borehole is marked in *yellow*. At the borehole, the bedrock elevation is 110 m a.s.l. and the ice-sheet surface reaches 2892 m a.s.l. (cf. Table 2.2). After Steinhage (2001). Courtesy of D. Steinhage (AWI)

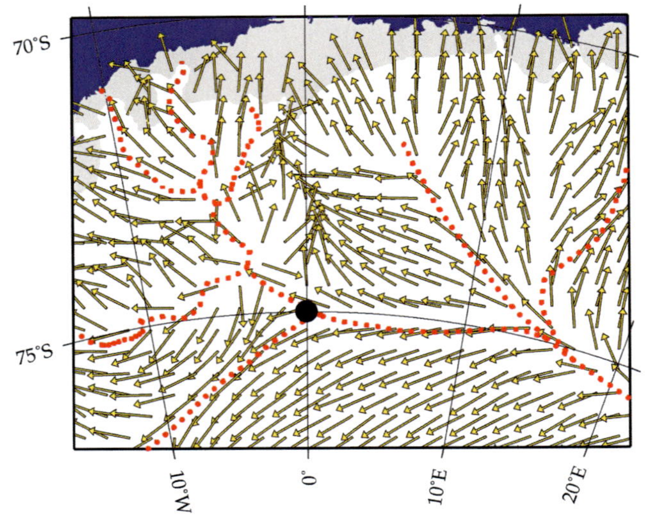

Fig. 3.7 Flow directions (*yellow arrows*) in the neighbourhood of the EDML drilling site (*black disc*), computed with the Budd–Warner numerical model (Budd and Warner, 1996) and data compiled by Steinhage (2001). *Red dotted lines* define the contours of the ice divide (i.e. lines of divergent flow). From these results we conclude that the dominant stress regime in the *upper part* of the ice sheet at EDML is horizontal extension transverse to the direction of flow. This conclusion has recently been corroborated by Weikusat et al. (2017). After Steinhage (2001). Courtesy of D. Steinhage (AWI)

horizontal position, therefore facilitating the core extraction and drill maintenance, but requiring in turn the excavation of an additional narrow trench (called inclined borehole trench) for accommodating the lower half of the drill tower when it is in vertical (drilling) position.

On 11 January 2001 the EDML base was inaugurated. The topping-out ceremony of the drill & science trench was celebrated two weeks later, on 24 January 2001. The base was named *Kohnen Station* in tribute to the German geophysicist Heinz Kohnen (1938–1997), who was the first Head of Logistics at AWI. Kohnen played a decisive role in the construction of the first West-German Antarctic base (Georg von Neumayer Station, at Atka Bay, Ekström Ice Shelf) and he strongly promoted EPICA from the very beginning.

3.3 Drilling and Processing

After the EDML drill & science trench was finished, drilling started with a dry access hole down to 98 m depth below the bottom of the inclined borehole trench (corresponding to 113 m below the original snow surface, cf. Fig. 3.10) and the hole was cased. Such a casing is necessary to prevent the

Fig. 3.8 Kohnen Station and the EDML drill & science trench. *Top* Kohnen Station (the drill trench is behind the observer), with a cylindrical fuel tank on the *right* (*blue*) and a white food storage refrigerator on the *left*. Further left there are different kinds of dormitories (several types of cabins and a large tent). In the main building, the two blocks of containers on the left (*darker orange*) were salvaged from the former Filchner Station, while the block of containers on the right (*lighter orange*) was added to complete Kohnen Station. *Bottom* entrance ramp to the EDML drill & science trench (Kohnen Station is on the *left*, not shown). Above the trench is the lift cover (*red*), with the drilling-fluid tank to its right and the drillers' workshop behind it (both in *blue*). Further to the right there are two science tents in *red* and a small "Scott" polar pyramid tent in *orange*

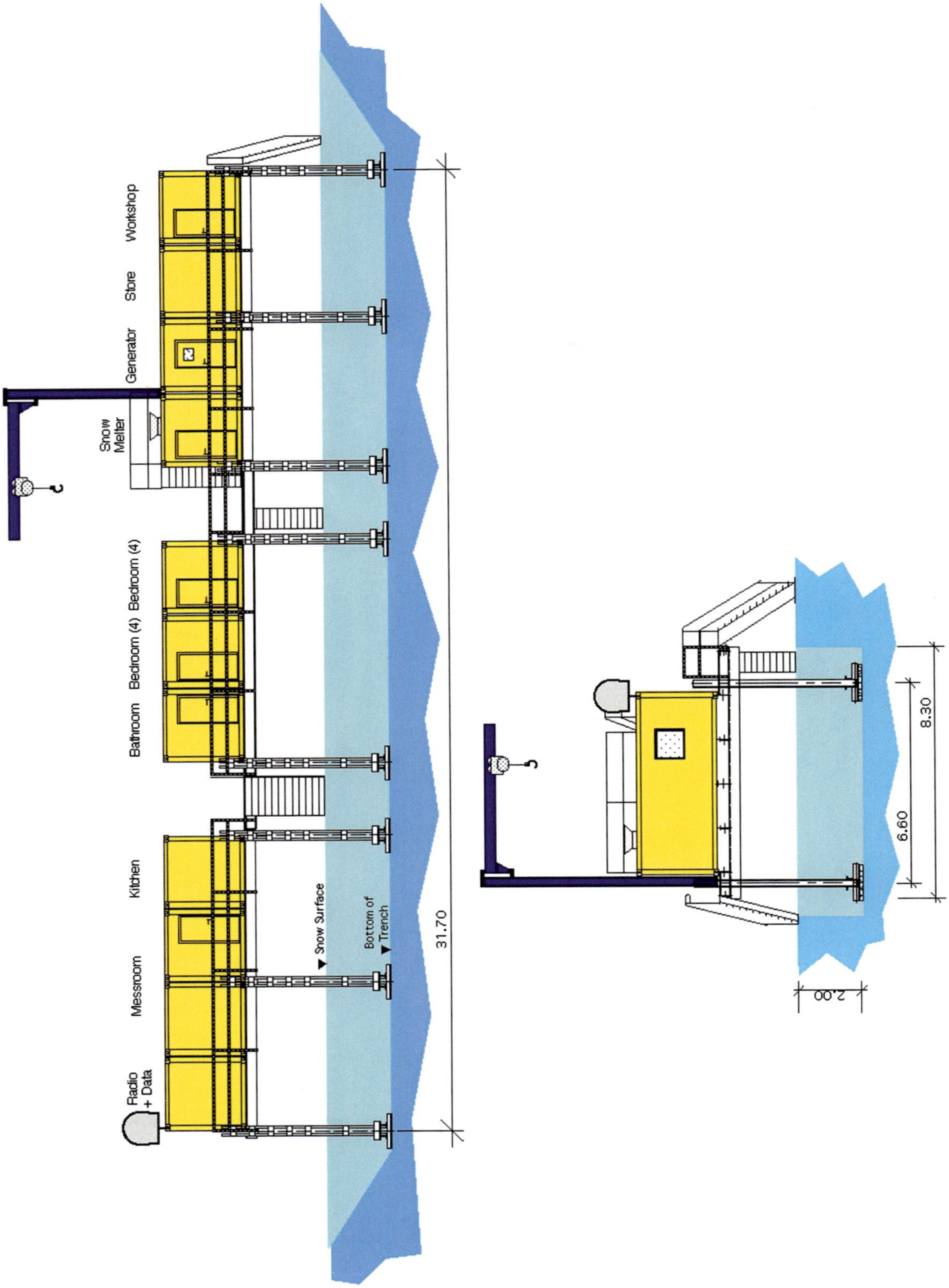

Fig. 3.9 Layout of Kohnen Station central building. This type of elevated design has proven to be very practical and robust, from the experience with other Antarctic stations (e.g. Filchner, Halley V, SANAE IV). The adjustable elevation prevents the building from being buried in snow, while the gap between the building and the snow surface allows the blown snow to pass underneath, therefore inhibiting the formation of undesirable snowdrifts and sastrugi by persistent katabatic winds. From Oerter et al. (2009). Courtesy of C. Drücker (AWI)

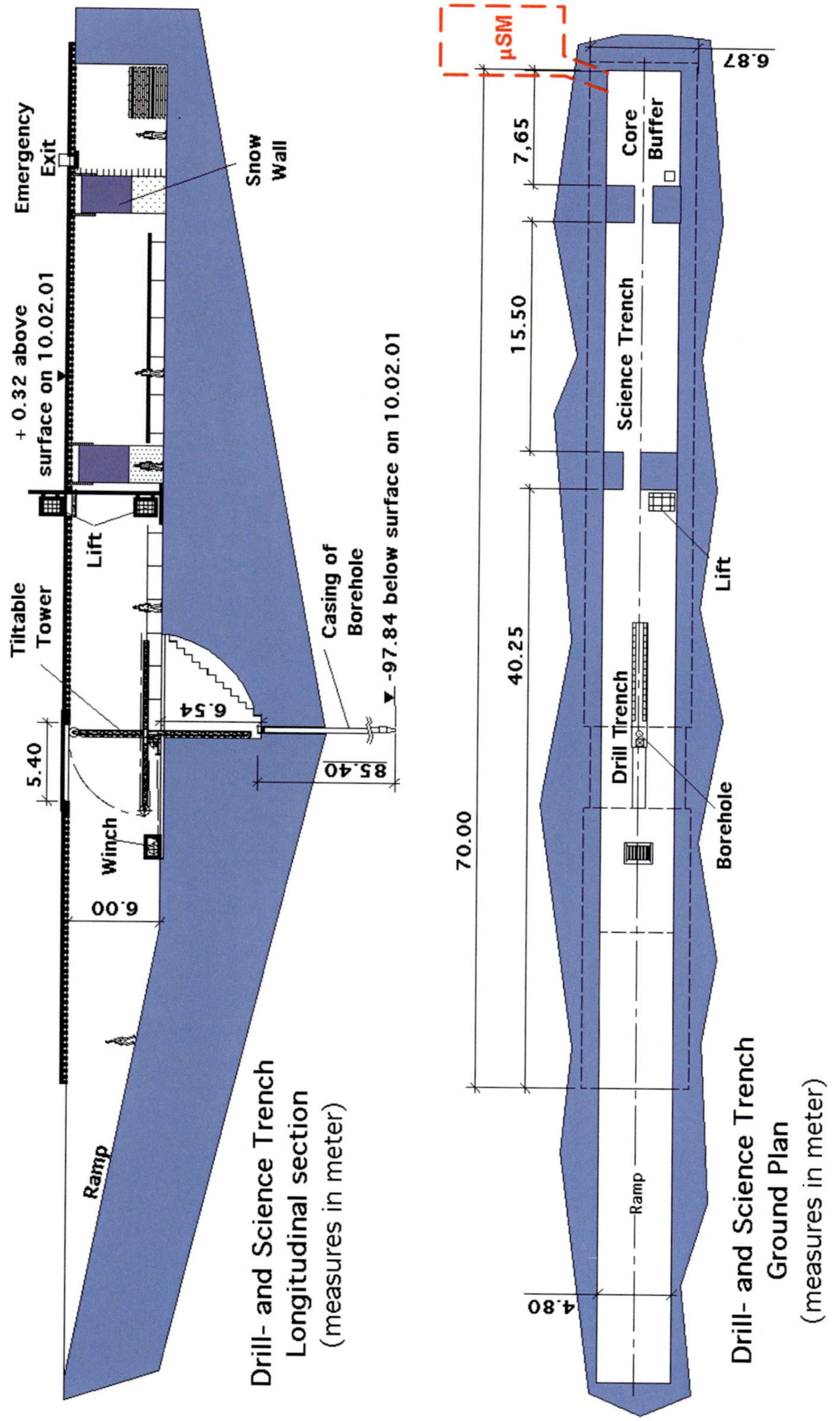

Fig. 3.10 Layout of the EDML drill & science trench. Added to the original layout is the sketch (*red dashed outline*) of the hand-excavated cave at the left side of the core buffer, which became the laboratory for microstructure mapping (µSM) studies. After Oerter et al. (2009). Original layout courtesy of C. Drücker (AWI); *red dashed outline* added by the Authors

drilling fluid from penetrating into the porous firn in the upper part of the ice sheet.

Deep ice-core drilling started in the EDML site on 25 January 2002 (season 2001/2002) and continued during four austral summer seasons (no drilling in the season 2004/2005)

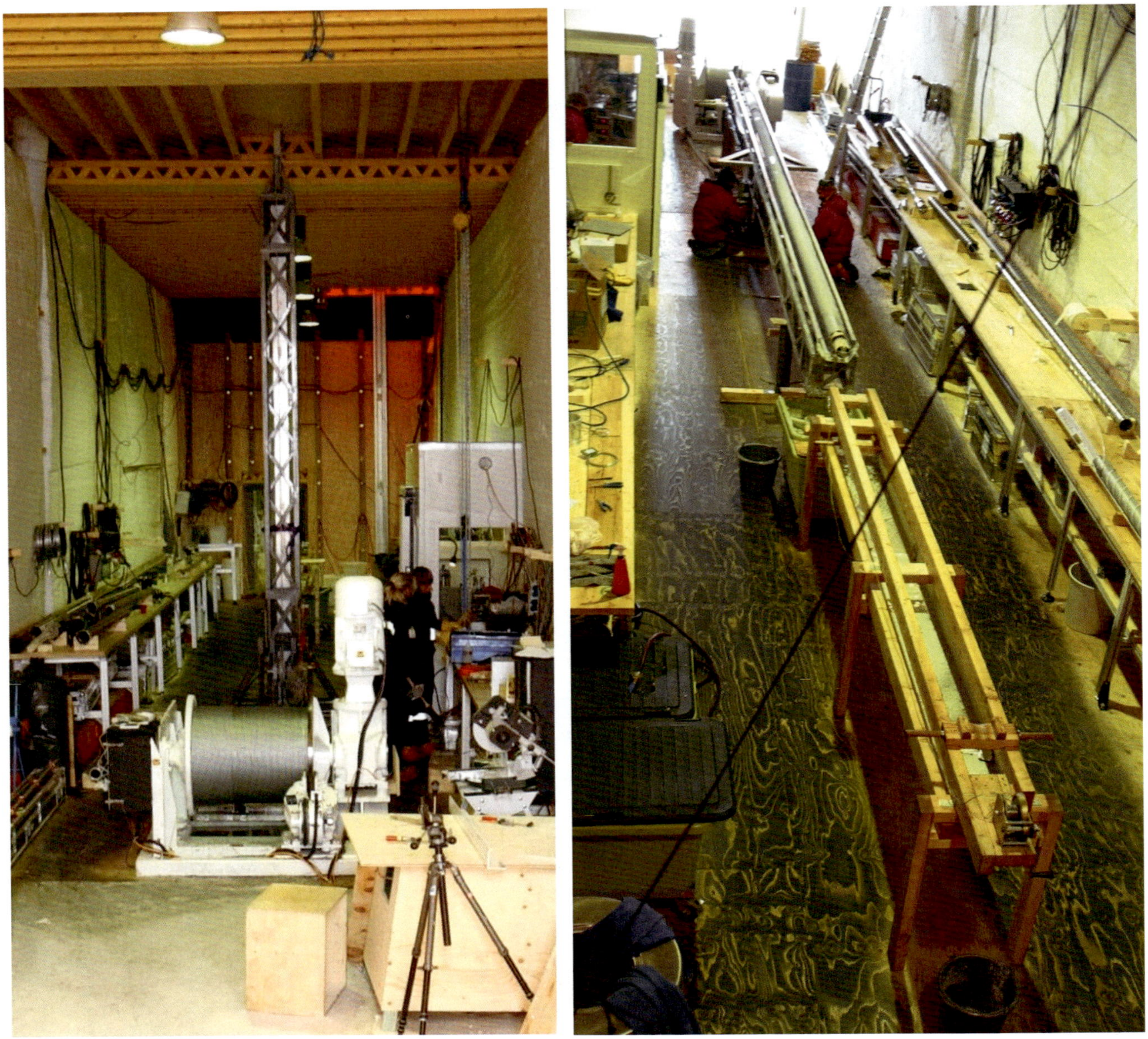

Fig. 3.11 Interior of the EDML drill trench. *Left*: Drill in vertical (drilling) position. Only half of the drill tower is seen; the lower half goes into the borehole trench, which is a coverable, narrow and inclined trench that houses the actual borehole under the drill trench's ground (cf. Fig. 3.10). Also visible are the drill winch in the foreground, the drill control cabin (*white*) on the right, and the entrance door to the science trench and core buffer far in the background. *Right*: Drill in horizontal position, enabling the easy extraction of the ice core. The whole 12 m long tiltable drill tower is now exposed. For safety reasons, the trapdoor of the inclined borehole trench remains closed when the drill lies horizontal. The drill control cabin (*white*) is seen on the left. The brightness on the top of the image comes from the snow outside, because the main entrance to the trench is wide open

until 16 January 2006, when it reached the final logged depth of 2774.15 m below snow surface. The drilling operation was concluded after detection of subglacial water entering the borehole (Fig. 3.16).

Post-drilling operations usually consisted of the following steps. Recovered ice-core pieces ranging from some centimetres to more than three metres in length were first stored in the core buffer (Fig. 3.12-bottom). This initial storage period ranged from minutes to two weeks, depending on the progress of drilling and core processing, the brittleness of the core, and special needs of physical studies. After being removed from the core buffer, the core pieces experienced the first stage of processing in the science trench (Fig. 3.12-top), which consisted of cleaning, logging, and cutting (and/or assembling) into one-metre ice-core units. The dielectric profile (DEP) of each unit was then determined, in order to obtain a preliminary

Fig. 3.12 EDML science trench and core buffer. *Top*: Science trench seen from the passage to the drill trench (cf. Fig. 3.10). Various stages of ice-core logging and initial processing can be distinguished from back-left to front-right: ice-core cleaning and logging table, horizontal band saw (for cutting the logged cores into one-metre long pieces), dielectric profiling (DEP) rail frame and controller (the latter in the metallic boxes), table for preliminary visual stratigraphy analysis and packing, and finally, on the bottom-right, the white corner of an expanded polystyrene box for storage and transportation of packed ice cores. *Bottom*: The core buffer room. After drilling, ice cores are temporarily stored here for later logging, cutting, DEP, visual stratigraphic analysis, and packing. The empty metallic boxes are used to access the higher parts of the buffer. An open expanded polystyrene box for storage and transportation of packed ice cores can be seen on the right

Fig. 3.13 Somc features of the hand-excavated cave that hosts the EDML microstructure mapping (μSM) laboratory (cf. Fig. 3.10). *Top*: Unsheltered part of the μSM laboratory, showing computer-controlled microscopes for manual and automatic scanning analyses of ice sections. Behind the microscopes are two windows of the heated cabin that shelters the controlling computers and other delicate equipment. *Bottom*: Interior of the heated cabin (ca. 15 °C). The same windows shown in the top photograph are now seen from inside the cabin, on the right

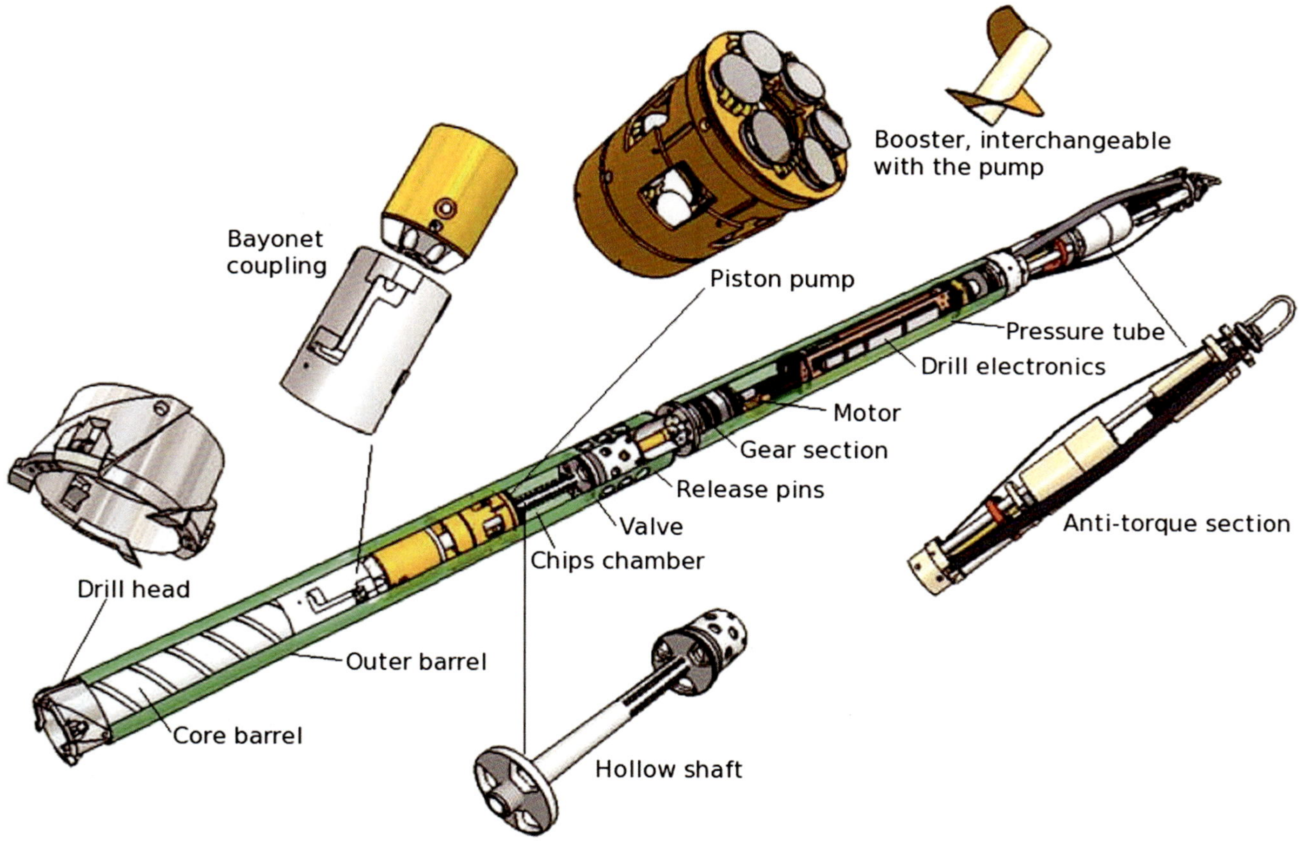

Fig. 3.14 Main components of the down-hole unit of the Hans Tausen (HT) drill. This was the shorter (6 m long) prototype version of the deep NorthGRIP and EPICA drills. The latter are mechanically identical to the HT drill, except for a much longer core barrel and chips chamber, in order to deliver in one run an ice core up to 4 m long. The HT drill bears its name from the Hans Tausen Ice Cap in Greenland, where it was first tested in 1995. From Johnsen et al. (2007)

estimate of the age of the core.[2] In addition, every ten metres a 1 cm thick stripe was cut lengthwise from the core unit (cf. physical properties cut in Fig. 3.17) for the microstructure mapping (μSM) of fresh ice-core samples in the microscopy cave[3] (Fig. 3.13).

The processed core units were thus packed in insulated polypropylene-foam boxes, six units in each box (summing up to a weight of 48 kg/box in the deeper parts of the core) and flown in small aircraft to Neumayer Station II at Atka Bay, where they were stored in large refrigerated containers. These containers were shipped to Cape Town, South Africa, by the Research Vessel Polarstern, or another supply vessel available. Finally, a commercial cargo vessel transported them from Cape Town to Bremerhaven, where the cores experienced the second stage of processing.

[2]Dielectric properties of polar ice vary with its impurity content (mainly acid and neutral-salt concentrations), which is closely related to climate proxies. Consequently, the non-destructive DEP analysis provides a first glimpse into key climatic events recorded in the ice core, which can be used to estimate the age of the ice piece.

[3]It is very important to record high-quality images of the microstructure of the core while the ice is still fresh, because relaxation and recrystallization effects, caused by the ageing of the ice core during storage, may alter or disguise its original structure.

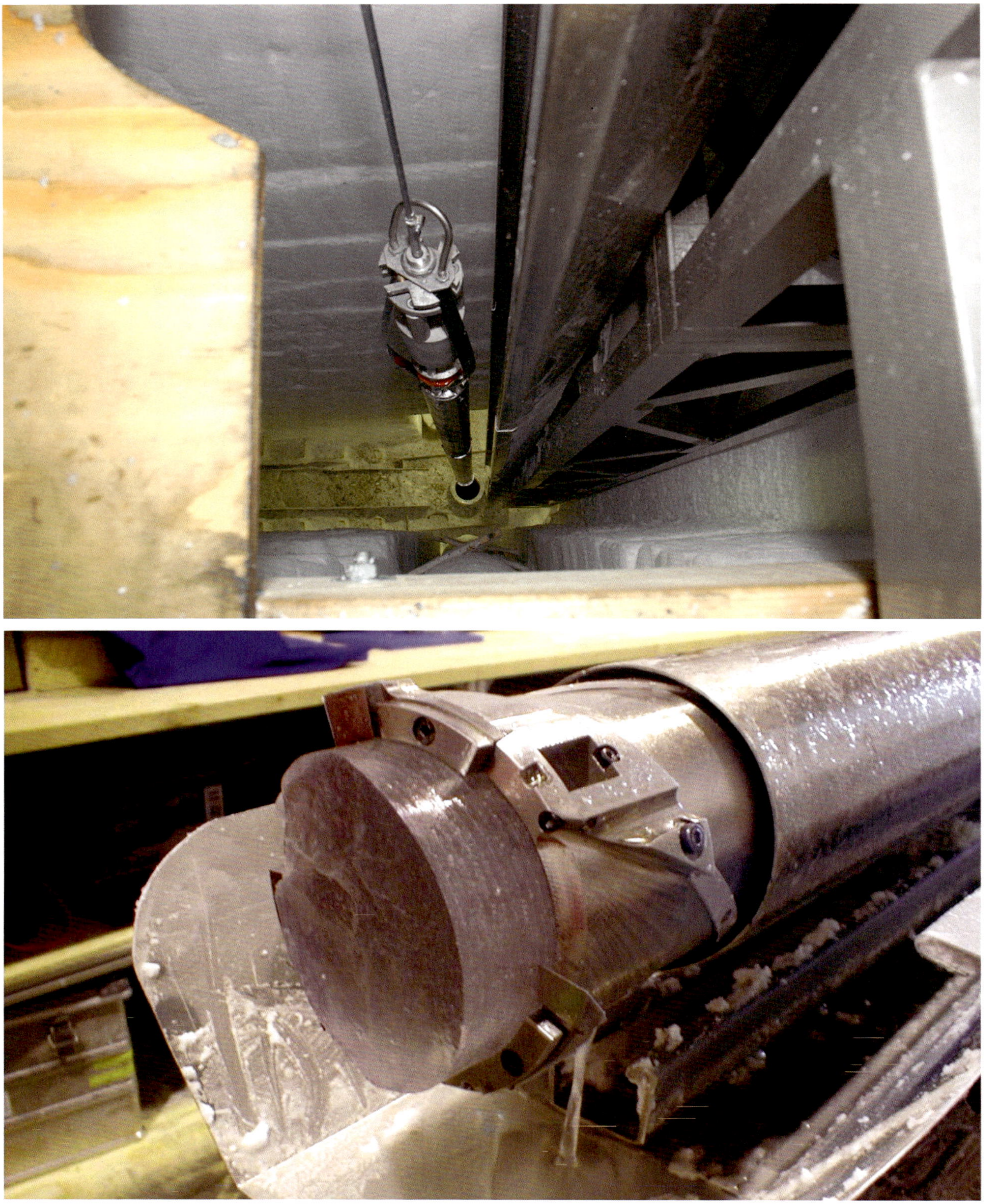

Fig. 3.15 EDML borehole and ice core. *Top*: Down-hole drill unit (suspended by the steel wire armoured cable) descending into the EDML borehole, circa 6.5 m below. Visible are the lower half of the drill tower on the right, as well as the anti-torque section at the top of the down-hole unit. The wood board on the left is part of the trapdoor that covers the inclined borehole trench (cf. Fig. 3.10). *Bottom*: A piece of the EDML ice core, still inside the core barrel (core diameter $\leq$10 cm). The drill is in horizontal position and the ice core and core barrel have been slightly shifted outside the outer barrel for better visualization. The liquid flowing out of the drill is the drilling fluid used to counterbalance the pressure inside the borehole

Fig. 3.16 Portrait of the international drilling team on the occasion of the end of the EDML deep ice-core drilling (photograph dated 17 January 2006). Mostly composed of German members, the team includes the logistics, drilling, and science groups active in the season 2005/2006. The drill is completely covered by frozen subglacial water from the bottom of the borehole. From bottom-left to top-right, *first row*: Karin Weiler (Switzerland), Tobjörn Karlin (Sweden; *dark-blue* overall), Sepp Kipfstuhl (*yellow jacket*), Jens Köhler (*dark-grey tuque*), Birthe Twarloh (no head cover), Fernando Valero (*white gloves*), and Sérgio H. Faria (*dark-blue trousers*); *second row*: Gunter Lawer (*light-blue tuque*), Dorothee Dick (no head cover), Anja Lambrecht (*red-yellow tuque*), Diedrich Fritzsche (*black ear-flap hat*), Frank Wilhelms (*multicoloured tuque*), Klaus Trimborn (*dark-blue tuque*), Heinz Miller (*grey tuque*), Hans Beiersdorf (no head cover), Johannes Freitag (*dark-blue ear-flap hat*), and Patrik Kaufmann (Switzerland; *blue jacket*); *third row*: Jochen Krischat, Marc Blattner (*black tuque*), Andreas Frenzel, Adi Ackermann (with *sun-glasses*), Cord Drücker (gripping the drill tower), Hans Oerter (*yellow tuque*), Gerit Birnbaum (*grey-black tuque*), Andreas Brehme (sun-glasses on the forehead), and Günter Stoff (with a Hein Blöd stuffed toy on the chest)

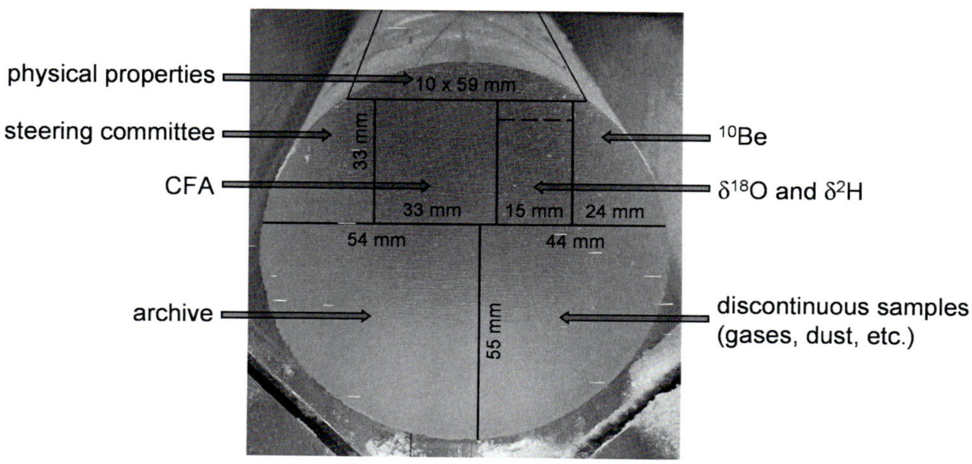

Fig. 3.17 The EDML cutting plan (courtesy of H. Oerter, AWI)

To accomplish this second stage, representatives of all EPICA partner laboratories met five times at AWI's cold room in Bremerhaven (June 2002, July and September 2003, June 2004 and April 2006). During this second processing the core was cut for diverse investigations, including stable isotope measurements ($\delta^{18}O$, $\delta^{2}H$), chemical analyses of dust, gases and other impurities, as well as physical and structural properties, following an agreed cutting scheme (Fig. 3.17). As a rule, the first cuts were performed for the analyses of physical and structural properties described in this book, namely visual stratigraphy (line-scanning) and microstructure mapping (cf. Chap. 4). After line-scanning, the core was finally cut for isotope studies, CFA, and special measurements subjected to prior approval by the EPICA Steering Committee.

4 Multiscale Structure of the EDML Core

But how can I explain, how can I explain to you? You will understand less after I have explained it. All that I could hope to make you understand is only events: not what has happened.
Harry, Lord Monchensey (fictional character),
T. S. Eliot (1960), Part I, Scene I, p. 28

4.1 Understanding Events

As explained in Sects. 2.1 and 2.3, whenever a snow layer is formed and buried by subsequent snow falls, a new portion of the ice-sheet stratigraphy and climate record begins to take shape. Besides climate records, deep ice layers also preserve microscopic structures that provide important clues about the ancient snow cover, the old pore structure of the firn layer, and past variations in the depth and thickness of the firn–ice transition zone (Bendel et al. 2013; Ueltzhöffer et al. 2010).

In the last decades, glaciologists have been measuring the signals of a wide variety of climate proxies in polar ice cores with increasing accuracy. These signals constitute fundamental evidence of climate events occurred in the past million years. Nevertheless, the bare detection of such signals is not sufficient to allow a reliable reconstruction of the paleoclimate history. The integrity of the ice-core record must also be verified. It has been commented in Sect. 2.1 that the same ice that preserves paleoclimate records can also operate on them, e.g. through the ice flow. Therefore, it is crucial to understand how these climate signals have been formed —and possibly transformed— within the flowing ice over the millennia. In other words, it is not enough to simply detect events in an ice-core record: we must understand the history of these signals, how they came into being and into their current state. Our knowledge of the processes of formation and modification of climate signals is still far from complete, therefore limiting also our ability to discriminate true climate events from mere artefacts.

Since the early times of ice-core drilling, the integrity of ice-core climate records has been assessed through a variety of methods, the most common of them being direct visual inspection of the stratigraphy, combined with microstructural analysis of selected ice sections via optical microscopy. Until recently, these methods were quite laborious and fatiguing for the scientist, who had to stay in the cold documenting the observations in written notes or photographic films. Fortunately, in the last two decades computing power and digital-imaging techniques underwent a striking progress. Modern digital cameras are today robust enough to operate under polar conditions and precise enough to successfully replace traditional film cameras. Additionally, they may be controlled by a computer that can optimize the image quality and store long series of high-resolution digital photographs automatically. New methods of stratigraphy analysis and optical microscopy that explore these modern digital technologies are presented in Sects. 4.2 and 4.4.

It should be remarked that the production of high-quality images not only facilitates the work of the glaciologist, but also opens forth new perspectives in the research of ice-core structures that could not be explored before. On the one hand, these images reveal new facts about the complexity of physical processes taking place in glaciers and ice sheets, therefore forcing us to review fundamental glaciological concepts (Faria et al. 2014b; Kipfstuhl et al. 2009). On the other hand, the investigation and quantification of such new facts demand an increasing effort towards the development of specialized methods of image analysis for glaciology, which in turn create new opportunities for multidisciplinary research. To sum up, image analysis is now ready to revolutionize ice-core physics in the same way as it has revolutionized other research fields, like medicine or engineering.

S.H. Faria et al., *The EPICA-DML Deep Ice Core*, Frontiers in Earth Sciences,
https://doi.org/10.1007/978-3-662-55308-4_4

4.2 Visualizing the Microstructure of Polar Ice

The concept of *structure*, as introduced by Thomson (1917) and Smith (1954), is understood as *the arrangement and interrelation of parts with reference to a whole*. Such "parts" are often called *structural features or elements*. Sometimes these structural features manifest themselves on various scales, giving rise to what is known as a *multiscale structure* (cf. Sects. 2.2 and 2.3). In this context, the term *microstructure* can be found in the literature with two different meanings, depending on the relative or absolute sense of the description:

- The *relative sense* is most commonly invoked when the multiscale structure is characterized by only two or three distinct scales (often called "micro-", "meso-" and "macro-scale"). In this situation the term "microstructure" refers simply to those structural features occurring on the smallest scale, irrespective of its absolute size.
- In contrast, the *absolute sense* is best suited for dealing with multiple scales, when the term "microstructure" encompasses all structural features occurring on the *microscopic scale* (loosely defined as the size scale of features or events that can be observed with a standard optical microscope, viz. between 0.2 cm and 0.2 μm, approximately).

Owing to the wide range of size scales relevant for the study of ice sheets, in this work we adopt the term "microstructure" in its *absolute sense*, as the collection of all microscopic structural features related to the crystalline lattice (e.g. grain boundaries, lattice orientations, dislocation walls) and microscopic inclusions (air bubbles and hydrates, microinclusions, etc.), cf. Fig. 4.1 and Sect. 1.2.

In general aspects, polar ice has a microstructure typical of polycrystalline materials creeping at high temperatures (recall that temperatures naturally occurring on Earth normally lie above 70% of the melting point of ice), which is characterized by curved (bulged, pinned, etc.) grain boundaries, subgrain boundaries and dislocation walls, as well as slip bands in some grains well-oriented for basal slip (cf. Fig. 4.2 and Sect. 1.2). As a consequence, even the cold, slow-deforming ice of Antarctica undergoes microstructural changes over the years under the action of continual deformation and thermally activated processes, like *dynamic recovery and recrystallization*, among others.

These microstructural changes may potentially affect the ice-sheet flow and paleoclimate records through several types of multiscale effects. For instance, on the micro- and mesoscopic scales, changes in the microstructure may alter detailed paleoclimate signals, e.g. via interactions between grain boundaries and microinclusions, transformation of air bubbles into clathrate hydrates, or diffusion and segregation of soluble impurities into grain boundaries (Barnes et al. 2003; Faria et al. 2010, 2009; Johnsen et al. 2000; Lüthi et al. 2010; Ohno et al. 2006). On macroscopic scales, microstructural changes can modify the mechanical properties of polar ice (including its rheology), causing flow instabilities that may result in stratigraphic disturbances and eventual disruption of the ice-sheet stratigraphy and related paleoclimate records (Alley et al. 1997; Azuma and Goto-Azuma 1996; Faria et al. 2010, 2009; NEEM Community Members 2013; Ruth et al. 2007; Thorsteinsson and Waddington 2002).

A number of experimental techniques have been in use to study ice-core microstructures, including optical and electron microscopy, X-ray diffraction and tomography, among others (Barnes and Wolff 2004; Faria et al. 2010, 2009; Freitag et al. 2008; Iliescu et al. 2004; Obbard and Baker 2007; Ohno et al. 2006; Wang et al. 2003; Weikusat et al. 2011a, b). In particular, optical microscopy quickly established itself as the primary method for natural ice microstructure studies, because of its simplicity, versatility, robustness, and above all: portability. Two optical microscopy techniques have made their way from post-war glaciology into the digital age (Faria et al. 2014a, b) and are commonly used today for the study of the microstructure of natural ice:

- Automatic Fabric Analysis (AFA),
- Microstructure Mapping (μSM).

Automatic Fabric Analysis (AFA) is the modern, automatized counterpart of the traditional, manually-operated analysis of *c*-axis orientations in polycrystalline ice via polarization microscopy (Wilen et al. 2003; Wilson et al. 2003). It is based on the phenomenon of birefringence (double refraction) in ice, in which incident light is split by the ice lattice into two polarized rays with mutually perpendicular planes of polarization (Fig. 4.1). The orientation of this orthogonal pair of polarization planes is determined by the incident ray and the *optic axis*. The latter coincides with the *c*-axis of the ice lattice, and defines an extraordinary refraction index for radiation polarized parallel to it.

Today, AFA is a powerful technique for studying *c*-axis orientations and strain-induced anisotropy in natural and artificial ices (Montagnat et al. 2014; Svensson et al. 2009; Weikusat et al. 2017; Wilson and Peternell 2011). It provides within minutes a wide variety of statistical data about *c*-axis orientations in an ice sample, as well as some information about its grain stereology (cf. Fig. 4.3 and Sect. 1.2). Owing to its underlying physical principles, however, the AFA method has also some limitations. In particular, the quality of the measurement depends on the optical path difference between the ordinary and extraordinary rays, which varies with the sample thickness. This means that AFA needs delicate thin sections of ice (thickness usually around 300–500 μm) with uniform thickness, which are somewhat troublesome to prepare and handle. Furthermore, the results obtained with AFA are actually *thickness-integrated*, in the sense that the observed microstructures emerge from the superposition of all features occurring through the thickness of the sample. As a consequence, the spatial resolution of AFA is constrained by the thickness of the thin section, so that microstructural features smaller than the section thickness (viz. several tenths of millimetre) cannot be properly identified and may give rise to large experimental errors. For instance, it has been shown (Faria et al. 2009, 2014b; Kipfstuhl et al. 2009, 2006; Weikusat et al. 2011b) that the disregard of small grains and localized microstructural features imposes serious cut-offs in the distributions of grain size and stored strain energy, which handicap interpretations of grain growth and recrystallization in natural ice.

From the above reasons, it follows that the analysis of c-axis orientations via AFA must be complemented by other methods capable of identifying a greater variety of microstructural features in higher resolution, in order to enable a comprehensive investigation of microstructure evolution and its effects on climate proxies of polar ice. It turns out that a recent technique of digital optical microscopy called *Microstructure Mapping* (μSM) is especially suitable for this purpose (Faria and Kipfstuhl 2004; Kipfstuhl et al. 2006; Wang et al. 2003; Weikusat et al. 2009b). It is sufficiently portable and robust to be carried with to the field, and versatile enough to be applied to a variety of ice samples, including thin and thick sections of arbitrary shape and non-uniform thickness. In fact, the possibility of analysing thick sections implies that even samples of fragile firn can be investigated via μSM (Faria et al. 2010, 2014b; Kipfstuhl et al. 2009).

The μSM method is essentially a digital form of optical microscopy. It consists of a digital video camera with automatic gain control mounted on an optical microscope equipped with a computer-controlled xy-stage (Fig. 3.13). The microscope automatically scans the whole sample, mapping a variety of microstructural features inside the ice (ranging from microinclusions and dislocation walls to air bubbles, clathrates and slip bands) with a microscopic resolution of up to ca. 3 μm/pixel. It is also very sensitive to grain boundaries and subgrain boundaries, being capable of detecting boundaries with misorientations smaller than 0.5° (Weikusat et al. 2011b).

Up to 1800 photomicrographs are needed to reconstruct a high-resolution digital mosaic image of a 50×100 mm section (Fig. 4.4). Micrographs are usually taken in transmitted light, with a standard size of 2.5×1.8 mm. Consecutive micrographs are taken every 2 mm in the x-direction. An overlapping of 0.5 mm is helpful for the later reconstruction of the full mosaic image through the matching of neighbouring micrographs. The portability of the method permits the mapping of fresh ice sections in the field, while drilling is ongoing. Under optimal conditions, the mapping of a complete ice section ($50 \times 100 \times 5$ mm, say) takes about one hour and can be accomplished as early as a few hours after core extraction, therefore avoiding extensive post-drilling relaxation of the microstructure.

Samples for μSM are prepared according to usual procedures for ice microscopy (Kipfstuhl et al. 2006; Weikusat et al. 2009b). Band saws and microtomes are respectively used for cutting and polishing the sections (extracted from the core portion assigned for *physical properties* in the cutting plan shown in Fig. 3.17). Clear surfaces are achieved by exposing the polished section to the free atmosphere: sublimation of the ice surface by thermal etching removes superficial imperfections (e.g. microtome scratches) and simultaneously highlights the sites where grain boundaries and other high-energy structural features meet the surface, through the formation of characteristic sublimation grooves and pits (Arnaud et al. 1998; Hobbs 1974; Kuroiwa and Hamilton 1963; Mullins 1957; Nishida and Narita 1996). A sublimation time varying between half an hour and half a day is usually necessary to obtain a clear surface, with well-developed grain-boundary grooves. This sublimation time depends very much on temperature and its gradient, humidity, and air circulation above the sample. The sublimation process is accelerated when the ice section lies under a light bulb, in direct sunlight, wind, or under vacuum. In general, the drier the air, the more distinct and clear the surface.

After the first (lower) surface of the sample is sufficiently clear, it is sealed off with a thin film of silicone oil (with a kinematic viscosity of ca. 10^{-4} m^2/s) and frozen onto a glass plate. The second (upper) surface is treated in the same manner, but it is sealed off with silicone oil and glass only after the first surface scan is completed, in order to optimize the quality of the μSM images. This first surface scan (e.g. Figs. 4.2, 4.5, and 4.6) aims mainly at mapping the thermally etched sublimation grooves produced by grain and subgrain boundaries, which can be qualitatively distinguished by their attributes: deep grooves with smooth traces are characteristic of high-angle grain boundaries, while subgrain boundaries and dislocation walls tend to produce shallower grooves, often with jagged, irregular or incomplete traces (Faria et al. 2014b; Saylor and Rohrer 1999; Weikusat et al. 2009b, 2011b).

In addition to this first surface scan, many ice sections are scanned again several times after sealing off the upper surface, with the microscope focused inside the sample, in order to map diverse microstructural features not related to the etched surface, like air bubbles and hydrates, microinclusions, slip bands, and most relaxation features (e.g. Figs. 2.6 and 2.7; notice that grain and subgrain boundaries are usually not visible inside the sample, mainly because they are too thin and their refraction index too similar to that of the ice matrix).

If the thermal etching is well done, the resolution of the μSM method becomes limited mainly by the resolution of the optical equipment and the digital image analysis software.

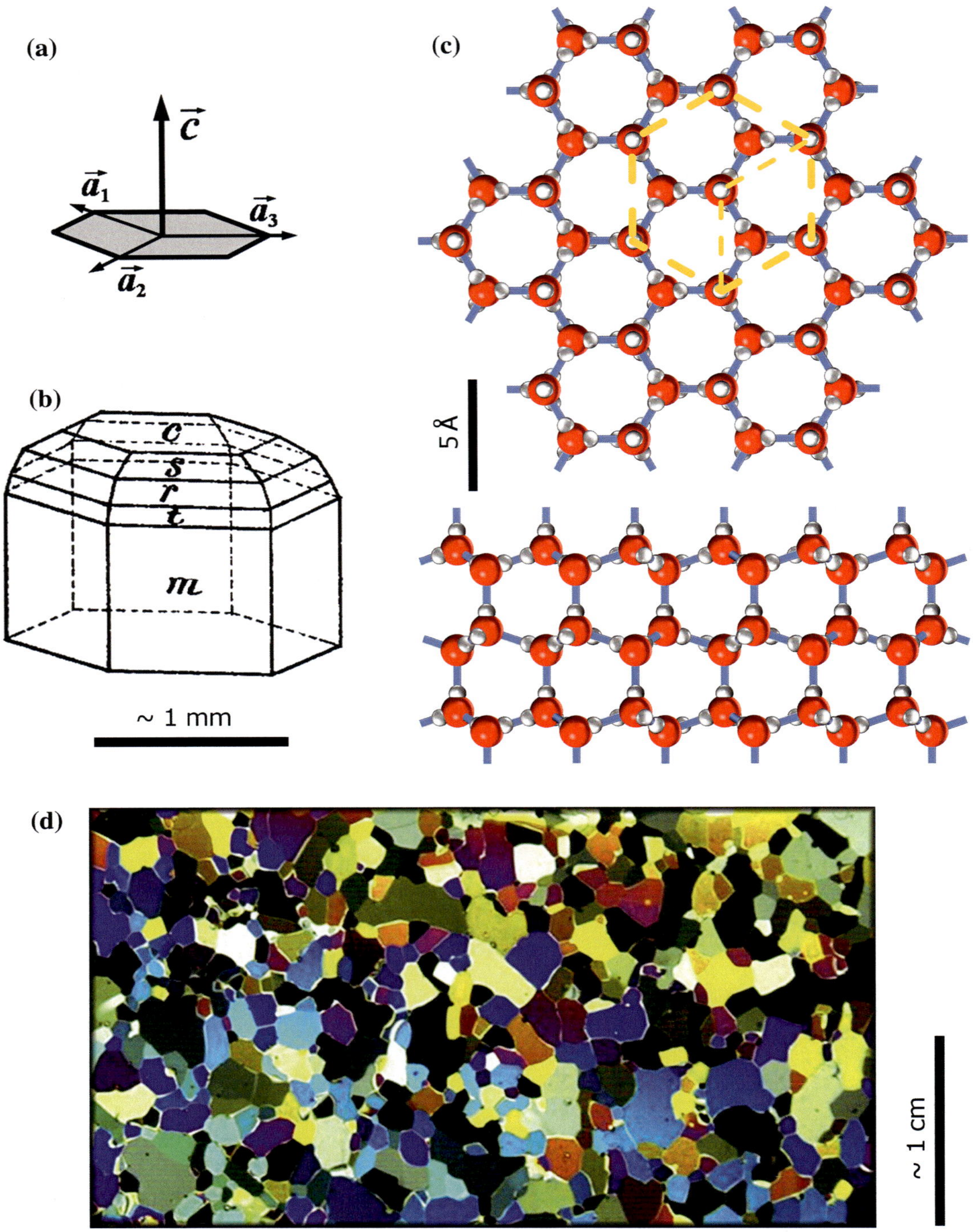

Fig. 4.1 The multiscale (poly-)crystalline structure of "hexagonal ice type I", concisely called *ice Ih*, which is the only naturally stable phase of ice on Earth. **(a)** Diagram of the main crystallographic axes of ice Ih. The c-axis is also called the *optic axis*, because it defines the uniaxial birefringence of the ice crystal. The hexagonal symmetry on the plane perpendicular to the c-axis is represented by three identical a-axes, symmetrically oriented at 120° from each other. **(b)** Facsimile of the original drawing by Nordenskjöld (1861) of an ice single crystal, illustrating the main crystallographic planes of ice Ih (cf. Fig. 2.2). Basal planes, denoted by c, are orthogonal to the c-axis. Prismatic planes (m) are parallel to the c-axis, while pyramidal planes (r, s, t) form different angles with the former (after Faria 2006). **(c)** Crystalline lattice structure of ice Ih. Water molecules are represented by red and white spheres that denote oxygen and hydrogen atoms, respectively, while *grey rods* are hydrogen bonds. Displayed are the views along the c-axis (*top*) and along an a-axis (*bottom*). The hexagonal symmetry of the lattice is highlighted by the yellow dashed line (after Faria et al. 2014b). **(d)** Thin section of polycrystalline ice between crossed polarizers, photographed in transmitted light. Incident light undergoes double refraction (birefringence) through the ice lattice, giving rise to interference colours that depend on the sample thickness and the optic-axis orientations of the various grains. Thus, each coloured domain in the image depicts an ice grain with a particular c-axis orientation, so that the interfaces between these domains are *grain boundaries*

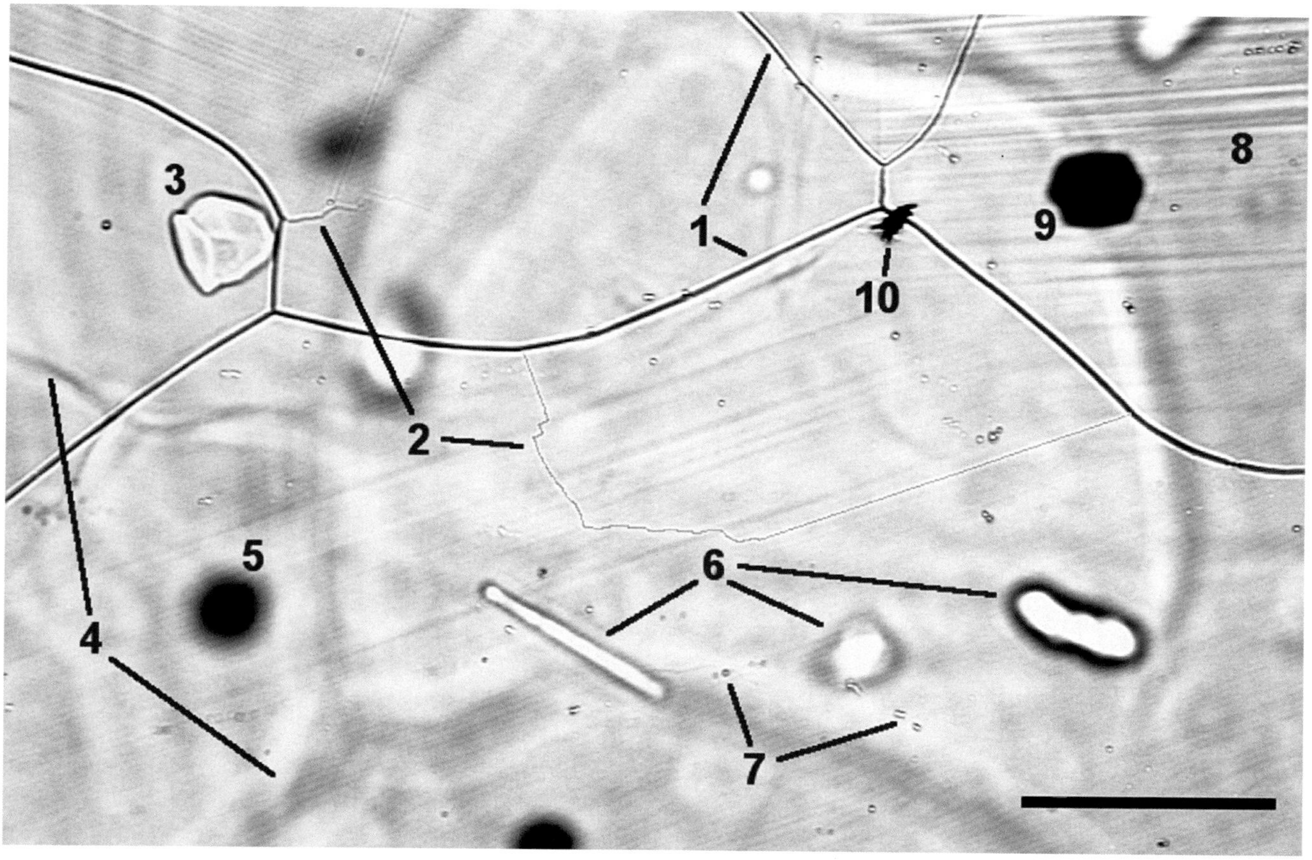

Fig. 4.2 Photomicrograph of an Antarctic ice thick section (EDML, 1155 m depth), which displays some of the most common microstructural features of polar ice. The image is focused on the thermally etched upper surface of the sample, which has been prepared according to the procedures for microstructure mapping (μSM; for more details about this microscopy method, see main text and the next figures in this section). **1** Grain boundaries meeting the upper surface (*sharp, thick black lines*). **2** Subgrain boundaries and dislocation walls meeting the upper surface (*sharp, thin grey lines*). **3** Rugged spheroidal hydrate close to the upper surface. **4** Grain boundaries inside the sample or meeting the lower surface (*blurred curvilinear shadows*). **5** Air bubble deep inside the sample (*blurred black disc*). **6** Several types of hydrates inside the sample. **7** Microinclusions close to the upper surface (*tiny spots*). **8** Slip bands (*blurred, parallel straight lines*). **9** Relaxed air bubble at the upper surface. **10** Extraneous imperfection or contamination of the sample's surface. Scale bar: 500 μm

As already mentioned, current set-ups work with optical resolutions down to 3 μm/pixel (Kipfstuhl et al. 2009, 2006). Another promising option with even higher resolution is *Electron Backscatter Diffraction* (EBSD; Iliescu et al. 2004; Piazolo et al. 2008; Prior et al. 2012; Weikusat et al. 2011a). The use of EBSD on ice poses many technical difficulties and is still in its infancy, but rapid technological and methodological developments suggest that it may become a powerful tool for high-resolution studies of ice microstructure in the near future.

4.3 EDML Ice Microstructure

Although the original purpose of the EDML Deep Ice Core was to provide answers to questions about past climate changes (e.g. EPICA community members 2006; Fischer et al. 2007a; Kaufmann et al. 2010; Siegenthaler et al. 2005; Stenni et al. 2010; cf. Sect. 2.4), the study of this core has also generated a major impact on the field of ice physics, by promoting a paradigm shift in our understanding of the genesis and evolution of polar ice microstructures. In this section we briefly review this shift.

For many decades, glaciologists had been attempting to devise a simple conceptual model for the microstructure evolution of polar ice, which could explain the microstructures observed in deep ice cores. Almost a half-century ago, such a conceptual model began to take form. As reviewed by Faria et al. (2014a, b), ideas about polar ice microstructure evolution devised in the 1970s and 1980s were gradually blended into what might be called the *Tripartite Paradigm* of polar ice microstructure (also known as the "Three-Stage Model"; Alley 1988, 1992), which many glaciologists still adopt today (cf. Fig. 4.7; the formulation below follows De la Chapelle et al. 1998; see Sect. 1.2 for definitions of the terms in *italics*):

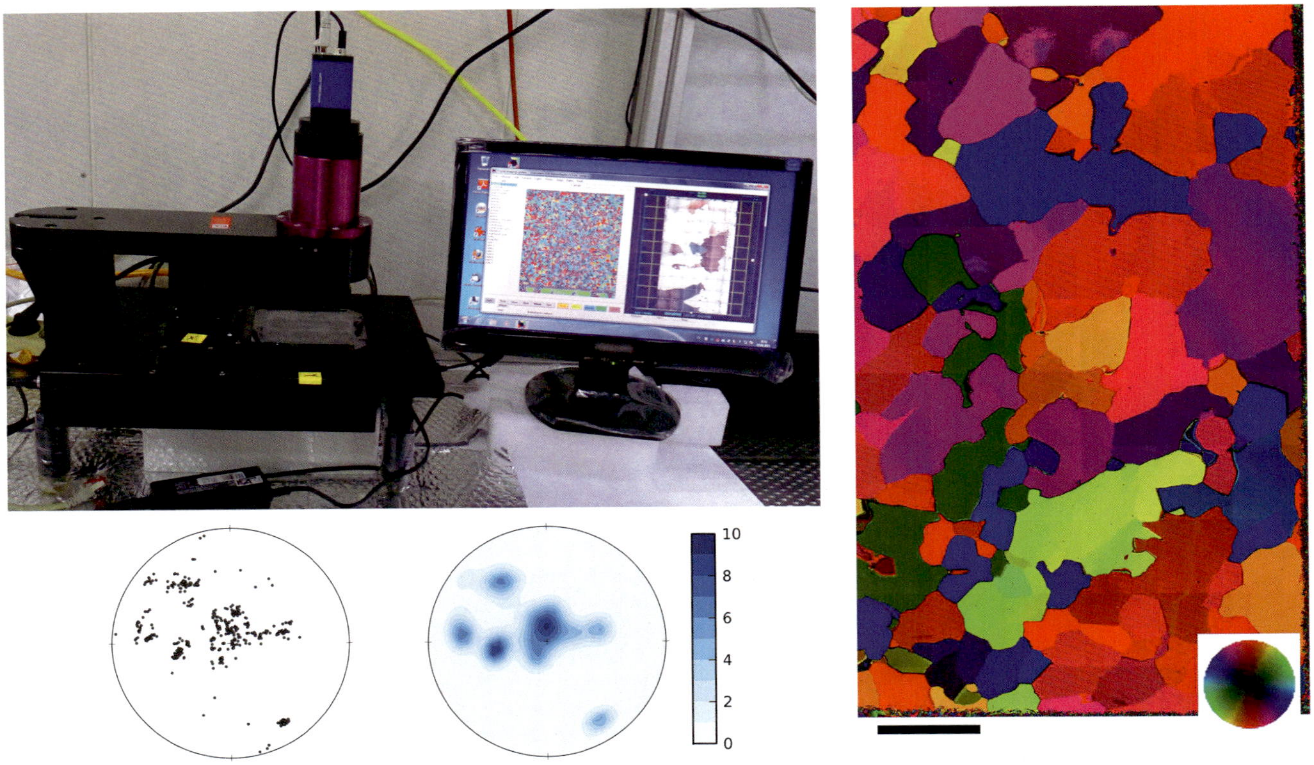

Fig. 4.3 Example of an Automatic Fabric Analysis (AFA). *Top-left*: AFA G50 by Russell-Head Instruments (www.russellheadinstruments.com). *Right*: Digital mosaic trend representation of the azimuth (colour) and colatitude (brightness) of *c*-axes in a vertical thin section of polar ice (EDML, 2306 m depth; scale bar: 1 cm). *Bottom-left*: Corresponding *c*-axis pole figures for this sample, in discrete and continuous versions. Each dot in the discrete version represents the *c*-axis orientation of one grain (or one point in a regular grid superposed on the section). The continuous version describes the respective *c*-axis orientational distribution density. The centre of the pole figures defines the vertical direction (nearly parallel to the core axis)

1. in the upper hundreds of metres of an ice sheet, grains grow in the regime of *Normal Grain Growth* (NGG; Gow 1969; Stephenson 1967);
2. in intermediate depths, NGG is counterbalanced by grain splitting via *polygonization* (Alley et al. 1995);
3. at the bottom of the ice sheet, where the ice temperature rises above circa $-10\,^{\circ}$C, *migration recrystallization* with *nucleation* of new grains (SIBM-N) markedly transforms the microstructure (Duval et al. 1983).

Since then, this paradigm has defined the status quo of our understanding of polar ice microstructures. It established the framework for interpreting the evolution of grain sizes (Alley et al. 1986a, b; Durand et al. 2006; Gow 1969; Stephenson 1967) and lattice preferred orientations (Alley 1992; Alley et al. 1995; Thorsteinsson et al. 1997), as well as the onset of dynamic recrystallization (Duval and Castelnau 1995). It set the basis for models of microstructure evolution of polar ice (De la Chapelle et al. 1998; Faria et al. 2002; Ktitarev et al. 2002; Montagnat and Duval 2000), as well as for anisotropic ice-sheet flow models (Azuma and Goto-Azuma 1996; Faria et al. 2002; Gödert and Hutter 1998; Montagnat and Duval 2000; Staroszczyk and Morland 2001; Thorsteinsson 2002; Van der Veen and Whillans 1994). It has even provided arguments in disputes about deformation mechanisms in polar ice (Duval and Montagnat 2002; Pimienta and Duval 1987).

Undoubtedly, the major virtue of the Tripartite Paradigm is its simplicity. It seems compatible with past observations of average grain size of most polar ice cores, and deems of an ice sheet consisting in great part (viz. excluding the messy bottom layers where recrystallization is supposedly active) of nice polyhedral grains growing like bubbles of froth, without being noticeably afflicted by internal stresses or localized strain inhomogeneities, except for the sporadic punishment of "polygonization" applied to those grains eager to grow too much. In spite of being so welcome and sought-after, today we know that the Tripartite Paradigm is fundamentally unsound, as revealed by a large body of evidence accumulated during the last decade, most of it derived from studies of deep ice cores like the EDML core (for a detailed discussion, see Faria et al. 2014b).

For instance, a cornerstone of the Tripartite Paradigm is the assumption that normal grain growth (NGG) dominates the microstructure evolution of polar ice in the upper hundreds of

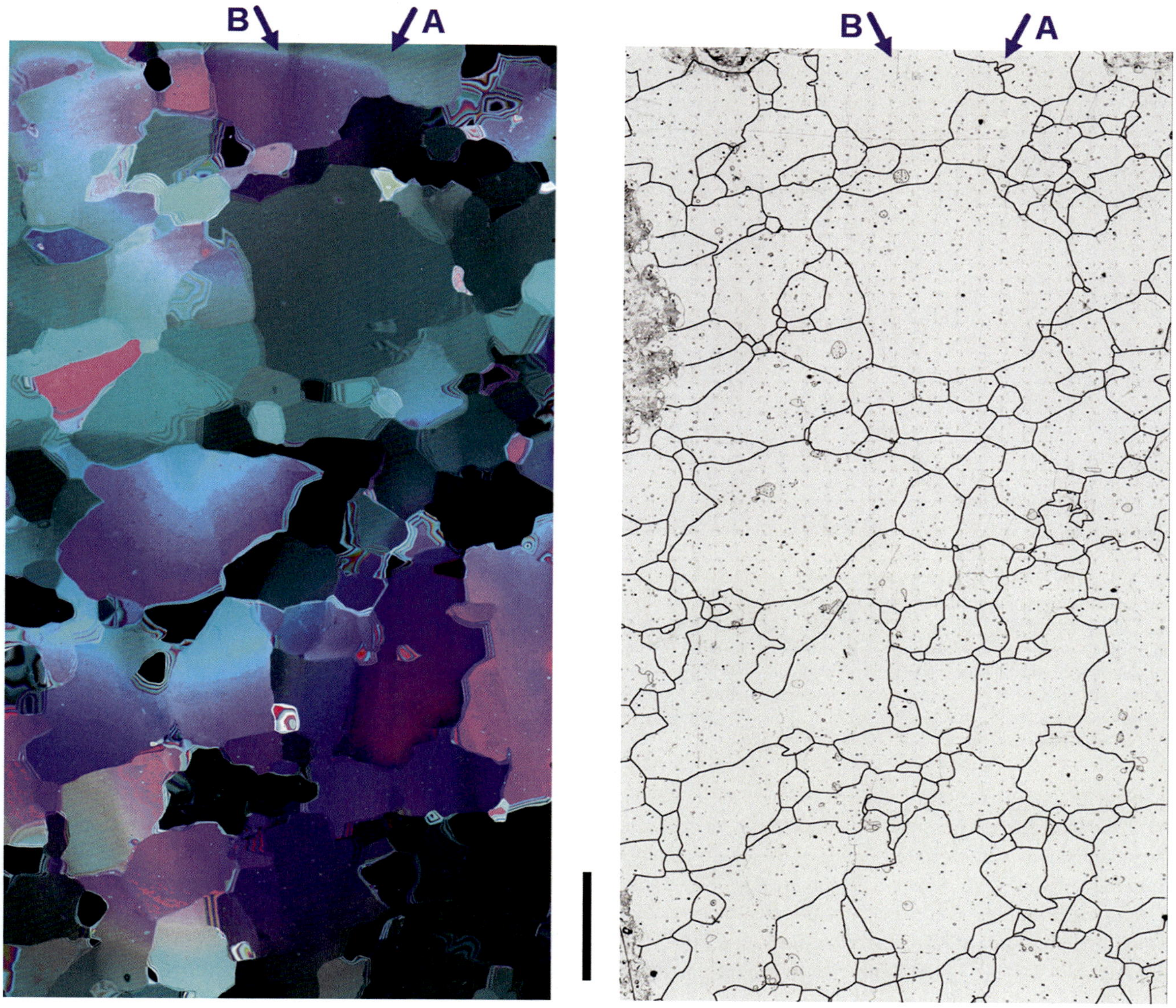

Fig. 4.4 Comparison between microstructure mapping (μSM) and a conventional method of microstructure visualization. Both images portray the same Antarctic ice thin section (EDML, 2266 m depth; scale bar: 1 cm). The *blue arrows* labelled A and B point to details displayed in high-resolution images in Figs. 4.5 and 4.6, respectively. *Left*: Conventional photographic visualization of ice microstructure with the sample sandwiched between crossed polarizers. Traditionally, images like this would be used to determine fundamental microstructural properties of the sample, like grain sizes and shapes. Each coloured domain may be identified as a particular grain (cf. Fig. 4.1) and slight colour variations within a grain suggest the presence of subgrain boundaries. This image is depth-integrated over the thickness of the section, so that inclined grain boundaries may appear blurry or generate interference fringes around certain grains. Tiny bright spots scattered over the image are mainly caused by air hydrates inside the section. *Right*: Microstructure map (μSM) mosaic image composed of more than 1800 high-resolution photomicrographs, here reduced to only a few percent of its original size. The image depicts the features of the section's bare upper surface (and some depth-integrated features as well). Dark lines (enhanced for better visualization) are grain boundary grooves on the surface, produced by thermal etching. Tiny dark spots scattered over the image are mainly caused by contours of air hydrates inside the section. Further microstructural features are hardly recognizable in this drastically reduced image, but they can be identified in the full-resolution micrographs shown in Figs. 4.5 and 4.6. A few large grey blots are caused by bubbles in the protective silicone oil between the base glass plate and the sample's bottom surface. These bubbles may be produced by various causes, including loss of silicone oil at the sample's borders or air release from decomposing bubbles and hydrates

metres of the ice sheet, including the firn layer (Fig. 4.7; Alley et al. 1986a; De la Chapelle et al. 1998; Gow 1969; Paterson 1994; Stephenson 1967). However, recent observations have proved that rotation recrystallization (RRX) and migration recrystallization with and without nucleation (SIBM-N and SIBM-O, respectively, cf. Sect. 1.2) are widespread phenomena in polar ice sheets and take place already at very shallow depths (a few tens of metres at EDML; Faria et al. 2010,

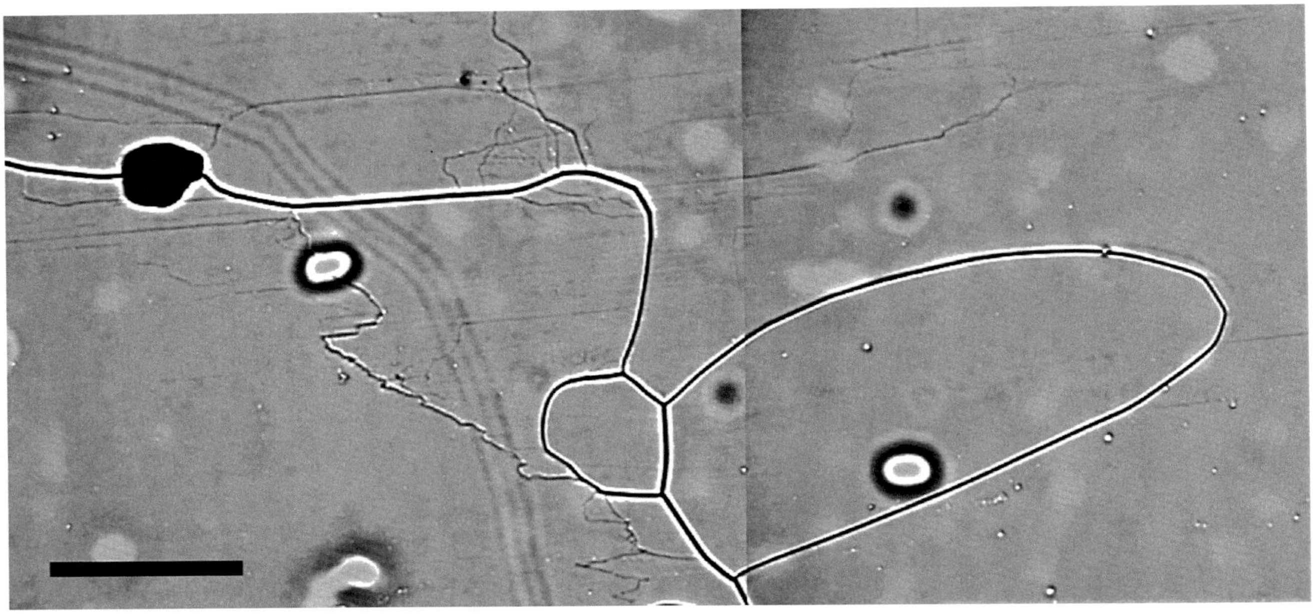

Fig. 4.5 High-resolution microstructure mapping (μSM) image of *Detail A* in Fig. 4.4 (EDML, 2266 m depth; scale bar: 500 μm). It comprises two matched micrographs showing two small grains, some strongly bulged grain boundaries and their interactions with a large number of *subgrain boundaries* and *dislocation walls* (cf. Sect. 1.2 for the definitions of the terms in *italics* in this caption). The curved, defocussed line on the left micrograph is the groove of the large grain boundary meeting the sample's lower surface. Large, round inclusions are *air hydrates* preserved inside the sample (bright) or decomposing at its surfaces (dark). Tiny spots, sharp or defocussed, are *microinclusions*. The degree of detail displayed in this image could never be achieved with traditional methods (cf. left image of Fig. 4.4) and it presents clear evidences of localized internal stresses, inhomogeneous strain, and *dynamic recrystallization*, including the formation of new grains via *rotation and migration recrystallization* (RRX and SIBM-N, respectively, cf. Sect. 1.2)

2009; Kipfstuhl et al. 2009, 2006; Weikusat et al. 2009a, b, 2011b). Nucleation is not predominant in polar ice, but newly nucleated grains can be found regularly in ice-core samples from any depth (Figs. 4.8 and 4.9), and are specially frequent in samples from the lower firn.[1] Additionally, theoretical and experimental studies of NGG in ice (Azuma et al. 2012; Roessiger et al. 2014, 2011) have shown that the microstructural evolution of shallow (bubbly) polar ice is seriously affected by processes other than NGG. These results are in accordance with the rigorous understanding that NGG cannot take place in a material undergoing deformation, since it is —by definition— a *static recrystallization mechanism* (Humphreys and Hatherly 2004; cf. Sect. 1.2). Thus, the habitual fitting of average grain size data of polar ice and firn with the standard NGG parabolic formula (the Burke–Turnbull–Hillert law) has in reality *no physical meaning*.

Considering that there is no reason to regard the EDML core as extraordinary (it has been extracted from a rather unexceptional site), we conclude that the structures observed in it are most likely representative of the typical structures and processes occurring elsewhere in the interior of polar ice sheets. Therefore, the Tripartite Paradigm needs serious reconsideration. The first steps towards the establishment of a new microstructure paradigm have already been made, as illustrated in the references cited in this chapter (see also Sect. 4.6), and this matter will certainly continue to be a prominent research topic within glaciology in the coming years. In this regard, the Tripartite Paradigm still has an important role to play as a *conceptual template* to be compared with real polar ice microstructures: the contrasts observed between the Tripartite Paradigm and reality may help us to grasp the complexity and relevance of the various microstructural processes taking place in polar ice (many of them not well understood yet), including internal stresses, inhomogeneous strains, dynamic recovery and recrystallization.

4.4 Visualizing the Stratigraphy of Polar Ice

We have already mentioned in Sect. 4.1 that a wealth of climate proxies and great detection accuracy are good but not enough to ensure a detailed and reliable reconstruction of past climate. The *integrity* of paleoclimate records must also be verified. Today, one of the simplest and most efficient ways

[1]This is a remarkable result, considering that firn at EDML has an annual average temperature of circa −45 °C. Formally, the lower firn is defined as the bottom part of the firn layer, where the firn mass density exceeds the critical value of 730 kg/m^3 (cf. Sect. 2.3). In EDML this occurs close to 60 m depth, at an overburden snow/firn load of 0.2 MPa (Kipfstuhl et al. 2009). According to Maeno and Ebinuma (1983), this critical mass density corresponds to the stage in which the firn particles reach optimal packing and densification becomes controlled by dislocation creep.

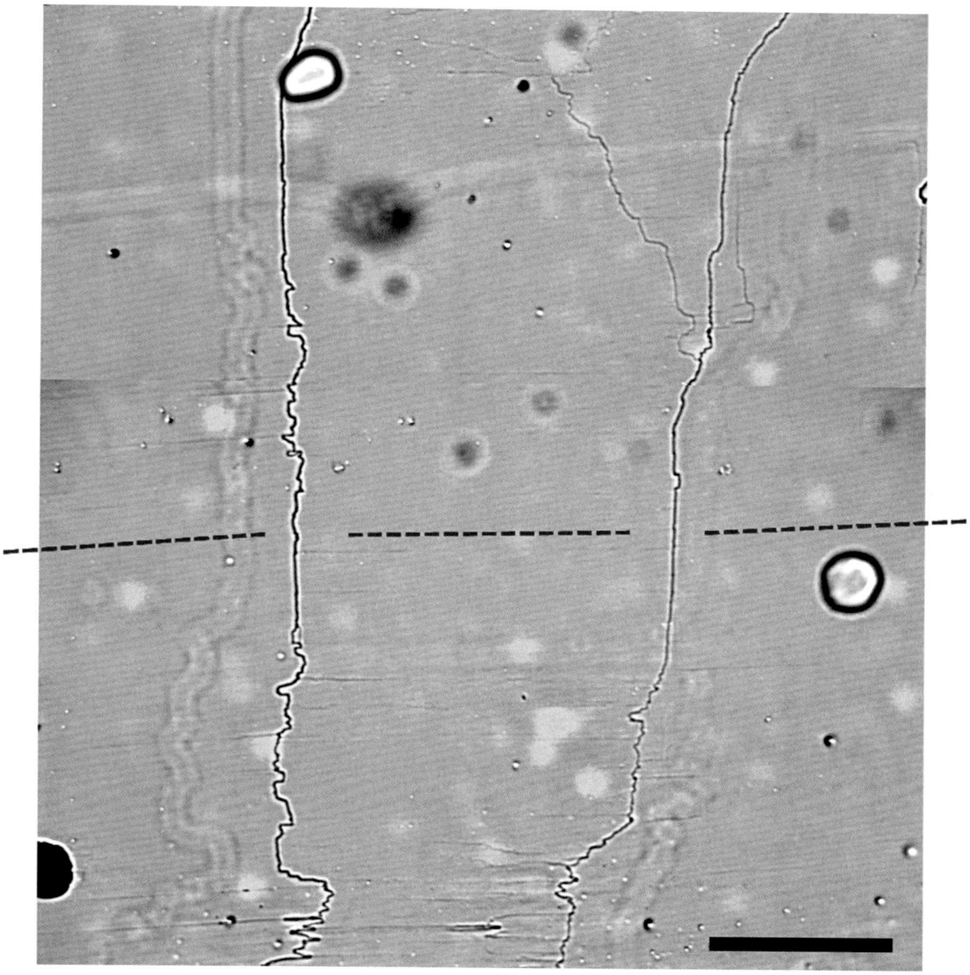

Fig. 4.6 High-resolution microstructure mapping (μSM) image of *Detail B* in Fig. 4.4 (EDML, 2266 m depth; scale bar: 500 μm). It comprises two matched micrographs showing the details of a typical *kink band* (cf. Sect. 1.2 for the definitions of the terms in *italics* in this caption). The *blue dashed lines* indicate the orientations of the visible *slip bands*, which are inclined with respect to horizontal by ca. 4.4°, 1.0° and 3.1° (left, centre, and right, respectively). Thus, the left *tilt boundary* has a slightly greater misorientation than the right one, as evidenced by its stronger trace (groove) etched on the section's upper surface. Also visible are the defocussed traces of the tilt-boundary grooves on the sample's lower surface, as well as many irregular *dislocation walls* emanating from the tilt boundaries. Large, round inclusions are *air hydrates* preserved inside the sample (*bright*) or decomposing at its surfaces (*dark*). Tiny spots, sharp or defocussed, are *microinclusions*

of performing a precise but comprehensive inspection of the paleoclimate-record integrity of an ice core is through the *optical scanning* of a lengthwise polished surface of the core. In the last decade, ice core line-scanners have been developed and tested (McGwire et al. 2008; Svensson et al. 2005; Takata et al. 2004). Contemporary devices can produce high-resolution digital images (ca. 0.1 mm/pixel) of the whole core length, which can later be enhanced and analysed digitally with all potentialities of modern image processing.

The operating principle of an *ice-core line-scanner* (LS) is very simple. First, an ice slab is cut lengthwise from the core (Fig. 4.10) and its upper and lower surfaces are polished, e.g. with a special microtome. The polished slab is then placed on a dark background and illuminated from below by two oppositely inclined planar light beams, while a digital line-scan camera scans the ice surface from above along a single line across the slab (Fig. 4.11). The whole optical system (light source and camera) is mounted on two trolleys, located below and above the ice core slab, which move synchronously along the core axis.

This particular configuration implies that the scanned region is just indirectly illuminated, in such a manner that only the light scattered by the slab is captured by the camera. Since ice is in principle transparent, it follows that the captured light must have been scattered by inclusions in the ice matrix, like microscopic particles or air bubbles. In other words, the degree of brightness in a particular portion of the LS image is proportional to the amount of inclusions in there, so that a clear piece of ice free of inclusions appears dark (Fig. 4.12). In this sense, ice-core line-scanning may be regarded as a macroscopic counterpart of dark-field microscopy.

Because of the high resolution and sensitivity of the digital line-scan camera, LS images display the visual stratigraphy of an ice core in great detail. Strata thinner than 1 mm can be easily detected with this method. Any small change in the optical properties of the ice matrix, usually caused by a change in the size or concentration of inclusions, gives rise to a new horizon. As a rule, such horizons can be regarded as *isochrones*, since they can be traced back to the chemical and physical conditions on the ice-sheet surface at the time of snow depo-

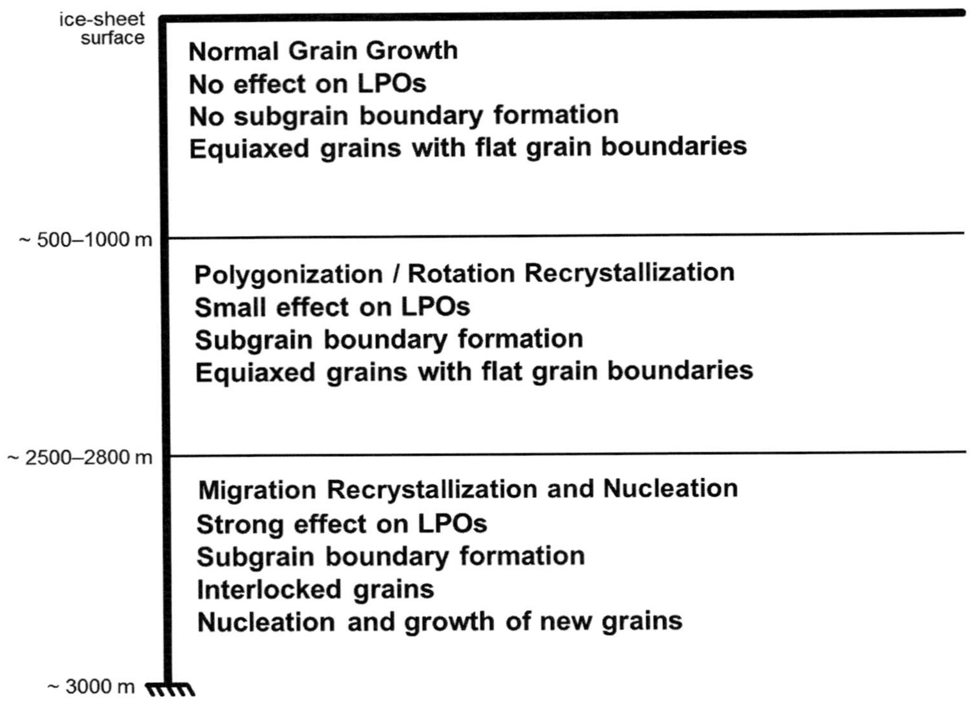

Fig. 4.7 Main features of the *Tripartite Paradigm* of polar ice microstructure. For illustration, approximate critical depths implied by the Paradigm for a site with 3 km of ice thickness are indicated on the *left*. The acronym "LPOs" stands for *Lattice Preferred Orientations* (often called *fabrics* in the glaciological literature; see Sect. 1.2 for definitions)

sition and transformation into firn and ice. Glaciologists have exploited this fact in attempts to estimate the thickness of past annual layers via spectral analysis (Svensson et al. 2005; Takata et al. 2004), which would be a valuable information for ice-core dating. However, the methods of identification and interpretation of LS horizons are not trivial, and still need improvement.

At this point it is important to remark that, to date, LS images actually offer an *optimistic estimate* of ice-core stratigraphy integrity, in the sense that they provide sufficient, but not necessary, evidence of stratigraphic disturbances. This stems from the fact that ice-core drilling has not yet succeeded in recording the azimuthal orientation of an entire ice core (relative to the ice flow or any other absolute reference). Admittedly, the mutual orientation of a series of ice-core pieces within a certain depth interval can sometimes be determined through the matching of core breaks (cf. *logging* in the Glossary of Sect. 1.2), but this method is doomed to failure after many metres, because of the accumulation of interferences and errors (e.g. matching ambiguity, human fault, missing core pieces, etc.). Therefore, as a rule, we must assume that the LS images are randomly oriented with respect to the ice flow. Any stratigraphic disturbance may be just as bad as, or worse than, that displayed in the LS image. Likewise, the true inclination of a layer with respect to horizontal may be equal to, or larger than, the inclination displayed in the LS image.

Future advances in ice-core drilling technology shall hopefully overcome the challenge of recording the absolute orientation of an entire ice core. This will greatly enhance the capabilities of the LS method, by allowing us to keep the LS images always oriented with the direction of local ice flow.

4.5 EDML Visual Stratigraphy

The visual stratigraphy of polar ice cores in general is dominated by air bubbles down to the beginning of the *bubble–hydrate transition zone* (BHT zone), which sets in at around 700–800 m depth in Antarctica, and ca. 300 m deeper in the warmer Greenland ice (cf. Sect. 2.3). Layering in the bubbly ice above the BHT zone is mainly governed by depth variations in the collective structure of bubbles (mainly number and size), which are by far the largest and most efficient light scatterers. These variations can sometimes be quite abrupt and may be traced back to diverse processes of snow deposition and firn metamorphism that took place when the material was originally deposited on the ice-sheet surface.

In the EDML core, visible layers are fairly horizontal and parallel to each other throughout the bubbly-ice zone. This observation is not uninteresting, in view of the diverse densification processes taking place in the snow and firn layers, as well as the multiscale roughness of the ice-sheet surface, which is dominated by a patchwork of snow types, sastrugi and snowdrifts (cf. Sect. 2.3).

Within the BHT zone, which in EDML covers the depth interval 800–1200 m, stratigraphic variations in the number and size (i.e., collective structure) of air bubbles gradually increase with depth. Noteworthy is the extreme case of certain

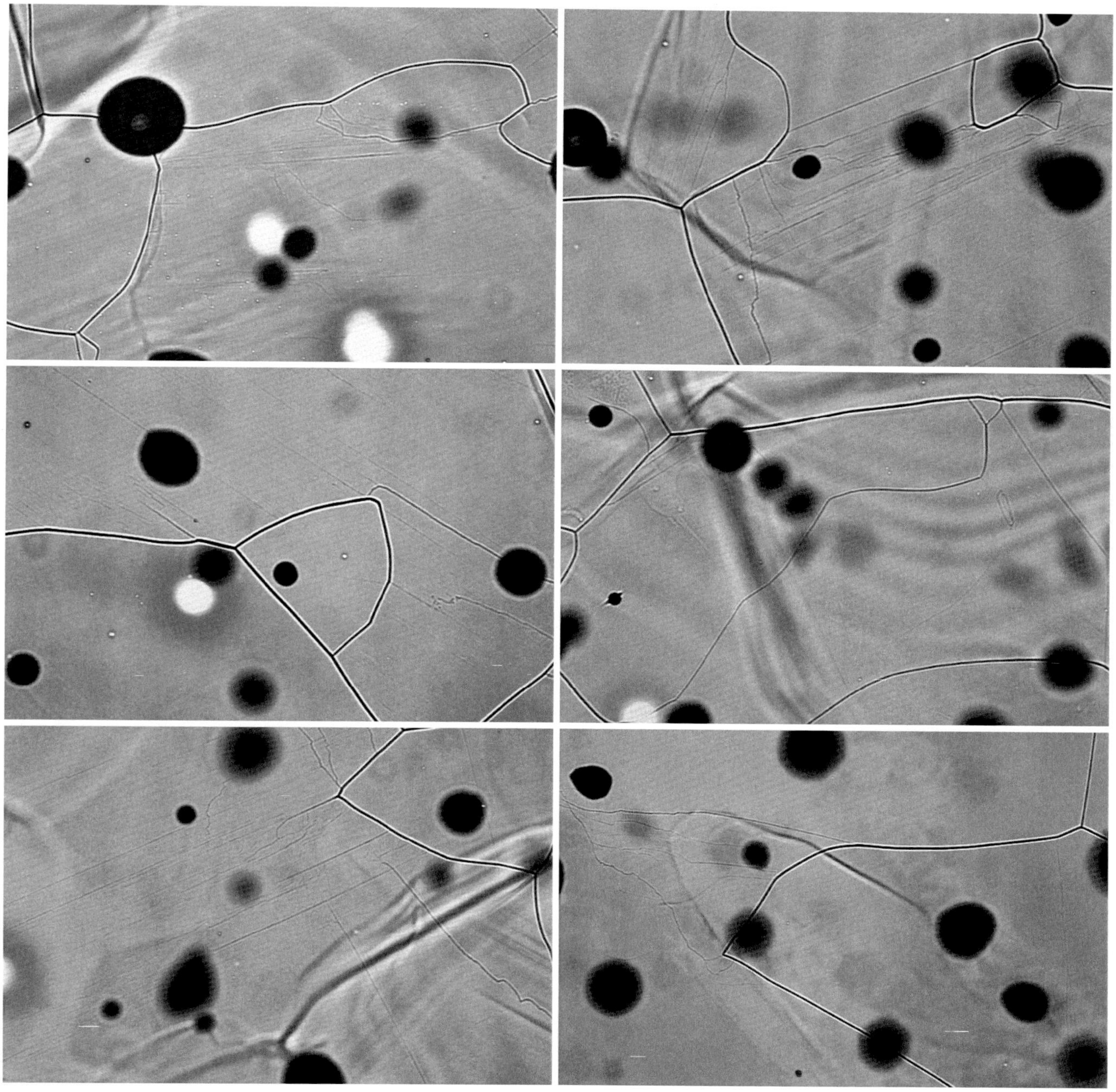

Fig. 4.8 Evidences of rotation and dynamic recrystallization in a thick section of EDML ice from 556 m depth (bubbly ice). The keys to interpret these images are provided in Figs. 4.2, 4.5 and 4.6. Air bubbles appear as black discs. A few glares and fringes in some images are caused by reflection and diffraction effects in the interior of the ice sample. Width of each micrograph: 2.5 mm. Definitions of the terms in *italics* are presented in Sect. 1.2. Further images of EDML ice microstructures are available in Fig. 4.9 and in the works by Faria et al. (2010, 2009, 2014b); Kipfstuhl et al. (2009, 2006); Weikusat et al. (2009b, 2011b). *Top-left and -right*: Typical examples of *migration recrystallization* with *pseudo-nucleation* (SIBM-N). A portion of the grain boundary bulges and advances over a region with high *stored strain energy* (revealed by many *dislocation walls*). Simultaneously, the material inside the bulge rotates, forming one or several subgrain boundaries that delimit the back of the bulge. *Centre-left*: If the the region inside the bulge continues to rotate with respect to the parent grain, the result is a new, two-sided little grain like this. *Centre-right*: Another mechanism of pseudo-nucleation, without the bulging of grain boundaries. It occurs through the suitable combination of subgrain boundaries inside the parent grain, forming a subgrain boundary loop either at a grain boundary (top-right side of the micrograph) or as a subgrain island inside the parent grain (centre-right side of the micrograph). *Bottom-left and -right*: Common examples of migration recrystallization without nucleation (SIBM-O). Notice the dislocation walls concentrated at the convex side of the bulged grain boundaries

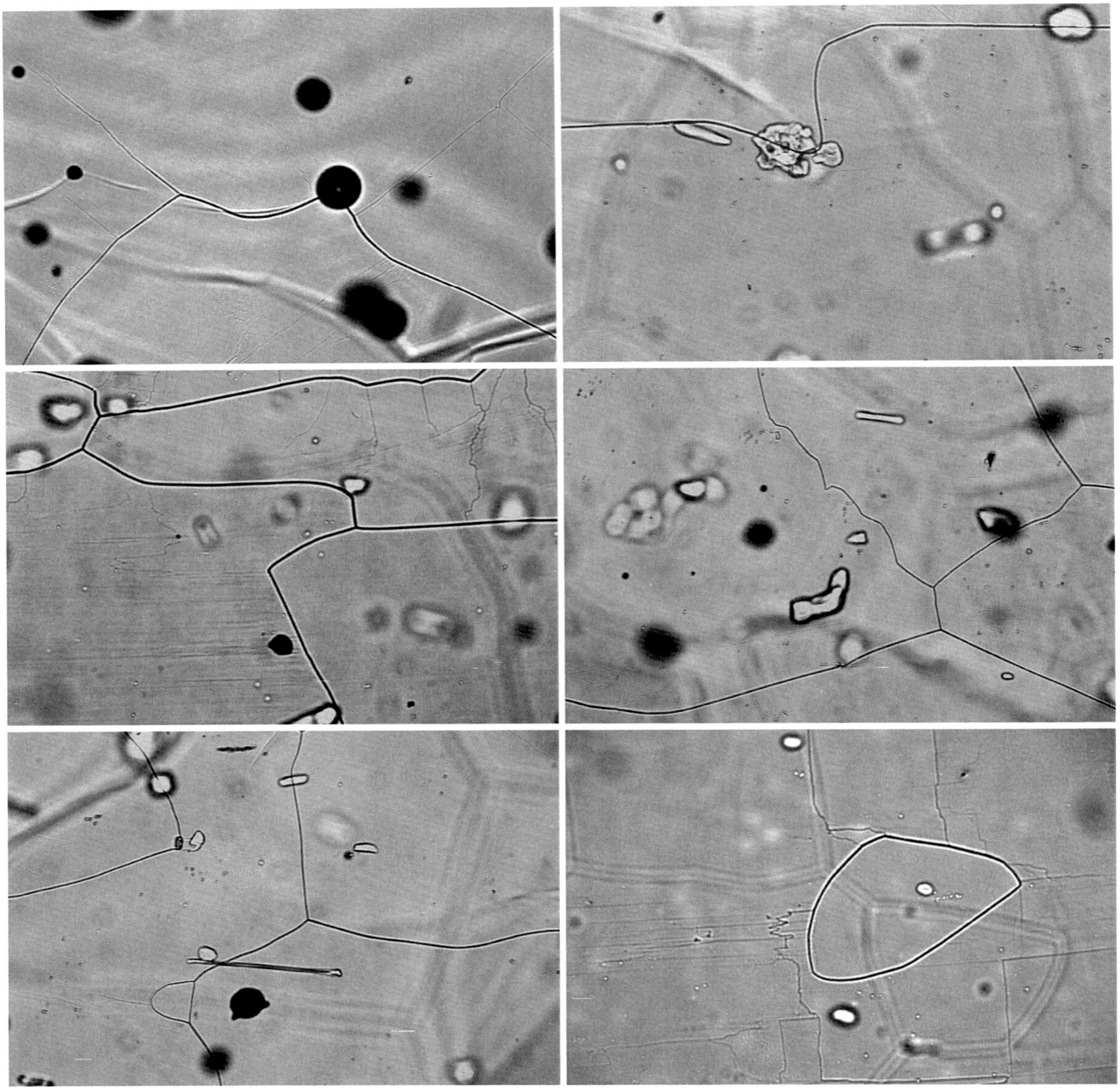

Fig. 4.9 Grain boundary pinning and dynamic recrystallization in EDML ice. The keys to interpret these images are provided in Figs. 4.2, 4.5 and 4.6; see also Fig. 4.8. Air bubbles appear as smooth black discs, while the variously shaped bright objects (from rough lumps to straight needles) are air hydrates. Tiny spots scattered over the samples are mostly microinclusions. Width of each micrograph: 2.5 mm. *Top-left*: Grain boundary pinning by air bubbles (556 m depth). Notice that the pinning of the grain boundary by the small bubble on the left created a visible corner in the grain boundary in the interior of the ice section. *Top-right*: Grain boundary pinning by an air hydrate (1156 m depth). *Centre-left*: Grain boundary pinning by subgrain boundaries (top-left side of the micrograph; 1156 m depth). *Centre-right*: Grain boundary pinning by clusters of microinclusions (1156 m depth). *Bottom-left*: Grain boundary pinning by an air hydrate (top-left side of the micrograph) and pseudo-nucleated two-sided grain (bottom-left side; 1156 m depth). *Bottom-right*: Grain boundary loop surrounded by subgrain boundaries and dislocation walls (2266 m depth). The defocussed lines of grain-boundary grooves on the lower surface of the section reveal that the loop is actually not a "grain island" fully immersed inside another grain, but rather just the cross-section of the protuberance of a small grain into a larger one

layers completely devoid of bubbles, called *bubble-free bands* (Bendel et al. 2013; Faria et al. 2010; Lüthi et al. 2010). With thicknesses ranging from a few millimetres to some centimetres, such bands first appear in the EDML core at ca. 950 m depth, and become remarkable stratigraphic features below 1000 m. Microstructure-mapping micrographs confirm that

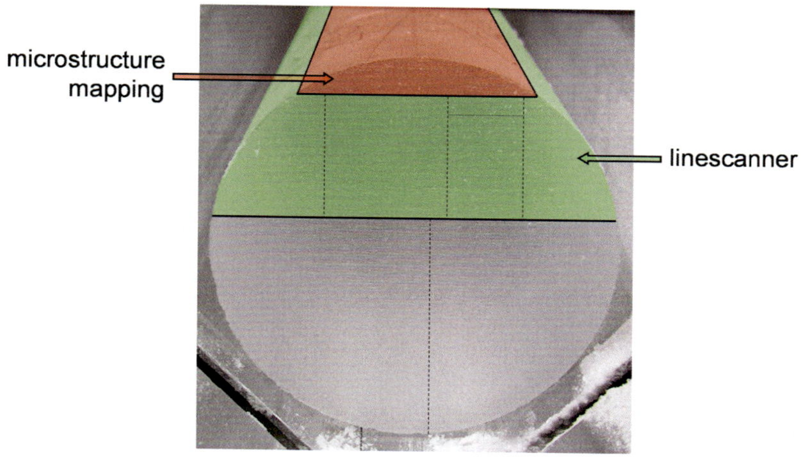

Fig. 4.10 Same as Fig. 3.17, but highlighting the first cuts to produce samples for microstructure mapping (*red*) and visual stratigraphy via line-scanning (*green*). The latter is placed on the line-scanner as shown in the figure, that is, the line-scanner camera is focused on the upper (smaller) surface of the ice slab (cf. Fig. 4.12)

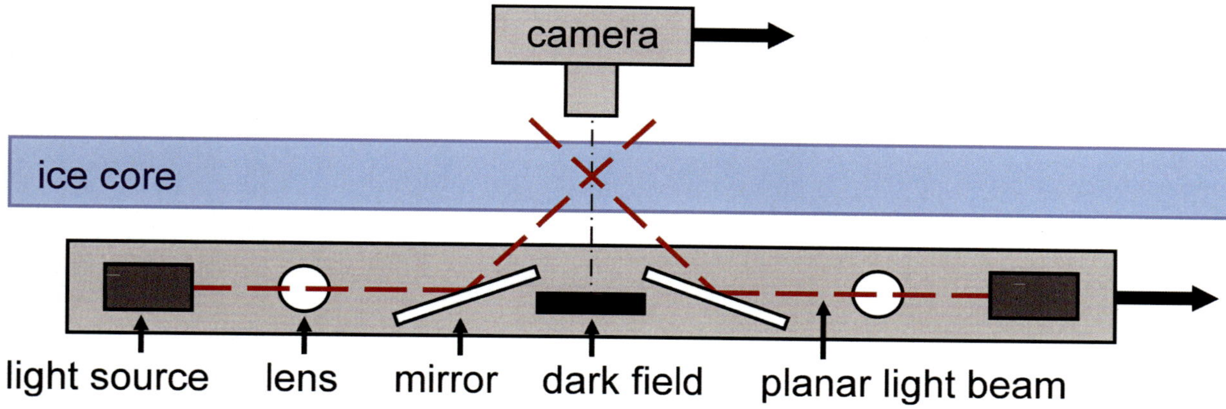

Fig. 4.11 Schematic drawing of a typical ice-core line-scanner. The polished ice-core slab is placed on a dark background and illuminated by planar light beams across the core, while a digital line-scan camera scans the upper surface from above along a single line. The light from the source enters the ice at an angle of 45°, so that the camera detects only light that is vertically scattered from inside the ice. Thus, transparent ice appears dark in the record (cf. Fig. 4.12). From Svensson et al. (2005), with modifications

all air bubbles in such layers have already been converted into hydrates, while the adjacent ice is still rich in bubbles. Also, they reveal that the ice in bubble-free bands is extremely fine-grained (average grain diameter < 1 mm) and that such bands form preferentially in impurity-rich layers that originally contained many small bubbles. The reason for the faster transformation of bubbles into hydrates in these layers is still an open question. A model of favoured hydrate nucleation in small air bubbles has been proposed by Lipenkov (2000), while Shimada and Hondoh (2004) and Ohno et al. (2010) suggest that chemical impurities and microinclusions could act as catalysts for the nucleation of air hydrates.

It is important to remark that the depth of the BHT zone at EDML coincides with that of the transition from the Holocene (chronologically back) to the last glacial period (MIS1–MIS2 transition), which is characterized by a noticeable increase in the concentration of soluble impurities and microinclusions (Faria et al. 2014a). Accordingly, the intensity and frequency of cloudy bands in the BHT zone is also enhanced. An unfortunate consequence of this depth coincidence is that it becomes very difficult to discriminate BHT effects from climate-change effects on the EDML ice structure in the BHT zone.

Below the BHT zone, all bubbles have transformed into air hydrates, which possess a refractive index comparable to that of ice. As a consequence, bubble-free deep ice is in principle quite transparent, even when it is rich in hydrates. This fact leads to the observation that the stratigraphy below the BHT zone is dictated neither by bubbles nor by hydrates; rather, it is characterized by variations in the concentration of *microinclusions*. In sufficiently high concentrations, microinclusions can scatter the incident light and make the ice seem

Fig. 4.12 The EDML line-scanner. *Left*: The ice-core line-scanner (LS) used for recording the visual stratigraphy of the EDML Deep Ice Core was developed at the Alfred Wegener Institute in Bremerhaven, and later improved at the Niels Bohr Institute of the University of Copenhagen. It is designed to scan ice cores up to 170 cm long and 10 cm wide, with a linear resolution of 115 pixels/cm. **(1)** Digital line-scan camera. **(2)** Trolley supporting the camera and light source. **(3)** Reference ruler. **(4)** One-metre long ice-core slab. *Right*: Example of an LS image of the EPICA Deep Ice Core. Depth interval: 1813.32–1813.52 m. Clear ice appears dark, while microinclusion-rich ice strata (viz. *cloudy bands*, cf. Sect. 1.2) appear in light shades of grey. The bright speckles scattered over the image are produced by air hydrates. The reference ruler appears on the left

opaque (Gow and Williamson 1976), forming strata of light-grey appearance called *cloudy bands* (cf. Sect. 1.2).

Like its northern counterparts (e.g. GRIP, GISP2, NGRIP), also the EDML Deep Ice Core contains a large number of cloudy bands, ranging from 1 mm to 10 cm in thickness, which are sometimes so intense as to be recognized with bare eyes in the raw ice core (viz. before cutting). Owing to their intensity and frequency, cloudy bands highlight the stratigraphy of the ice sheet and can be used as macroscopic markers for verifying the integrity of the paleoclimate records of an ice core: as long as cloudy bands are horizontal and parallel to each other, there are good reasons to believe that the paleoclimate records are well-preserved and are not jeopardized by large-scale disturbances induced by the ice flow.

On the mesoscale of cloudy bands, however, it becomes more difficult to assess the consequences of centimetre-sized flow disturbances (like undulations or folds) for the integrity of paleoclimate records. The presence of such flow anomalies is certainly an important warning of possible record disturbances, but the quantification of these disturbances usually requires an individual examination of the problem, including the type of climate proxy under analysis and the methods of signal detection (Faria et al. 2010). For instance, it is evident that inclined layers, even when they are nicely flat and parallel to each other, do change the signals measured through cross-section-integrated methods (e.g. CFA, DEP, ECM; cf. Sect. 1.2) because the layer inclination mixes up the signals of neighbouring strata and affects the variability of the recorded data. Notwithstanding, the average values of the record are usually still preserved in such cases.

It should be remarked that cloudy bands are generally characterized by much smaller grains than their surrounding clearer ice. The cause of such small grains is still not completely understood, but it is believed to emerge from a combination of two main phenomena:

- The well-established correlation between high impurity concentrations (especially of microinclusions) and small grain sizes indicates that grain growth in such bands may be inhibited by impurities and/or microinclusion (Zener) pinning (Durand et al. 2006; Faria et al. 2010, 2009, 2014a; Paterson 1991, 1994).
- Enhanced LPOs (lattice preferred orientations), widespread undulose extinction, and characteristic grain shapes often observed in cloudy bands suggest that small grains are continually produced by rotation recrystallization and nucleation (RRX and SIBM-N; cf. Sect. 1.2) induced by some kind of softening mechanism (Faria et al. 2010, 2009, 2014b; Gow and Williamson 1976; Paterson 1991, 1994).

This is a crucial issue in polar glaciology, seeing that a significant softening of cloudy-band ice could potentially lead to inhomogeneous layer thinning in the ice sheet, which would cause severe complications for ice-core dating. Therefore we conclude that a clear understanding of the relation between polar-ice flow and stratigraphy on the meso- and macro-scales cannot be fully achieved without a detailed analysis of the ice microstructure and related processes on the micro-scale.

From the bottom of the BHT zone at circa 1200 m depth down to around 1600 m depth, the EDML stratigraphy remains pretty ordered and shows just minimal undulations in some cloudy bands, with amplitudes no larger than a few millimetres. Below 1600 m depth, these undulations gradually increase in intensity and frequency, to the extent that at 1800 m depth stratigraphic disturbances reach the centimetre scale and some flimsy cloudy bands become disrupted. Notwithstanding, the layering remains fairly horizontal and it is only below 2050 m depth that small folds (with amplitudes up to 1 cm) and sloped strata (inclined up to 15° from the horizontal) become evident. Further down, the intensity of such disturbances increases noticeably: cloudy bands appear ragged and fuzzy at diverse inclinations, sometimes up to 30° with respect to horizontal.

Remarkably, below the precise logged depth of 2385 m all cloudy bands become neatly parallel, straight, and well-defined again, with marked alternations between cloudy and clear ice. Faria et al. (2010, 2006, 2009, 2014a) have pointed out that this striking stratigraphy change accurately coincides with a sharp increase in impurity content that marks the transition (chronologically backwards) from the last interglacial (MIS5e) to the penultimate glacial period (MIS6), ca. 130,000 years ago. It coincides also with conspicuous changes in ice microstructure and rheology (e.g. borehole constriction) at that depth, including the most abrupt reduction in grain size in the whole core, from an average diameter of 5 mm to circa 1 mm within a depth interval of only 20 m.

Drastic variations in the intensity, inclination and folding of cloudy bands in the EDML core are recognized below 2405 m depth, which once again can be related to changes in the ice microstructure and rheology (Faria et al. 2010, 2006, 2009, 2014a). Intense cloudy bands, strata inclined up to 45°, larger folds (amplitudes > 0.1 m) and other serious stratigraphic disturbances are frequent in this deep zone. As a general trend, however, the intensity of the cloudy bands gradually reduces with depth until they completely disappear below 2600 m, where the ice temperature rises above -7 °C and the average grain size increases dramatically, with grains as large as 0.5 m in diameter.

4.6 Summary of the EDML Multiscale Structure

Based on the information presented in the previous sections and chapters, we can visualize the multiscale structure of the whole EDML Deep Ice Core as consisting of eight fundamental depth zones, which are schematically illustrated in Fig. 4.13 and summarized below.[2] Definitions of acronyms and technical terms are available in Sect. 1.2. Approximate ages (based on the EDML1 time scale published by Ruth et al. 2007) are shown within square brackets in *thousands of years ago* (ka).

1. *Firn zone: from the ice-sheet surface* [*present*] *down to the firn–ice transition depth at 88 m* [*0.9 ka*]. Granular snow (Fig. 2.2) transforms into porous firn (Figs. 2.4 and 2.5), which continues its metamorphism to bubbly ice with clear evidence of *dynamic recrystallization* (RRX, SIBM-O, SIBM-N) especially in the lower firn (60–88 m depth; Kipfstuhl et al. 2009). With an annual mean temperature of ca. -45 °C, this requires *very inhomogeneous deformation* and *high internal stresses* in the firn skeleton (Faria et al. 2014b). Ice grains originate from snow crystals and reach an average diameter of ca. 1.5 mm at the bottom of the firn zone. Initial surface irregularities (sastrugi, snow-drifts, wind crusts, etc.) are levelled out already in the first metres of firn (Figs. 2.3, 2.4 and 2.5), resulting in a visual stratigraphy that is nicely flat and ordered, governed by *firn porosity* variations with depth.
2. *Bubbly ice zone: from 88 m* [*0.9 ka*] *to 800 m depth* [*14.6 ka*]. Ice microstructure evolution continues to be strongly influenced by *rotation and migration recrystallization* (RRX and SIBM-O), with occasional *pseudo-nucleation* (SIBM-N; Fig. 4.8; Faria et al. 2010, 2009; Kipfstuhl et al. 2006). *Lattice (c-axis) preferred orientations* (LPOs) start to develop in the form of a *vertical girdle*, which is compatible with the dominant flow regime in the upper part of the EDML ice ridge, viz. *transversal elongation* (Fig. 3.7). Modest increase in grain size to circa 2.4 mm average diameter below 700 m depth. Variations in the *collective structure of air bubbles* (mainly number and size) dominate the visual stratigraphy of this zone, owing to their predominance and superior light-scattering power. First cloudy bands can barely be identified below 700 m depth, as air bubbles get compressed by the overburden pressure to 60% of their original sizes (average bubble diameter at 90 m depth around 300 μm, and ca. 180 μm at 700 m depth; Bendel et al. 2013; Ueltzhöffer et al. 2010).

[2] This summary is based on Faria et al. (2010). A comparable summary, but from the point of view of ice-sheet flow modelling, is presented by Weikusat et al. (2017).

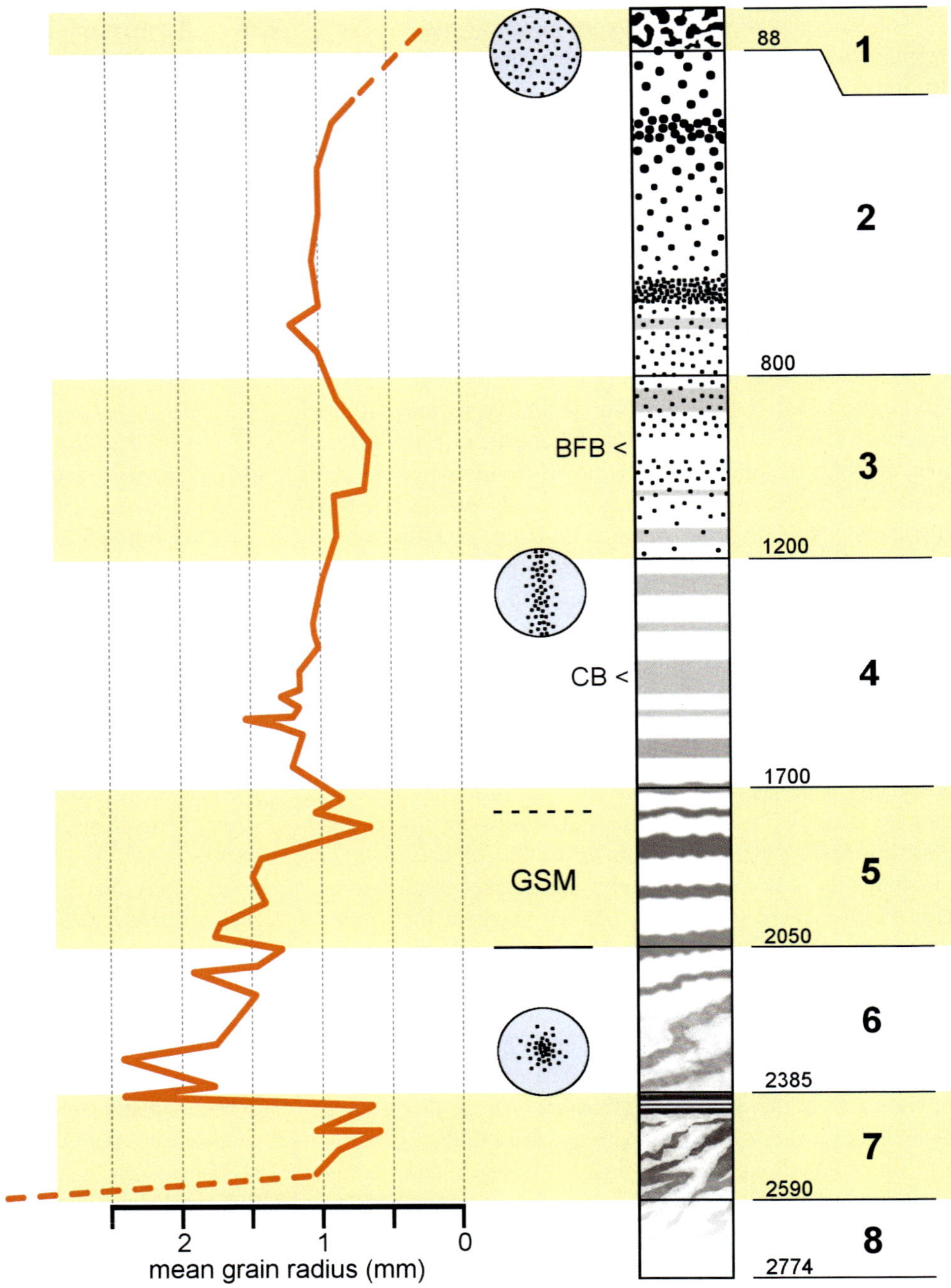

Fig. 4.13 Outline of the EDML visual stratigraphy, main depth zones, and basic microstructural information. *Left* mean-grain-size profile (Weikusat et al. 2009b) and selected *c*-axis pole figures describing LPOs. The centre of each pole figure coincides with the vertical core-axis direction. GSM stands for "girdle–single-maximum transition zone". Below circa 2550 m depth variations in grain size become so large that no reliable mean-grain-size estimate can be obtained from usual thin/think sections. *Centre* schematic outline of the visual stratigraphy (out of scale). Well-defined black patches and discs/dots represent pores and bubbles, respectively; grey stripes portray cloudy bands (CBs). A bubble-free band is also marked with BFB. *Right* The eight depth zones of the EDML Deep Ice Core, described in the text. The numbers on the left of the text denote depth in metres. After Faria et al. (2010)

3. *Bubble–hydrate transition (BHT) zone: from 800 m [14.6 ka] to 1200 m depth [31.7 ka].* Air bubbles start to convert into hydrates (Fig. 2.6), giving rise to strong variations in the concentration of bubbles with depth, which culminate with the occurrence of *bubble-free bands* (BFBs) below 950 m depth (Bendel et al. 2013; Faria et al. 2010; Lüthi et al. 2010). There is a noticeable increase in the concentrations of impurities and microinclusions, and therefore also in the frequency and intensity of cloudy bands, because the BHT zone in EDML coincides with the MIS1–MIS2 transition, i.e. the transition from the Holocene to the last glacial period. Consequently, *cloudy bands* become the prevailing features of the visual stratigraphy in this depth zone, while the mean grain size reduces markedly to just 1.0 mm at 900 m depth, and it remains small for further 150 m. Apart from this grain-size reduction, the ice microstructure and LPOs continue to evolve in the same fashion as in the bubbly ice zone (Fig. 4.9).
4. *MIS3 zone: from 1200 m [31.7 ka] to 1700 m depth [60.6 ka].* Bubble-free ice with lower impurity/microinclusion content than ice from the BHT

zone, characteristic of the warmer MIS3 period. Visual stratigraphy is marked by *nicely flat and parallel cloudy bands*. More or less steady increase in average grain diameter to almost 3.0 mm at 1700 m depth. The *strongest c-axis vertical girdle* in EDML is achieved in the lower part of this zone. *Dynamic recrystallization* continues to affect the ice microstructure evolution in the same fashion as in the upper zones.

5. *Girdle–Single-Maximum (GSM) transition zone: from 1700 m [60.6 ka] to 2050 m depth [87.0 ka]*. During the cooler MIS4 period (approx. 1700–1850 m depth) there occurs a peak in impurity/microinclusion concentrations that gives rise to intense cloudy bands, while average grain diameter gets as small as 0.8 mm. Below 1850 m depth (MIS4–MIS5 transition) impurity/microinclusion concentrations return to lower levels and mean grain diameter increases vigorously to ca. 3.5 mm at 2050 m depth. LPO evolution changes its trend for the first time, turning to a gradual *weakening of the c-axis vertical girdle*, which culminates in a somewhat abrupt transition to a vertical single maximum at ca. 2050 m depth (also detectable through radar sounding, Eisen et al. 2007). Such a girdle–single-maximum (GSM) transition points to the rise of *horizontal simple shearing* as the dominant flow regime in the lower third of the ice sheet at the EDML site. This change in flow regime is also suggested by the first signs of *stratigraphic disturbance* (Faria et al. 2010), in the form of *cloudy bands with millimetre-sized undulations*, which are the main feature of the visual stratigraphy in the GSM transition zone.
6. *MIS5 zone: from 2050 m [87.0 ka] to 2385 m depth [134 ka]*. Progressive deterioration of the stratigraphy with ever increasing undulations, tilted layers and folding. The stratigraphy at the bottom of this zone is characterized by *fuzzy cloudy bands*, tilted from horizontal by up to 30°, and *fully developed folds* with many centimetres in amplitude. Synchronization with EPICA Dome C Deep Ice Core is lost below 2366 m depth (Ruth et al. 2007). Gradual increase of mean grain size to a diameter of ca. 5 mm at 2370 m depth. Steady strengthening of the *c*-axis vertical single maximum.
7. *High-shearing zone: from 2385 m [134 ka] to 2590 m depth [no reliable dating]*. The top 20 m of this zone exhibit the most striking structural changes in the whole EDML Deep Ice Core, including a steep drop in the mean grain diameter to circa 1 mm and an almost 100-fold jump in impurity/microinclusion content to concentrations comparable with LGM levels. These remarkable changes are believed to be caused not only by the enclosed climate transition from the last interglacial to the penultimate glacial (MIS5e–MIS6 transition), but also by the activation of *enhanced shearing* in a many-metre thick *soft ice layer* between 2385 and 2405 m depth. Evidences of such a soft ice layer are diverse and range from the abrupt change in the stratigraphy just below 2385 m depth, characterized by intense, neatly flat and parallel cloudy bands, to the peculiar arrangement of grains in a "brick-wall pattern" that facilitates grain-boundary sliding by *microshear* (Faria et al. 2010, 2006, 2009, 2014b). Between 2400 and 2500 m depth the ice microstructure remains generally fine-grained, but mean grain size variability increases dramatically with the ultimate deterioration of the stratigraphy, which becomes *completely disrupted* below circa 2575 m depth.
8. *Basal ice zone: from 2590 m [no reliable dating] to 2774 m depth [bedrock]*. EDML basal ice appears exceptionally clear, with huge grains of up to a half metre in diameter. Local temperature rises with depth from circa −7 °C at 2590 m to the pressure melting point near the bedrock, at 2774 m depth. Stratigraphic structures like cloudy bands no longer exist and almost all microinclusions can be found collected at grain boundaries. Also the optical properties of the grain boundaries seem altered as they interact more intensely with the incident light of the line-scanner, becoming the dominant visible structures of the core in this deepest zone.

EDML Line-Scan Images

5

The motto here is always: take a wider *look round.*
Ludwig Wittgenstein (1983), p. 127

5.1 Key to Image Interpretation

As explained in Sect. 4.4, an *ice-core line-scanner* (LS) is a device for recording the visual stratigraphy of ice cores in high-resolution (≈0.1 mm/pixel) digital images. In the case of the EDML Deep Ice Core, the LS device scans polished ice slabs cut lengthwise from every full metre of the EDML core, according to the cutting plan shown in Fig. 4.10.

In a similar fashion as dark field microscopy, the LS system is designed to capture only the indirect light scattered by inclusions in the ice matrix, like microscopic particles or air bubbles, so that the amount of brightness in any point of an LS image is proportional to the concentration of inclusions in that location of the ice slab. Thus, a clear piece of ice, free from inclusions, appears dark, while another piece of ice full of inclusions (e.g. bubbly ice or cloudy-band ice) appears bright (Fig. 5.1).

Depending on the amount of light scattered by a given ice slab, a suitable aperture adjustment of the light-source diaphragm is needed to avoid over- or under-exposure of the LS image. Also the exposure settings of the LS digital camera are adjusted accordingly. Further optimization is obtained through digital image processing. All these adjustments are done automatically and/or manually during the scanning process. Therefore, the intensity of stratigraphic features recorded in different LS images may vary to some extent, because of distinct calibrations, and care should be taken when comparing individual images. The brightness of the ruler on the left side of each image may serve as a qualitative indication of the net exposure settings of a particular image.

The LS images of the EDML Deep Ice Core, from now on called EDML-LS images, are named after the core's *bag number*, which corresponds to the logged depth at the bottom of each one-metre piece of core (for a definition of *logged depth*, see the Glossary in Sect. 1.2). For instance, EDML-LS Image 672 corresponds to the 672th piece of core, which spans the logged depth interval 671–672 m. The EDML-LS image series covers every metre of the whole EDML Deep Ice Core, with exception of the upper 448 m (bag numbers 1–449), which comprise the firn and upper bubbly-ice zones. The reason for this omission is that the light scattering by air bubbles and pore space in this shallow depth zone is so intense that nothing useful can be seen in the LS images.

Occasionally, an LS image may not cover the full one-metre length of a core piece, for example because that part of the core consist of fragments too small to be cut into a slab. In particular, there are a few rare depths in the *brittle zone* (cf. Sect. 1.2), from which there is no LS image at all, simply because the entire ice-core piece from that particular depth was so fragmented that it was impossible to cut an ice slab out of it (e.g. EDML-LS Image 936).

As remarked in Sect. 4.4, the absolute azimuthal orientation of an ice core is generally not known, so that we have to assume that the LS images are randomly oriented with respect to the ice flow. This implies that the LS method currently provides an *optimistic estimate* of stratigraphy integrity and a *lower bound* of layer inclination. In other words, any stratigraphic disturbance may be just as bad as, or worse than, its depiction in the LS image, while the true inclination of a layer may be equal to, or larger than, the inclination displayed in the LS image. Future advances in ice-core drilling

Electronic supplementary material The online version of this chapter (doi:10.1007/978-3-662-55308-4_5) contains supplementary material, which is available to authorized users.

S.H. Faria et al., *The EPICA-DML Deep Ice Core*, Frontiers in Earth Sciences,
https://doi.org/10.1007/978-3-662-55308-4_5

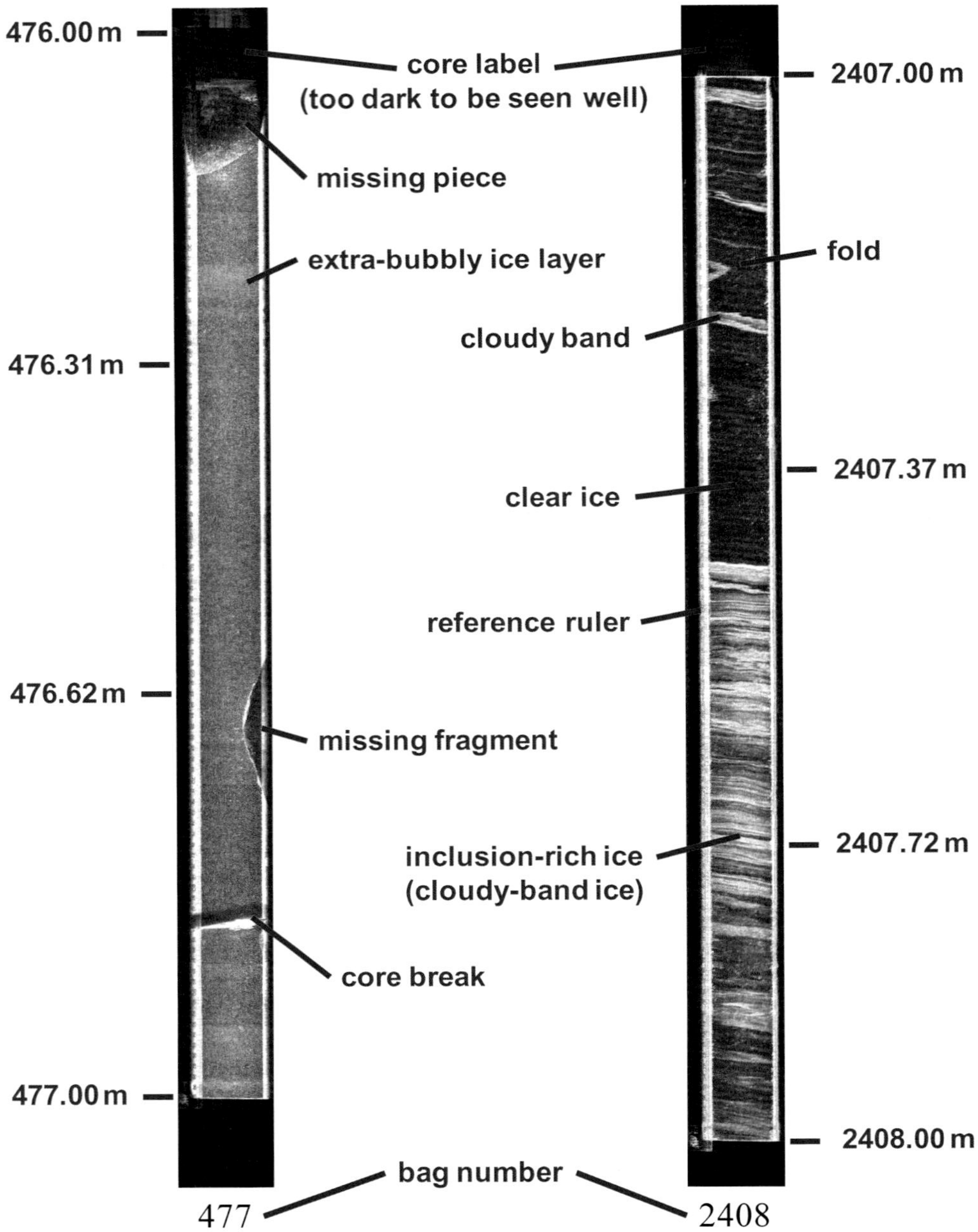

Fig. 5.1 Main features of the line-scanner images of the EDML Deep Ice Core. Two examples are shown (notice the distinctive brightness of the reference ruler on the *left-hand side* of each image, which indicates that they have been produced with different exposure settings). *Left*: EDML-LS Image 477 covering the depth interval 476–477 m (bubbly ice zone). Bright speckles distributed over the whole core piece are air bubbles. Depth variations in the collective structure of bubbles (mainly number and size) create layering (an extra-bubbly ice layer is marked). *Right*: EDML-LS Image 2408 covering the depth interval 2407–2408 m (high-shearing zone). Bubble-free ice with very strong contrast between clear ice and cloudy-band ice. A fully developed centimetre-sized fold is also indicated

technology shall hopefully overcome the challenge of recording the absolute orientation of an entire ice core, therefore enabling future LS images to be always oriented in the direction of the local ice flow.

5.2 The EDML Visual Stratigraphy Record

With the most essential features of the EDML-LS images being described in Fig. 5.1, we are now ready to explore the whole visual stratigraphy record of the EDML Deep Ice Core.

The ensuing pages display 2325 line-scan images covering the logged depth range 450–2774 m of the EDML Deep Ice Core. High-resolution versions of these images are also available for download from the link indicated on the copyright page at the beginning of this book.

This collection of line-scan images represents the most accurate visual record to date of the stratified structure of Antarctic ice from the Dronning Maud Land site. We hope that it may serve not only as a useful reference for glaciologists and climate modellers, but also as a unique compilation of ice-core images for enthusiasts of Antarctica.

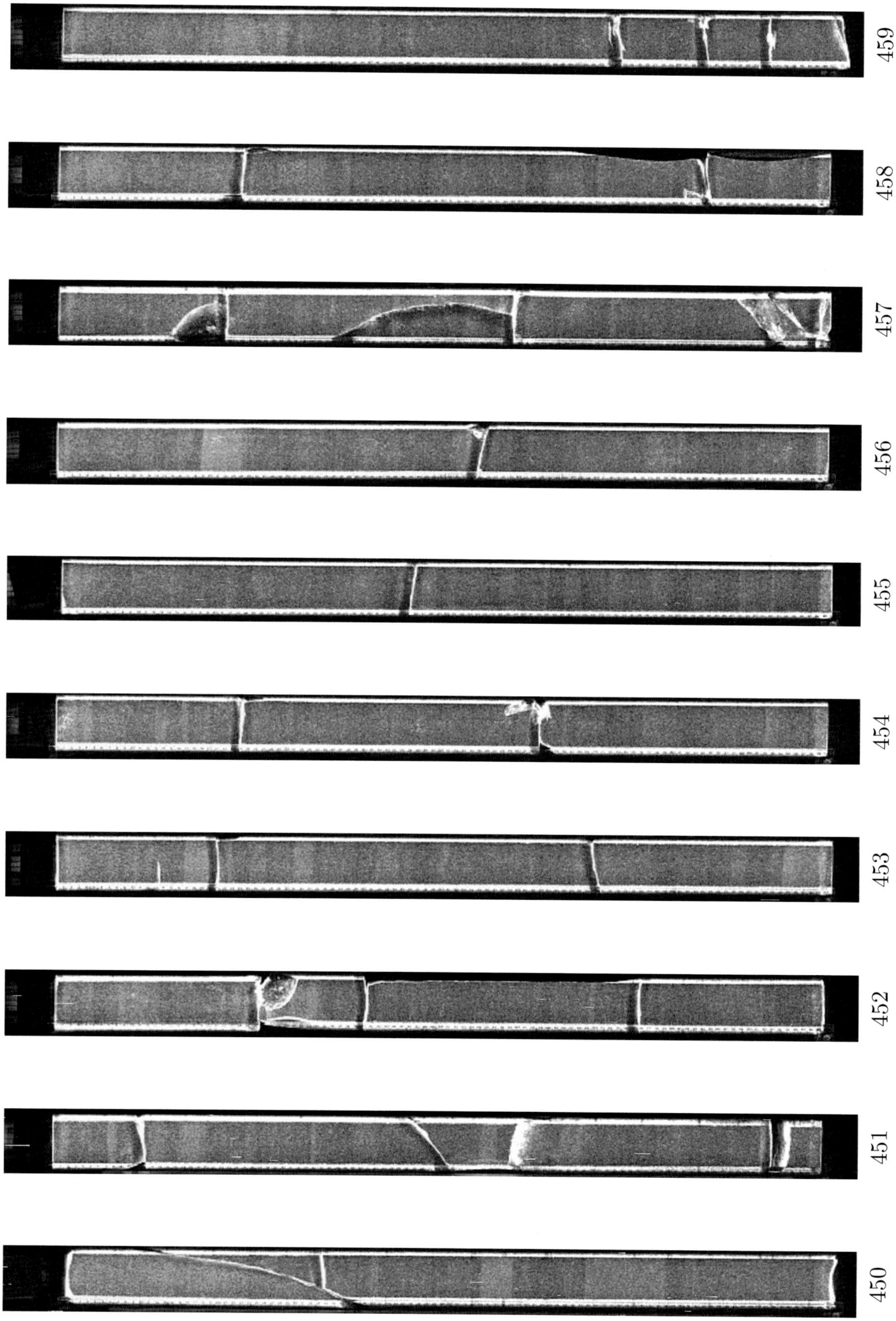
459
458
457
456
455
454
453
452
451
450

460

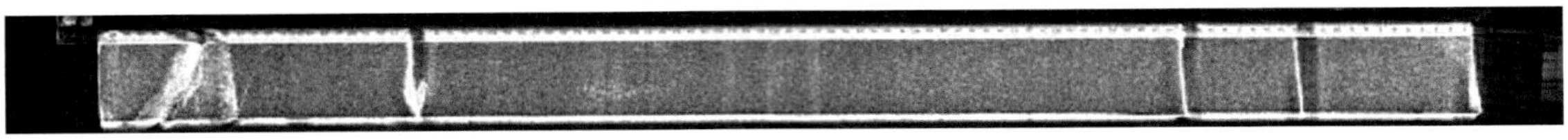

461

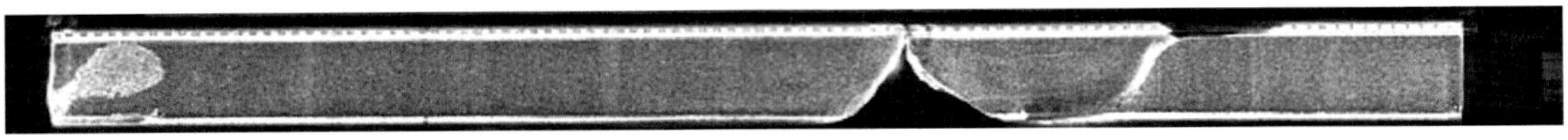

462

463

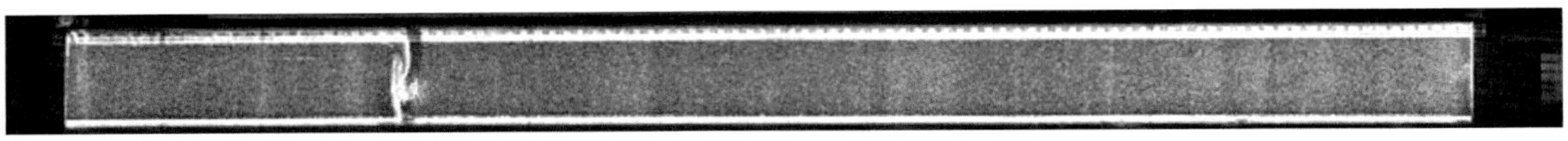

464

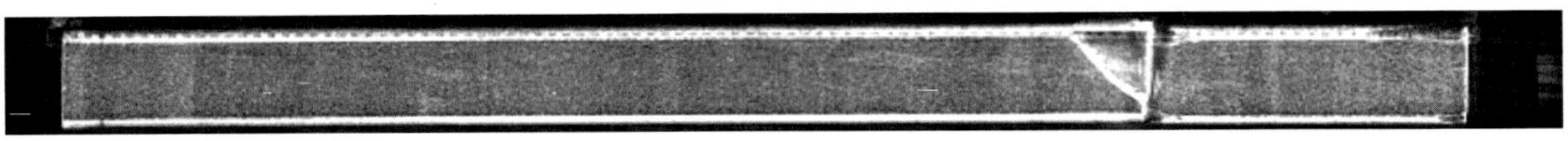

465

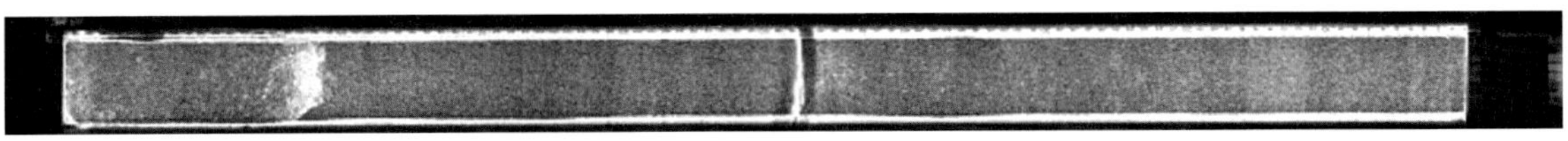

466

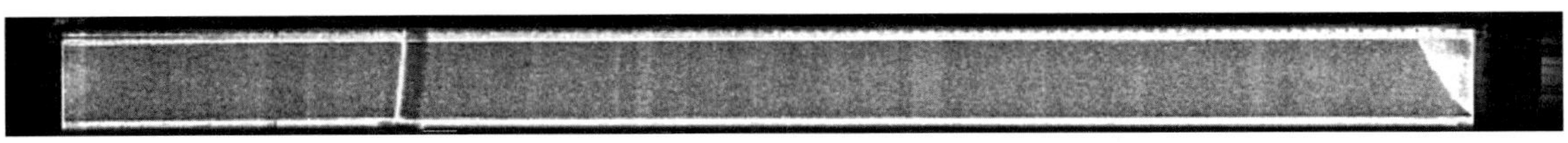

467

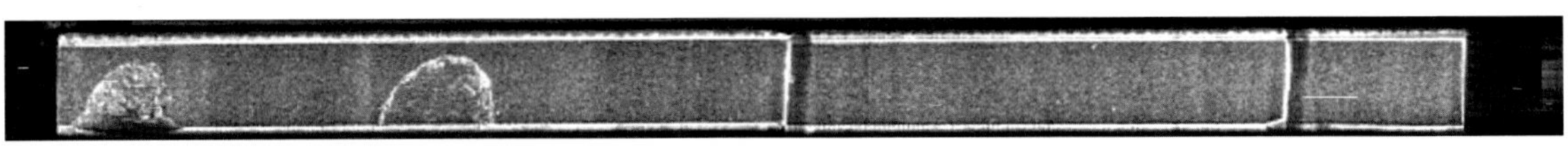

468

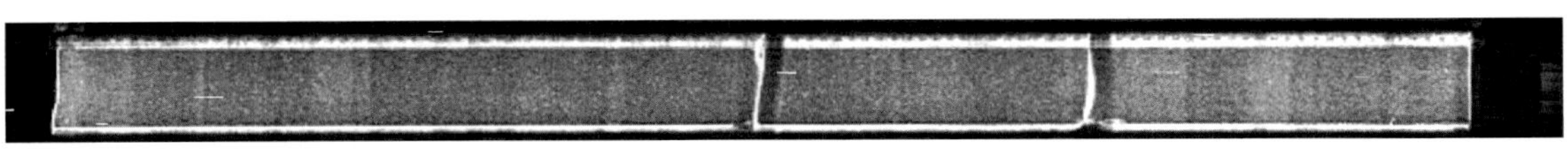

469

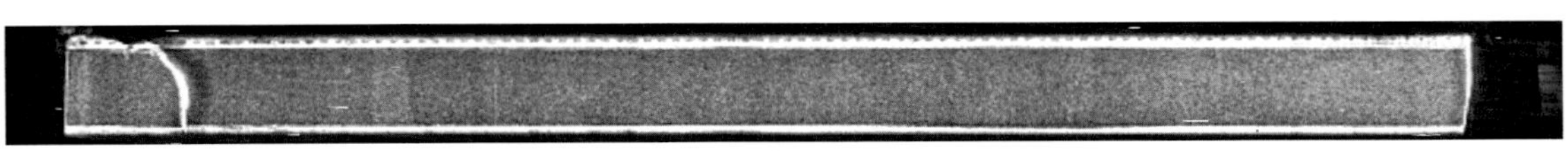

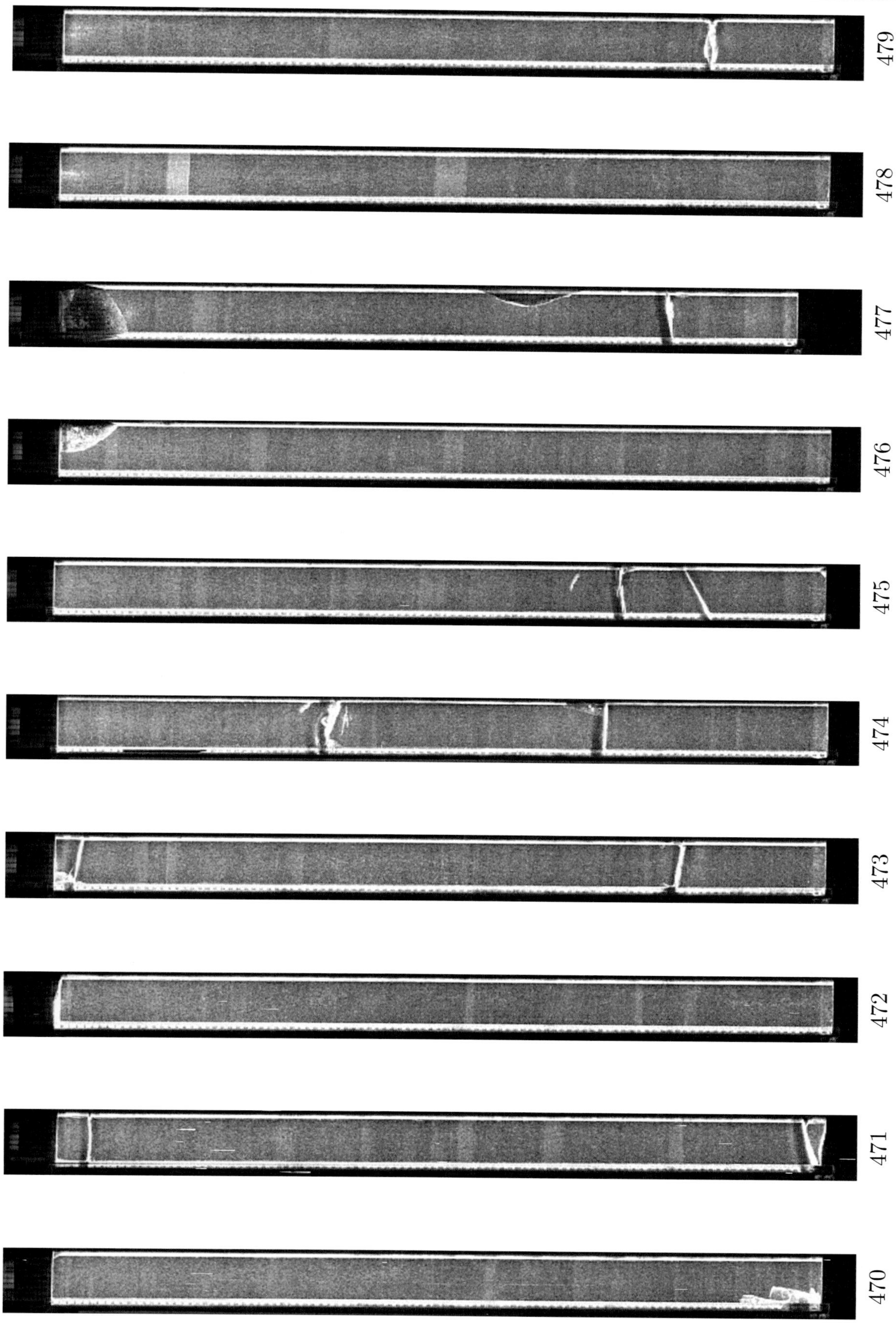
479
478
477
476
475
474
473
472
471
470

480

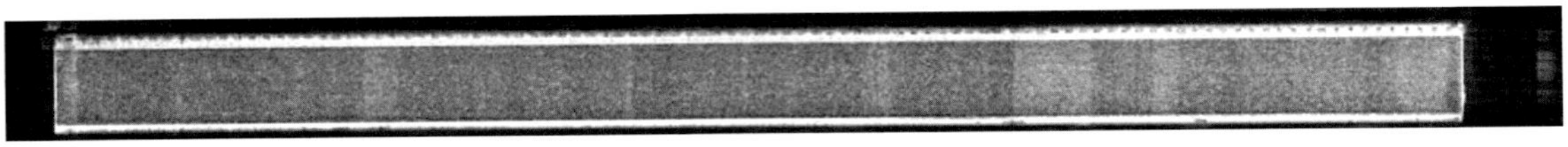

481

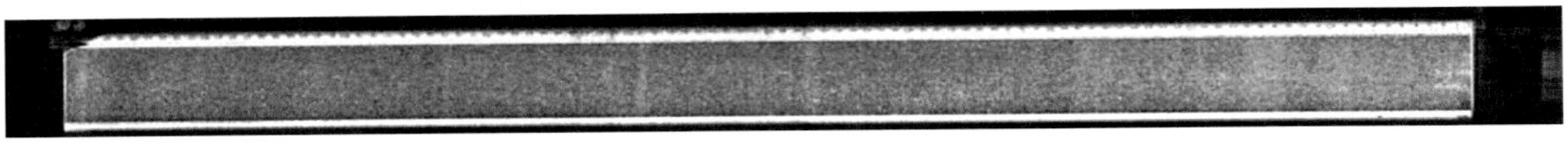

482

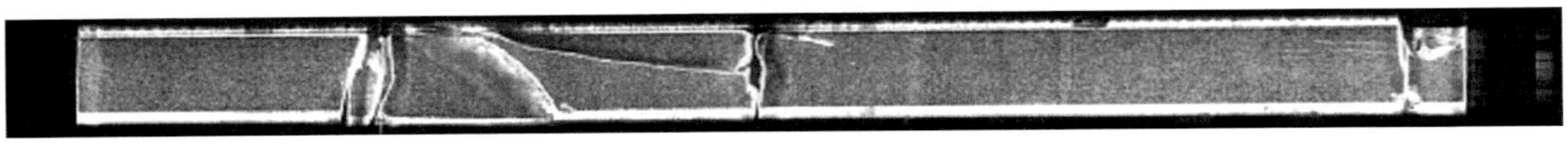

483

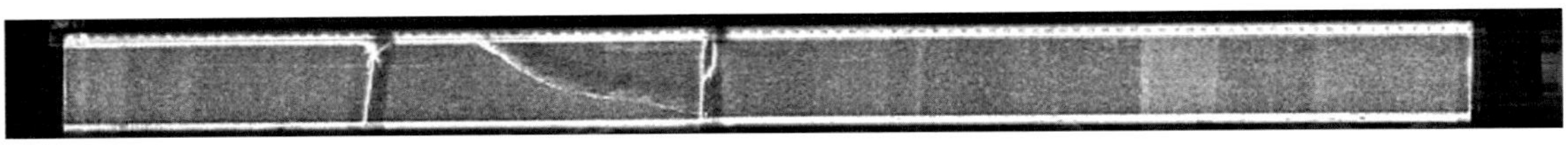

484

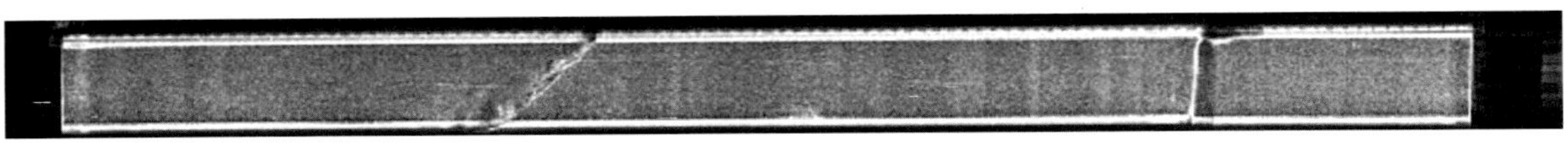

485

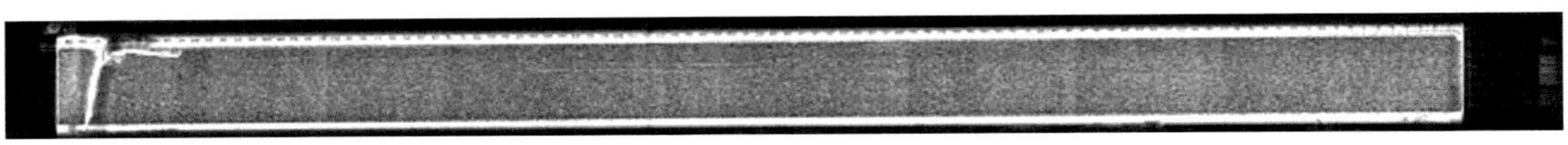

486

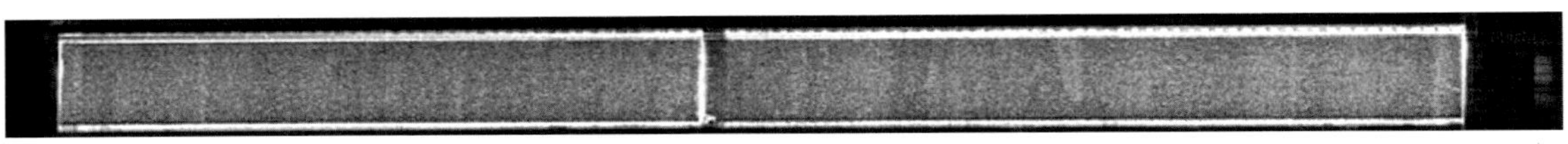

487

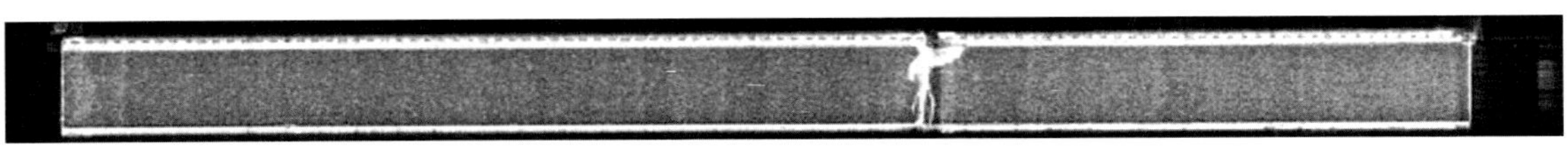

488

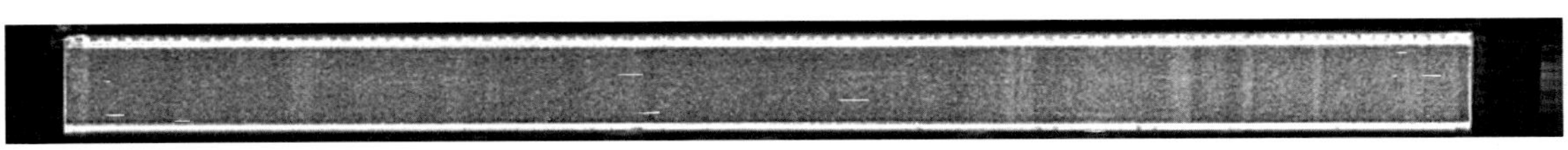

489

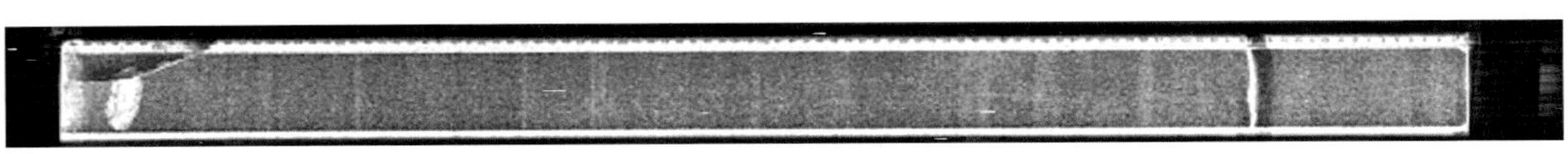

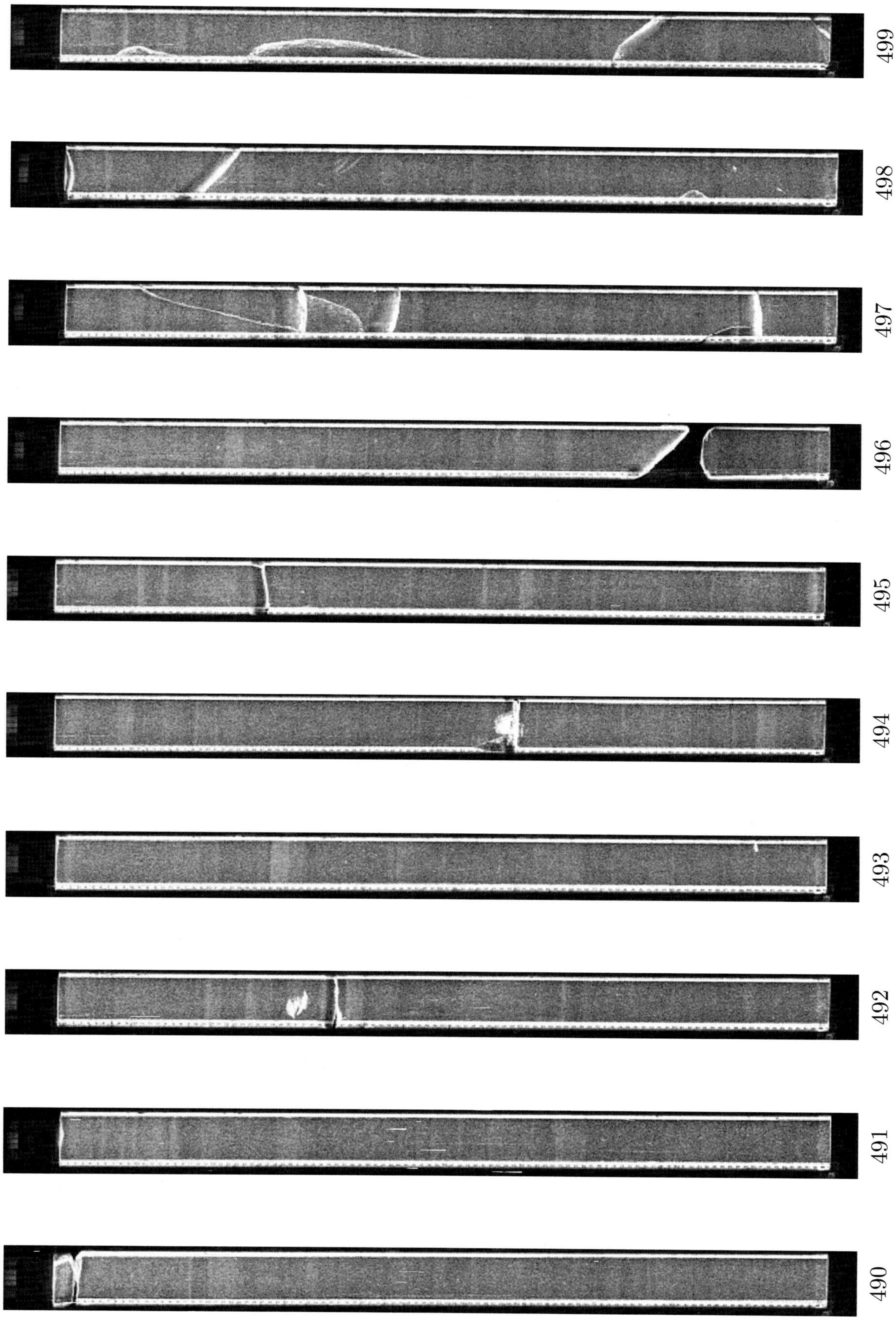
499
498
497
496
495
494
493
492
491
490

500
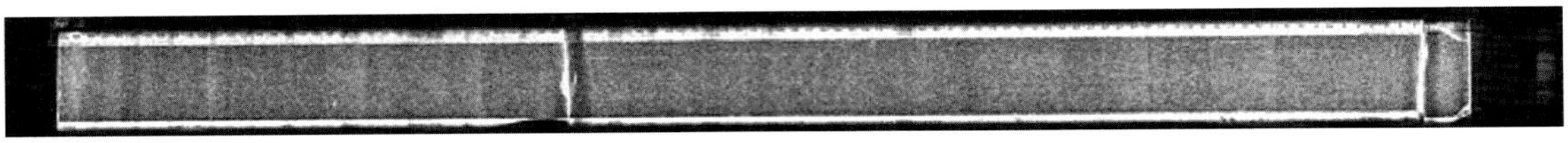

501
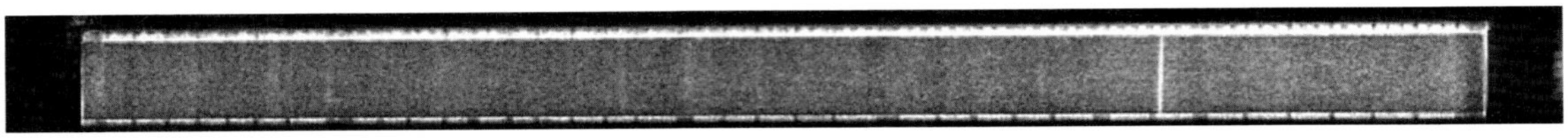

502
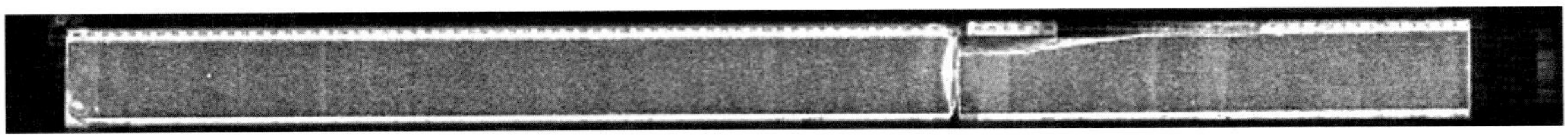

503
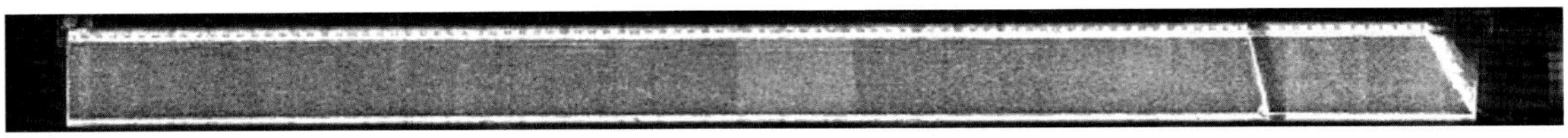

504
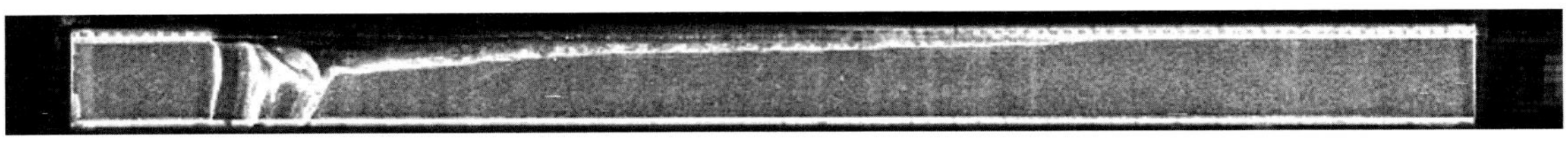

505
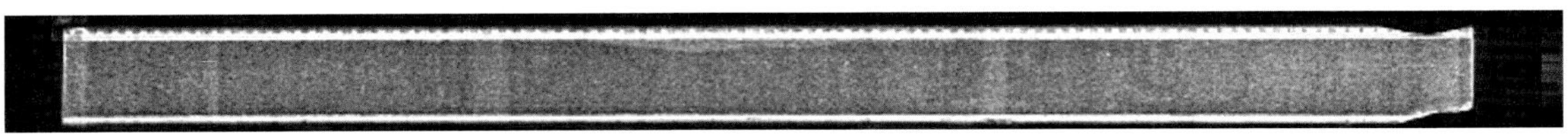

506
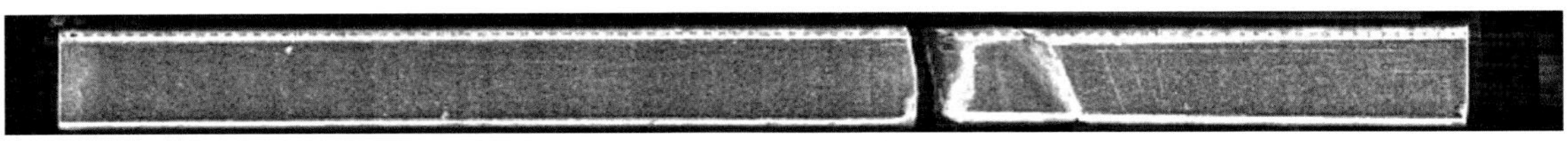

507
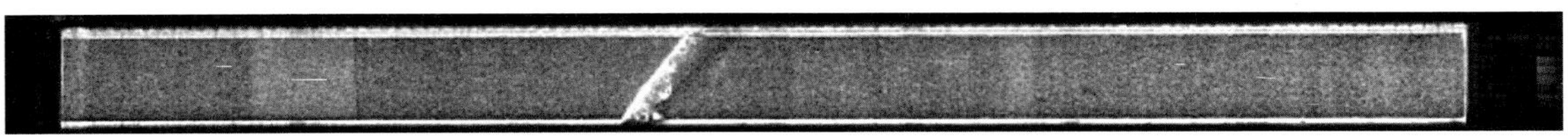

508
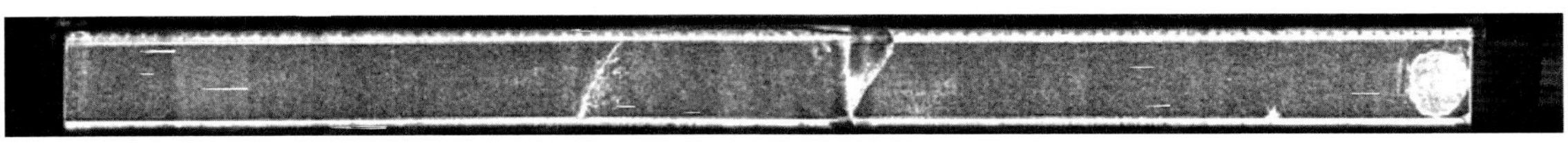

509
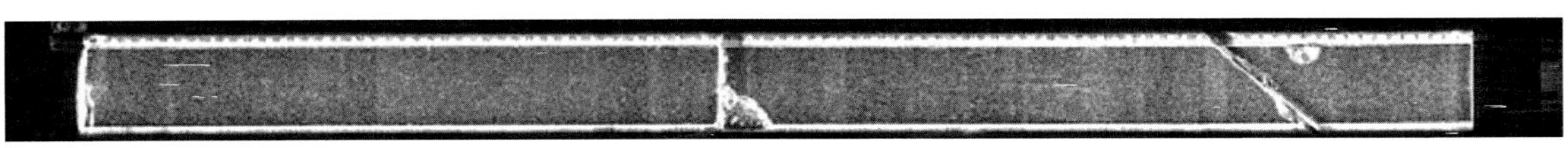

519

518

517

516

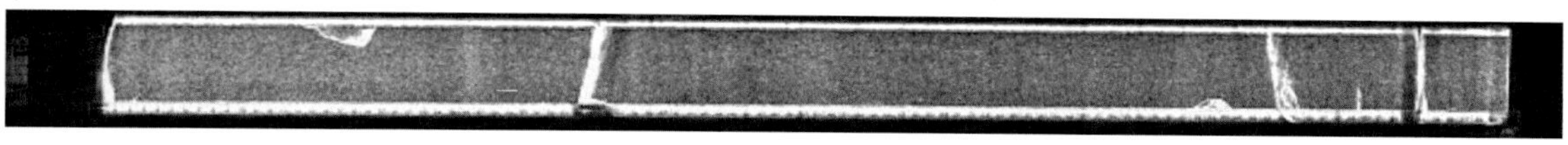
515

514

513

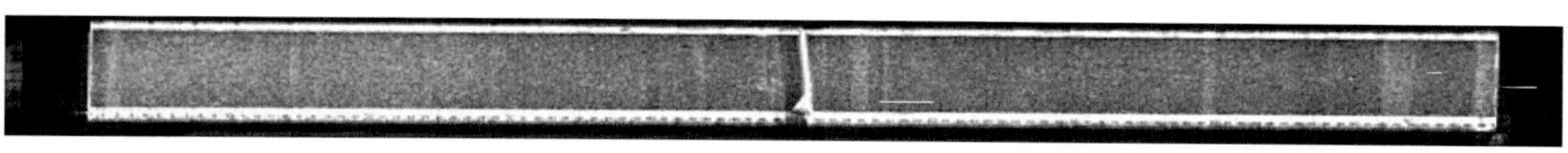
512

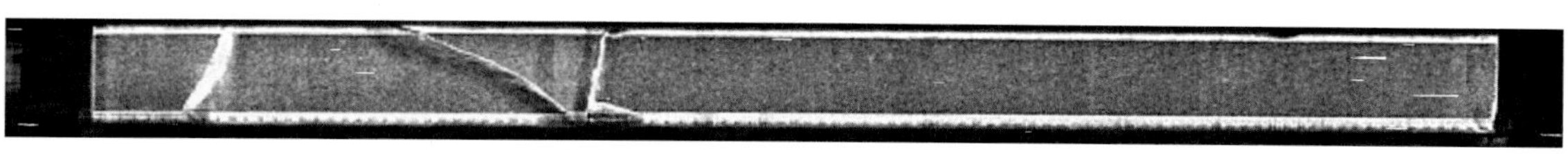
511

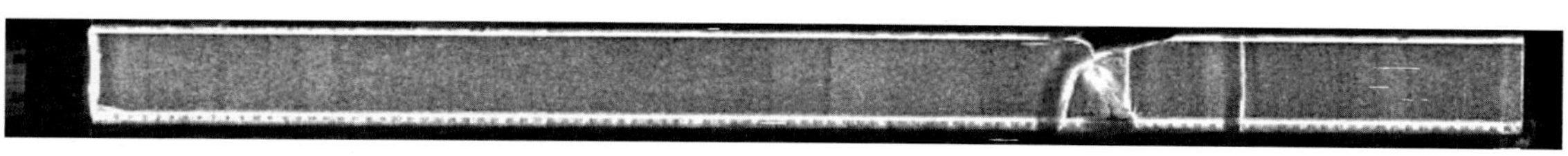
510

520

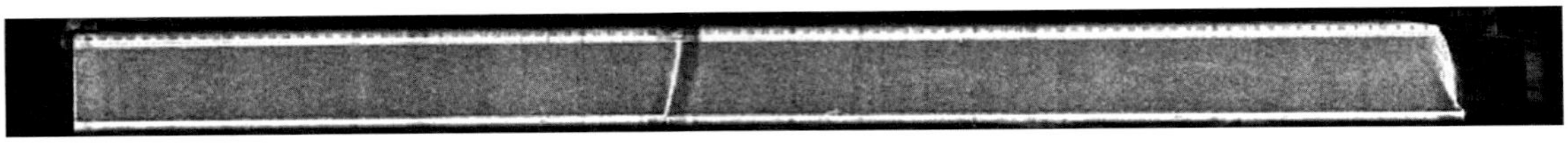

521

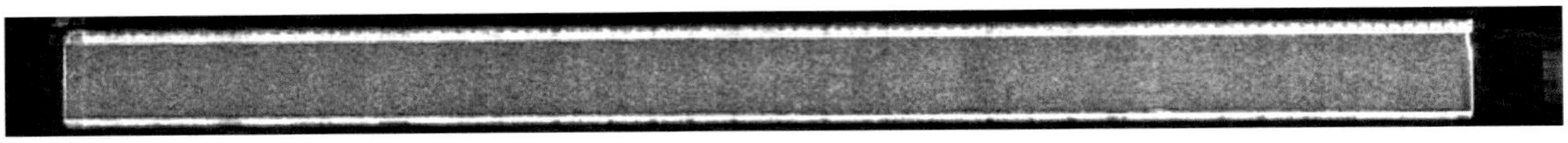

522

523

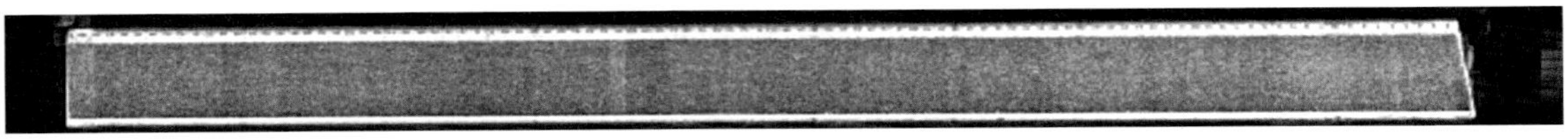

524

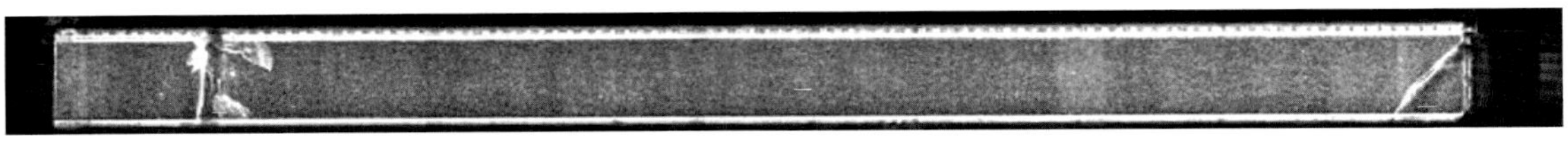

525

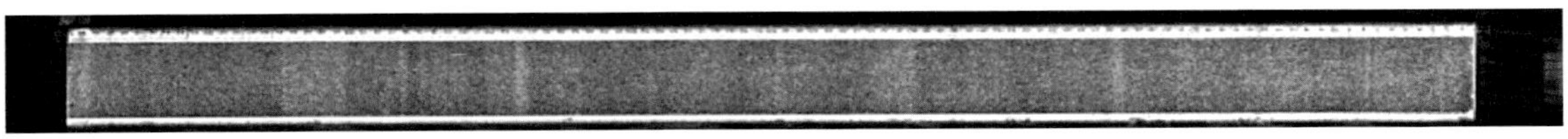

526

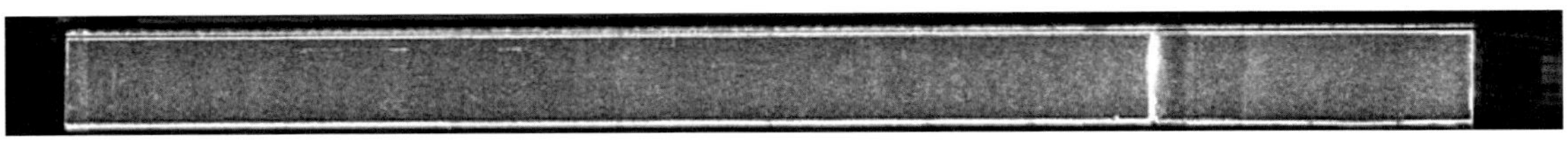

527

528

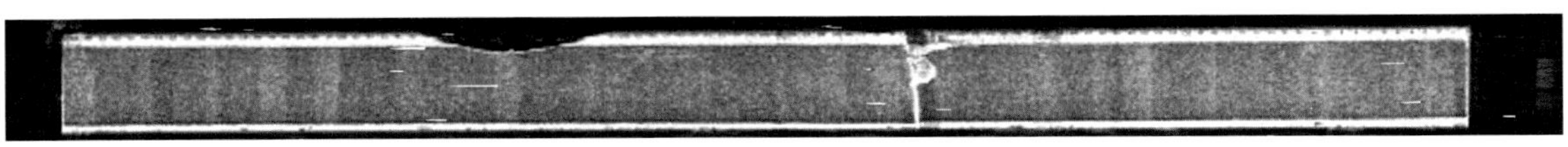

529

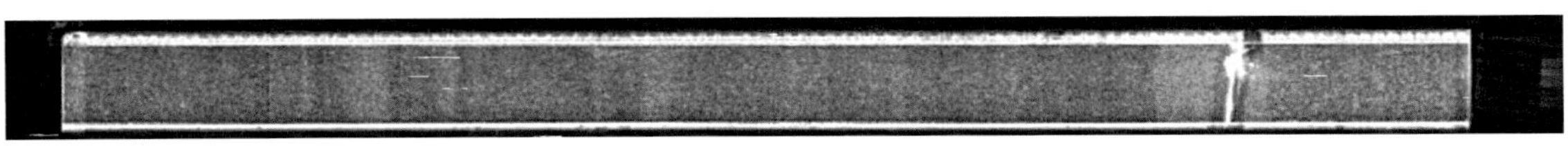

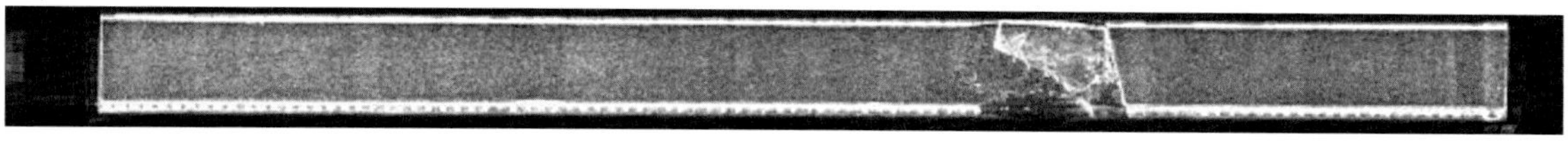
539

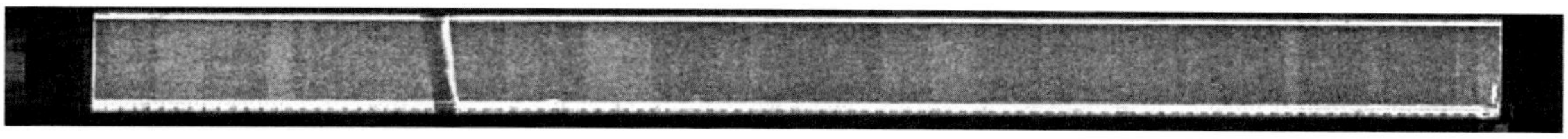
538

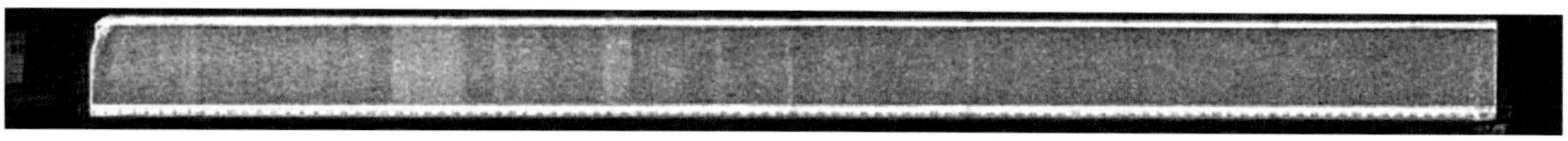
537

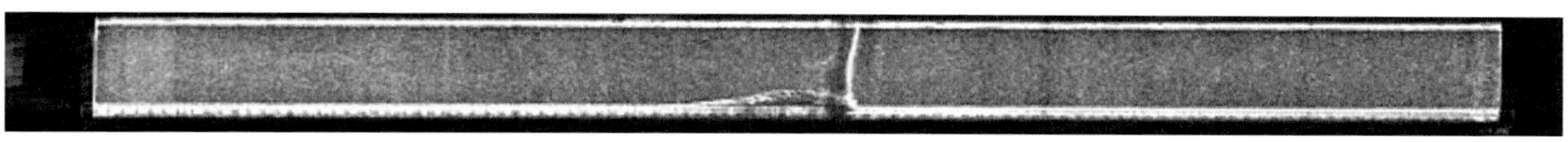
536

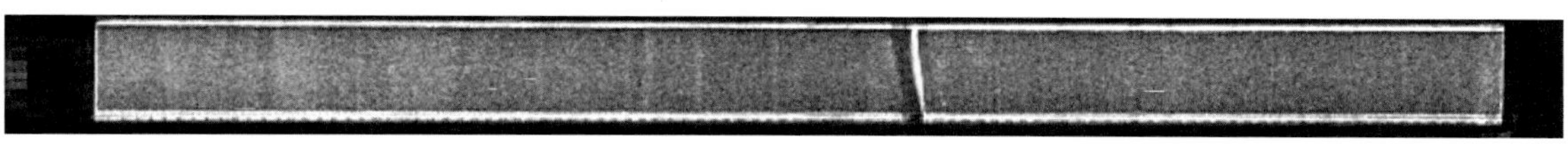
535

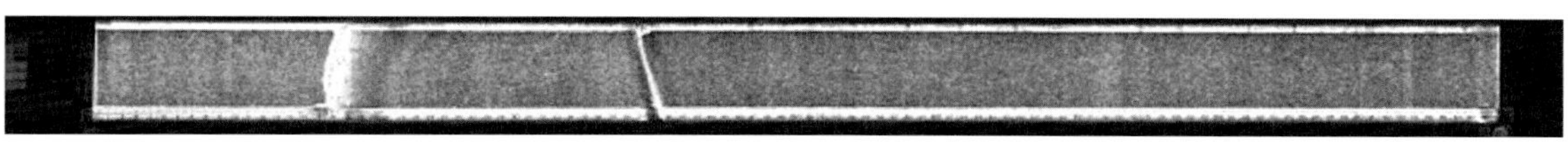
534

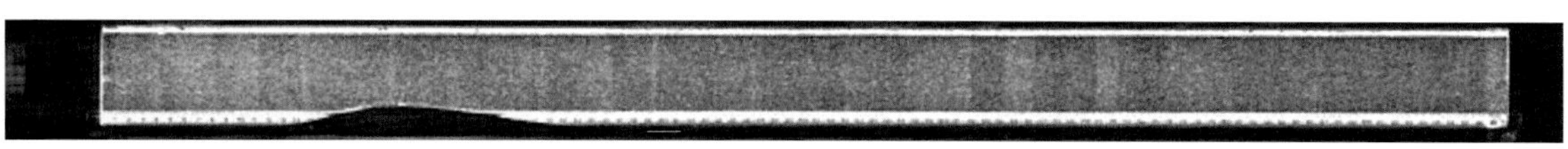
533

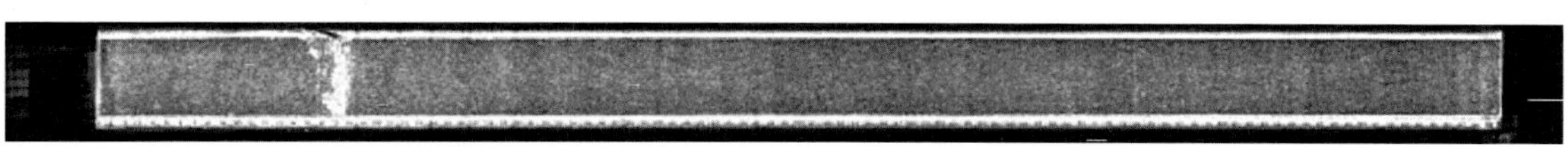
532

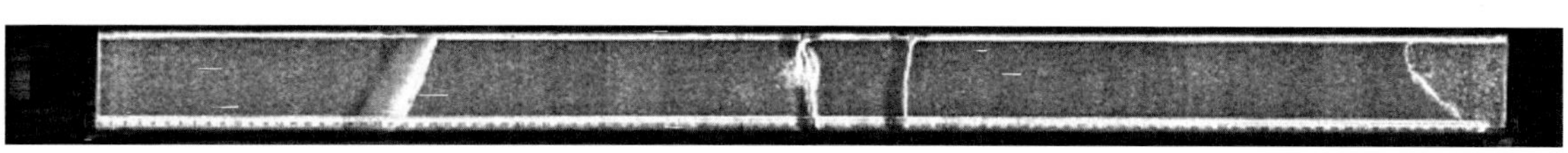
531

530

540
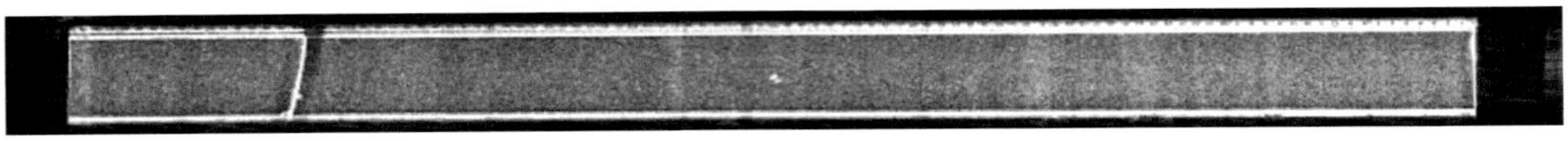

541
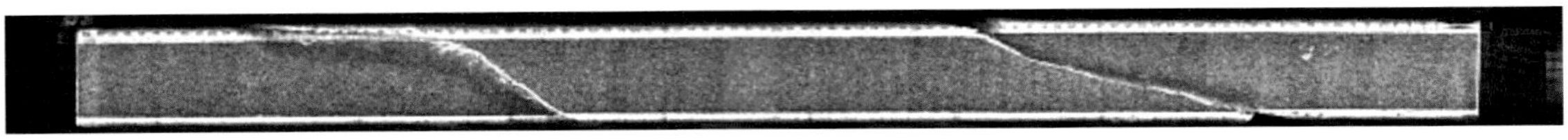

542
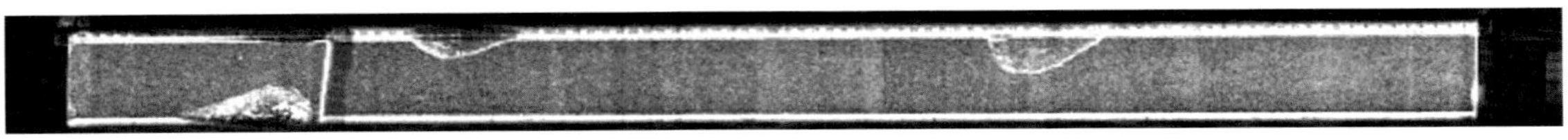

543
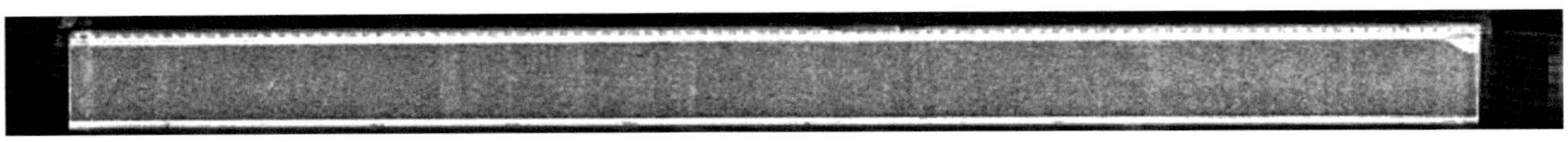

544
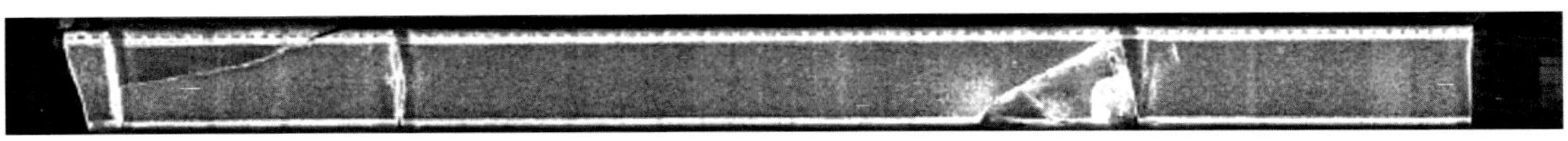

545
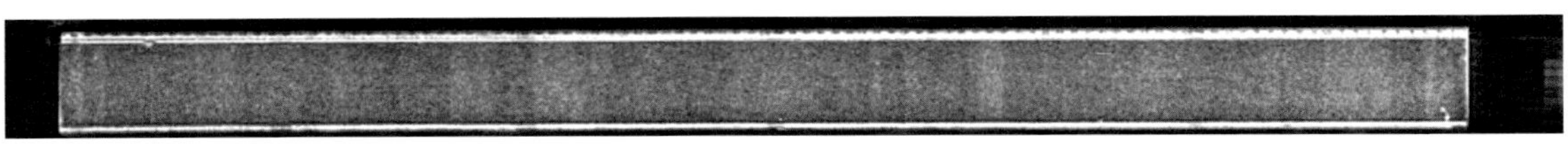

546
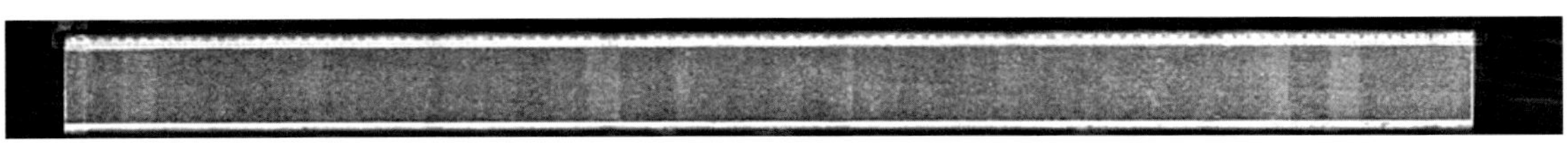

547
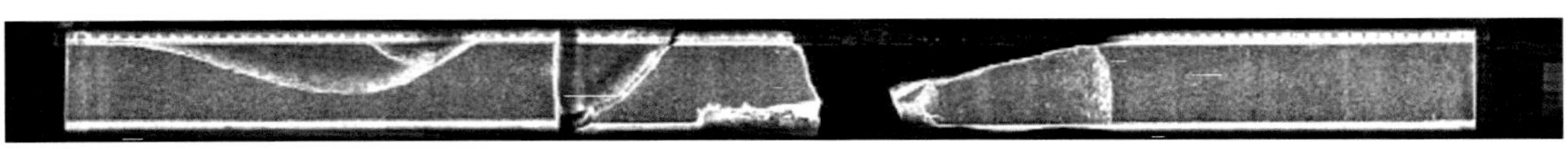

548

549
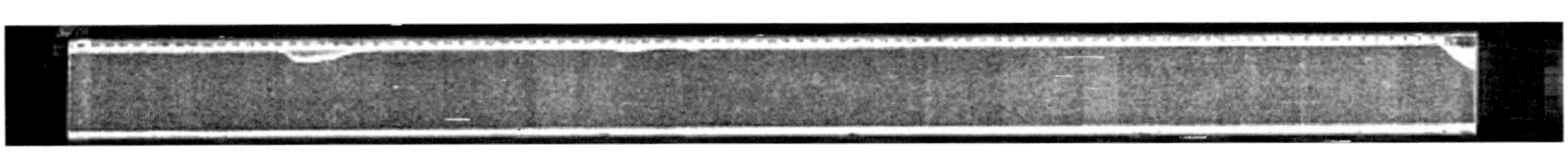

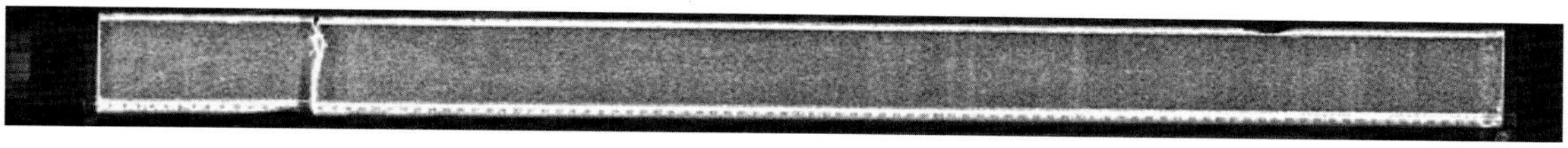

559

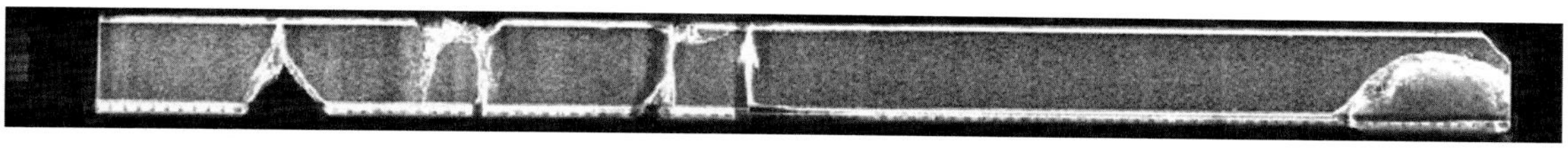

558

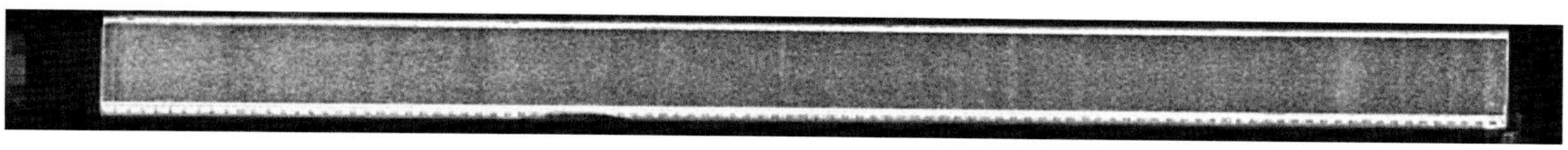

557

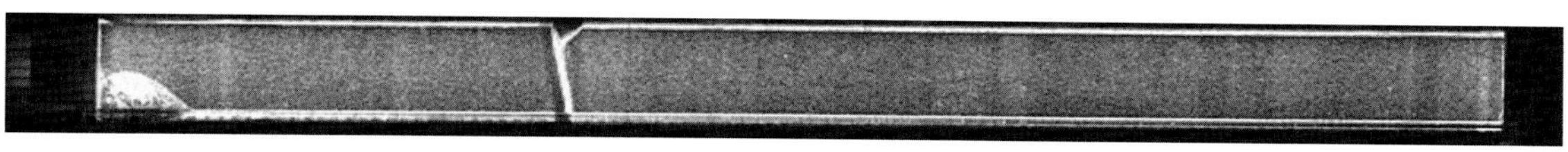

556

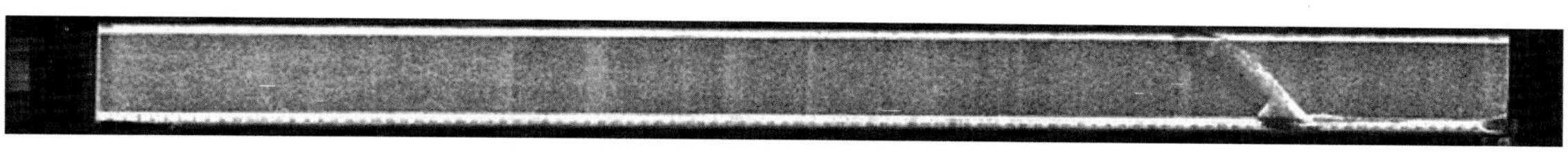

555

554

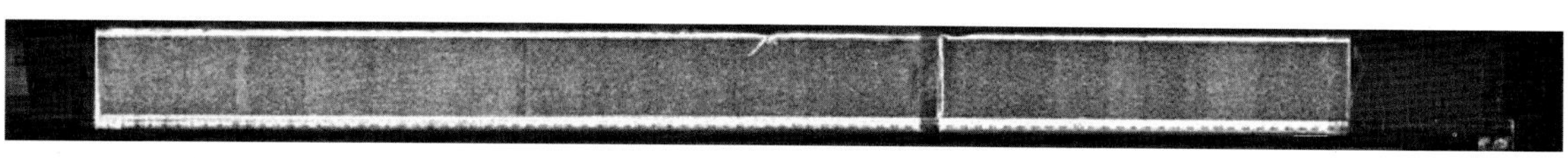

553

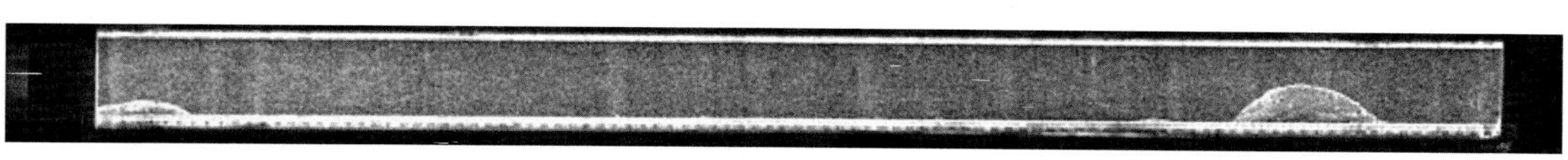

552

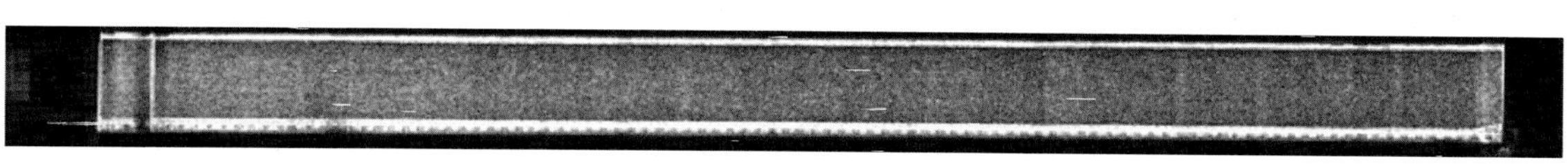

551

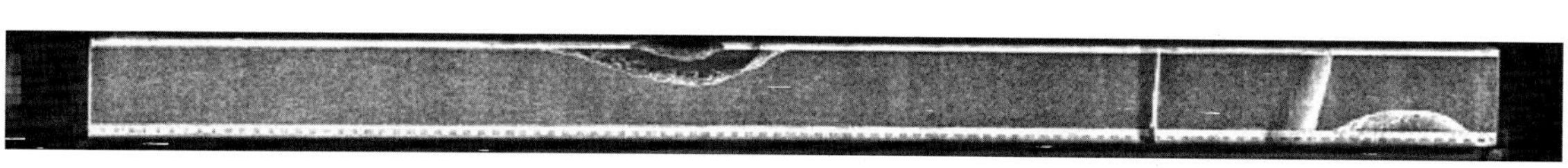

550

560
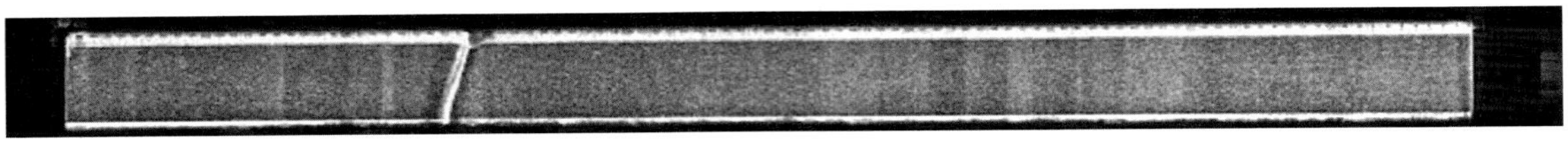

561
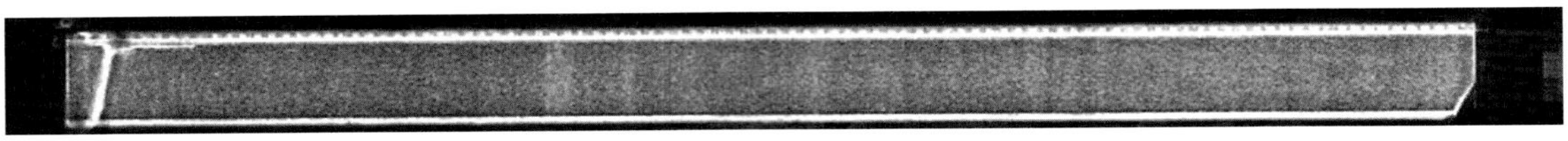

562
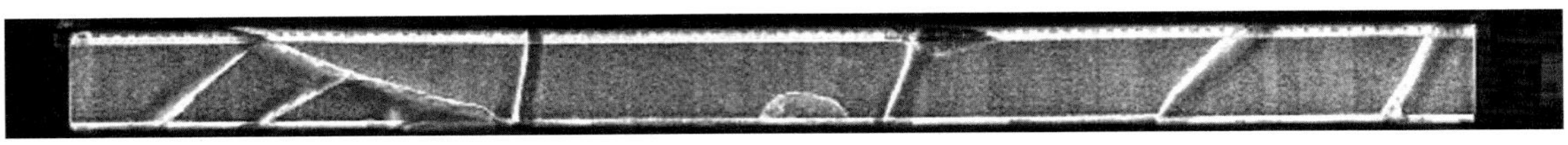

563
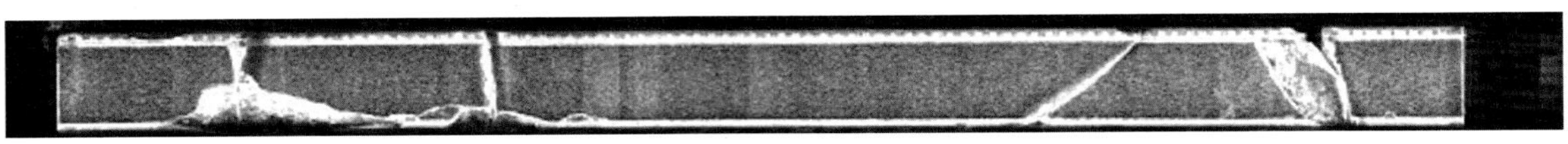

564
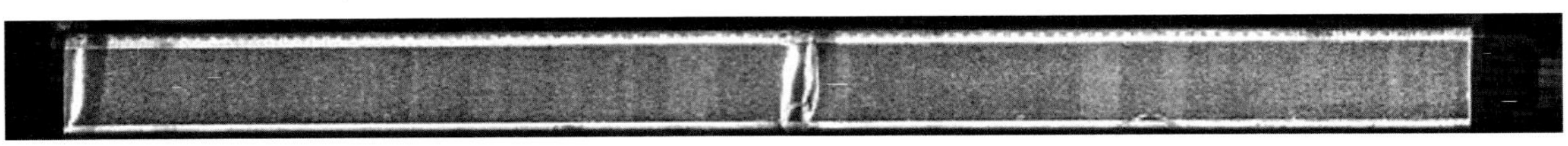

565
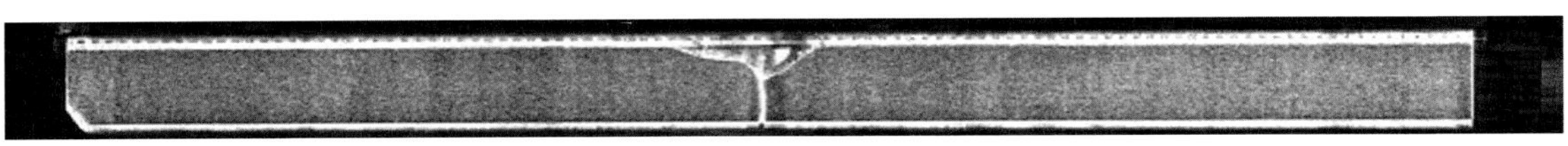

566
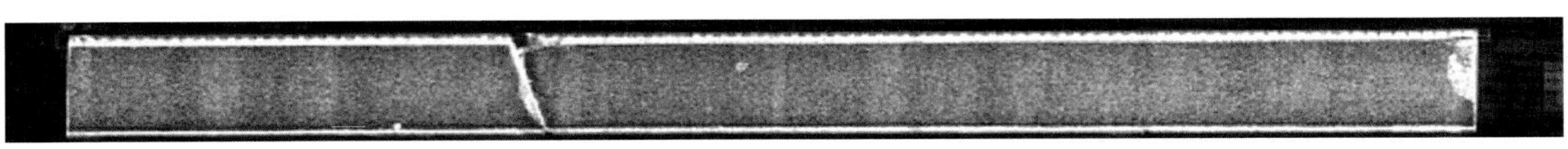

567
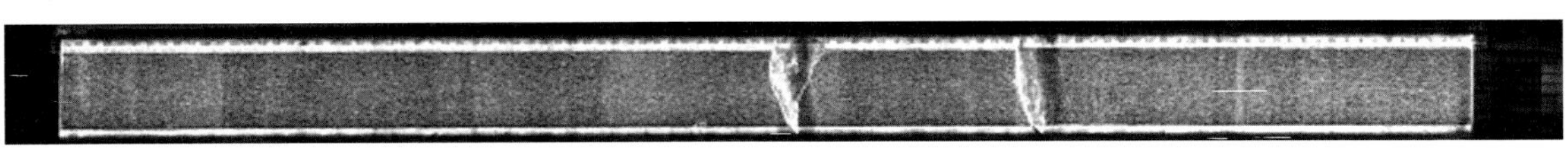

568
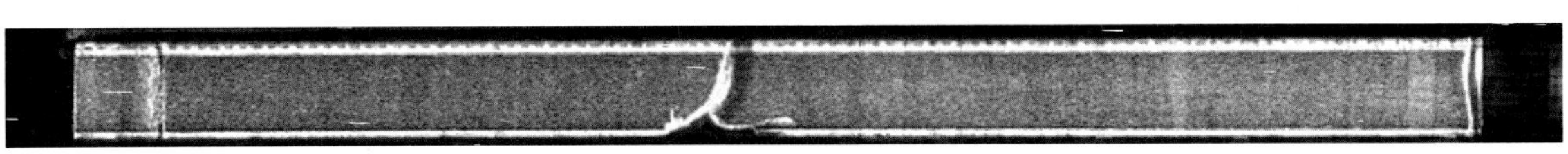

569
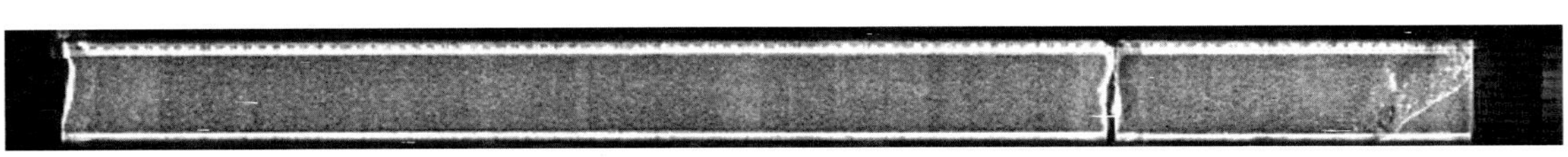

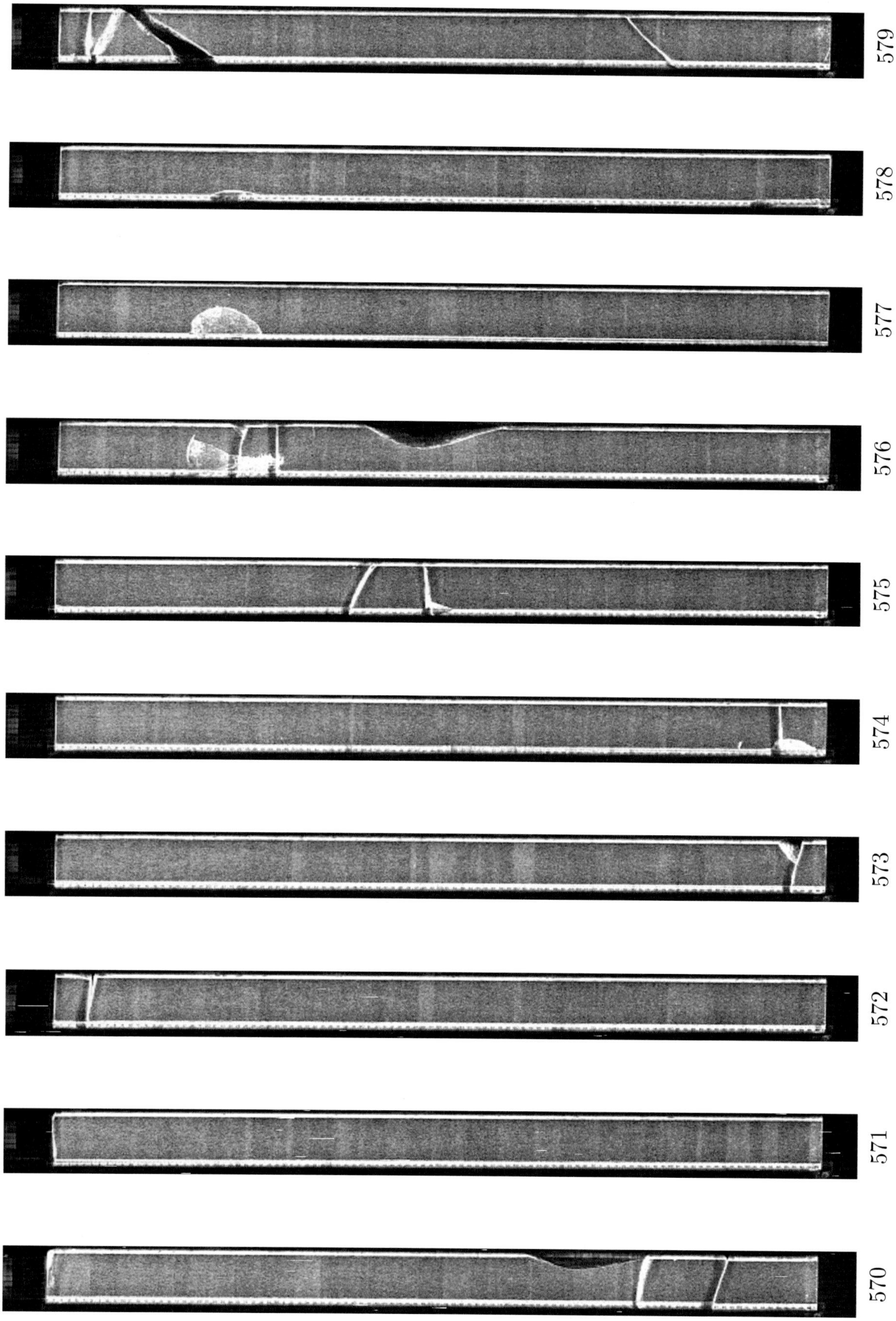
579
578
577
576
575
574
573
572
571
570

580

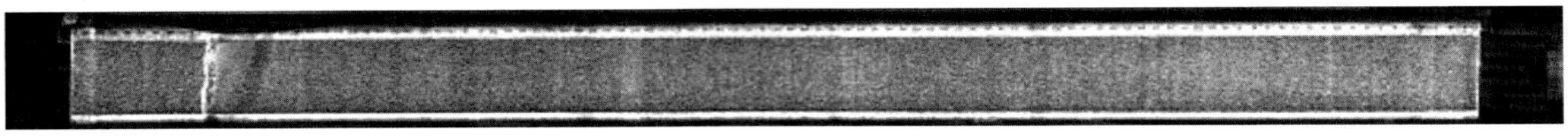

581

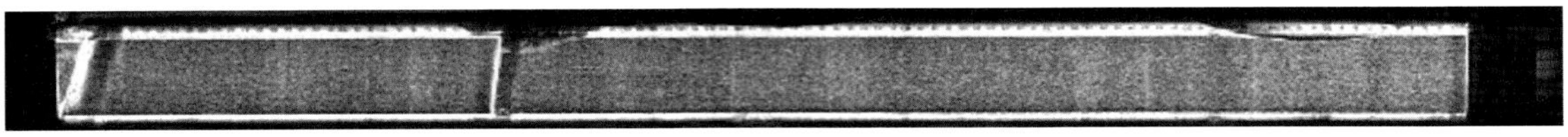

582

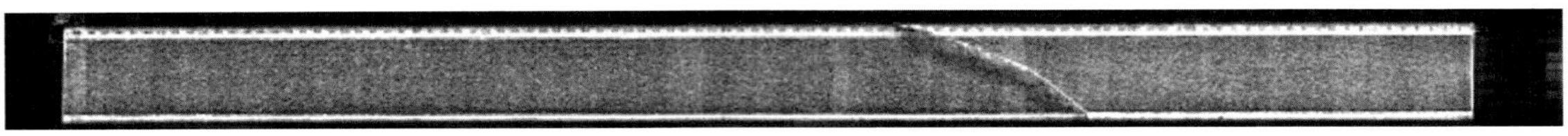

583

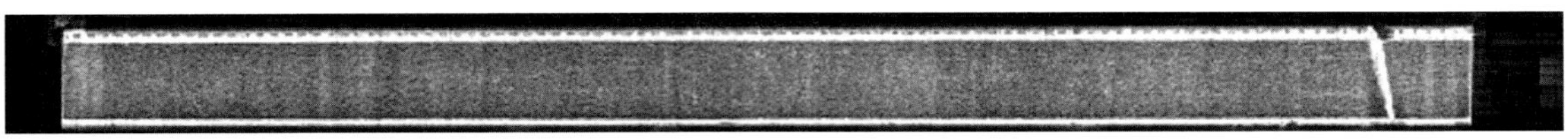

584

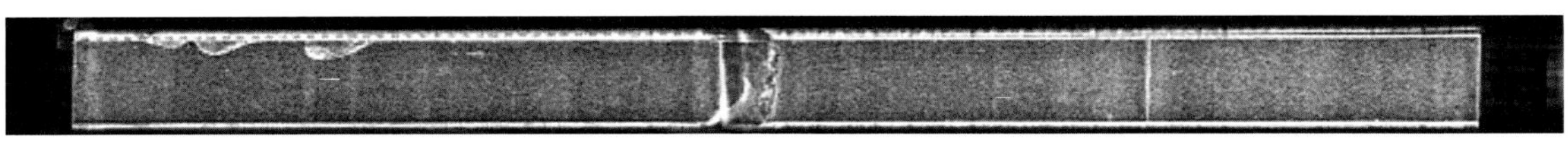

585

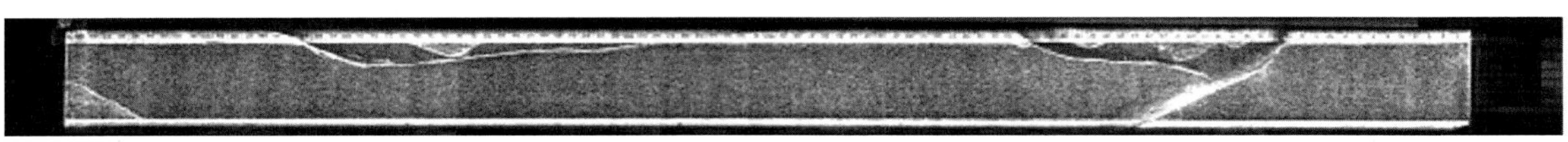

586

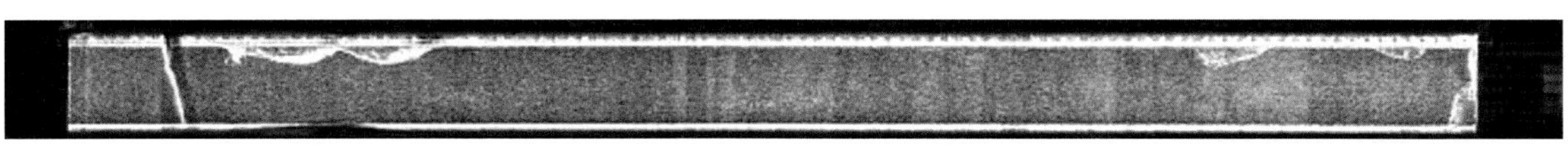

587

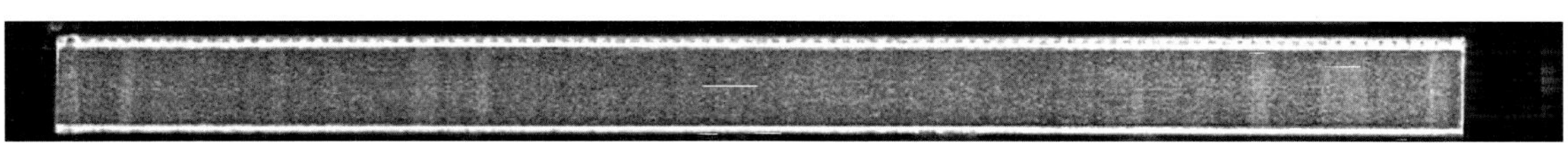

588

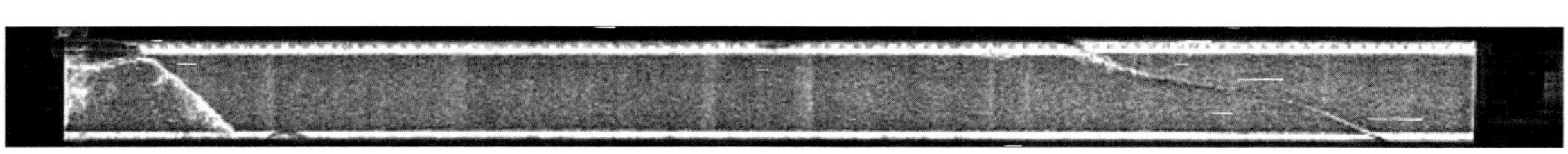

589

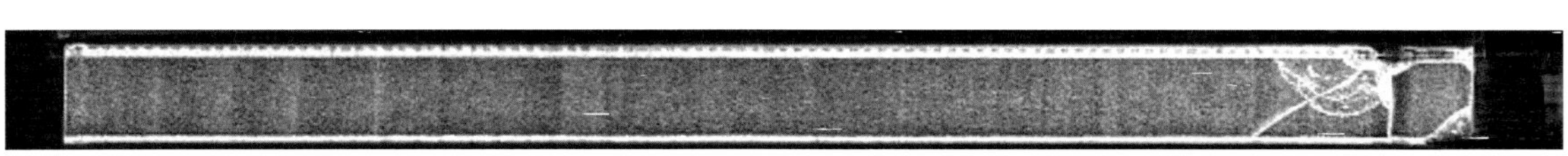

599

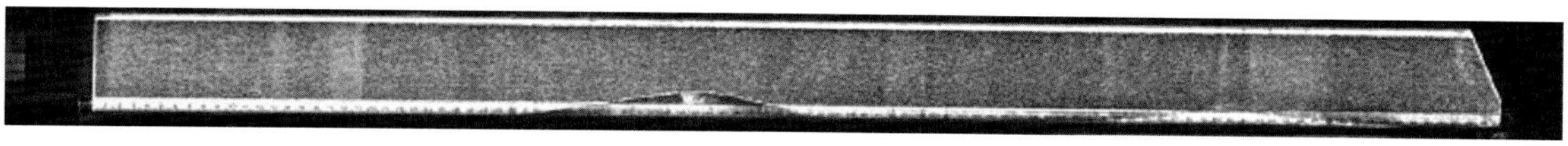
598

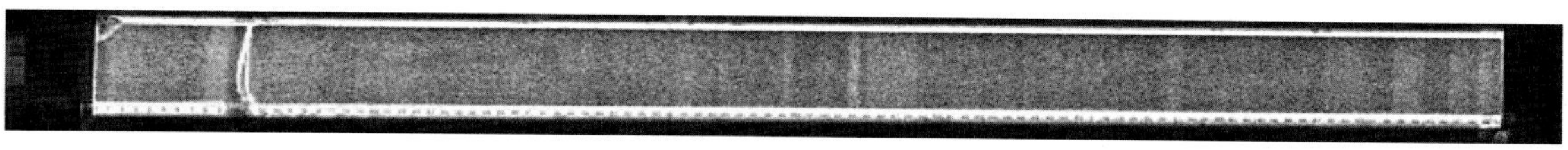
597

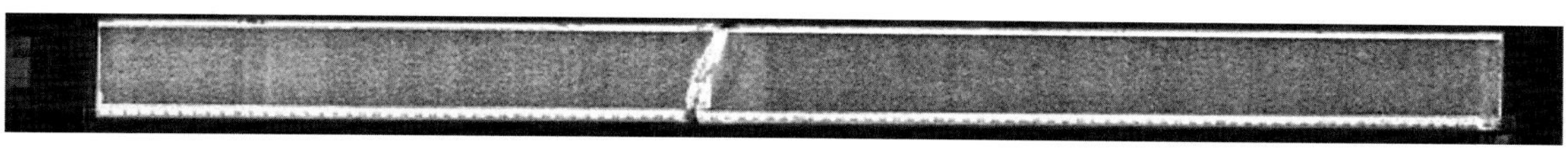
596

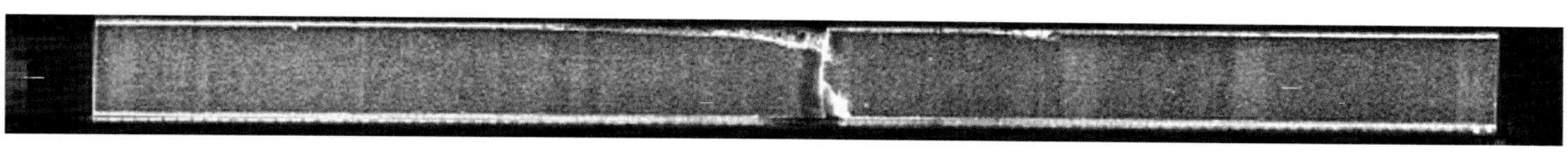
595

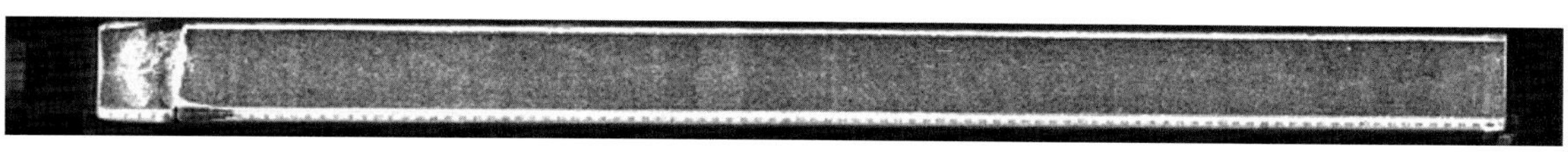
594

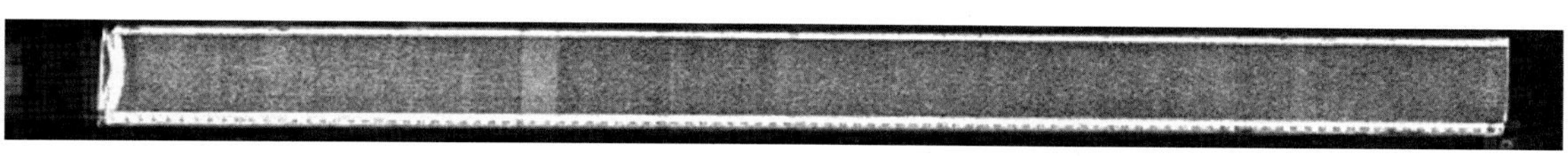
593

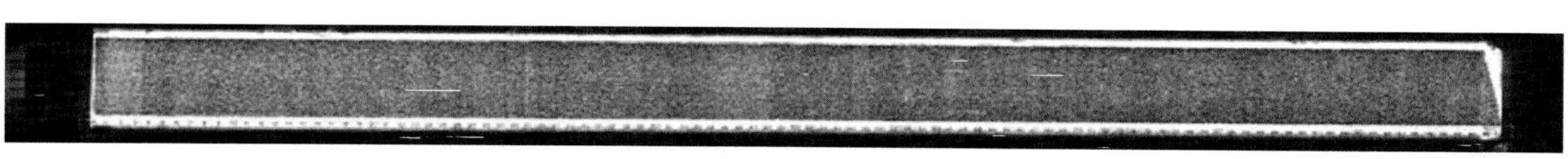
592

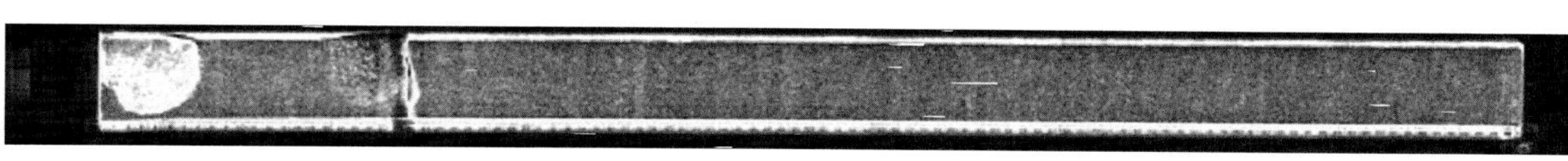
591

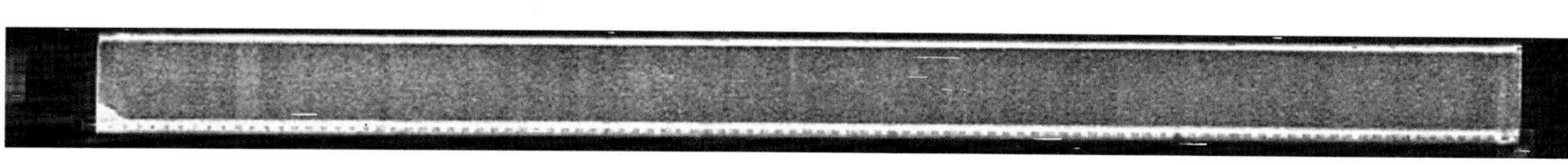
590

600
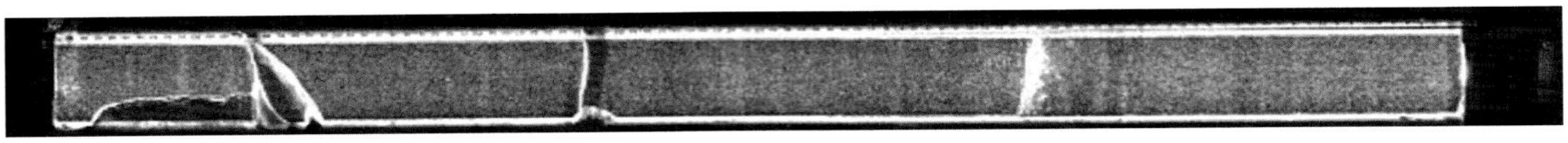

601
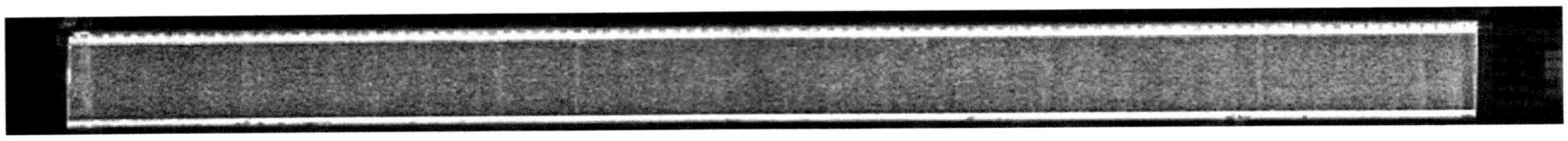

602
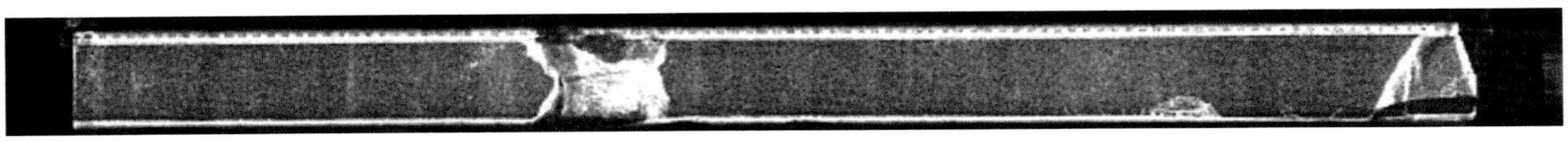

603
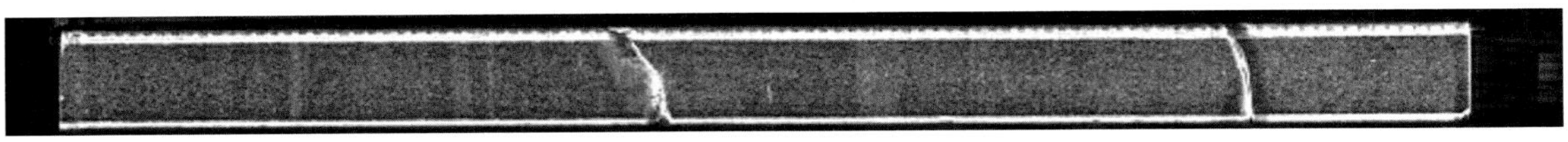

604
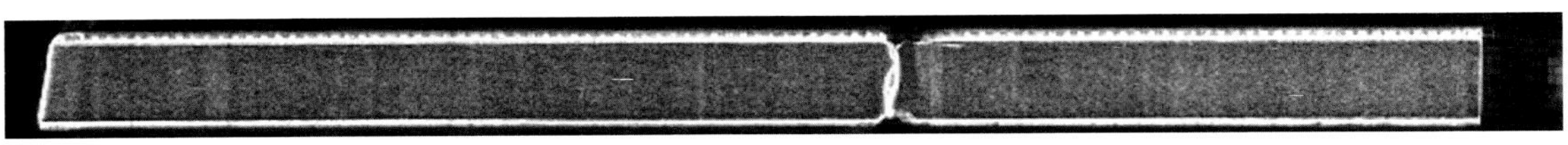

605
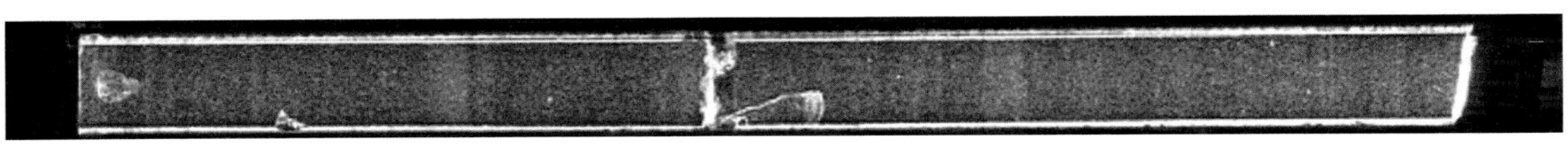

606
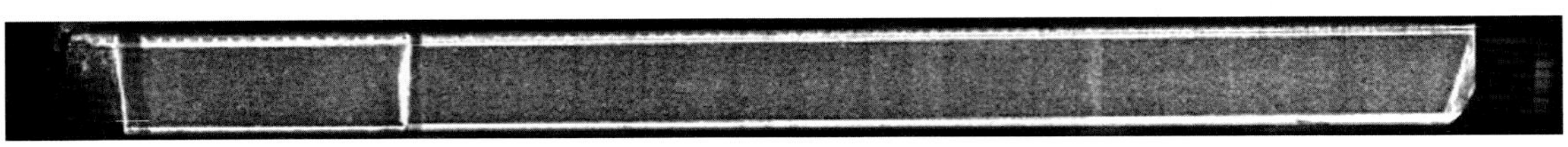

607
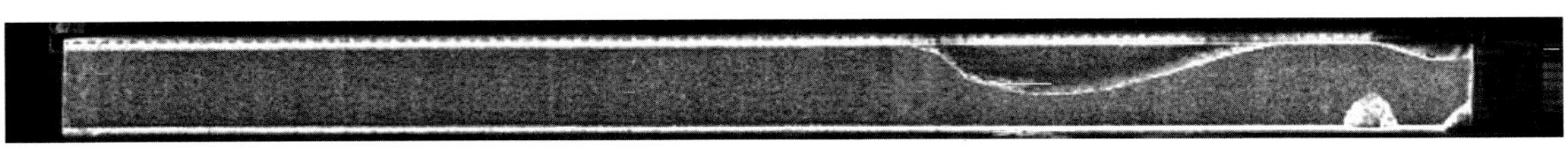

608
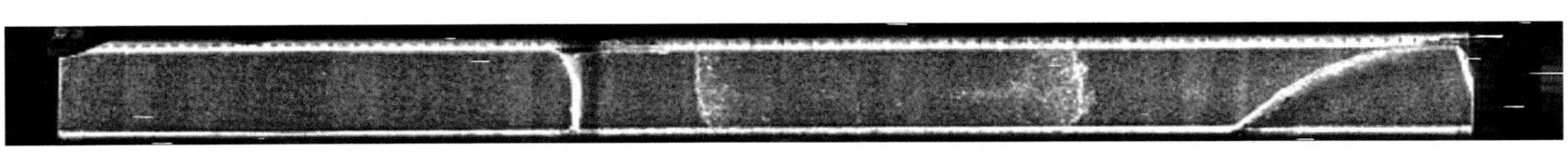

609
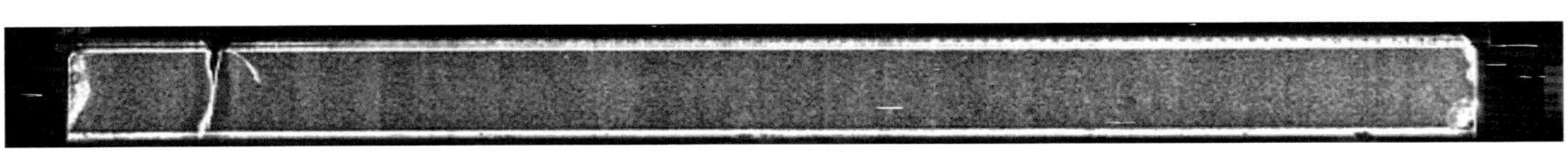

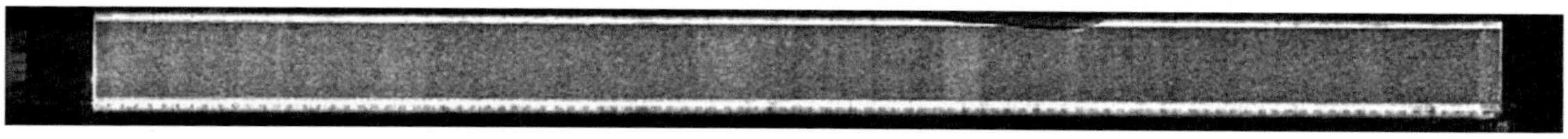
619

618

617

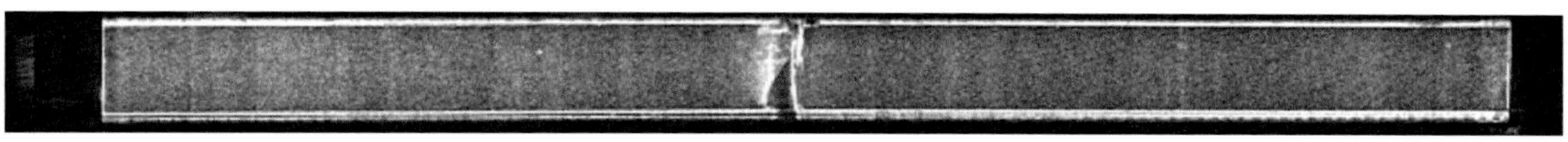
616

615

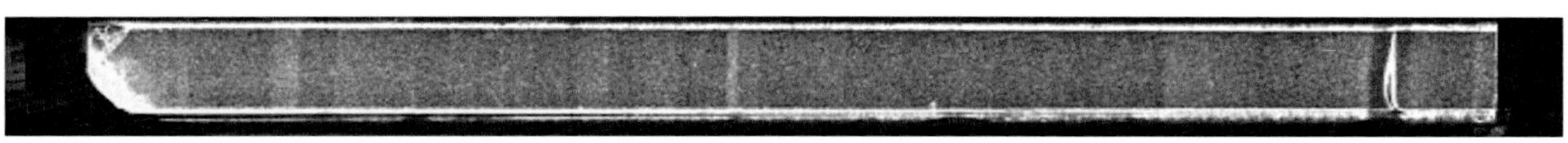
614

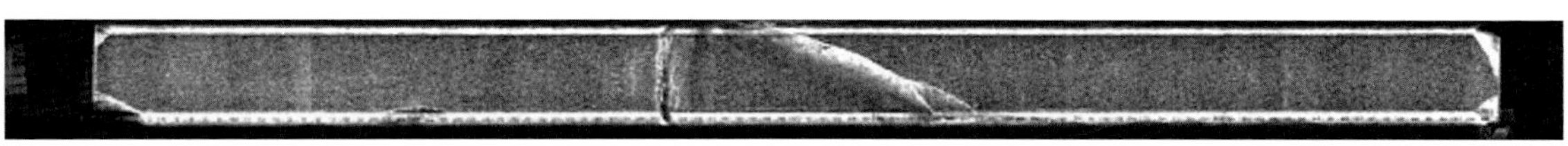
613

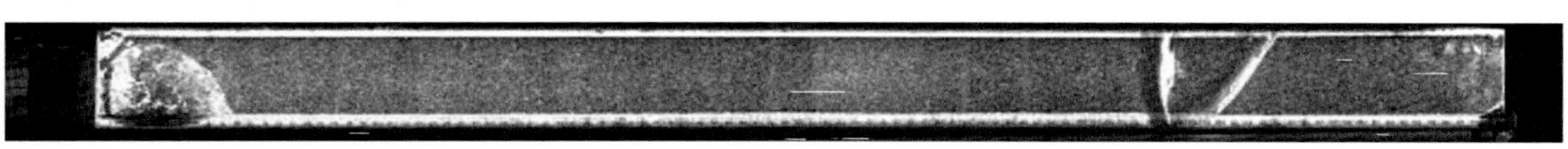
612

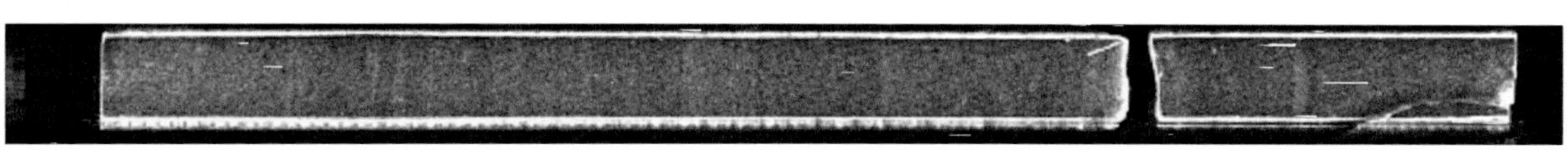
611

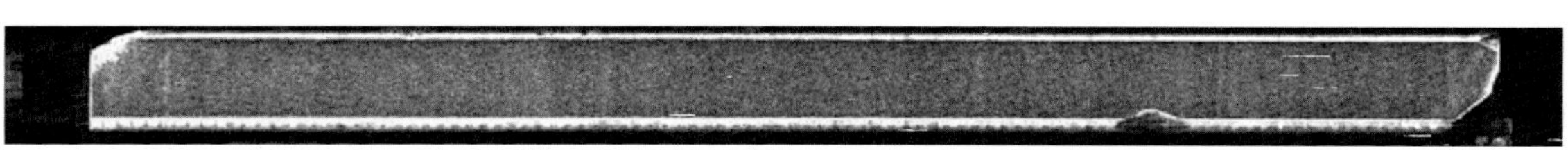
610

620
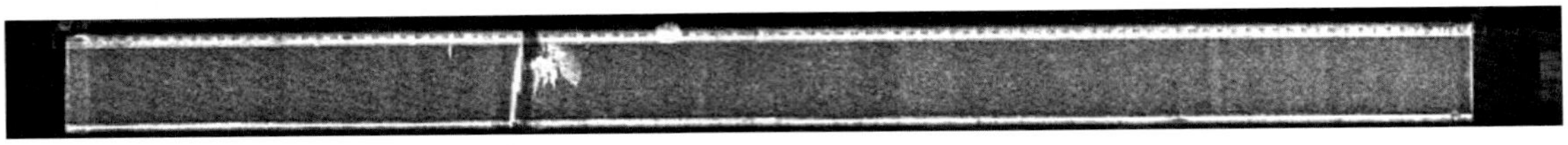

621
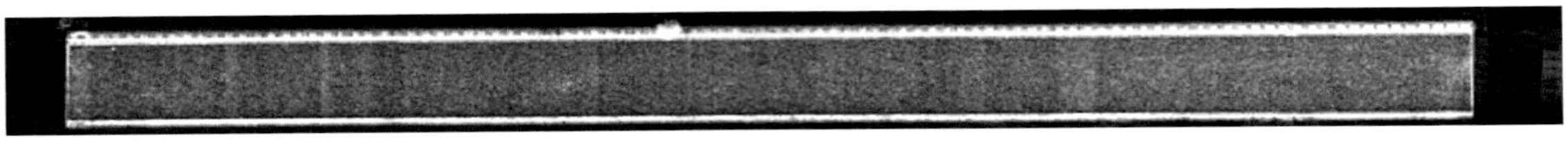

622
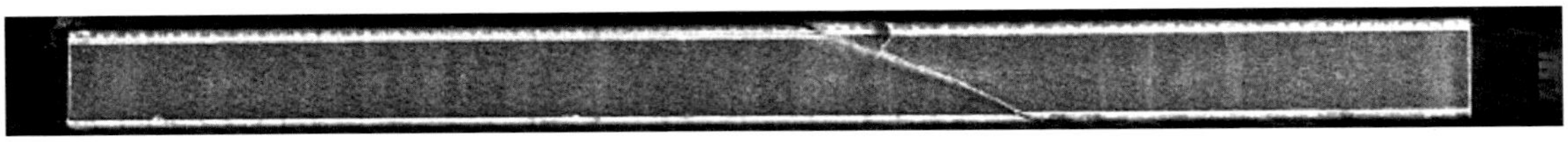

623

624
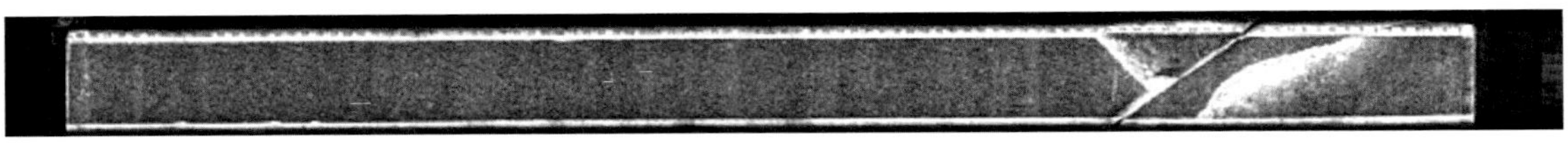

625
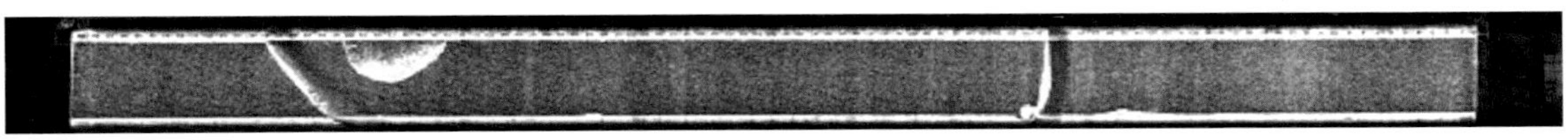

626
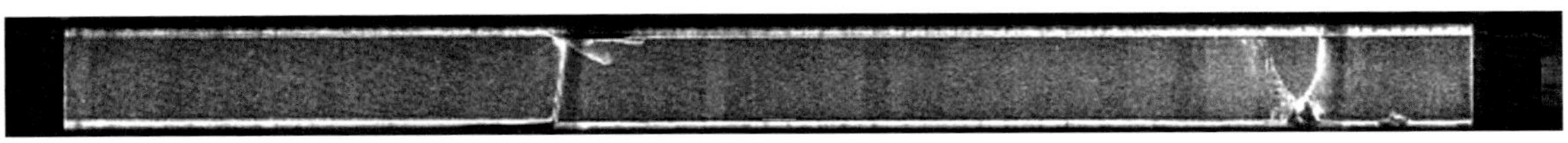

627
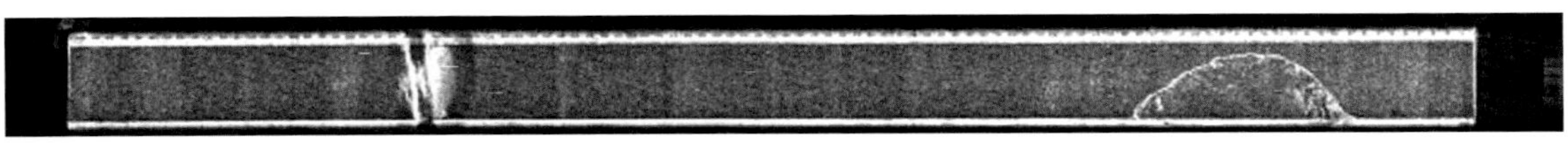

628
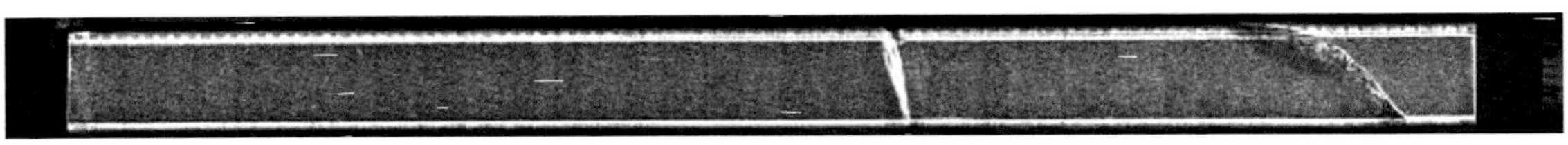

629

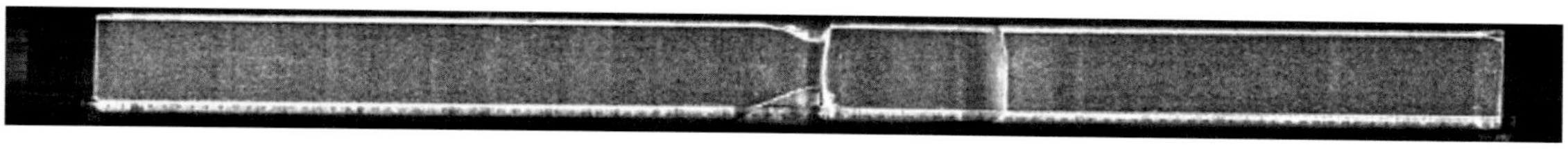

639

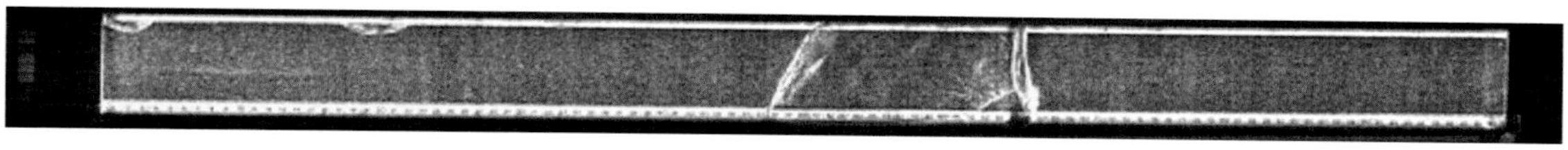

638

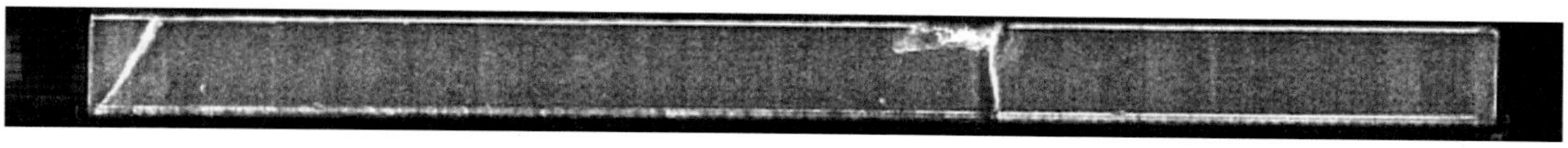

637

636

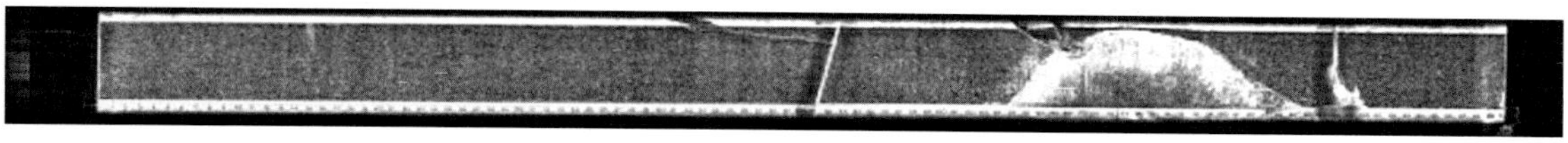

635

634

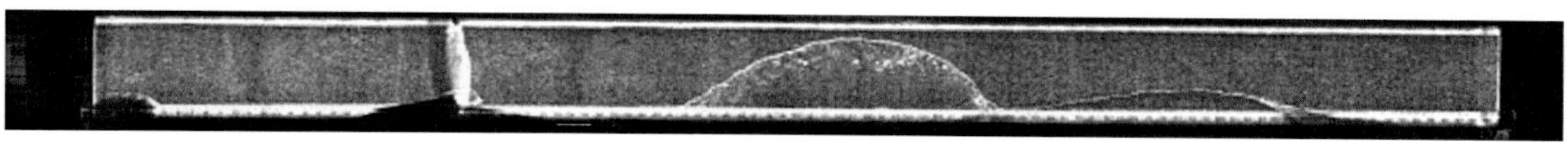

633

632

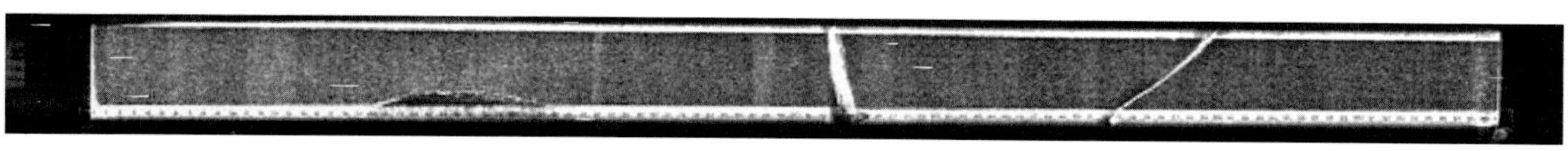

631

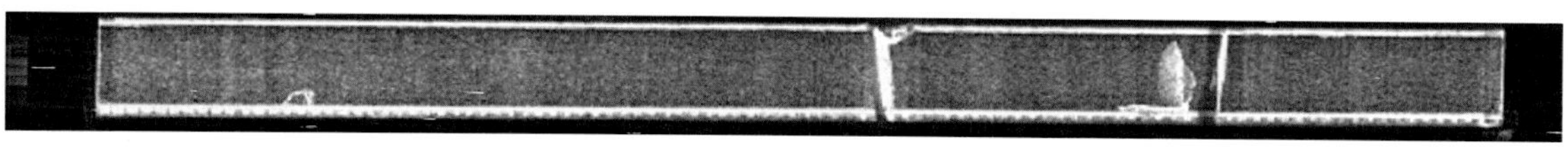

630

640
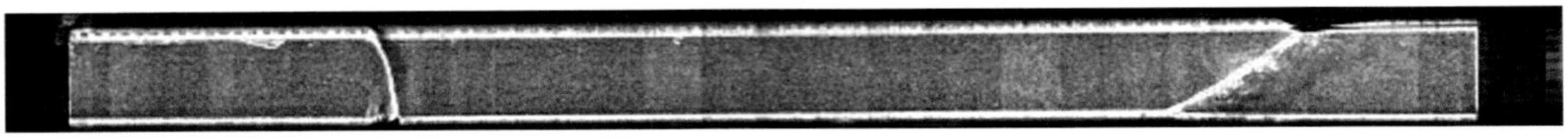

641
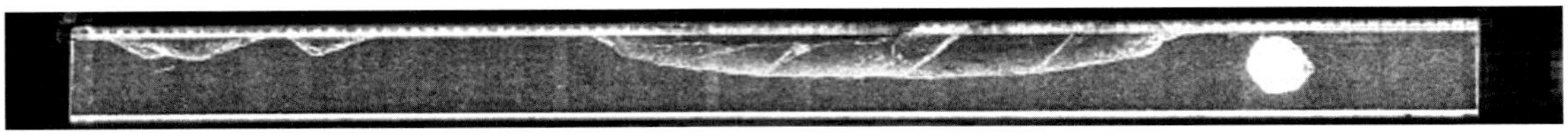

642
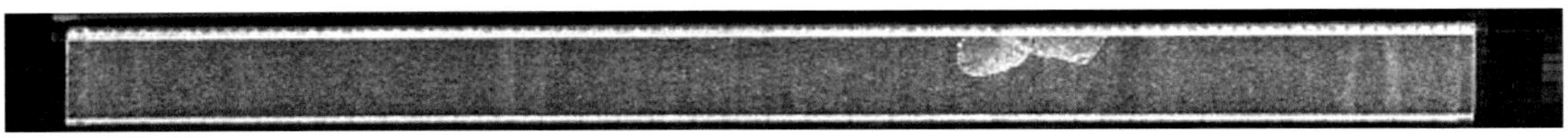

643
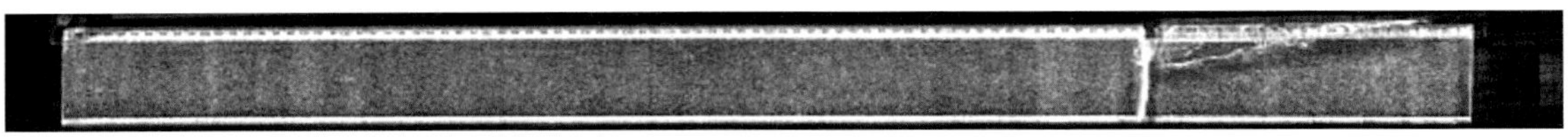

644
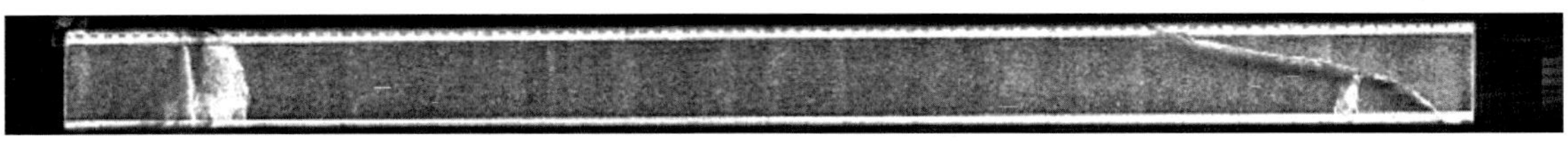

645
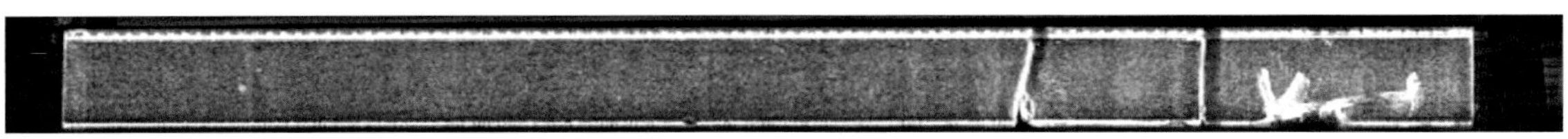

646
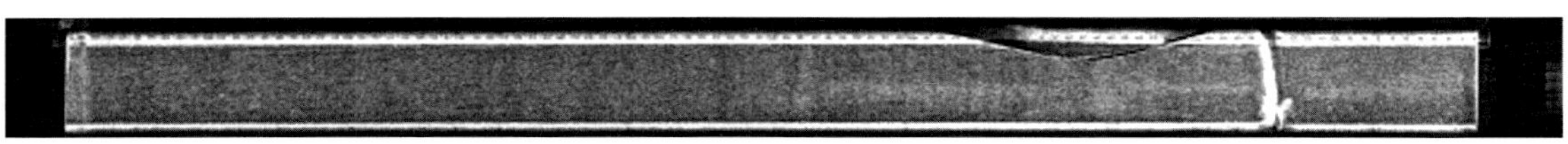

647
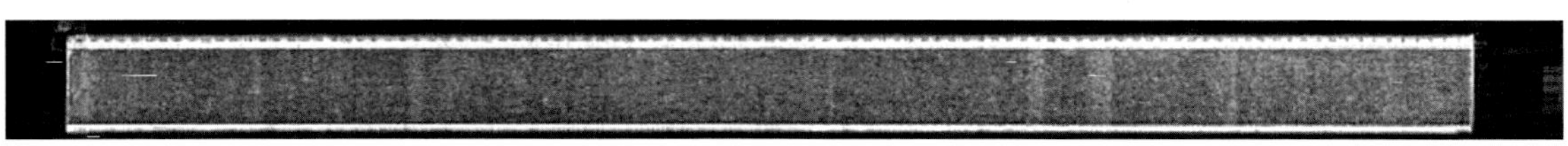

648
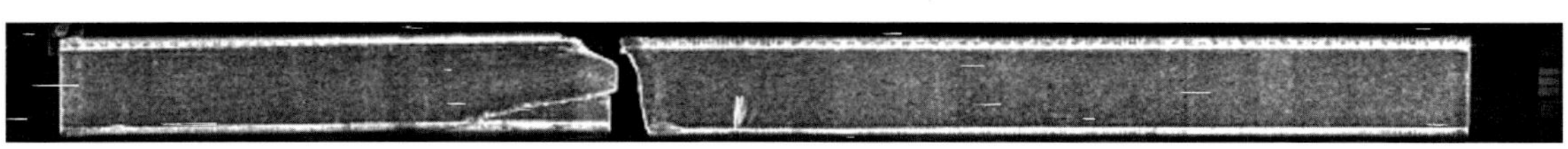

649
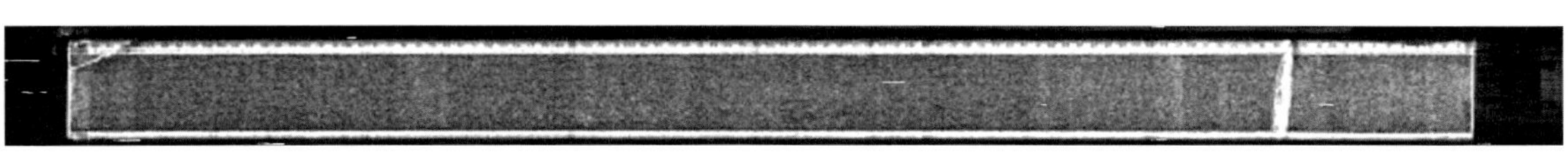

659

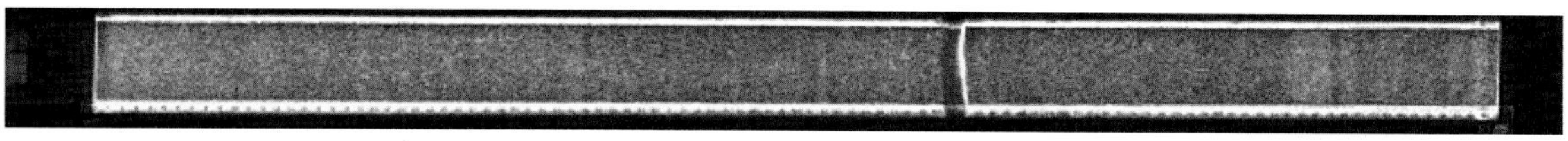
658

657

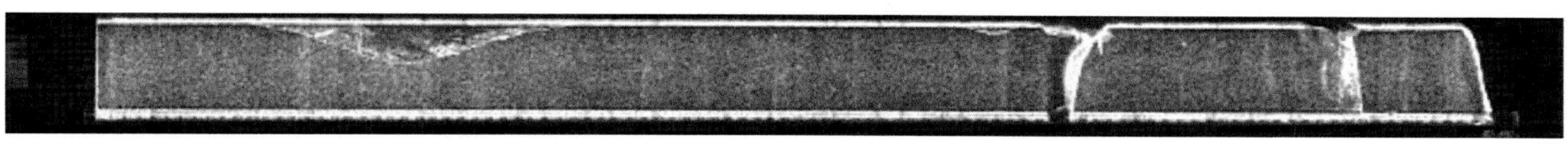
656

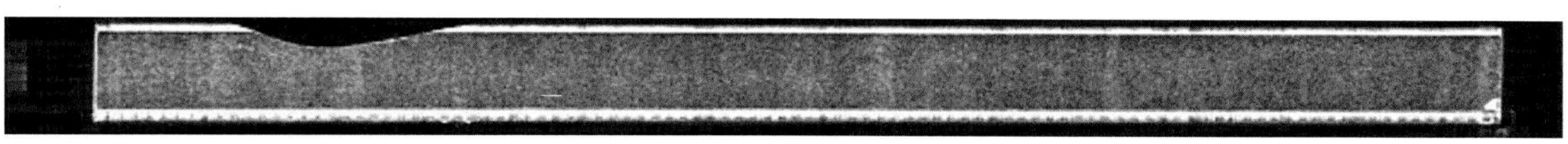
655

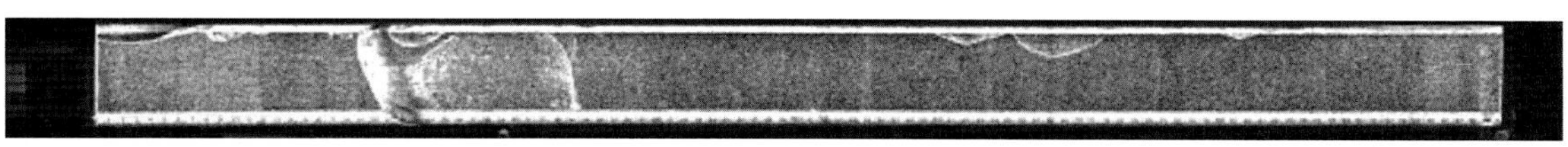
654

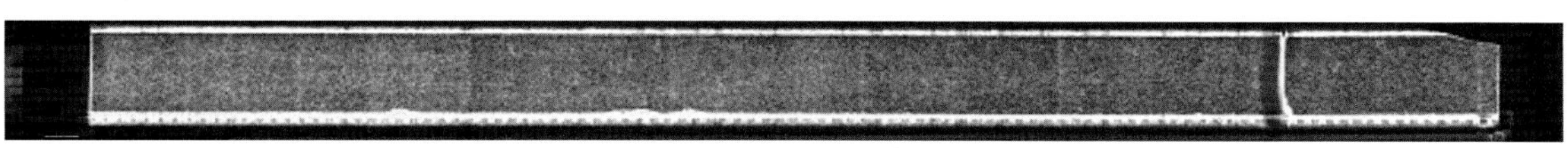
653

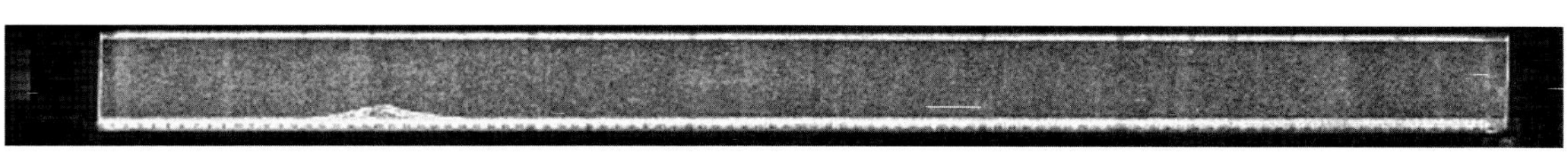
652

651

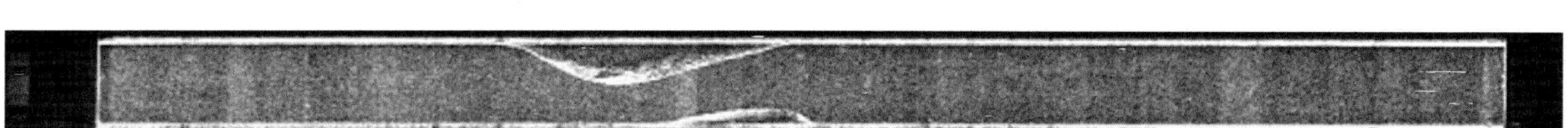
650

660
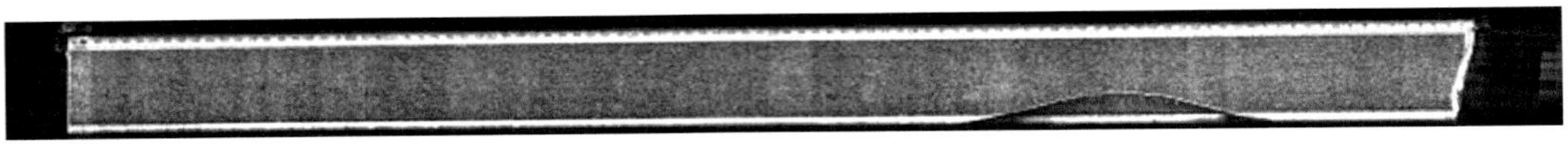

661

662
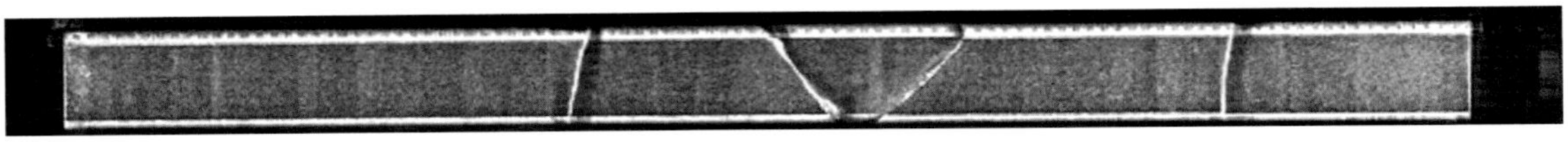

663
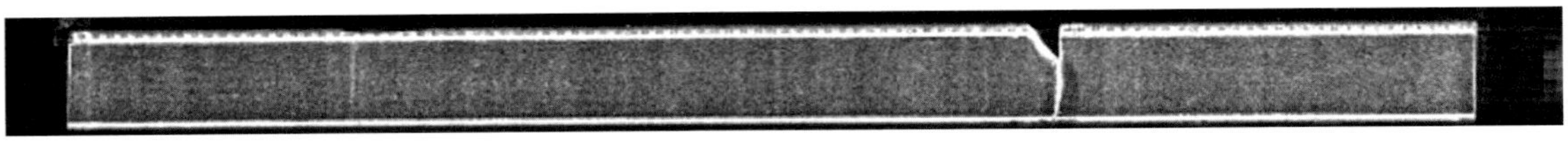

664
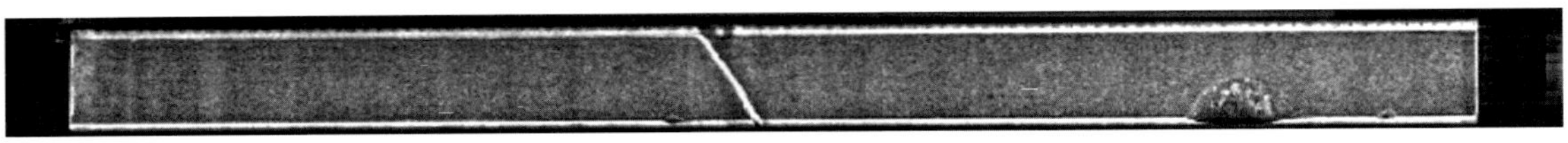

665
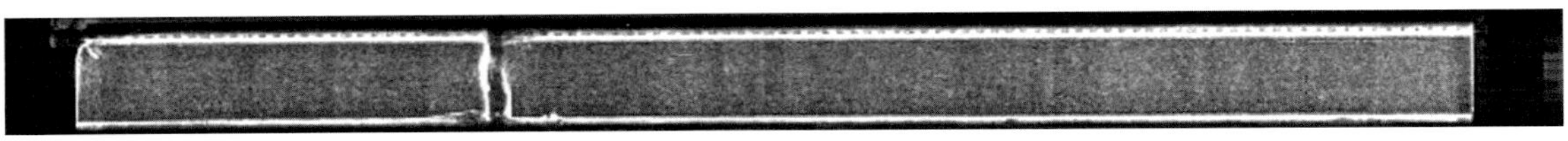

666
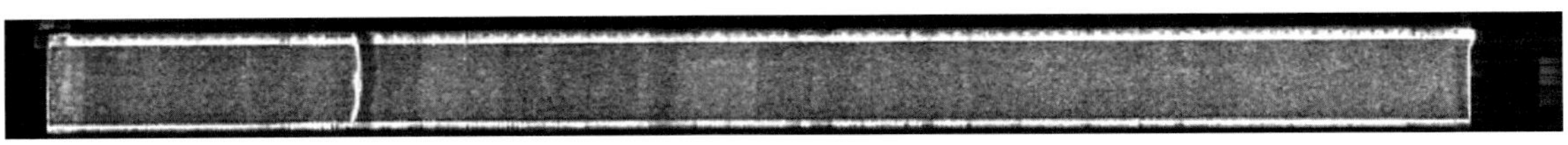

667
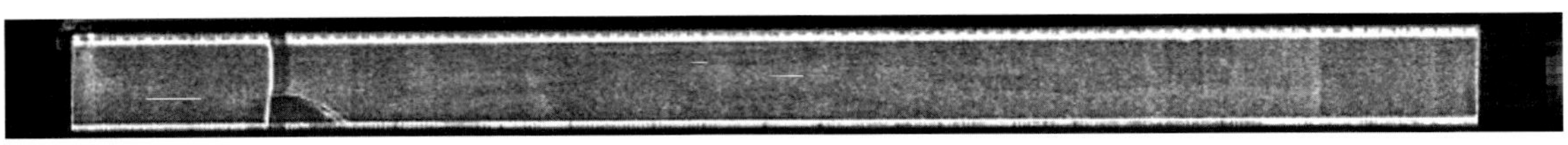

668
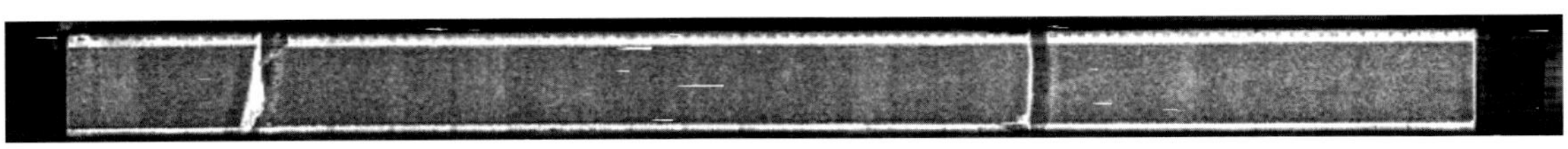

669
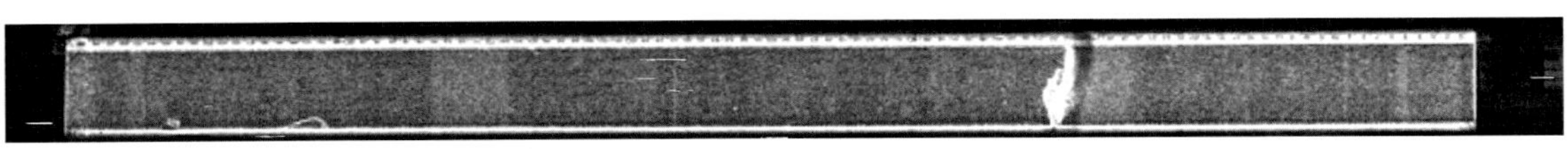

679

678

677

676

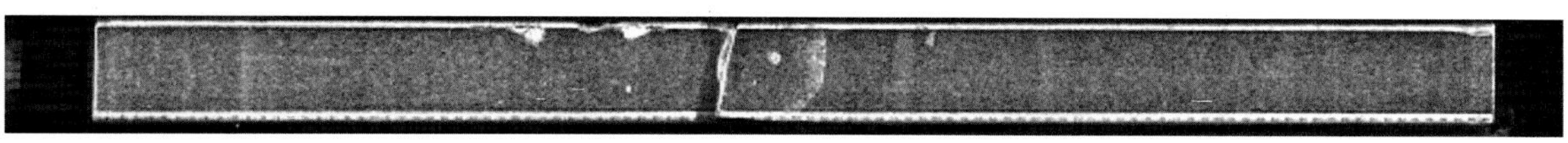
675

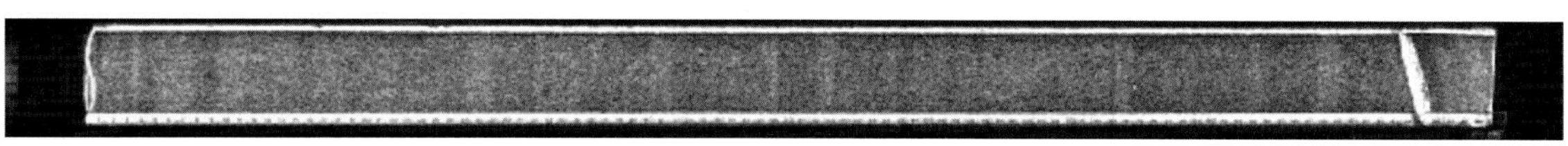
674

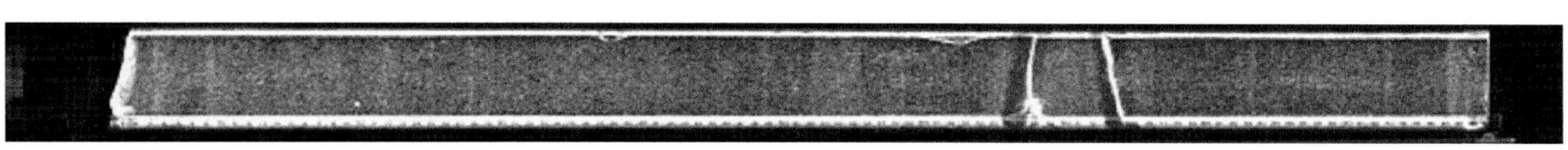
673

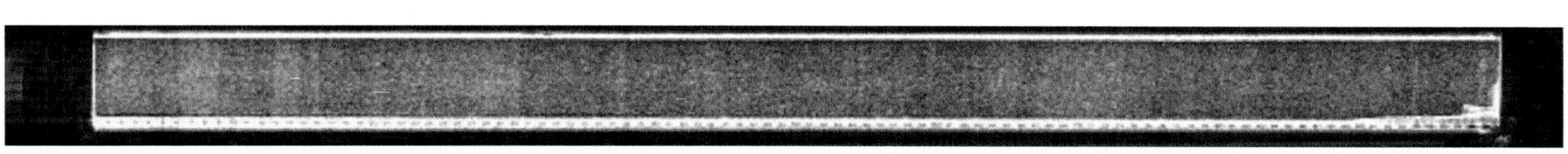
672

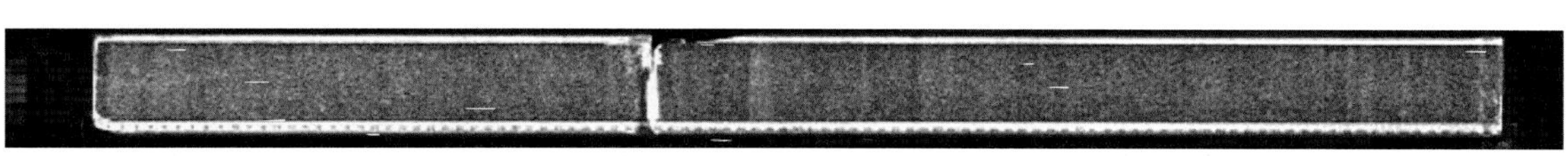
671

670

680

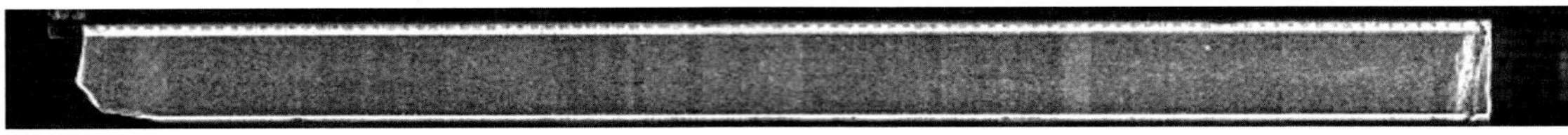

681

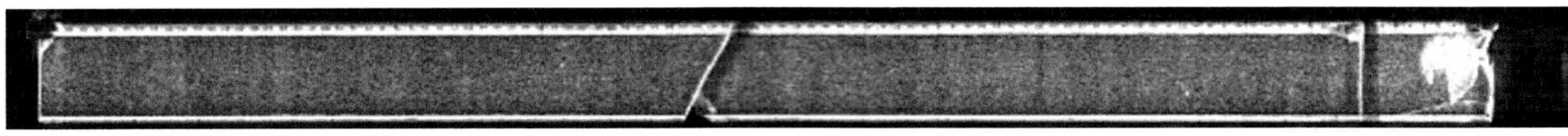

682

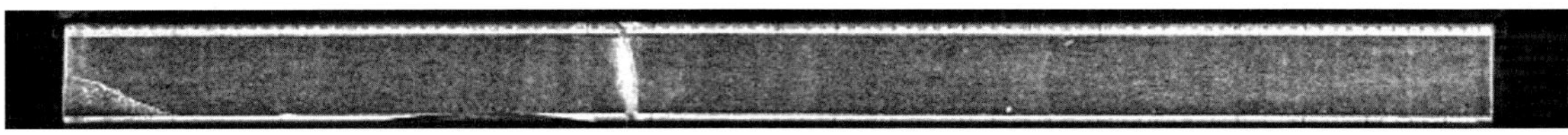

683

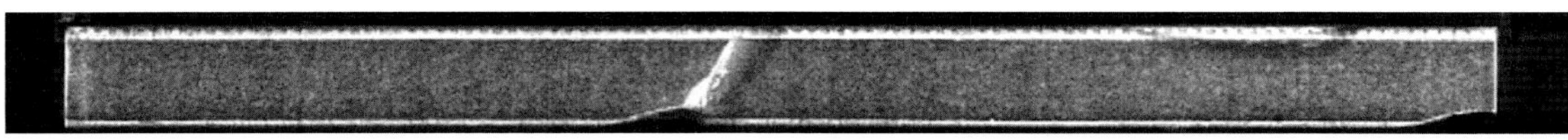

684

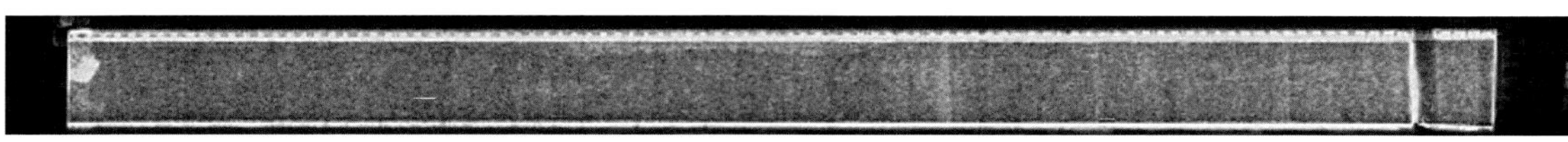

685

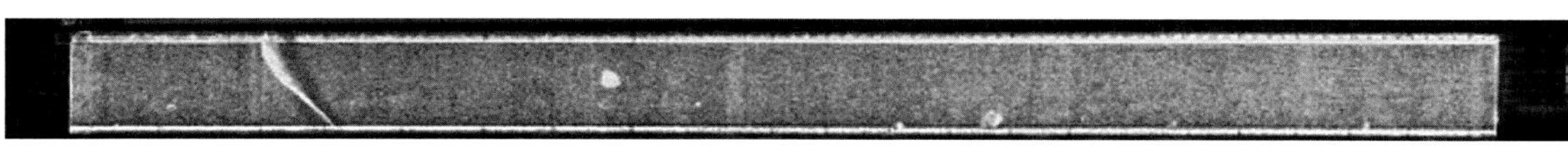

686

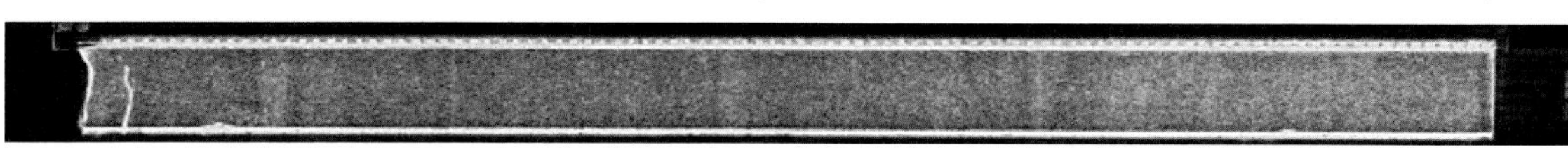

687

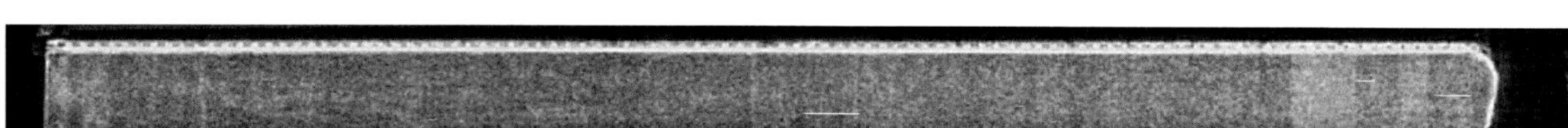

688

689

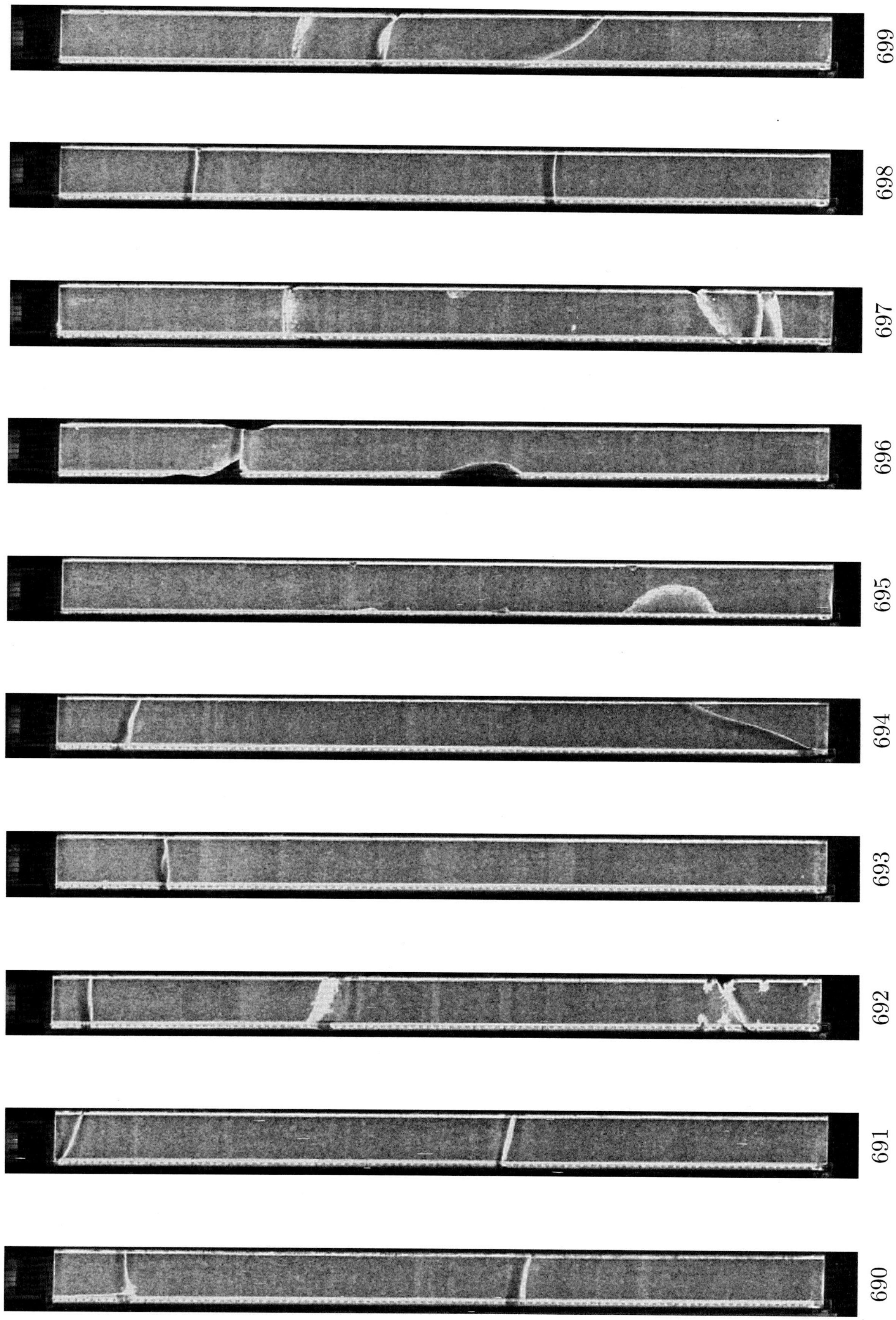
699
698
697
696
695
694
693
692
691
690

700

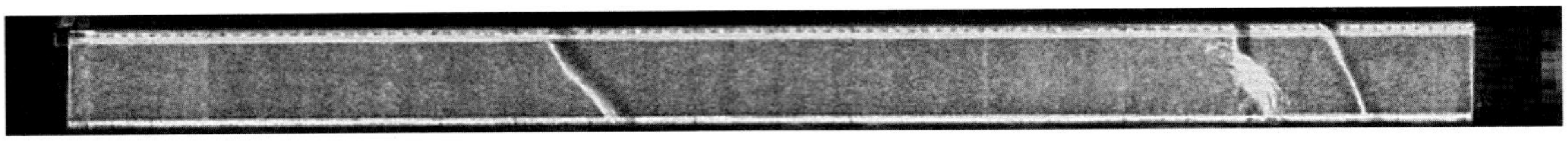

701

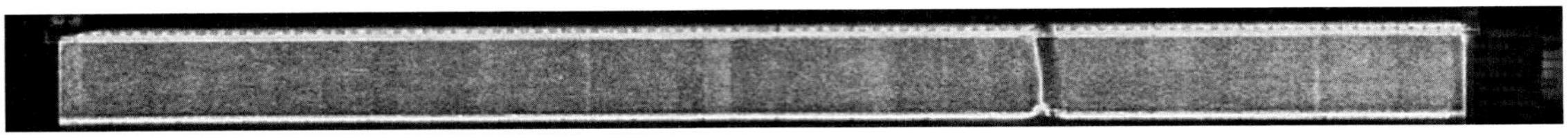

702

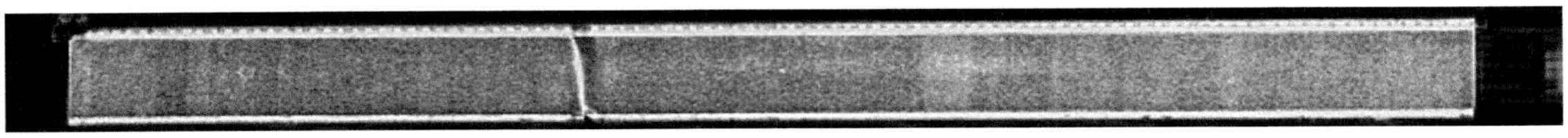

703

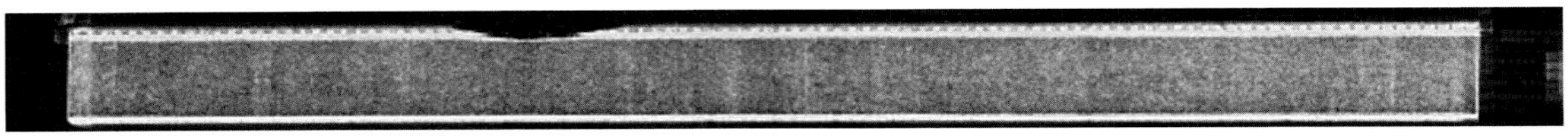

704

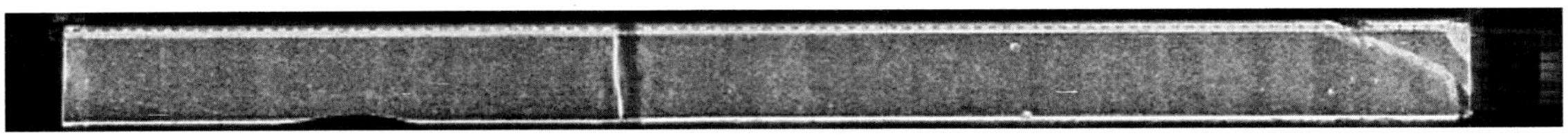

705

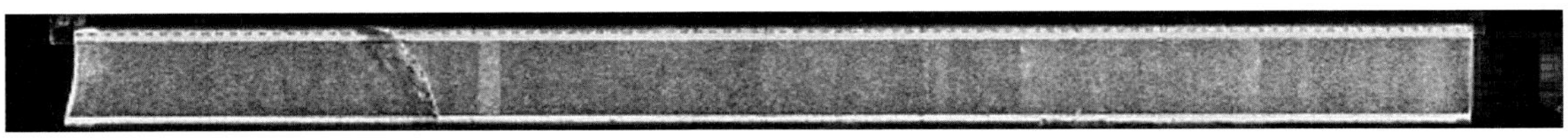

706

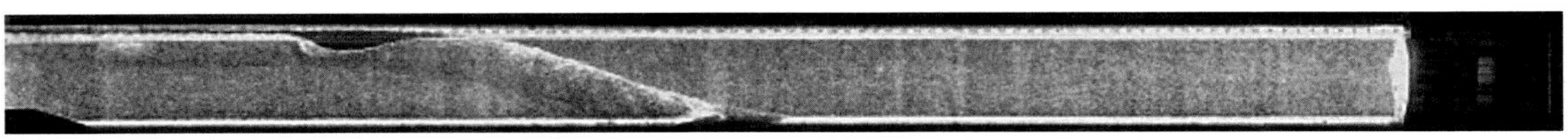

707

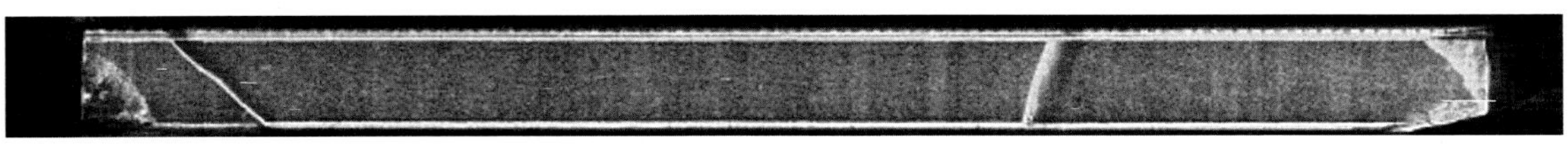

708

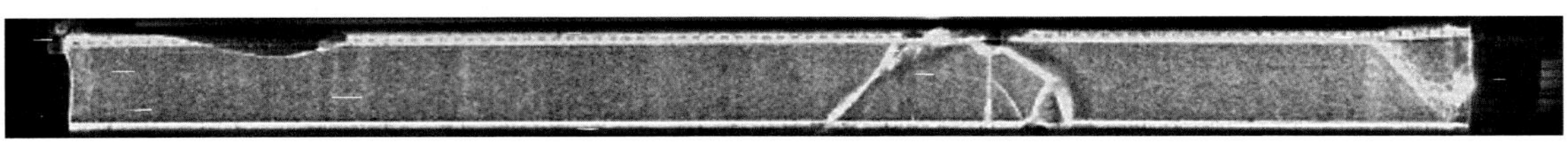

709

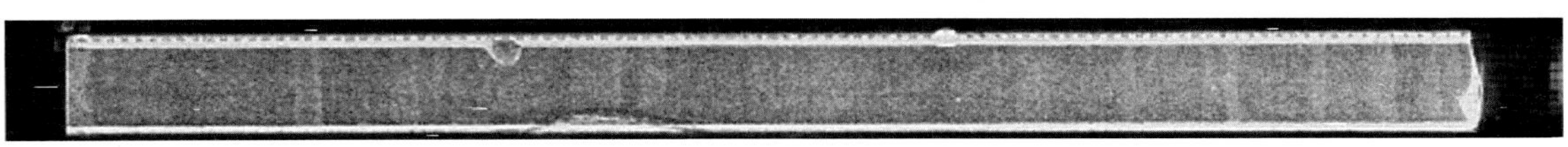

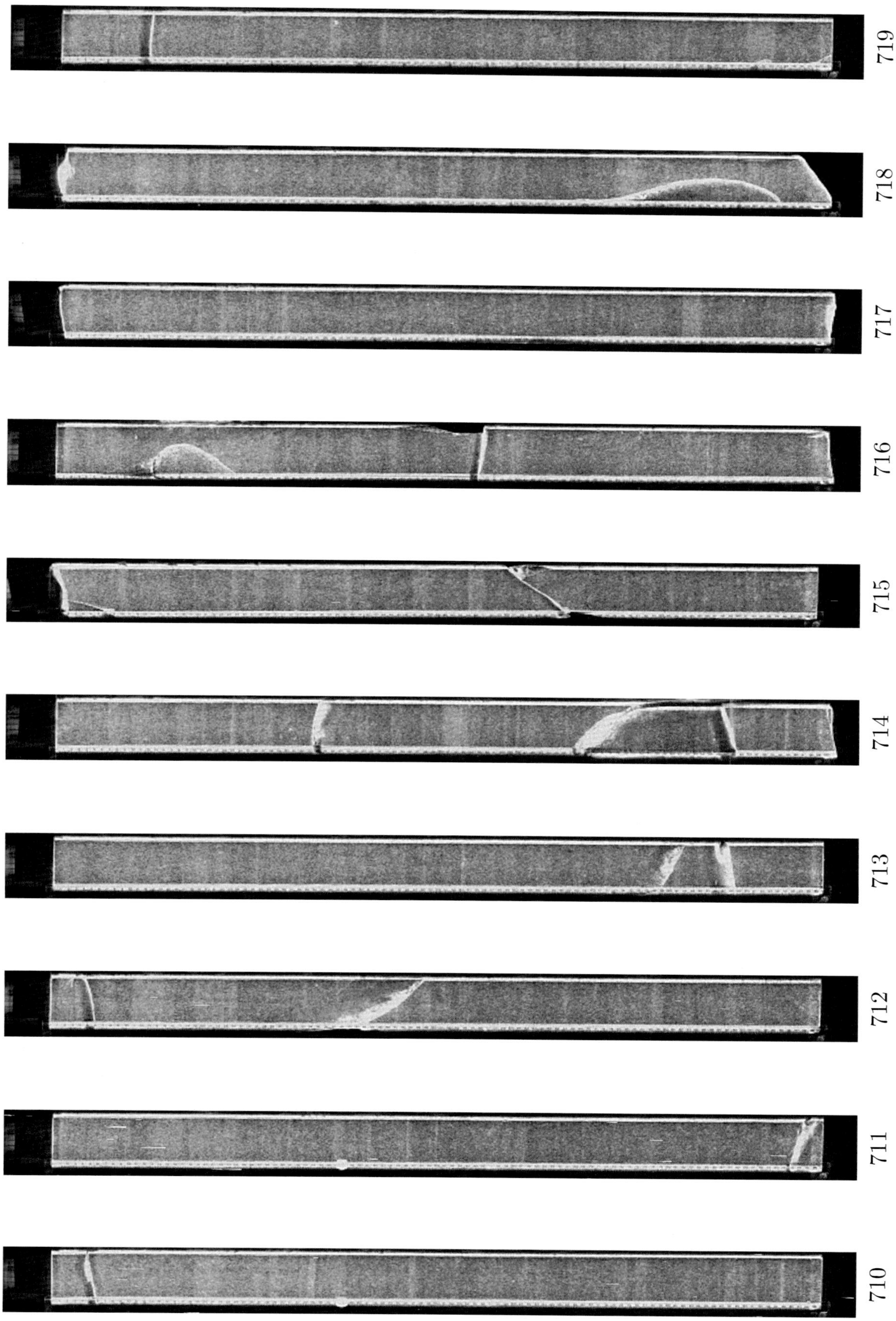
719
718
717
716
715
714
713
712
711
710

720

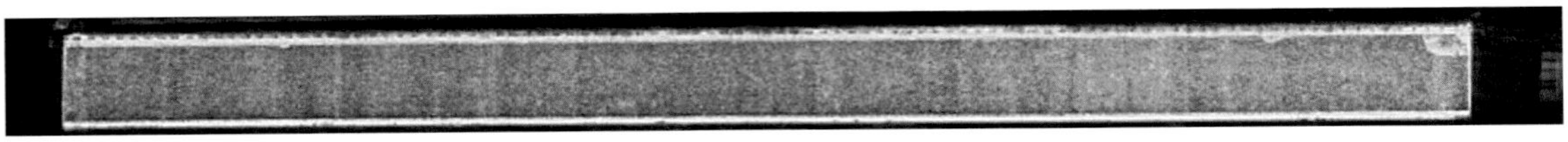

721

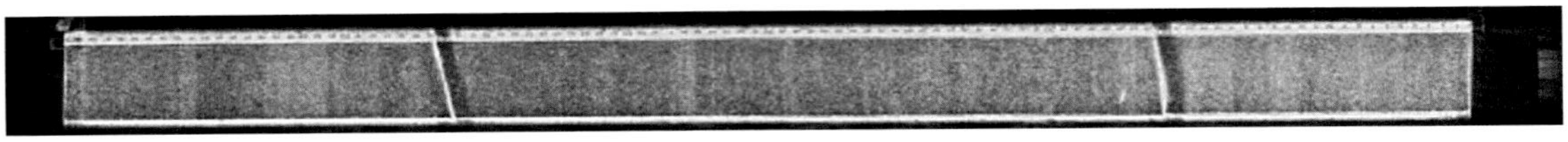

722

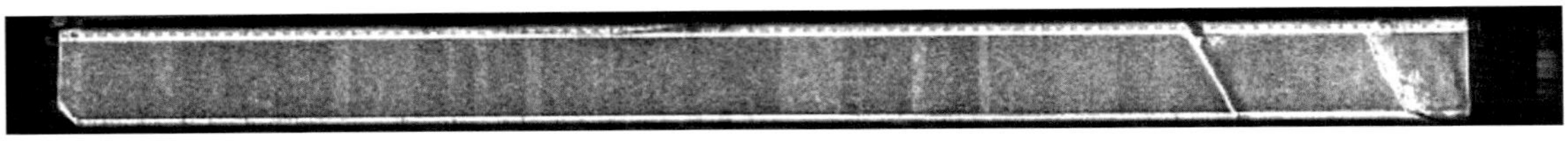

723

724

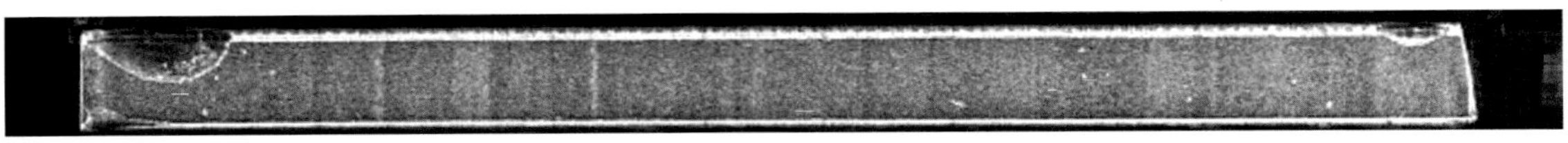

725

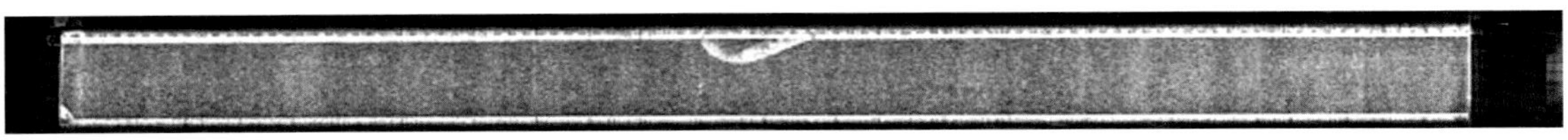

726

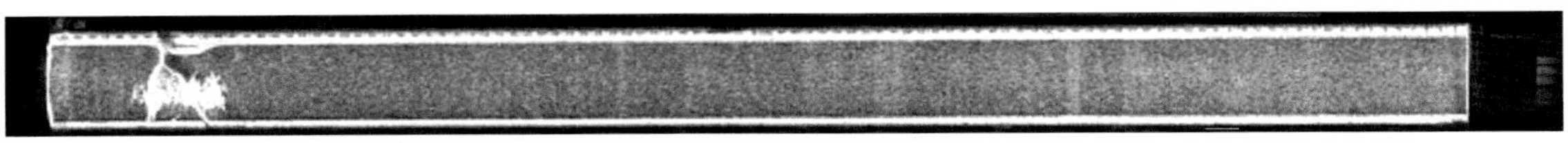

727

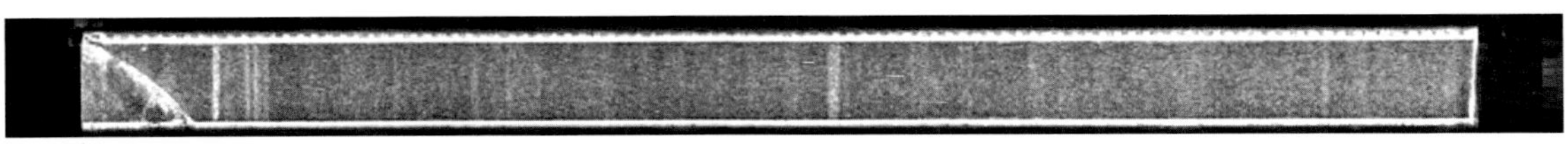

728

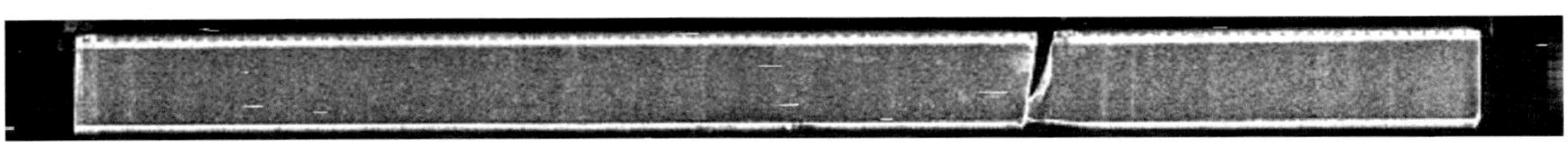

729

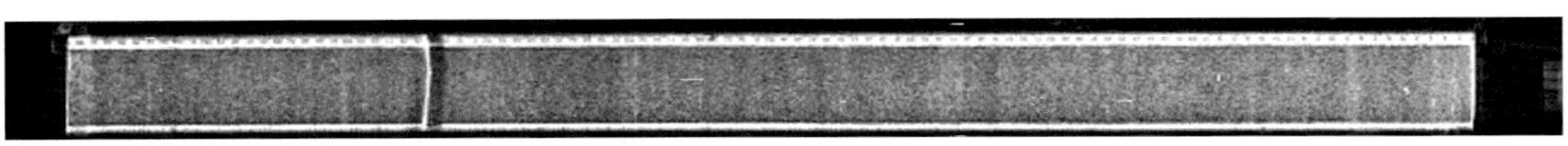

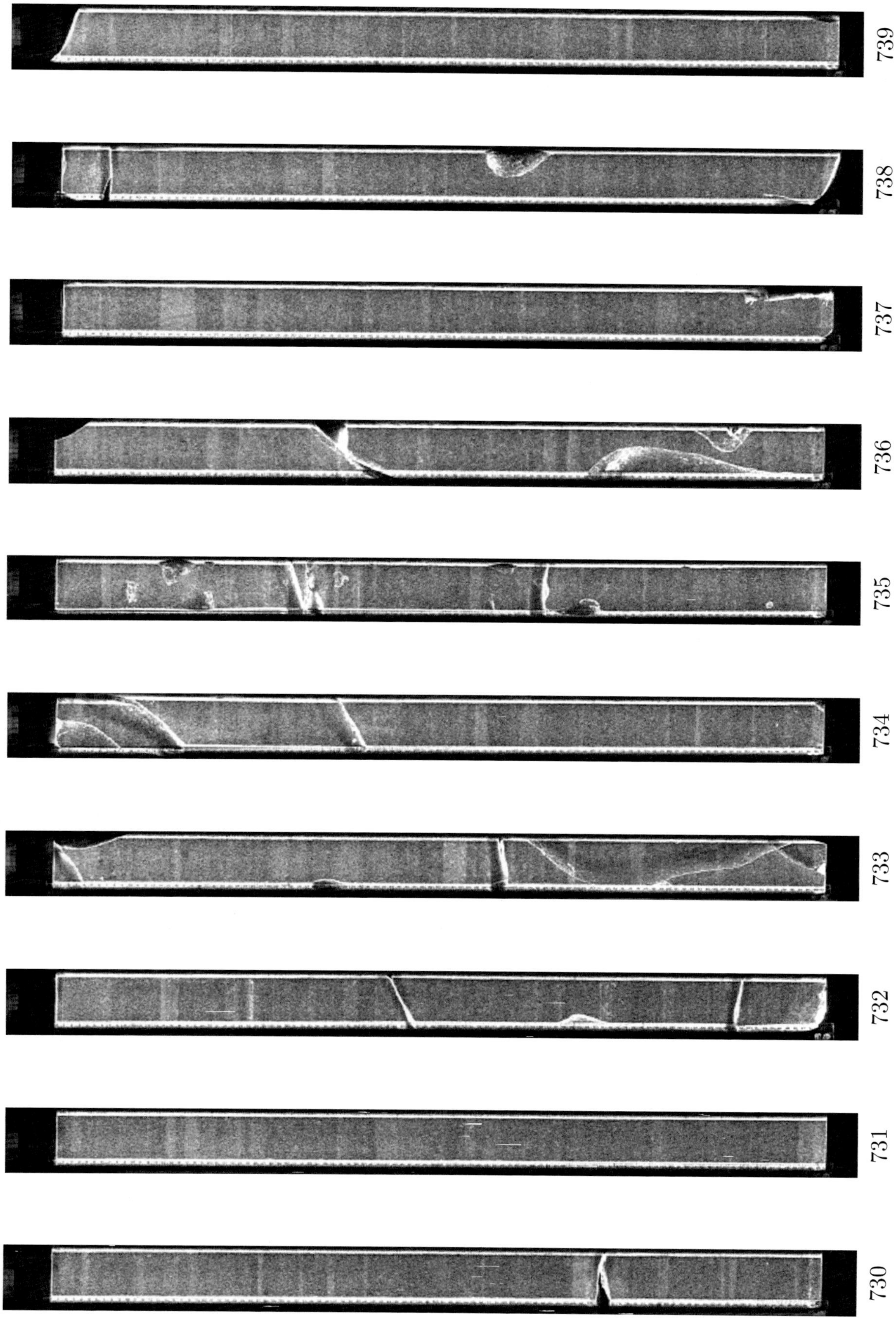
739
738
737
736
735
734
733
732
731
730

740
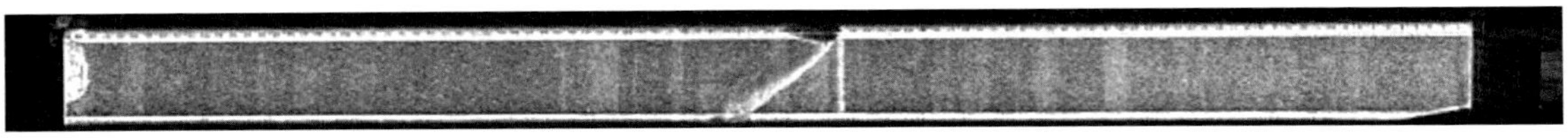

741
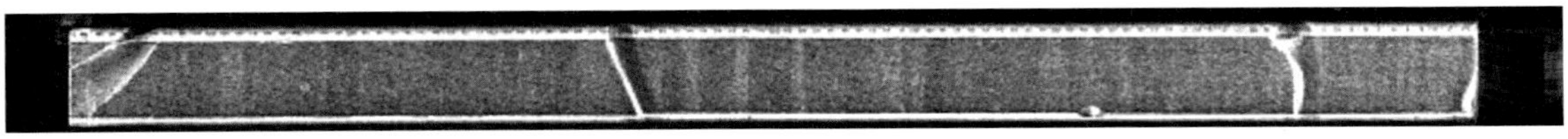

742
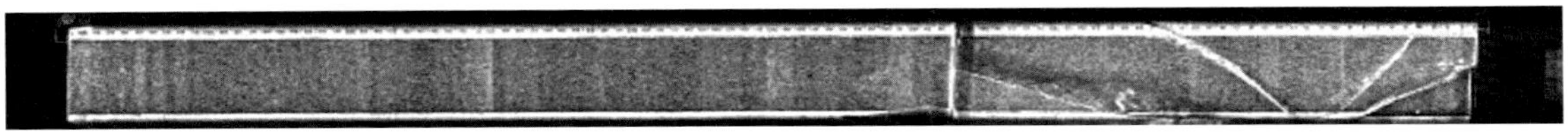

743
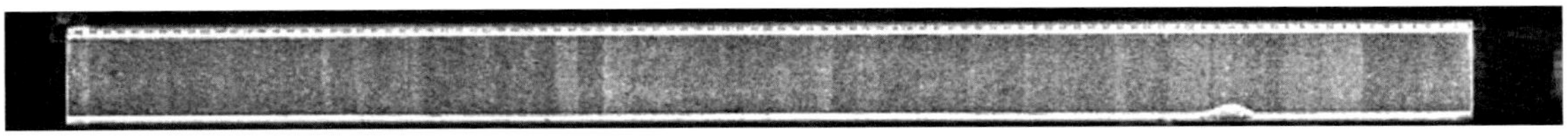

744
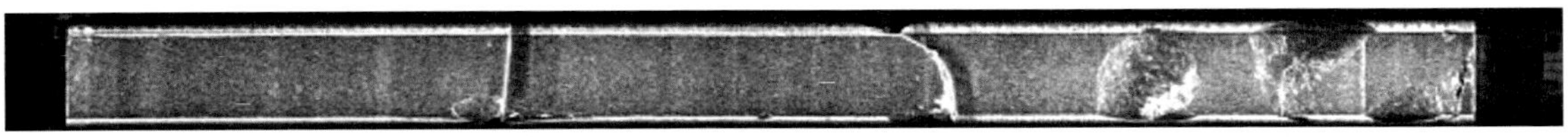

745
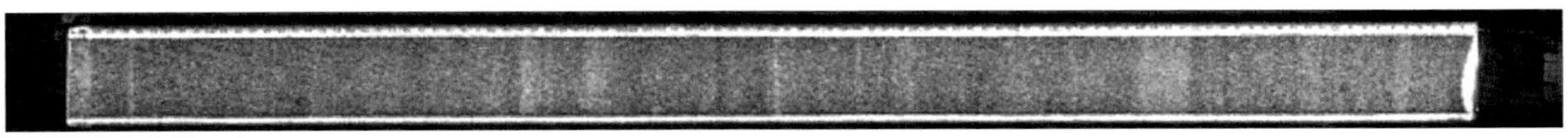

746

747
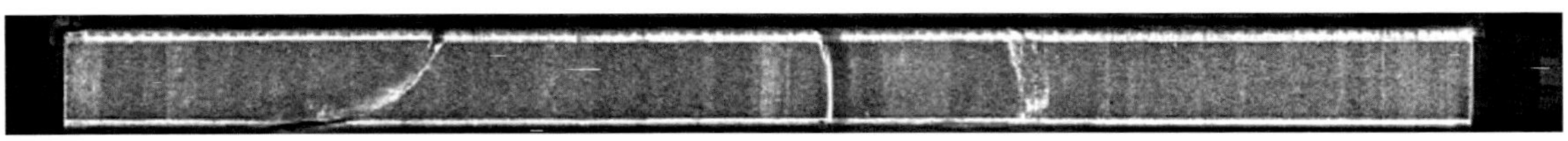

748
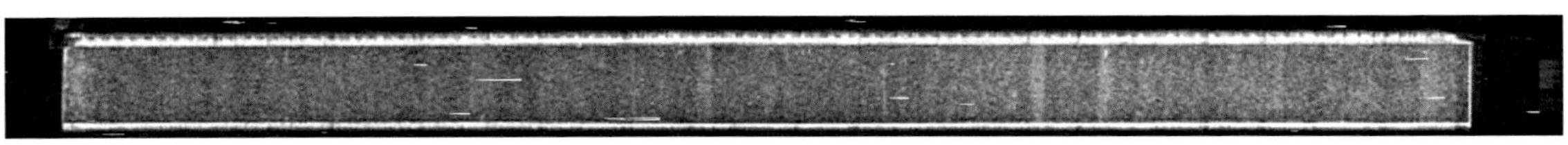

749
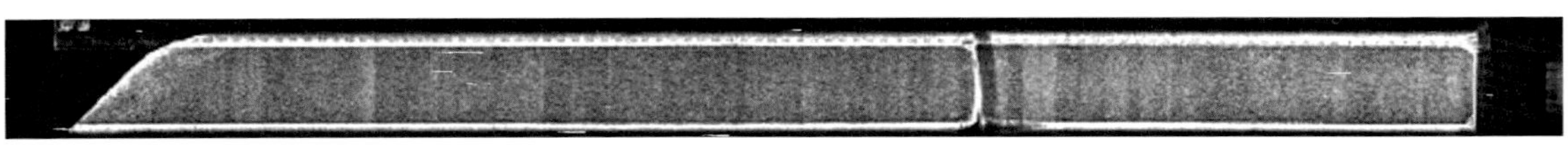

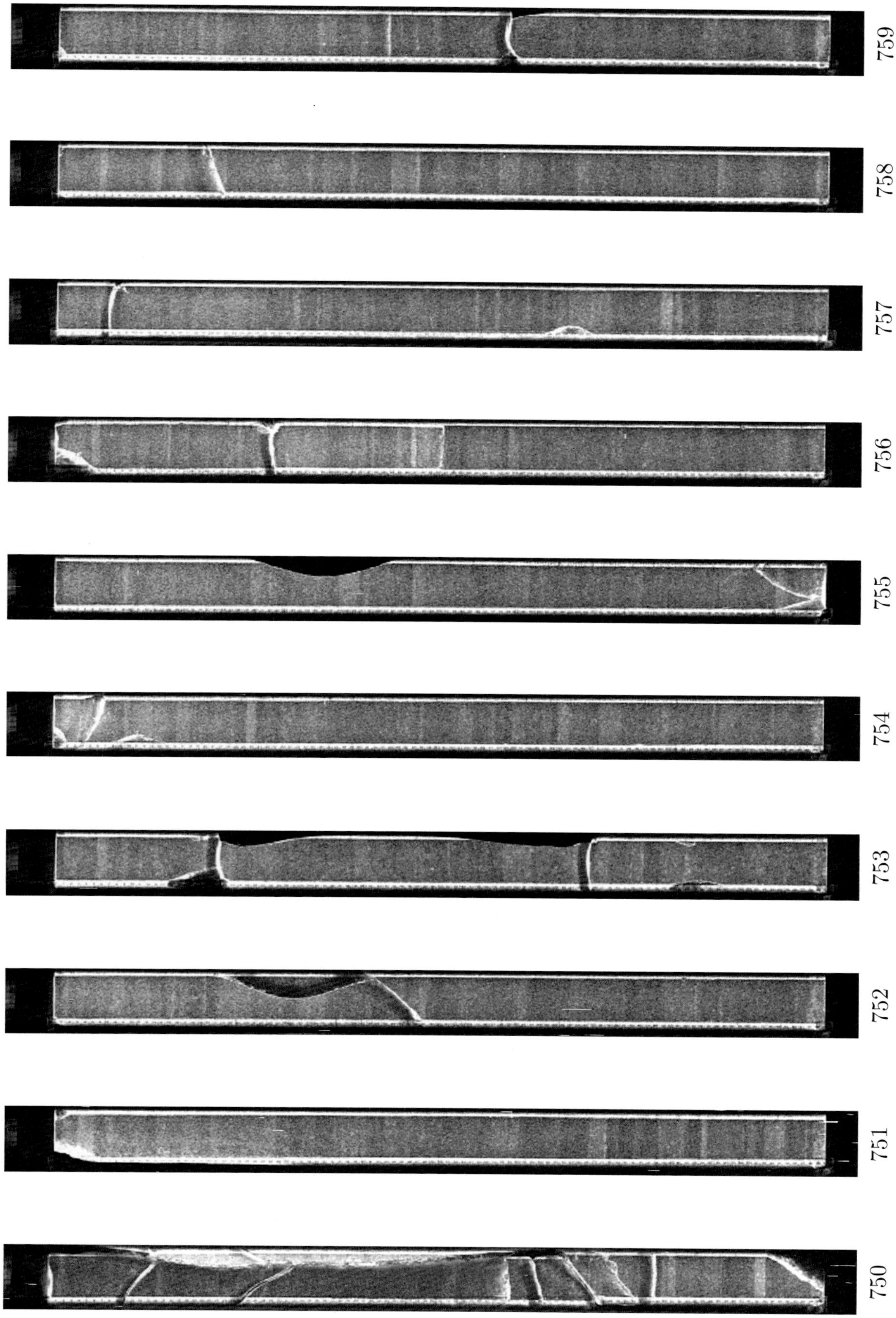
759
758
757
756
755
754
753
752
751
750

760
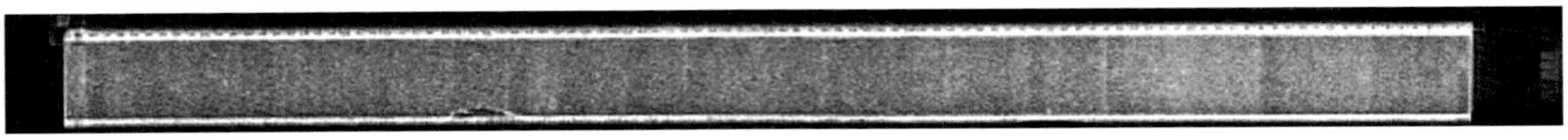

761
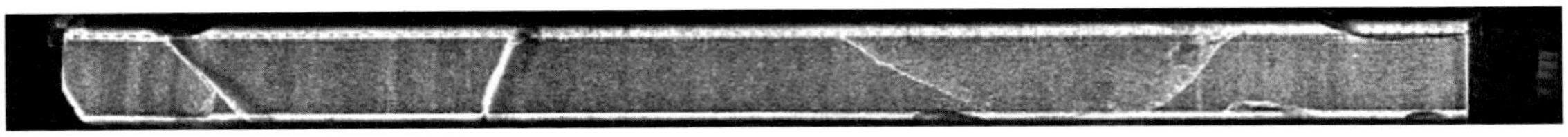

762
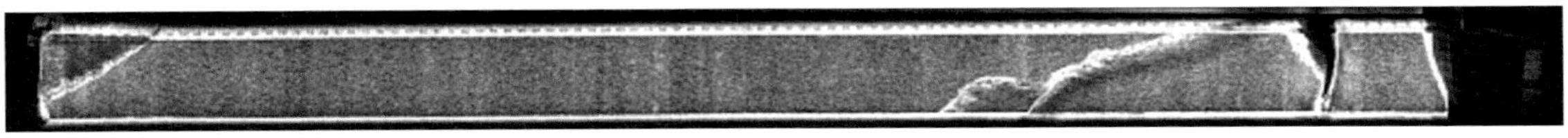

763
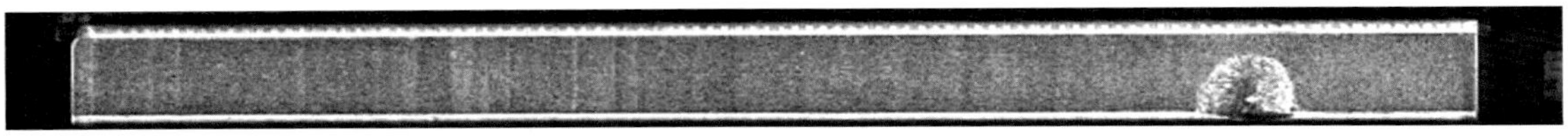

764
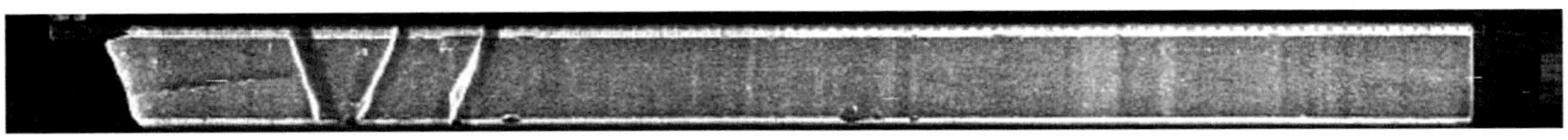

765
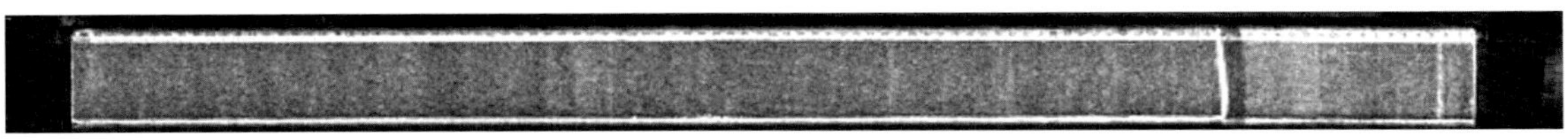

766
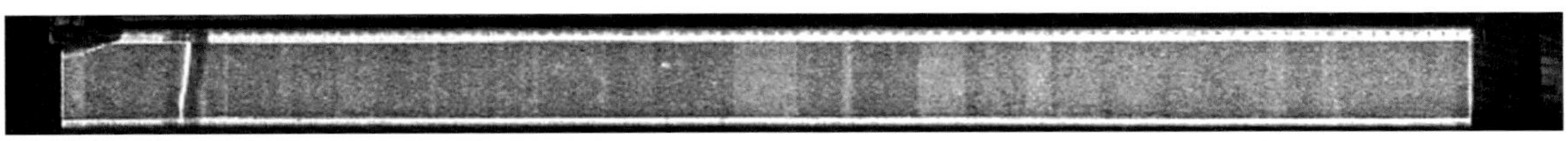

767
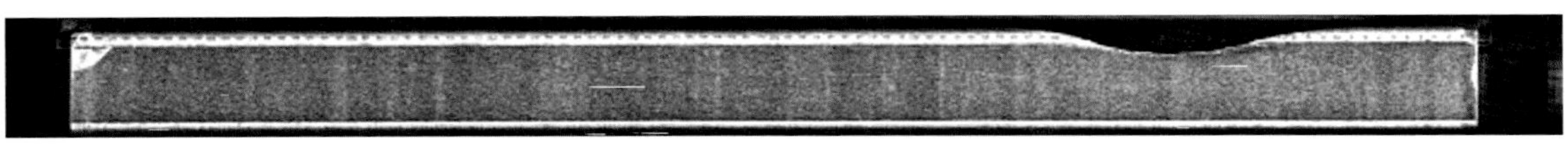

768
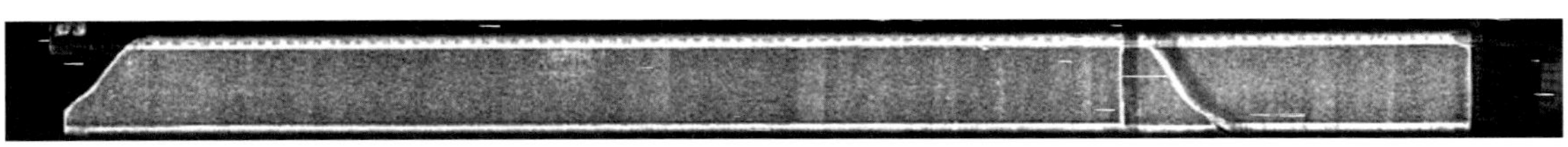

769
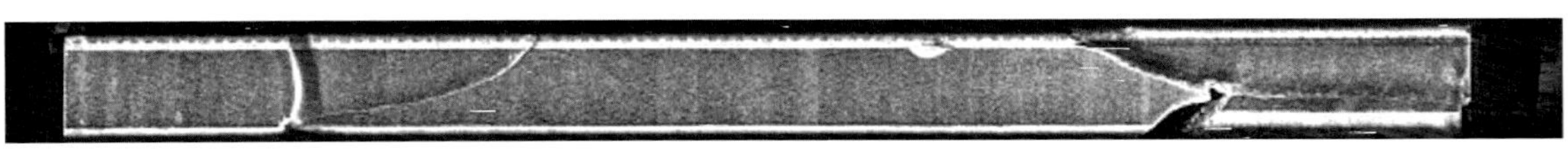

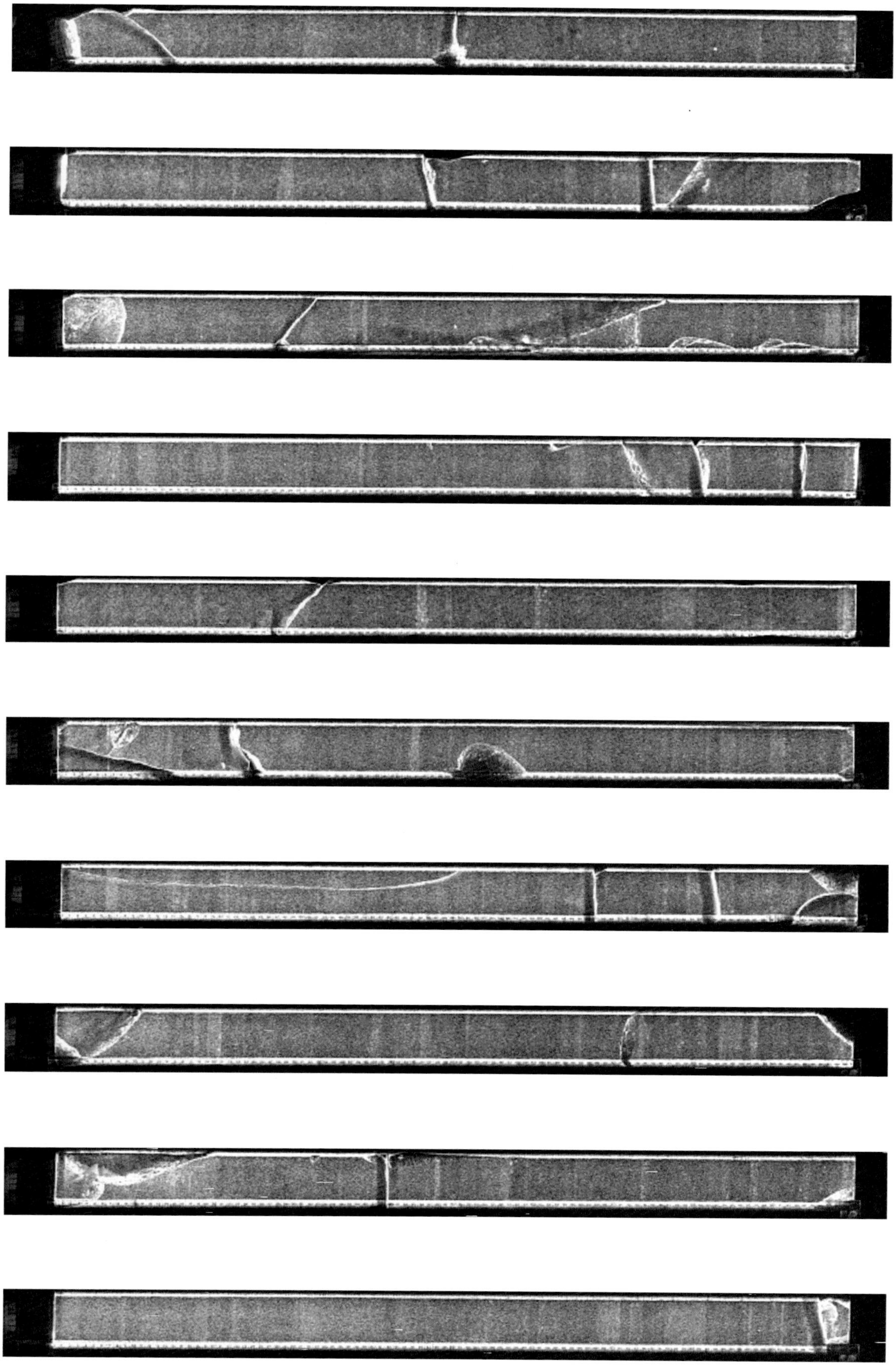

779

778

777

776

775

774

773

772

771

770

780
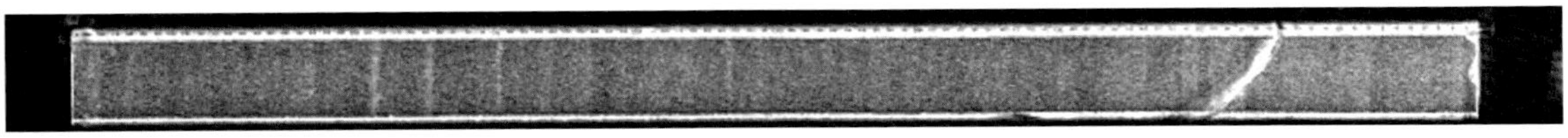

781
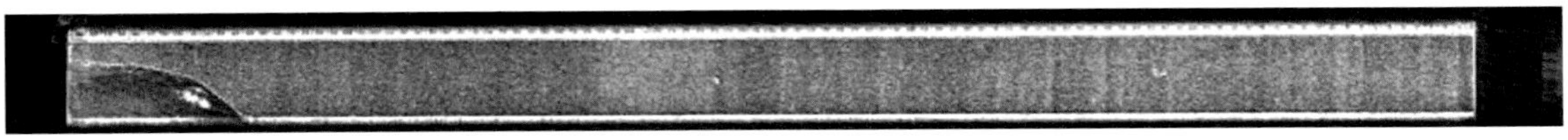

782
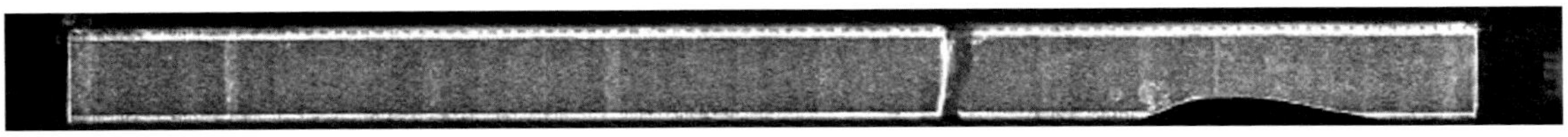

783
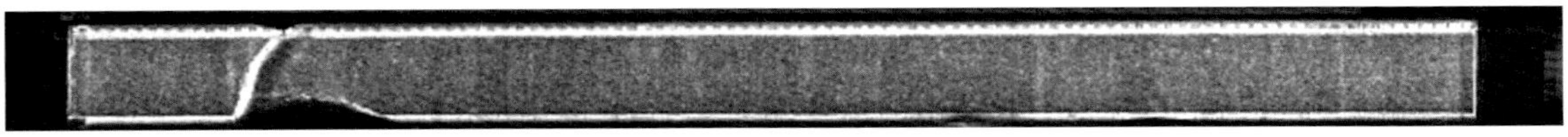

784
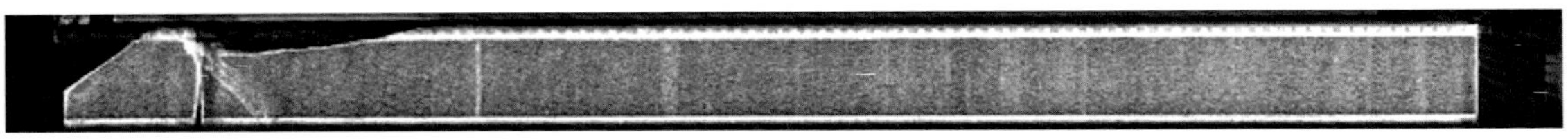

785
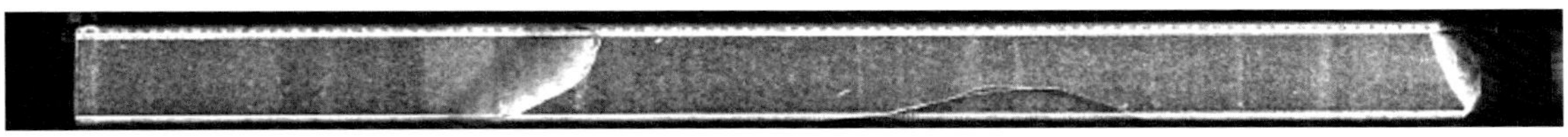

786
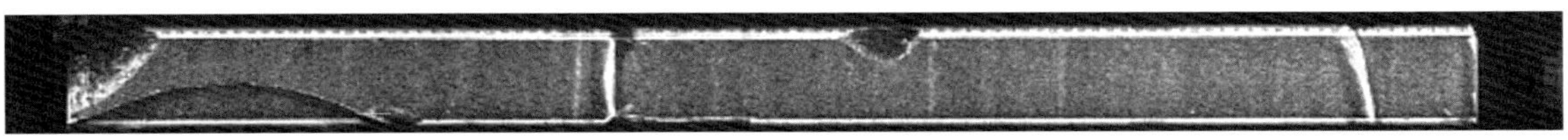

787
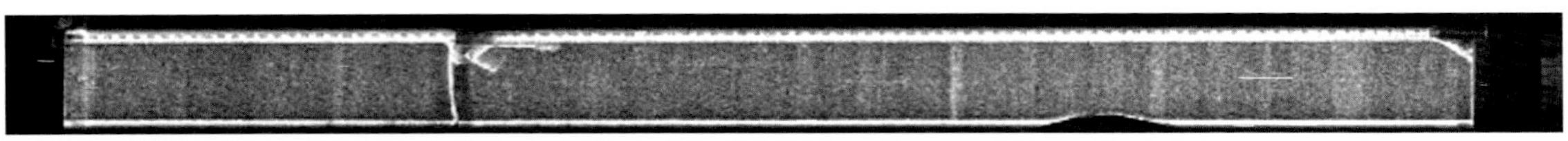

788
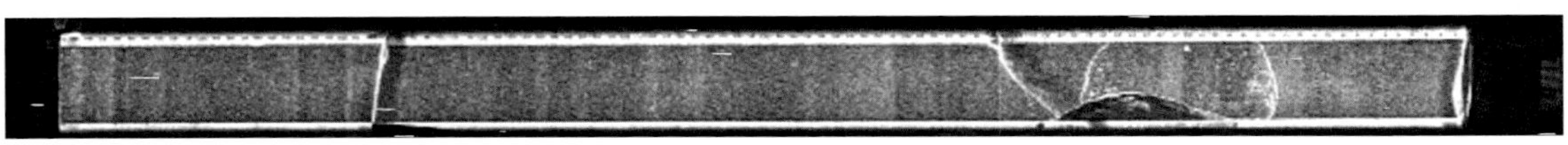

789
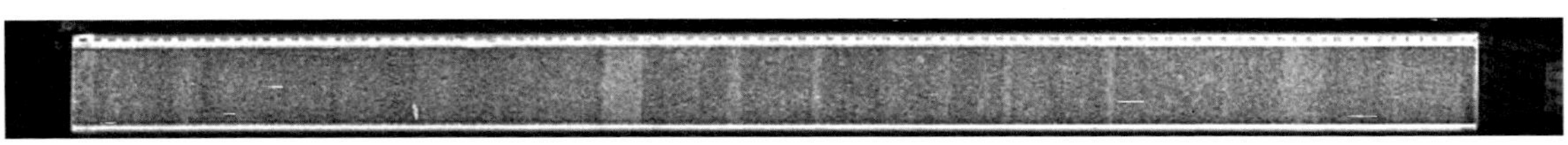

799

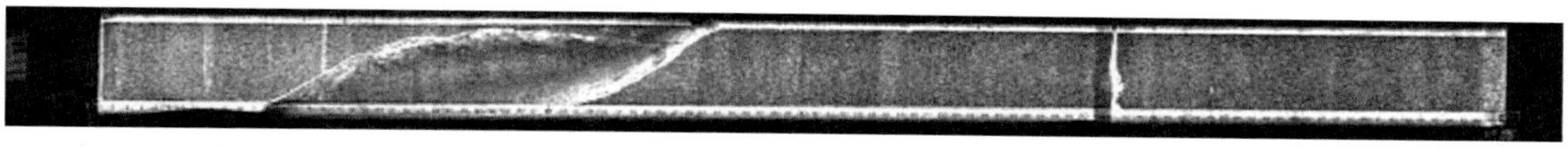

798

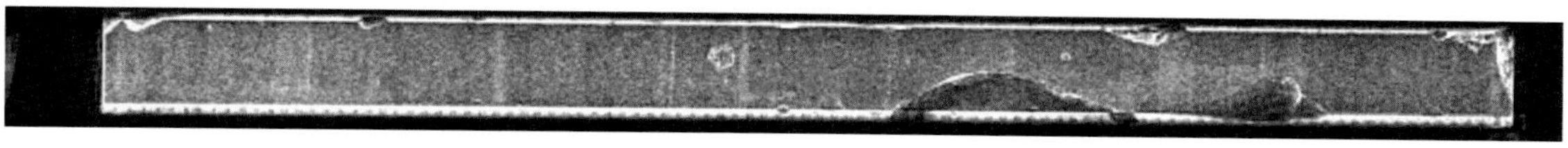

797

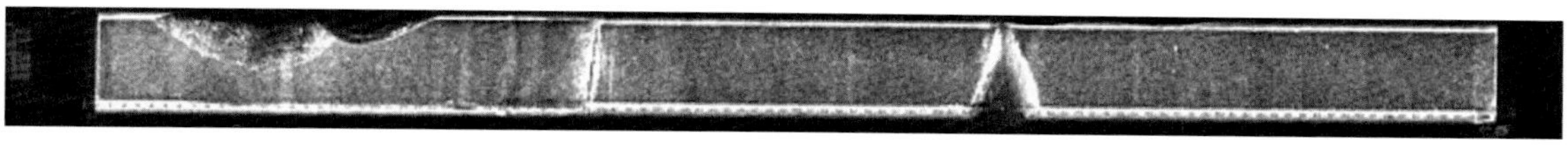

796

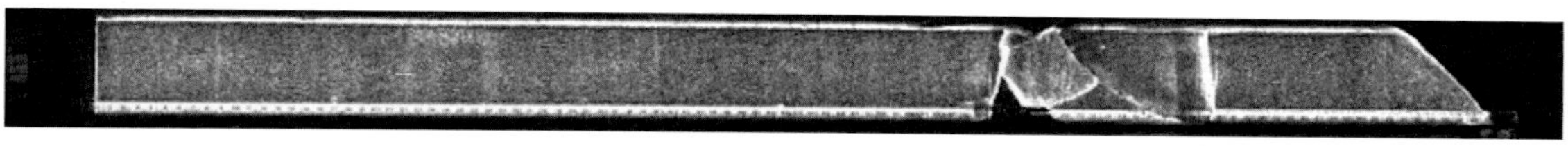

795

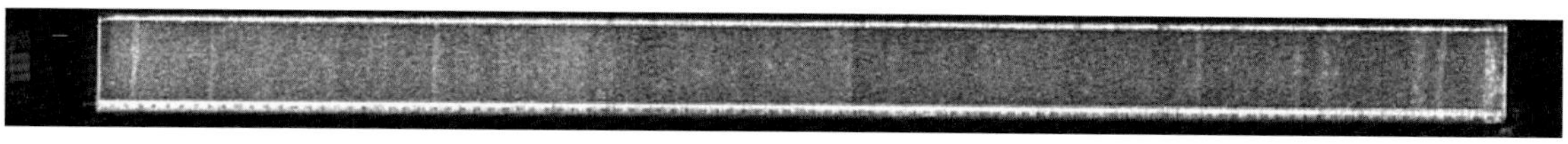

794

793

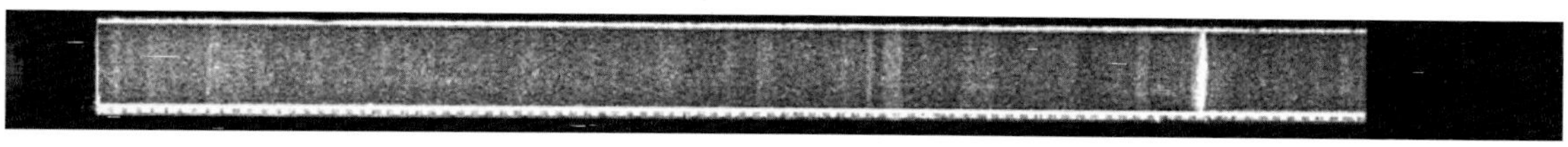

792

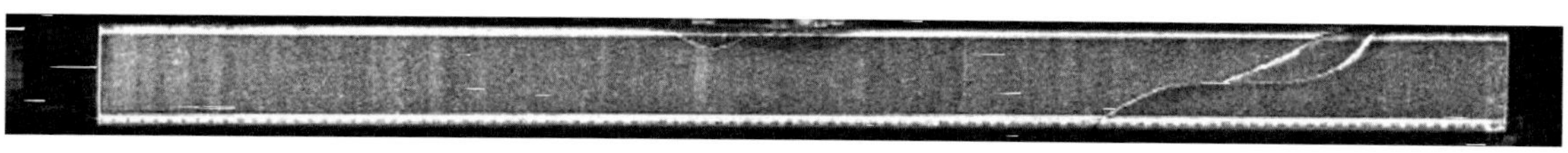

791

790

800
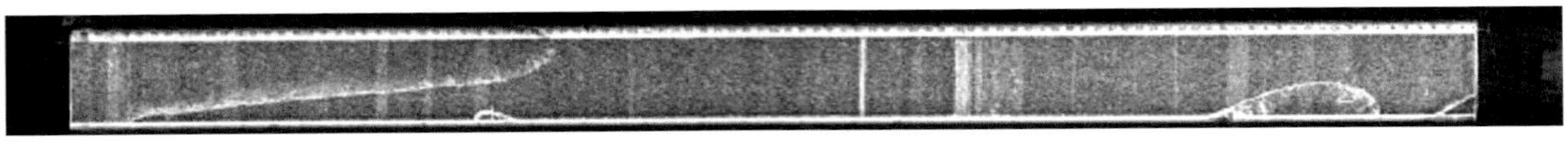

801

802
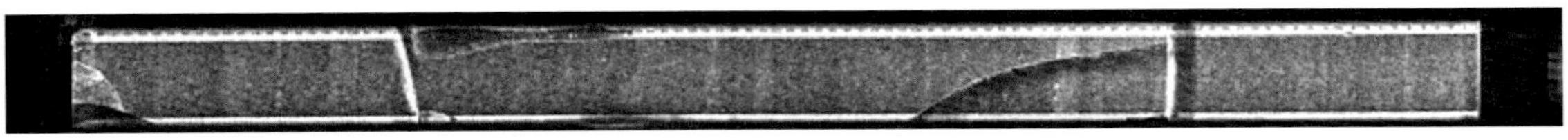

803
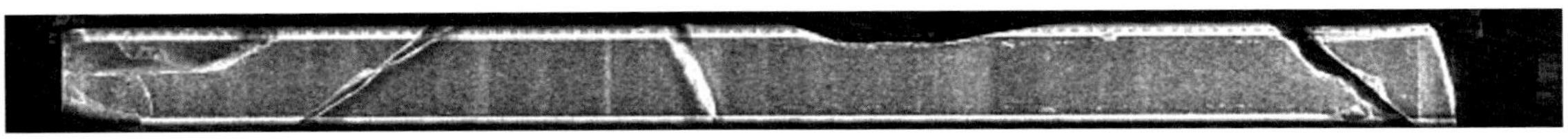

804
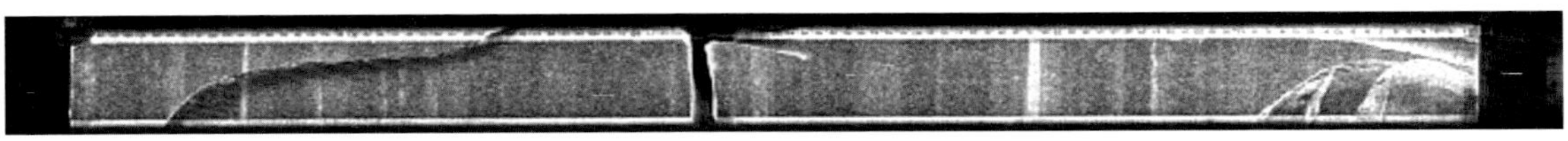

805
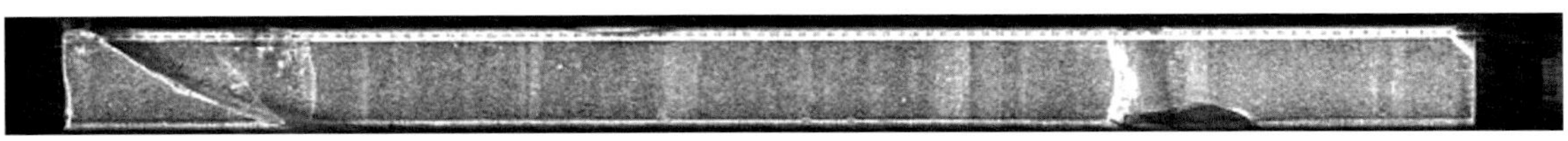

806
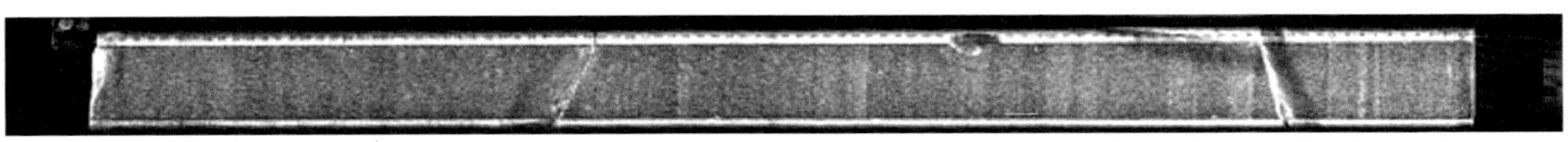

807
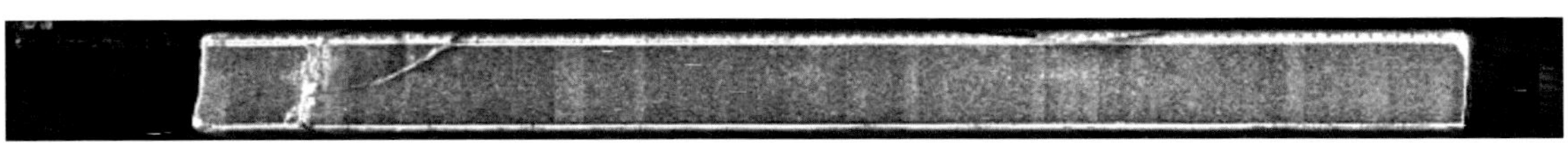

808
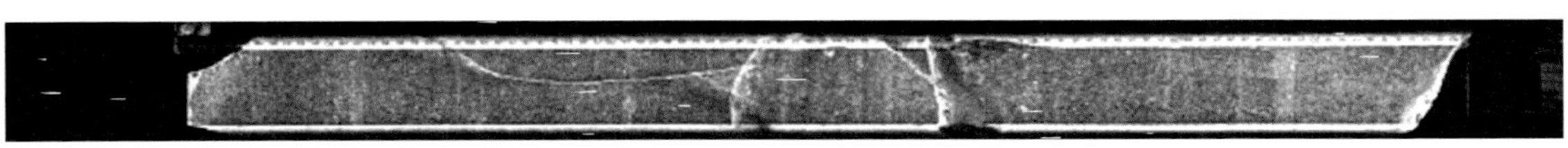

809
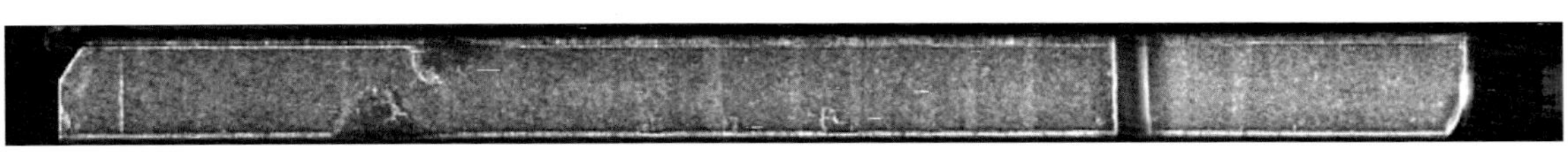

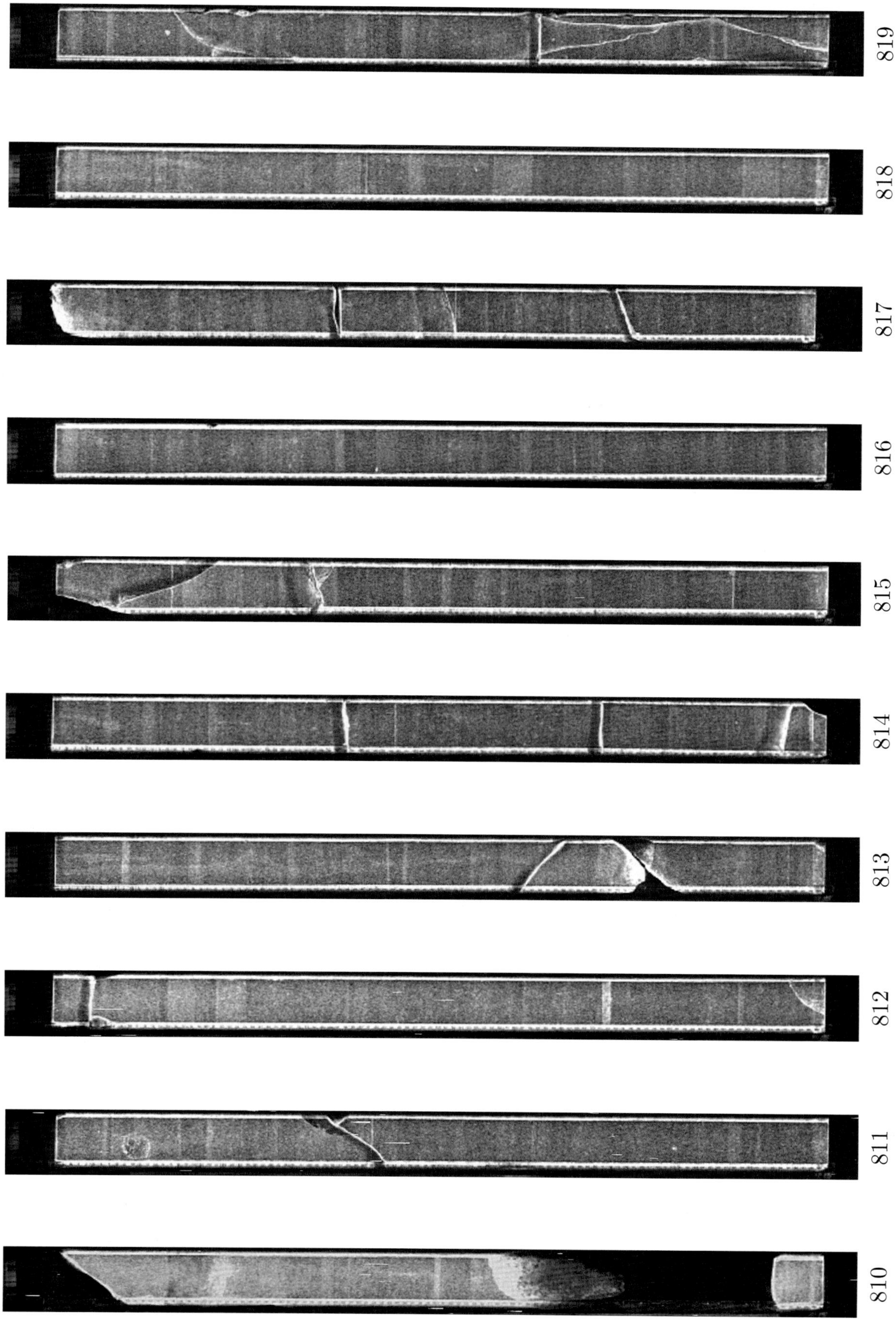
819
818
817
816
815
814
813
812
811
810

820

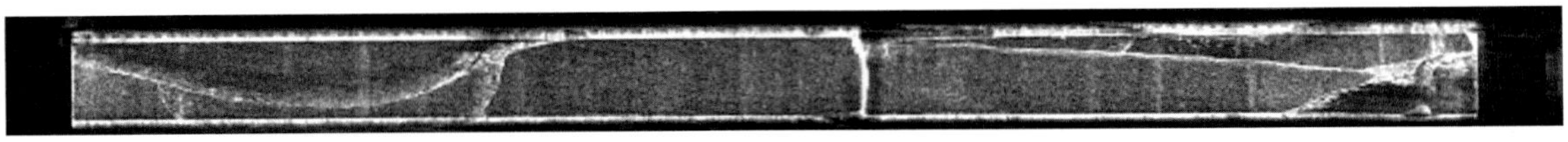

821

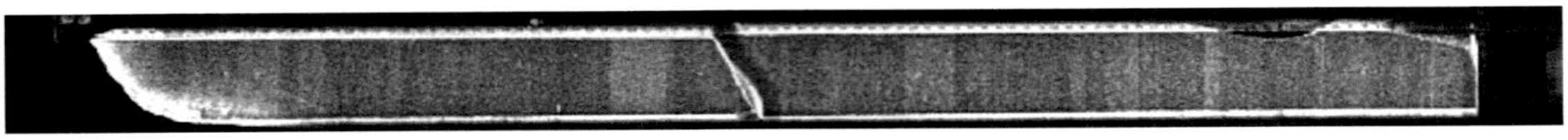

822

823

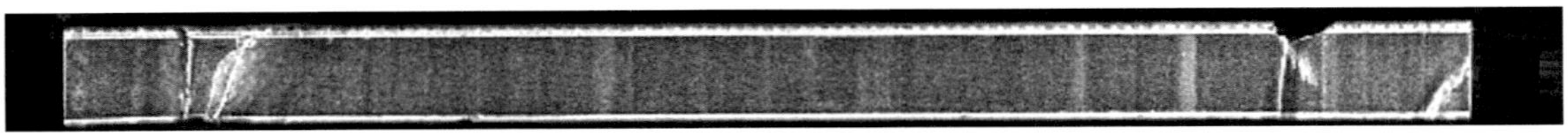

824

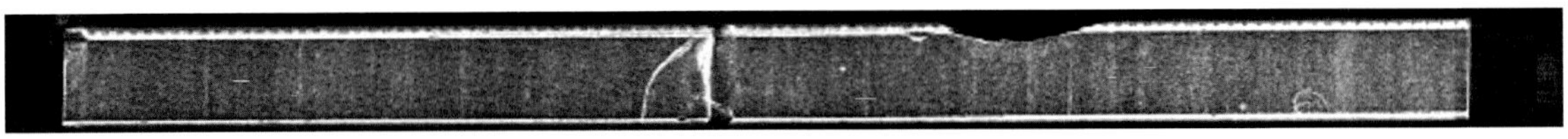

825

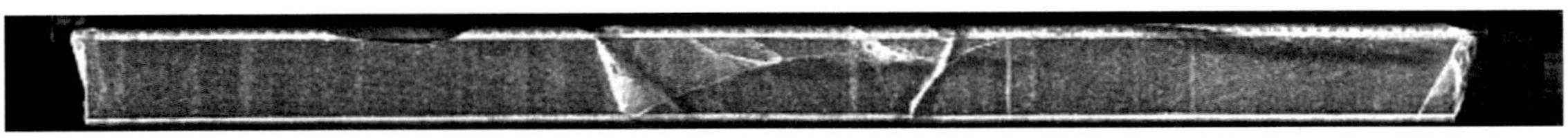

826

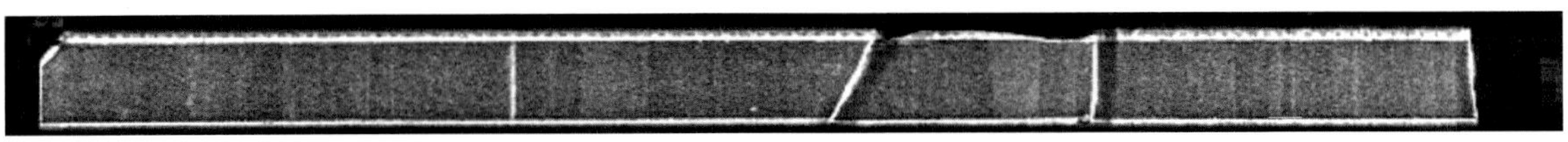

827

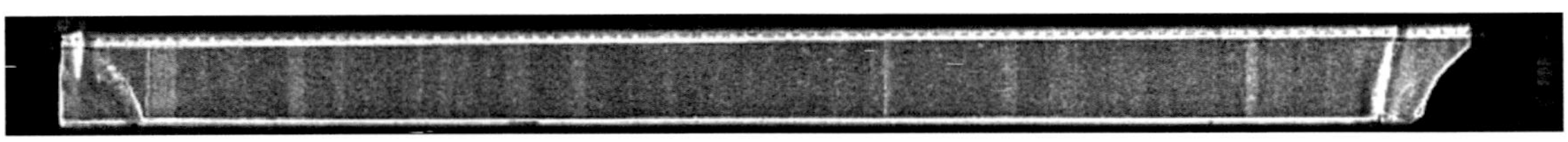

828

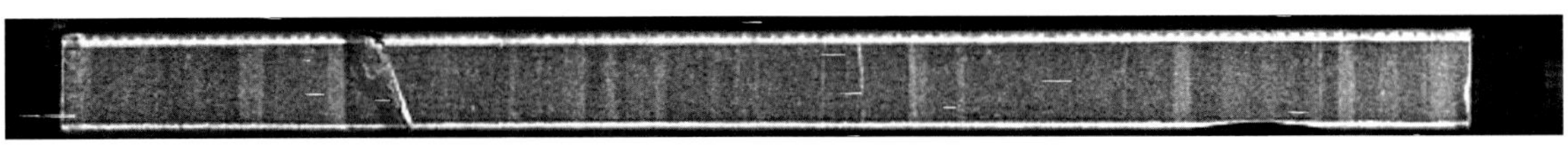

829

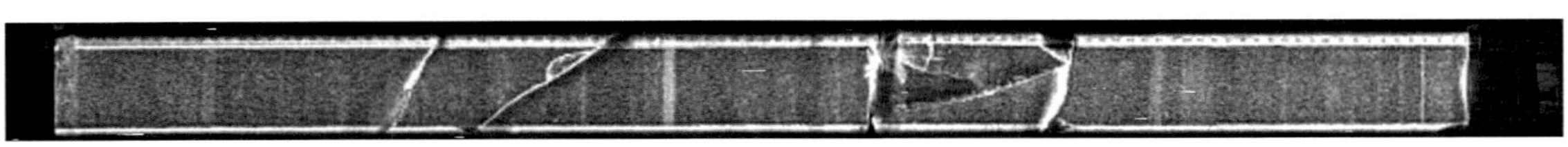

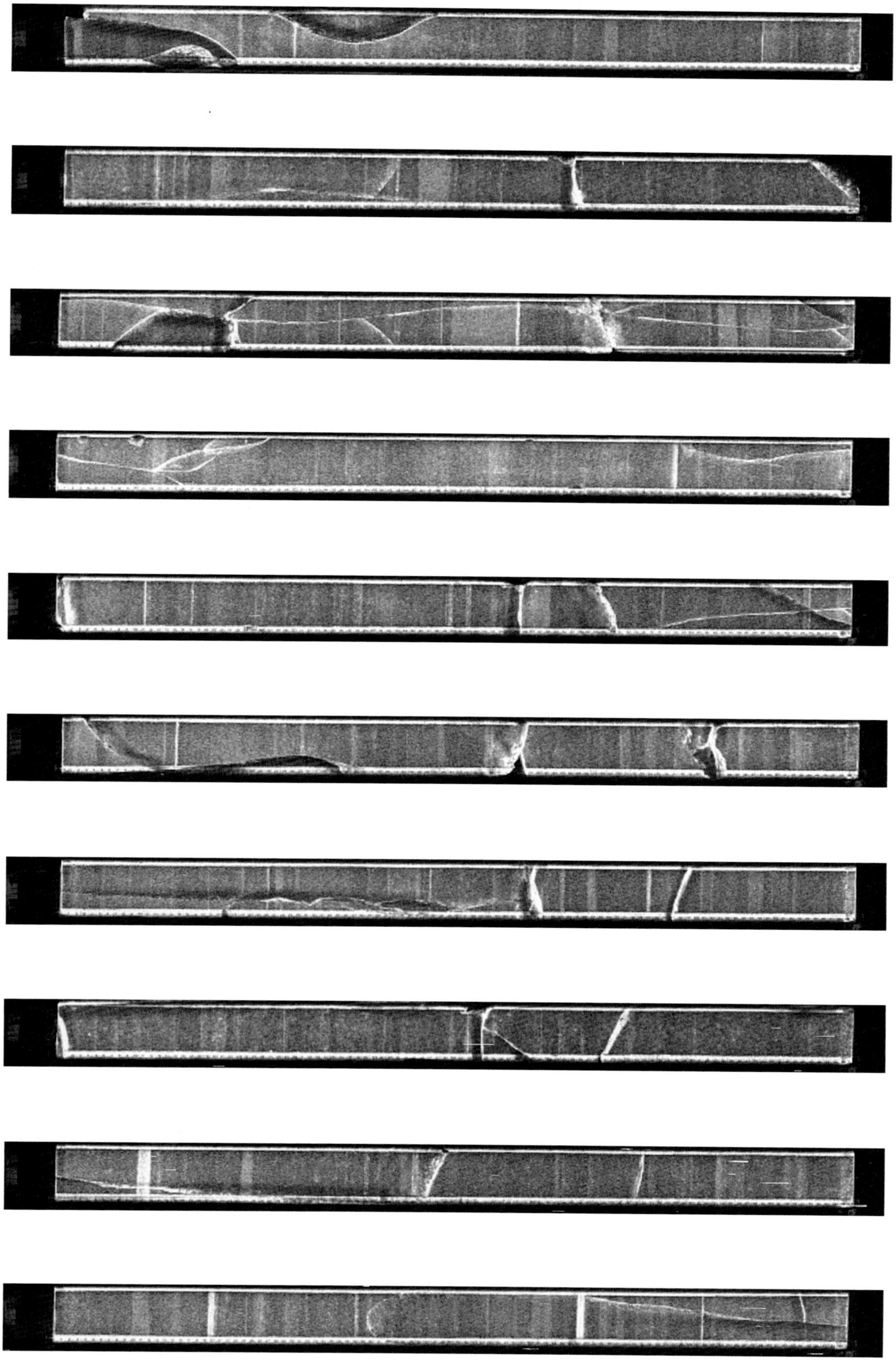

839
838
837
836
835
834
833
832
831
830

840

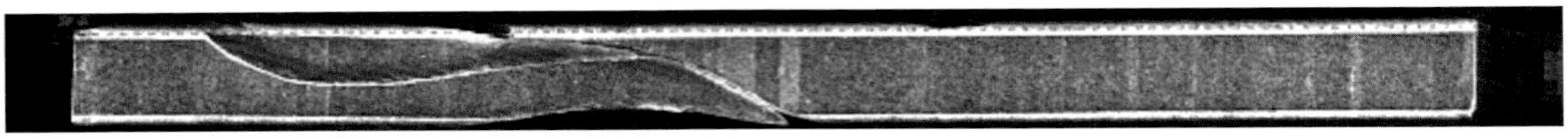

841

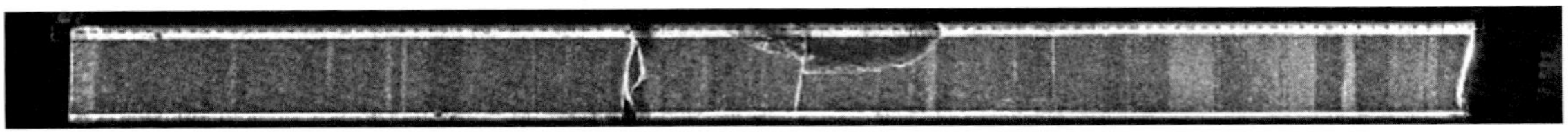

842

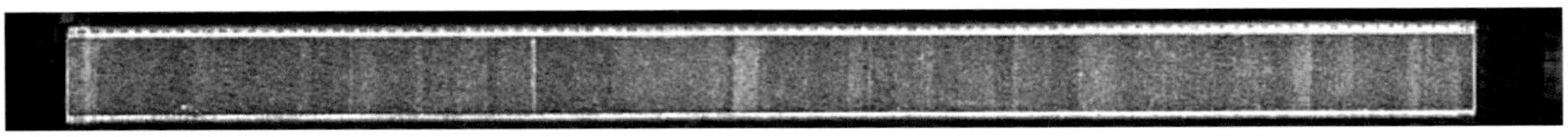

843

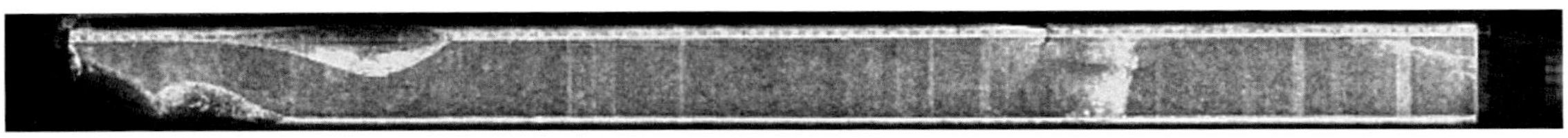

844

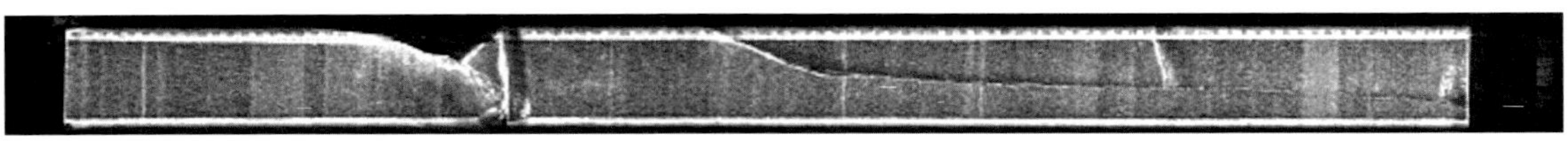

845

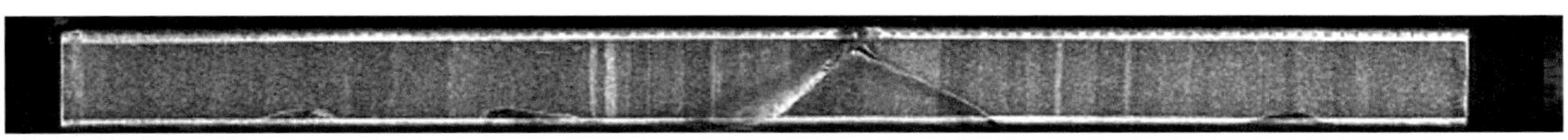

846

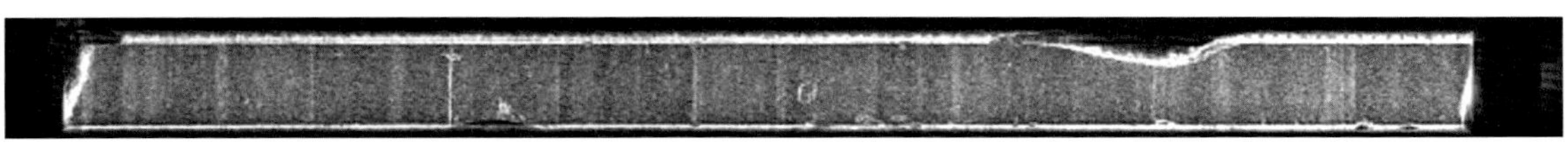

847

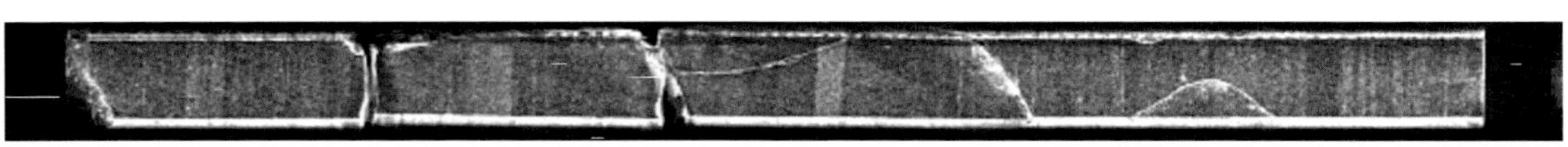

848

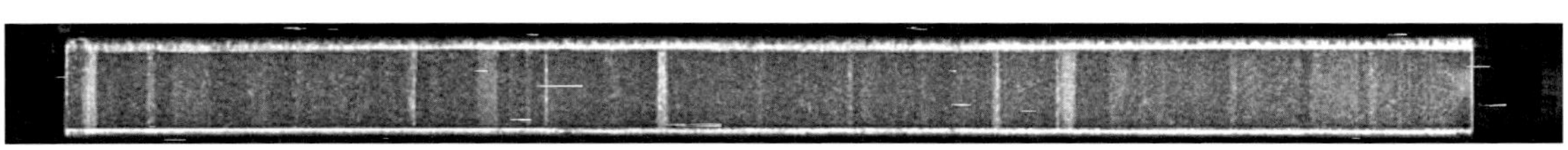

849

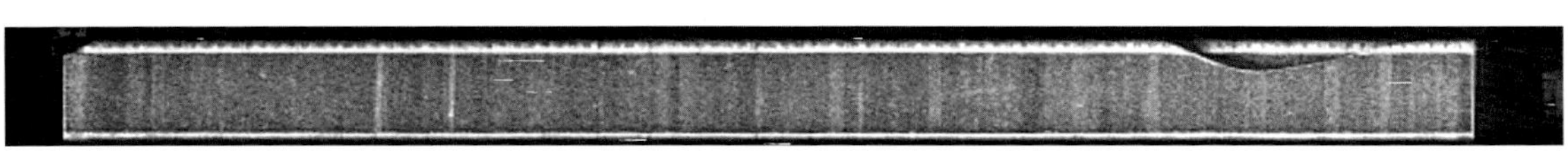

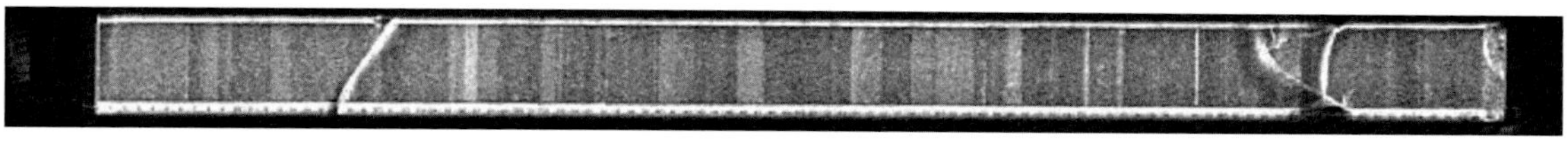

859

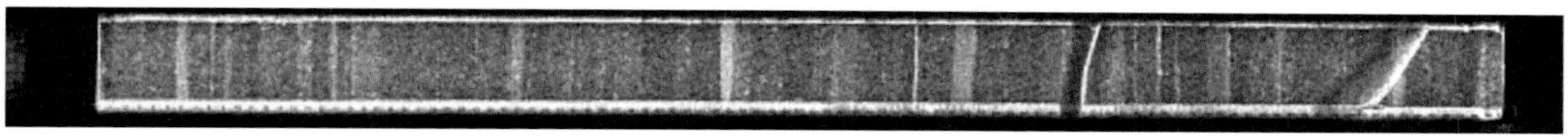

858

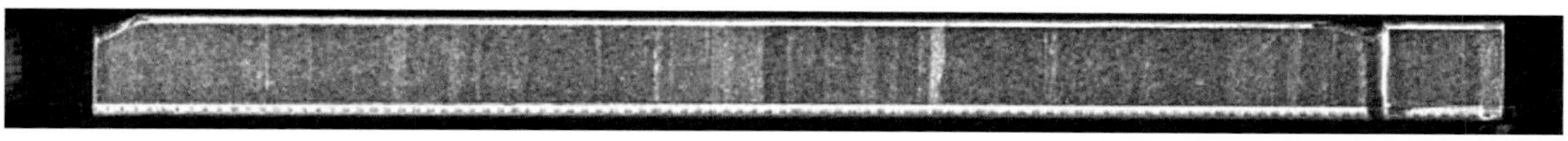

857

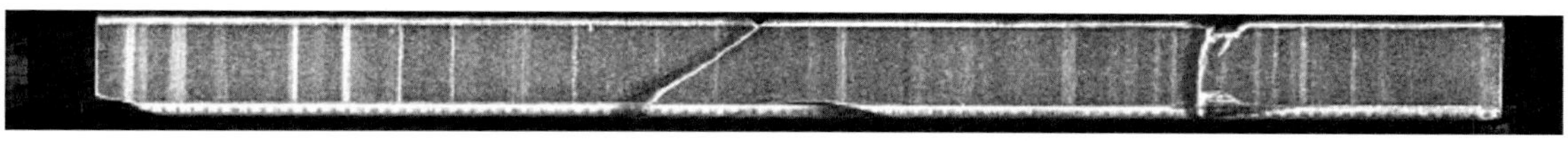

856

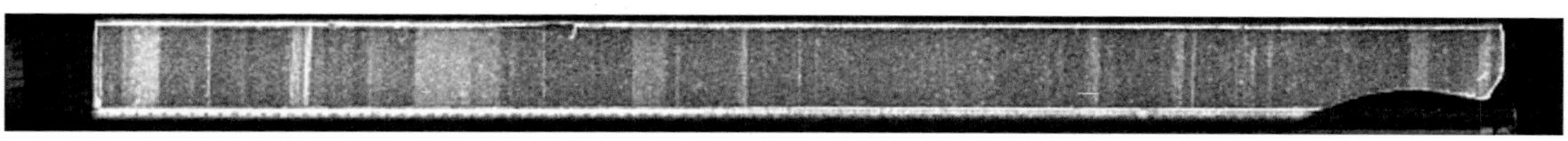

855

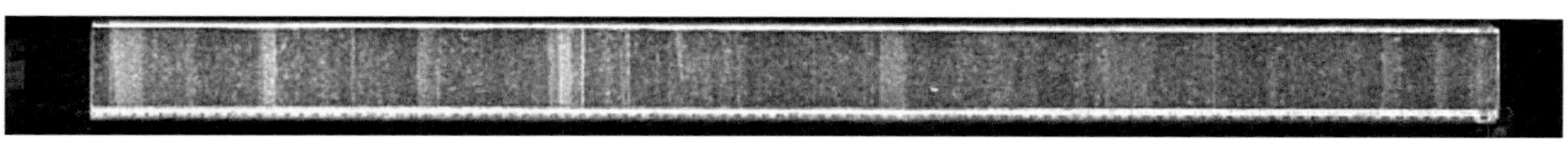

854

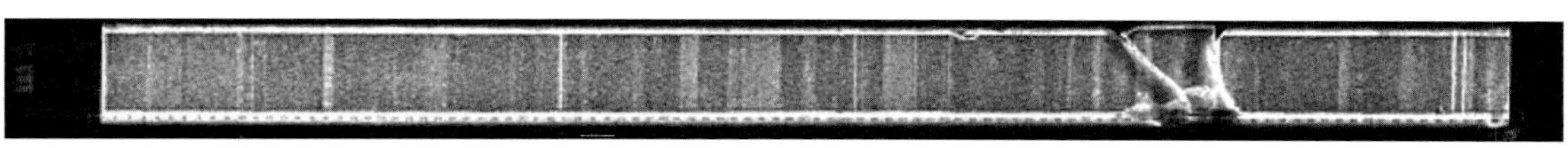

853

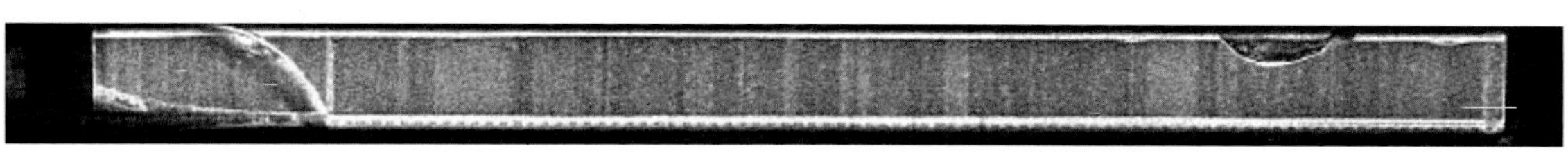

852

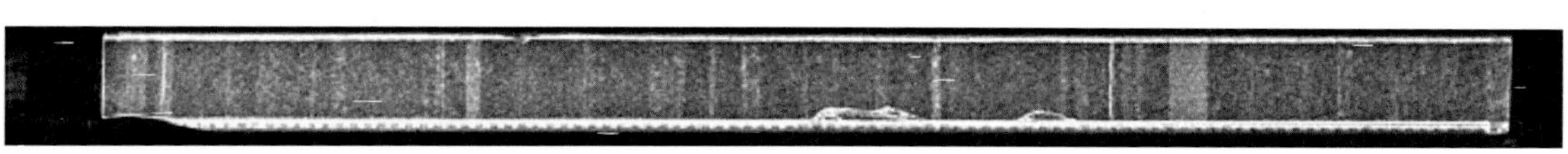

851

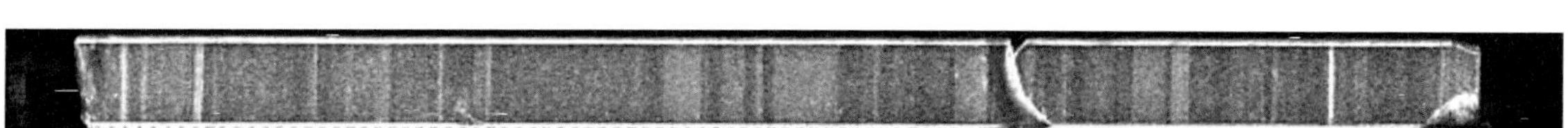

850

860

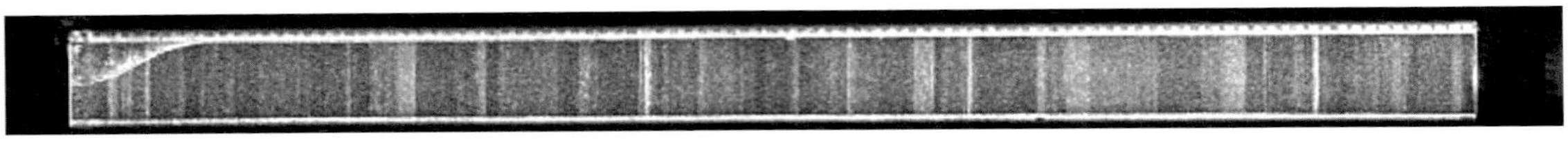

861

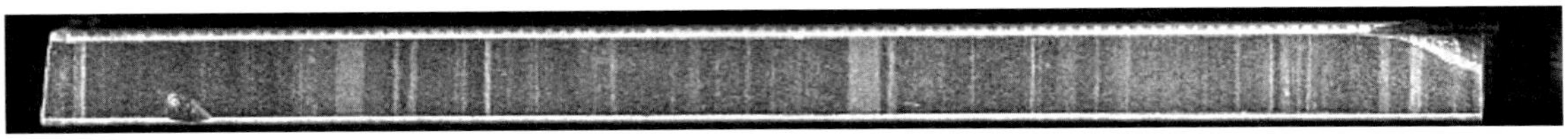

862

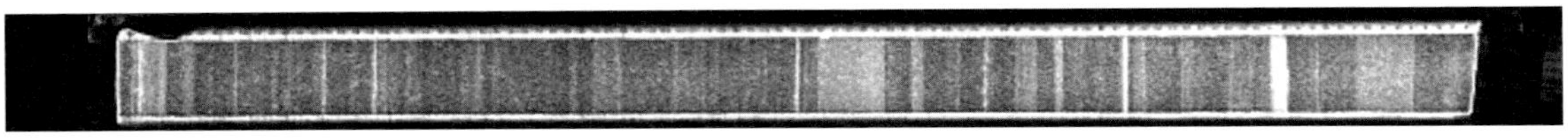

863

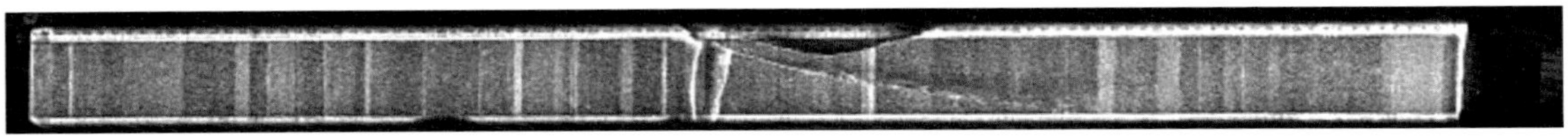

864

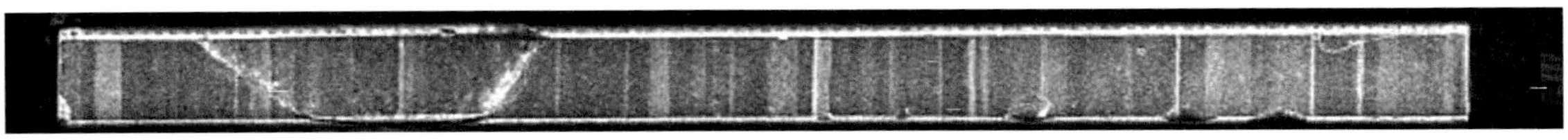

865

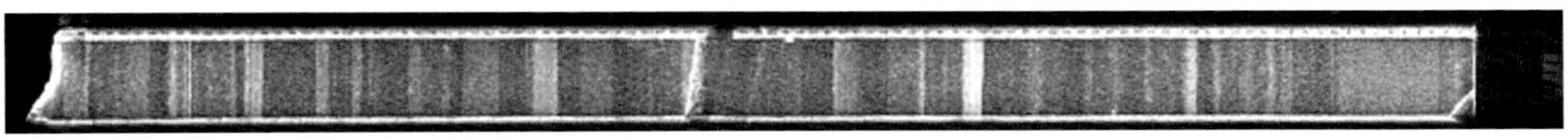

866

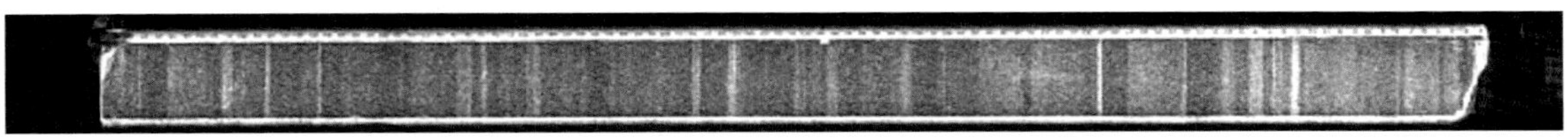

867

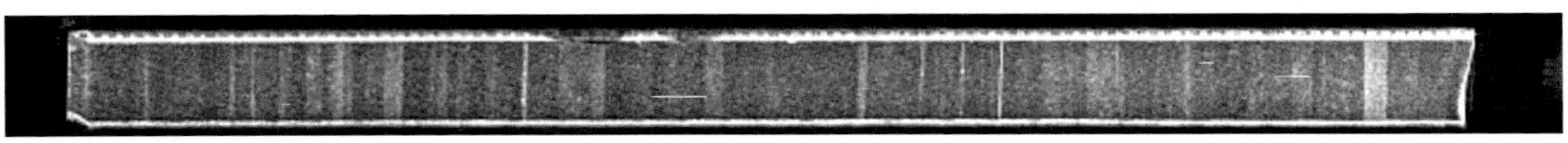

868

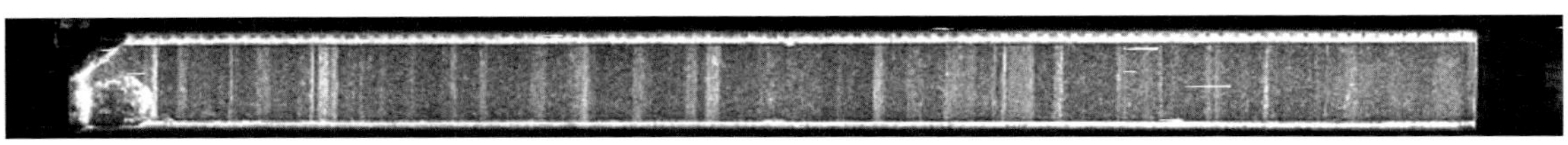

869

NO IMAGE AVAILABLE NO IMAGE AVAILABLE

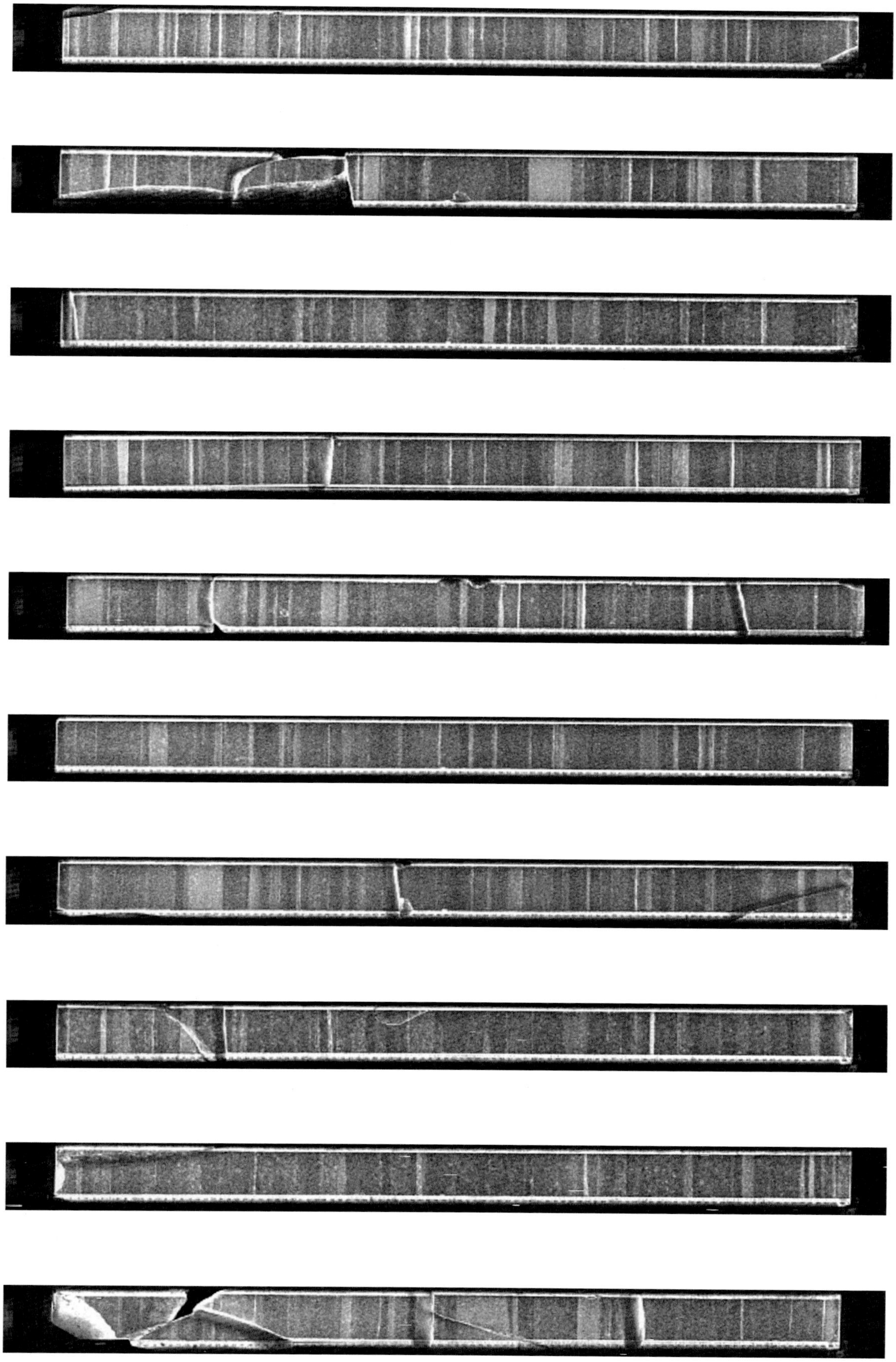

879

878

877

876

875

874

873

872

871

870

880
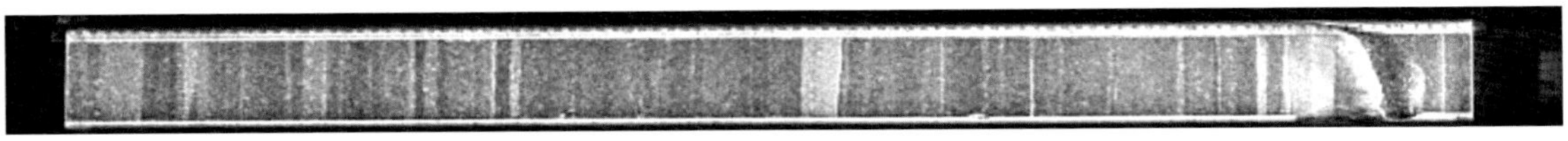

881
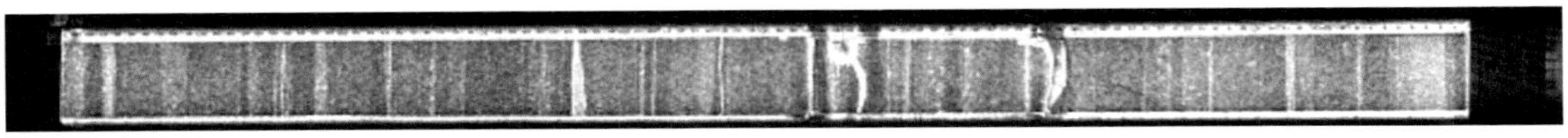

882
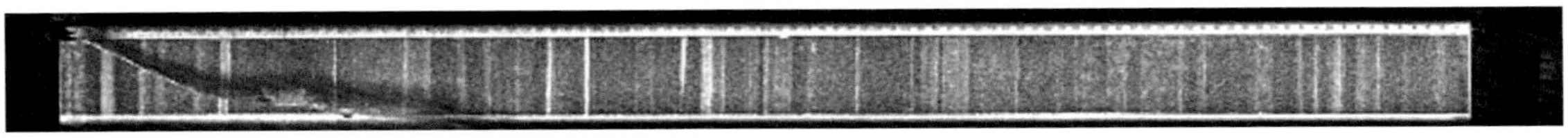

883
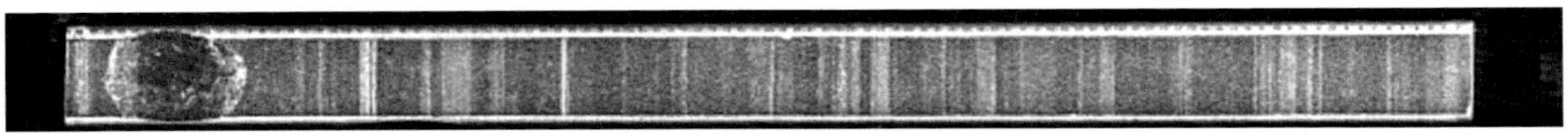

884
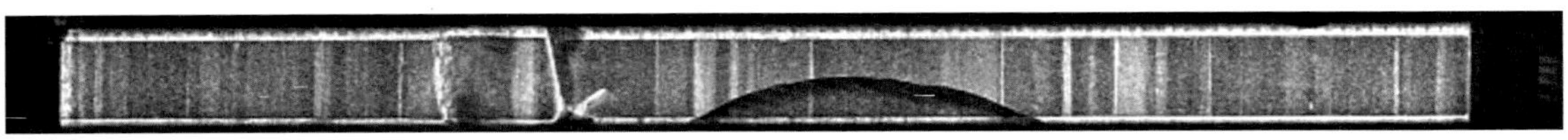

885
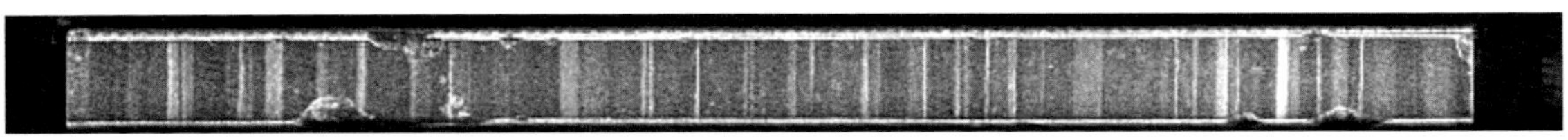

886
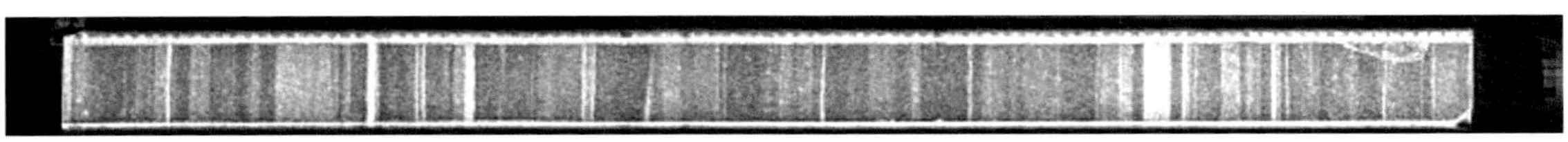

887
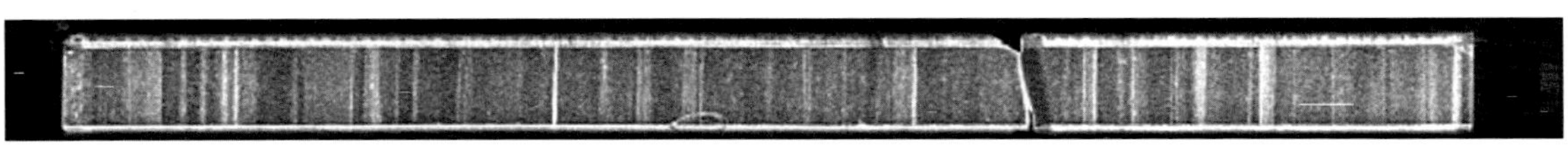

888
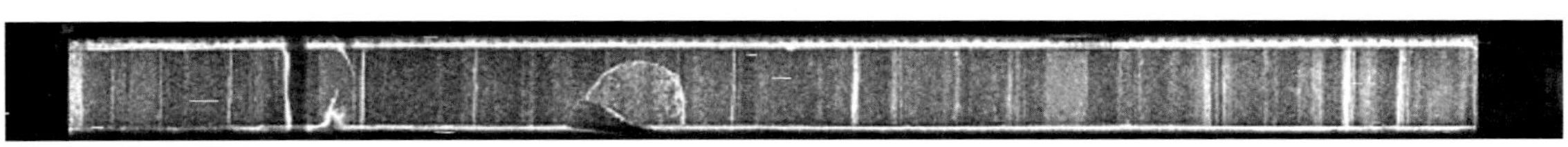

889
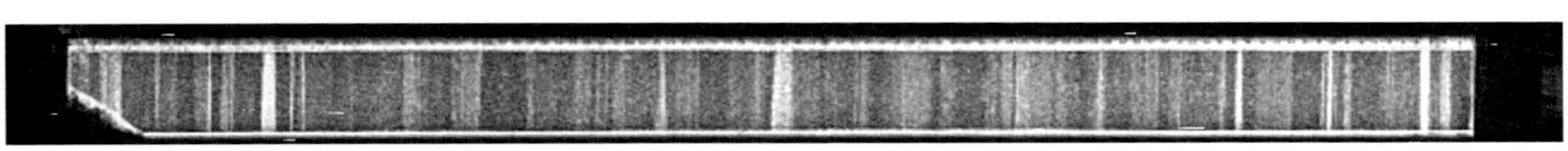

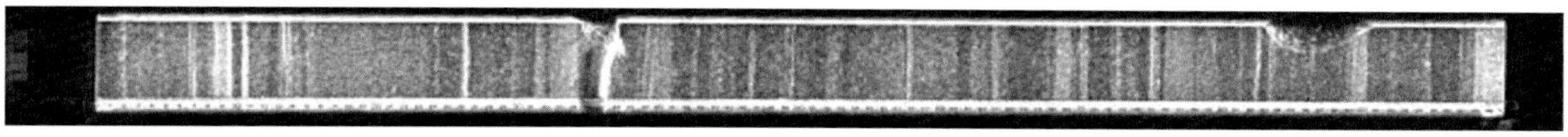

899

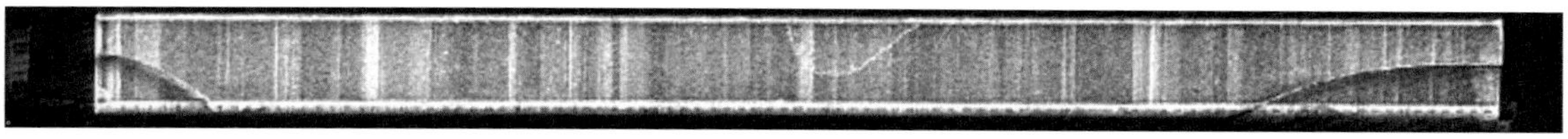

898

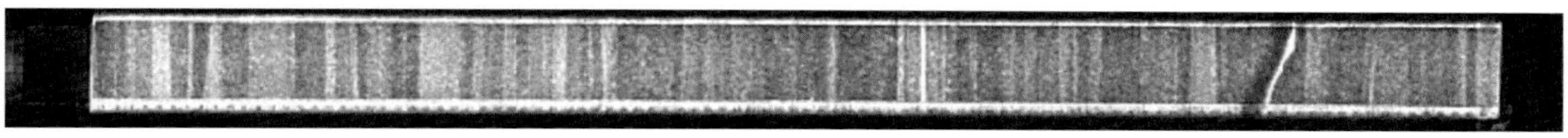

897

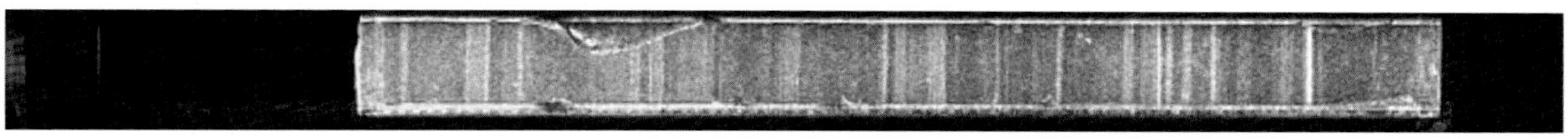

896

NO IMAGE AVAILABLE NO IMAGE AVAILABLE

895

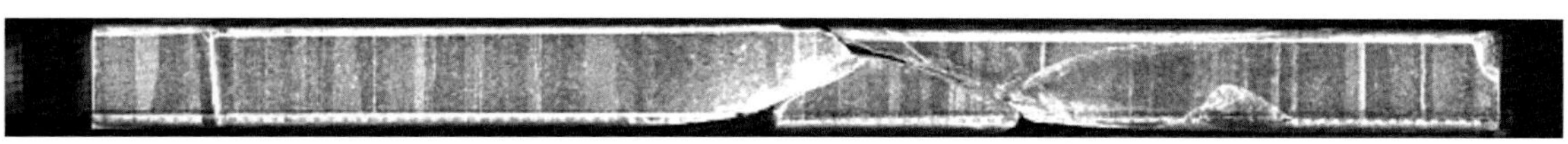

894

893

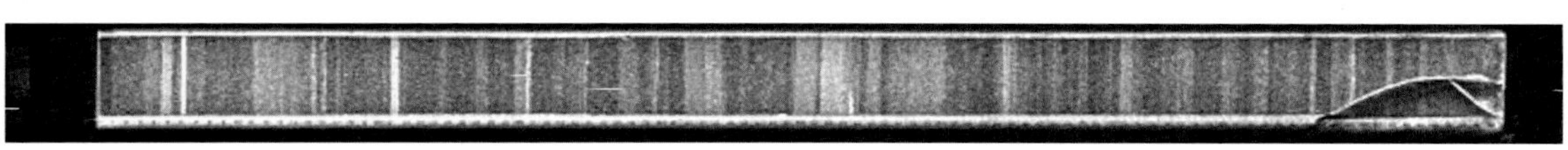

892

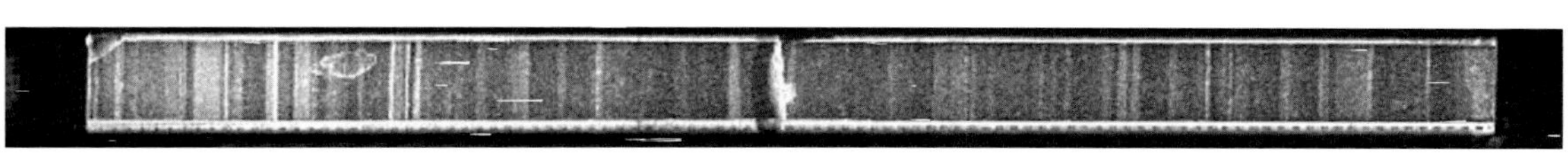

891

890

900
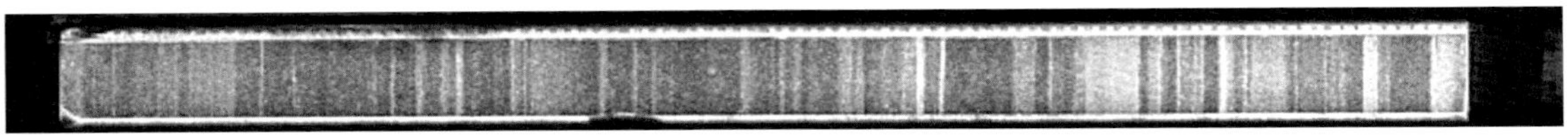

901
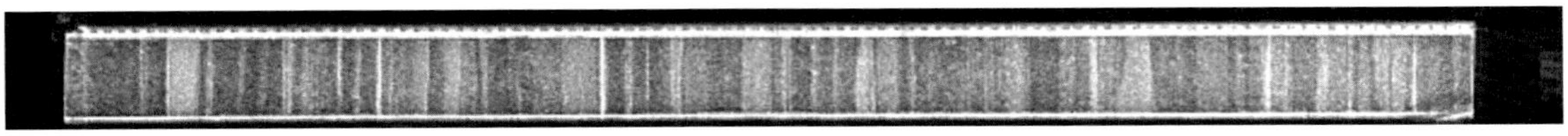

902
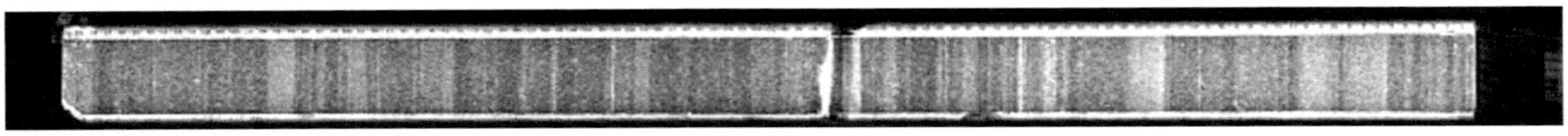

903
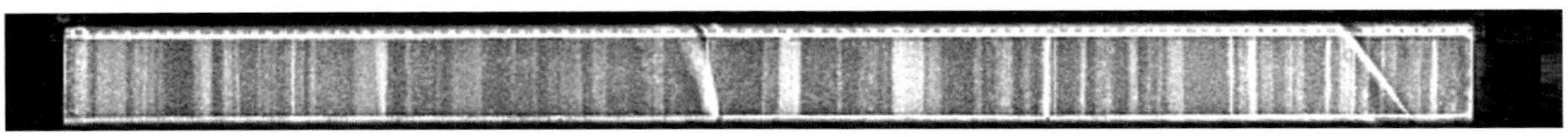

904
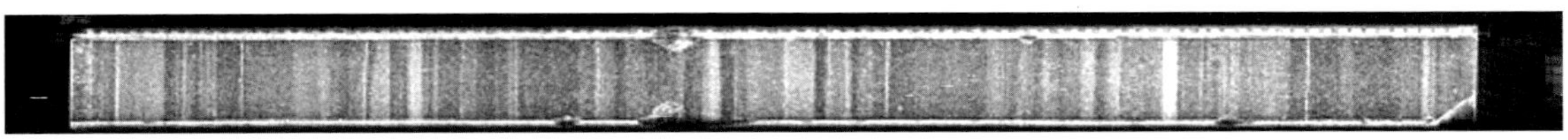

905
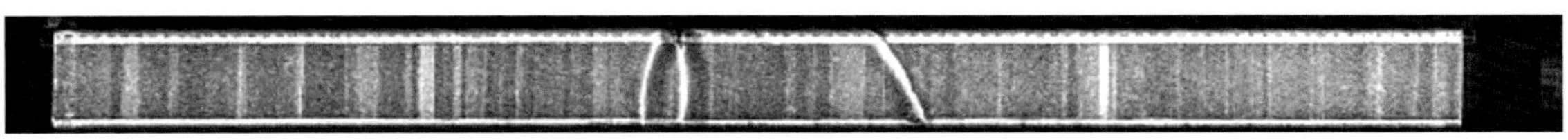

906
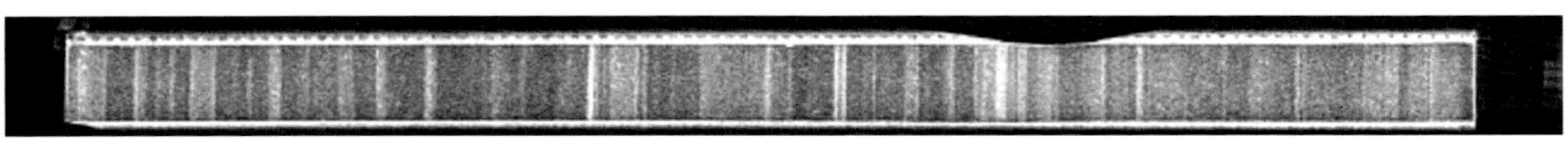

907
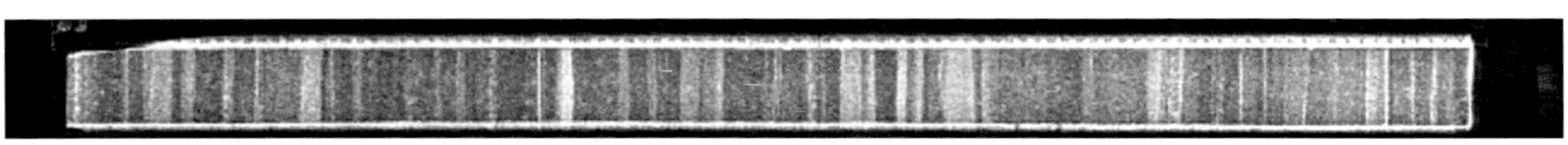

908
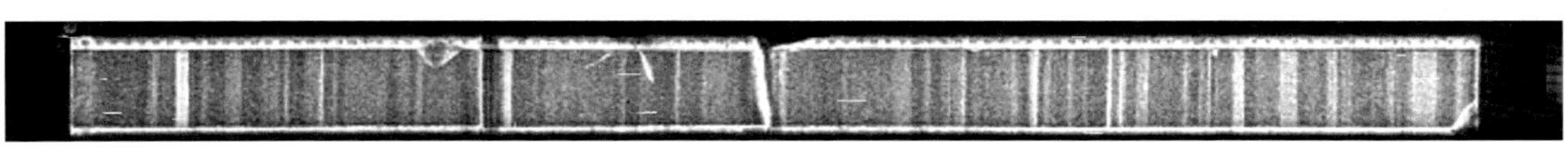

909
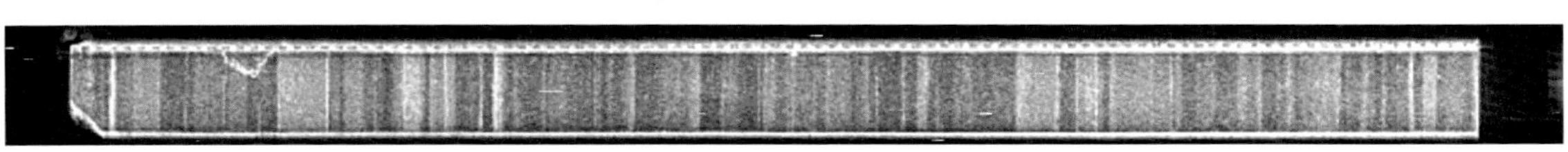

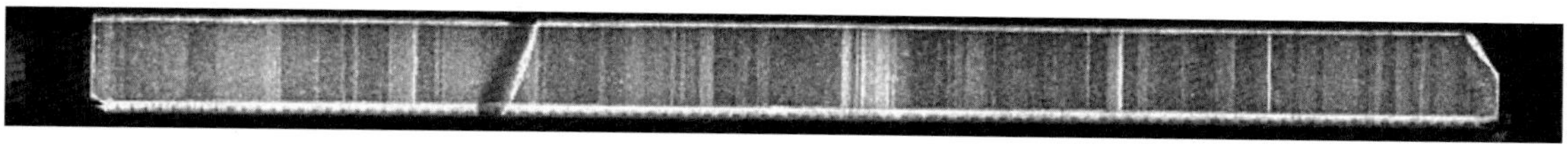

919

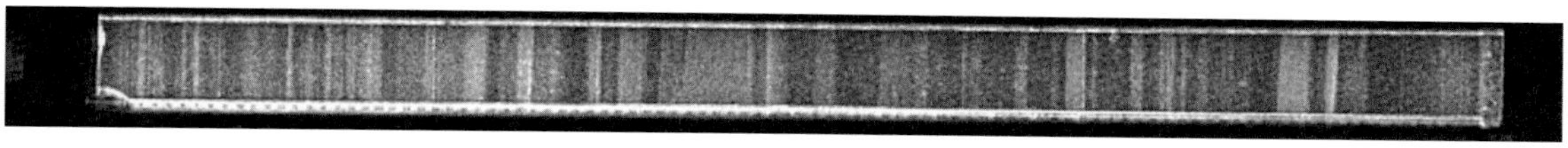

918

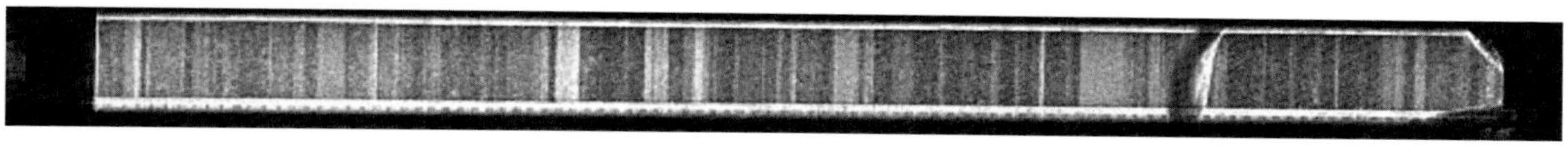

917

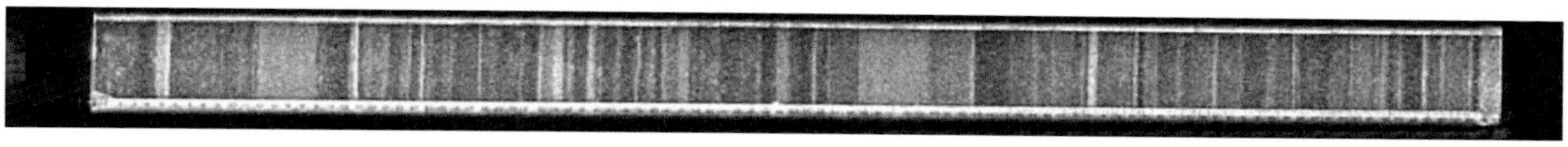

916

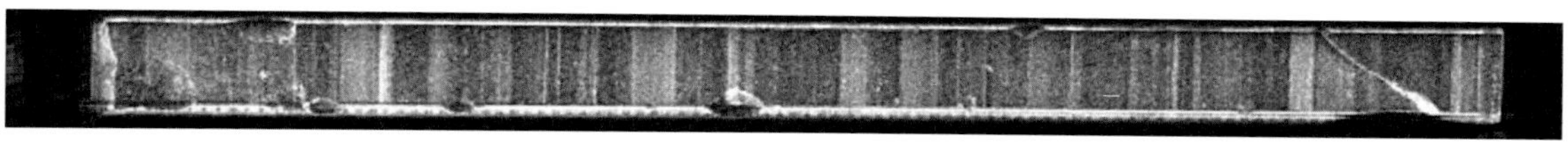

915

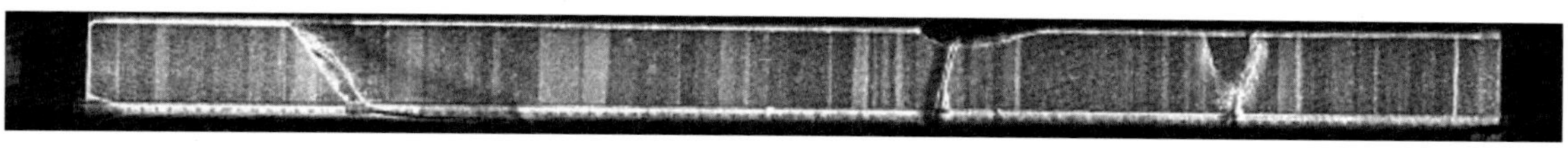

914

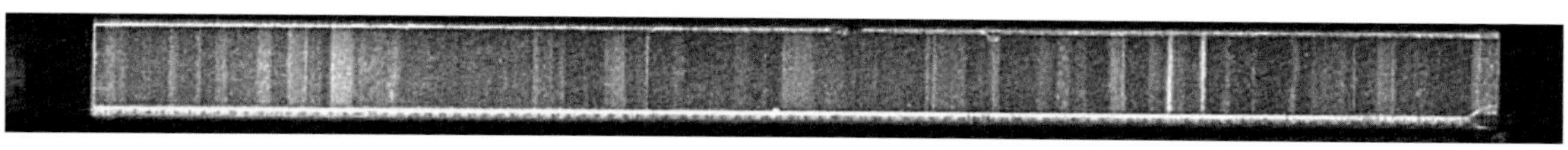

913

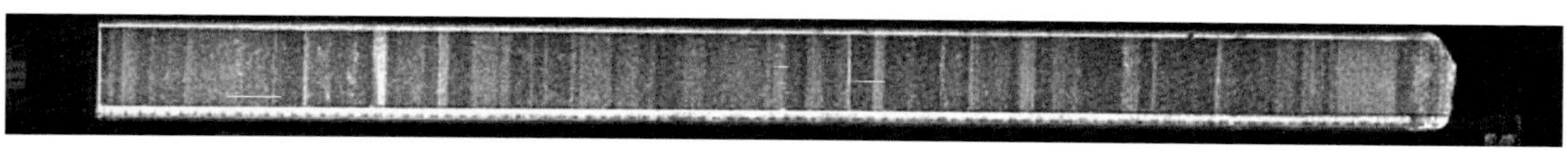

912

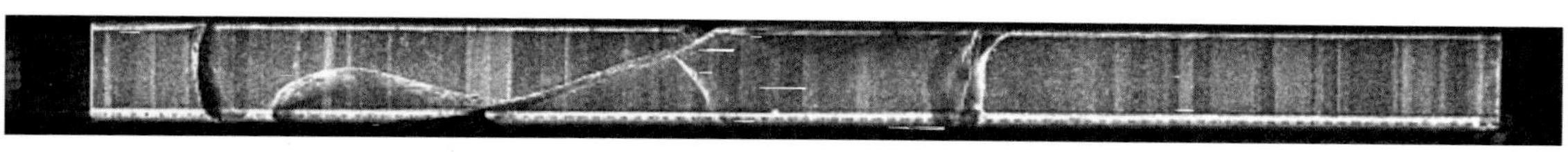

911

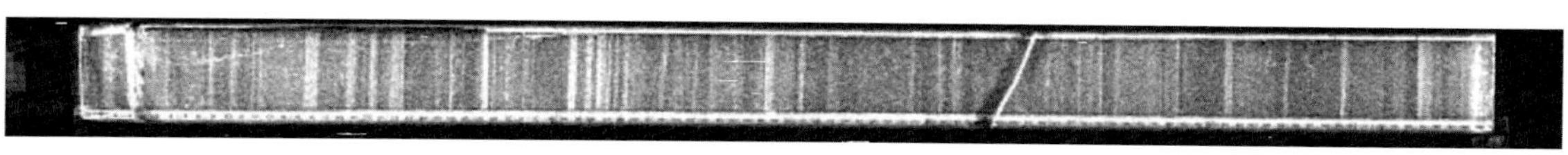

910

920

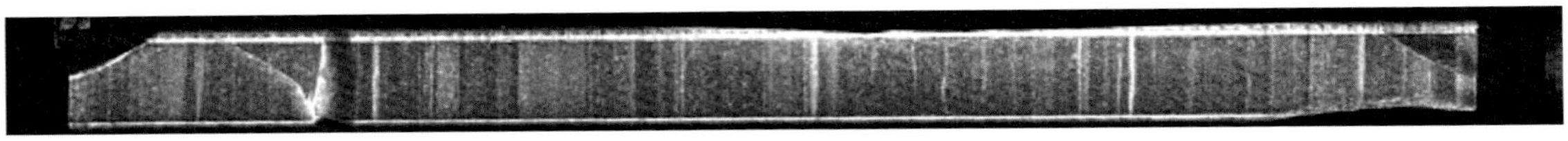

921

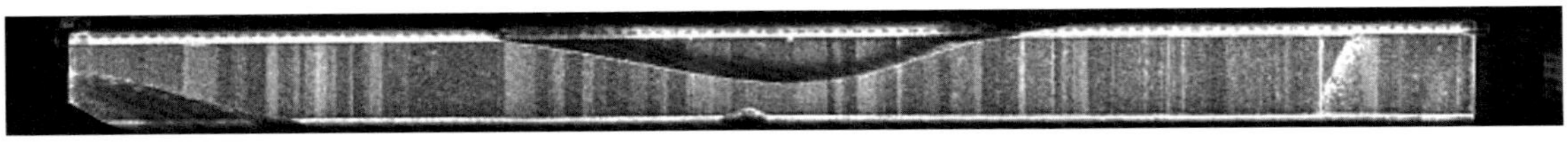

922

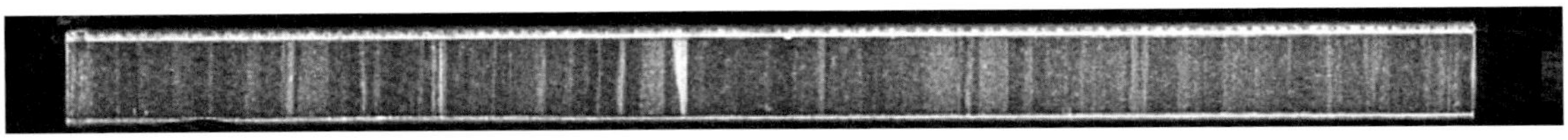

923

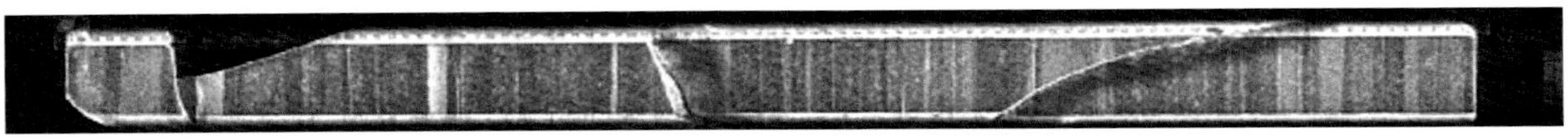

924

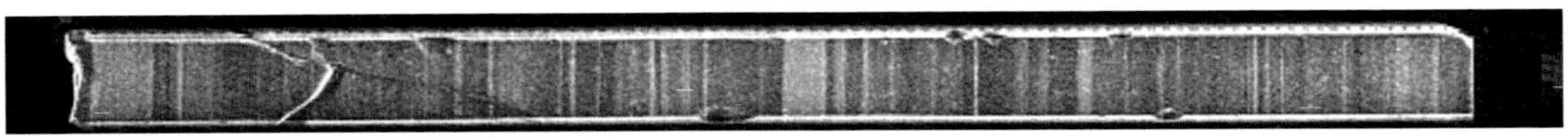

925

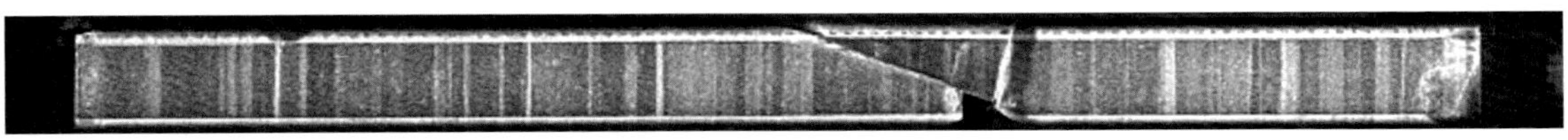

926

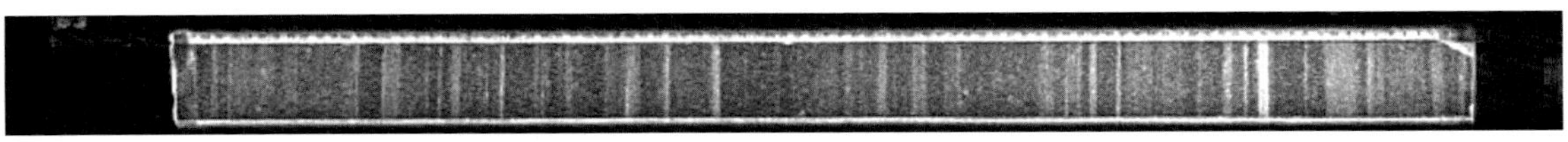

927

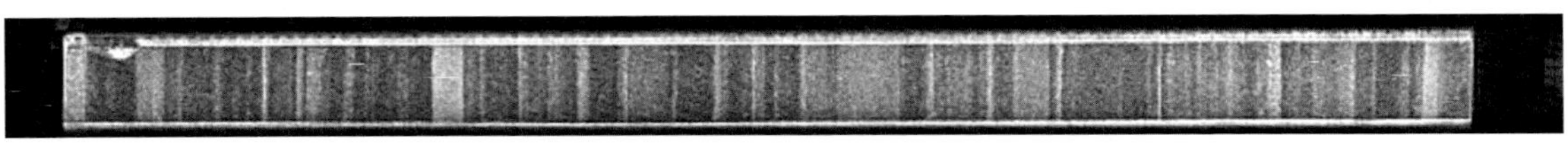

928

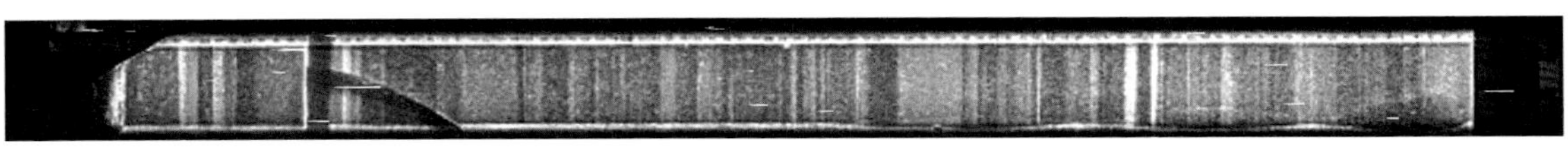

929

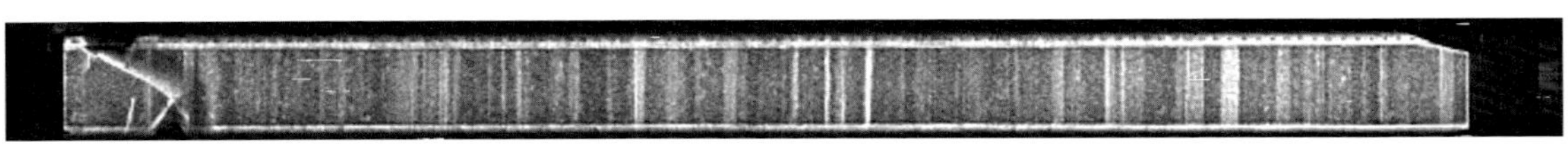

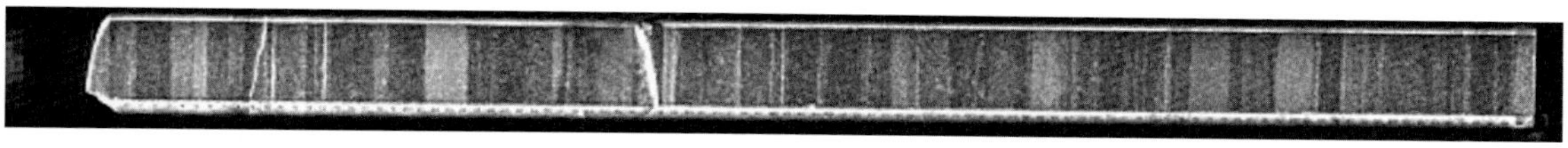
939

NO IMAGE AVAILABLE NO IMAGE AVAILABLE
938

NO IMAGE AVAILABLE NO IMAGE AVAILABLE
937

NO IMAGE AVAILABLE NO IMAGE AVAILABLE
936

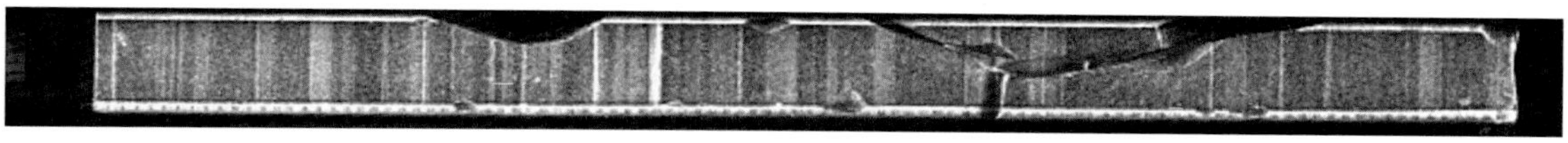
935

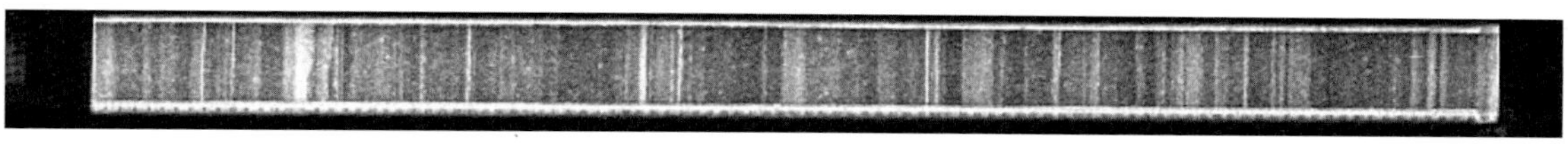
934

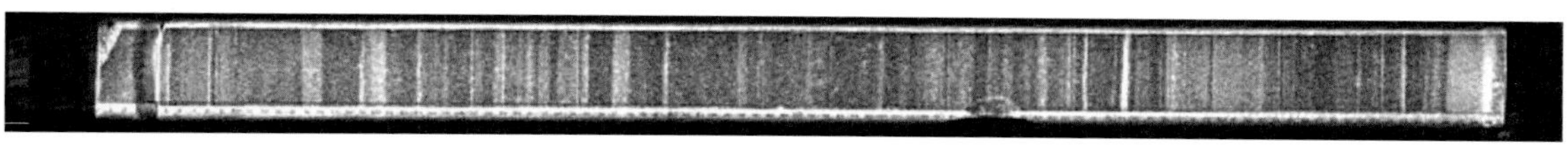
933

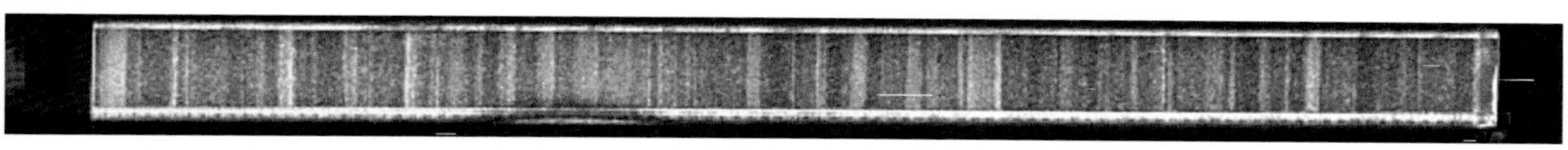
932

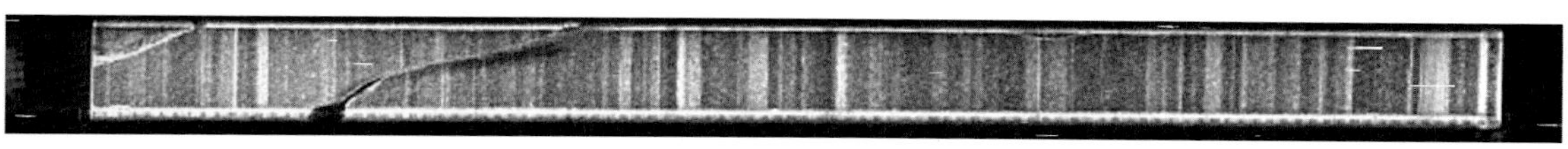
931

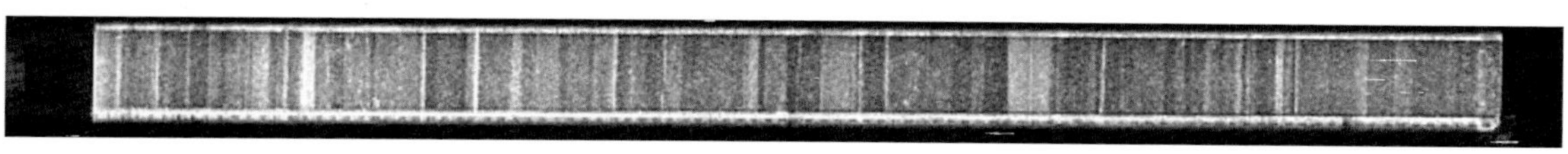
930

940
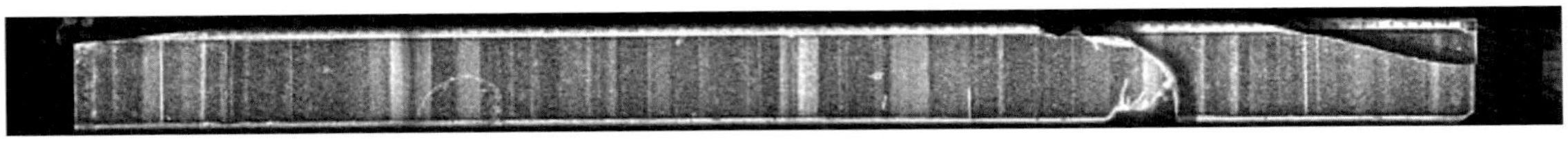

941
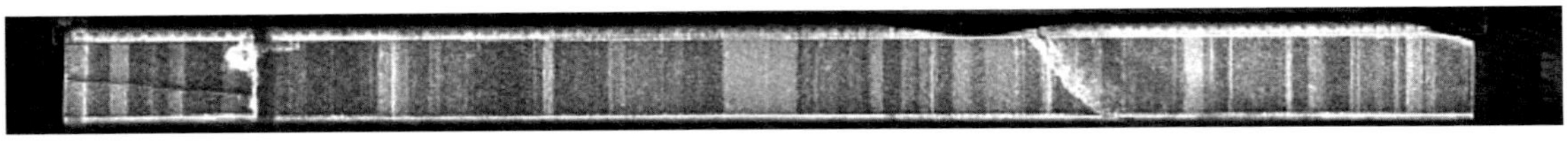

942
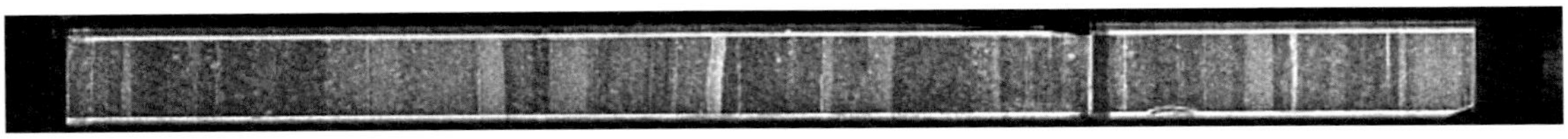

943
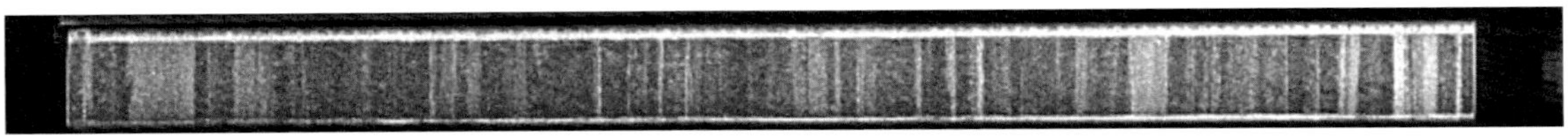

944
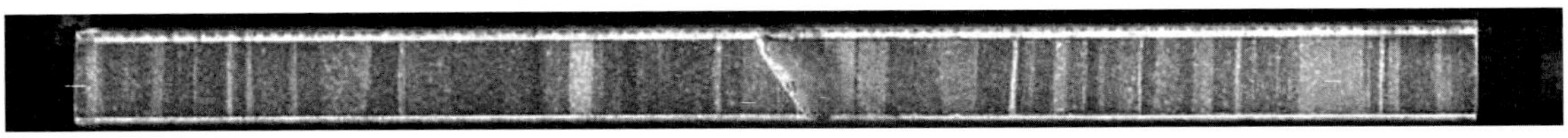

945
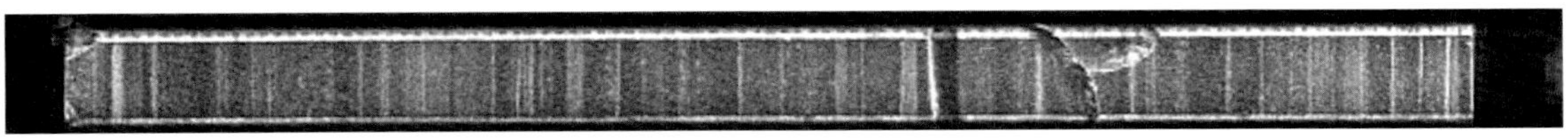

946
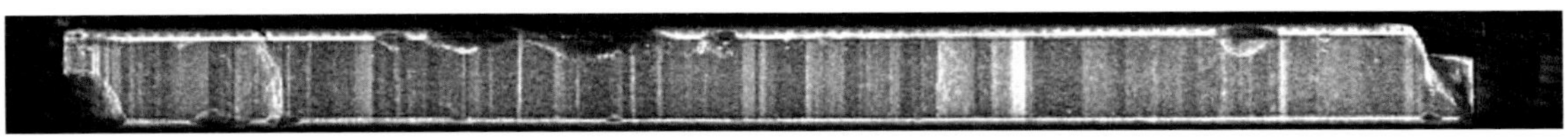

947
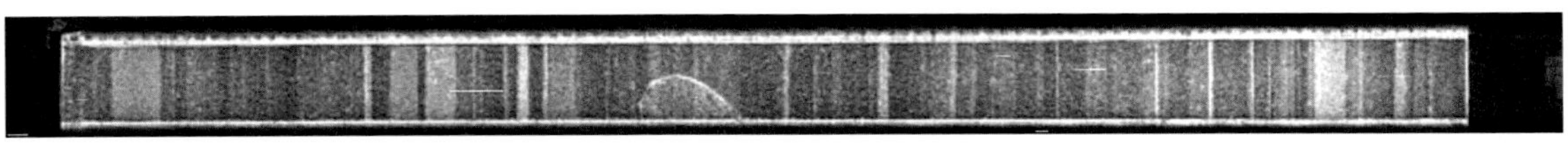

948

949
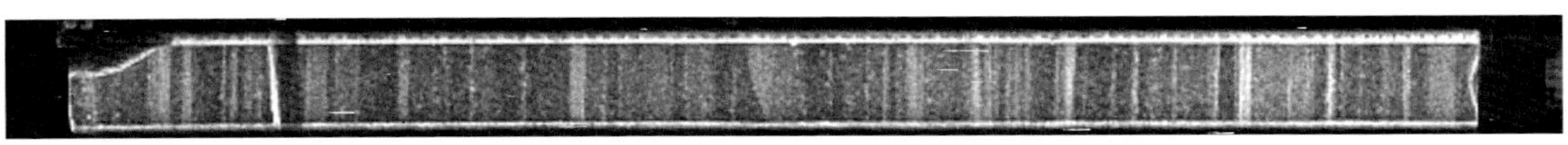

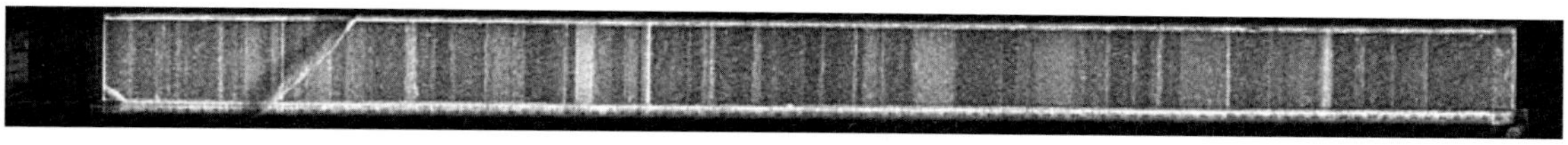
959

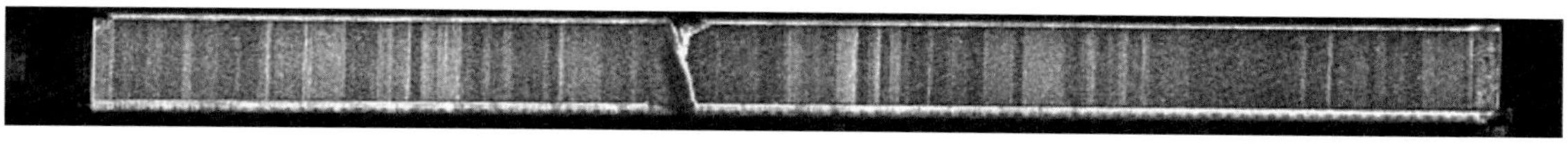
958

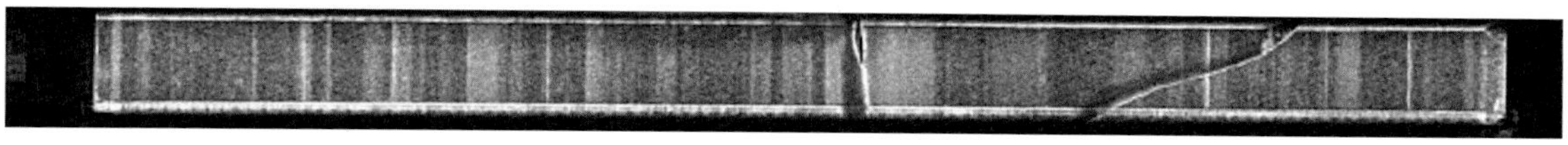
957

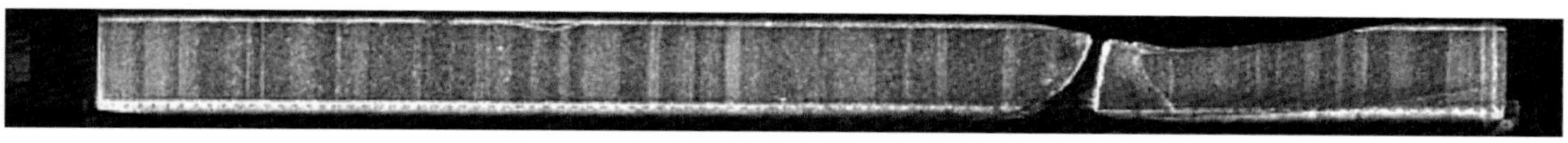
956

955

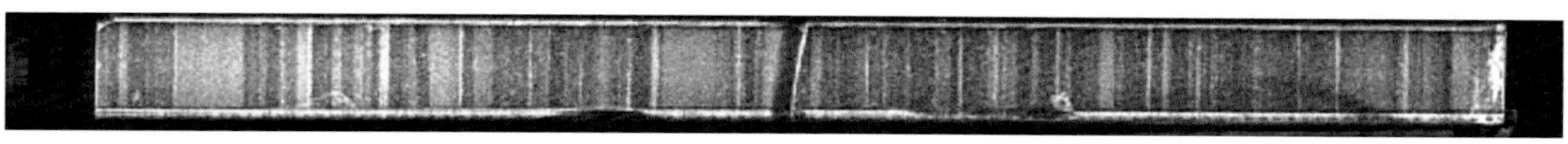
954

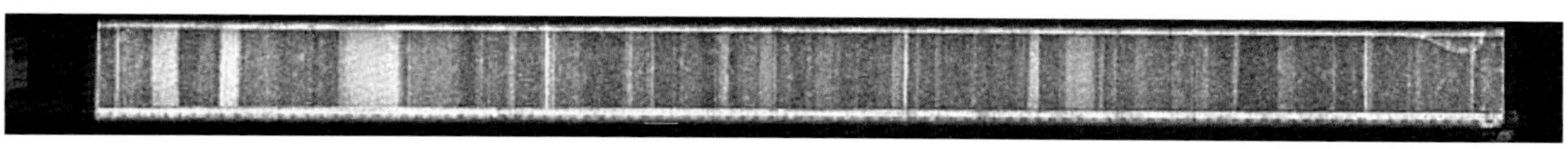
953

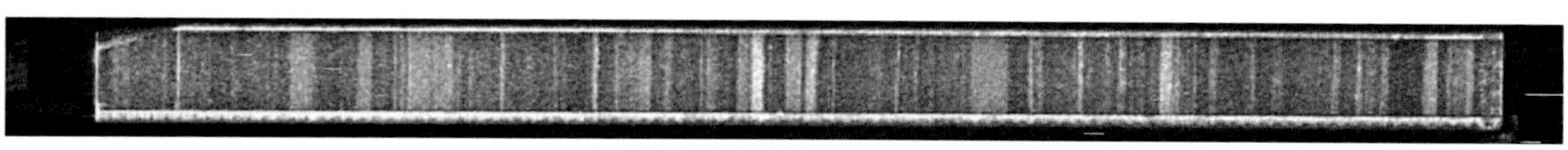
952

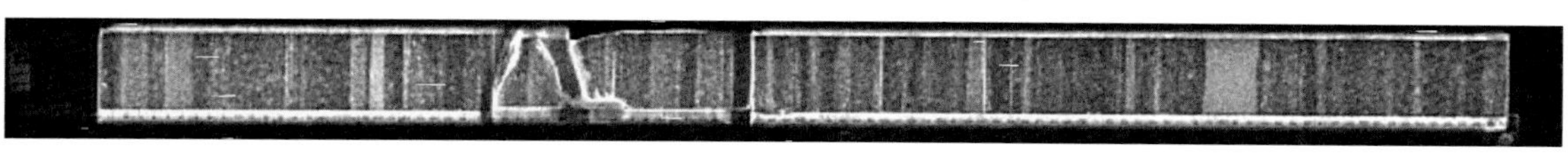
951

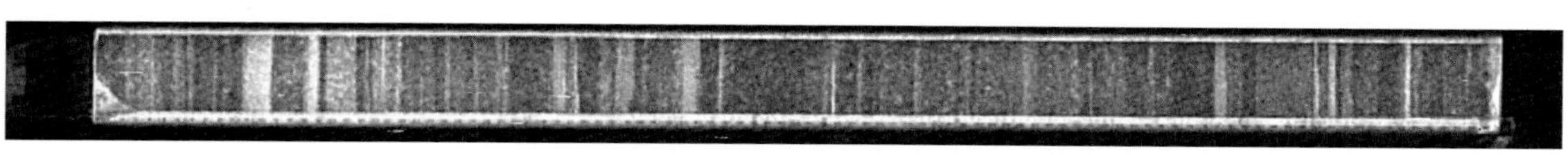
950

960
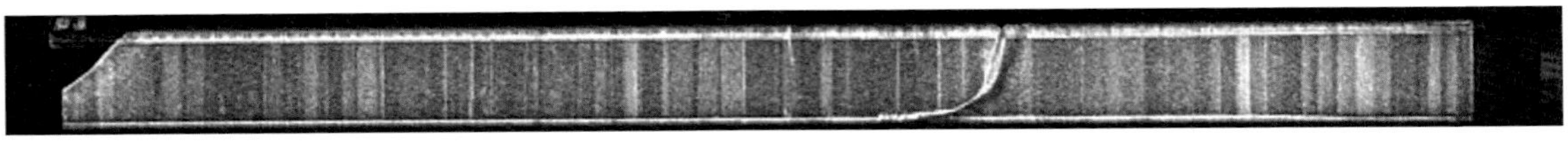

961
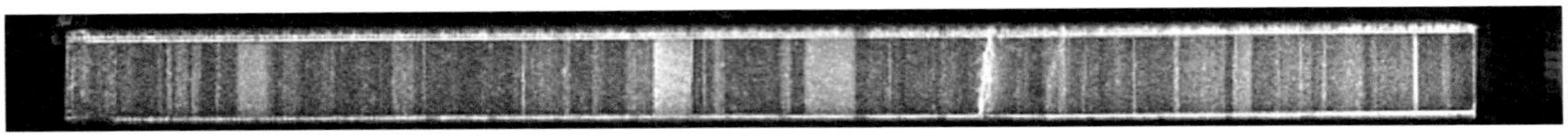

962
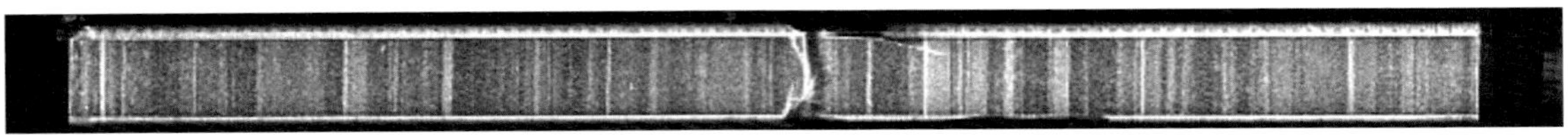

963
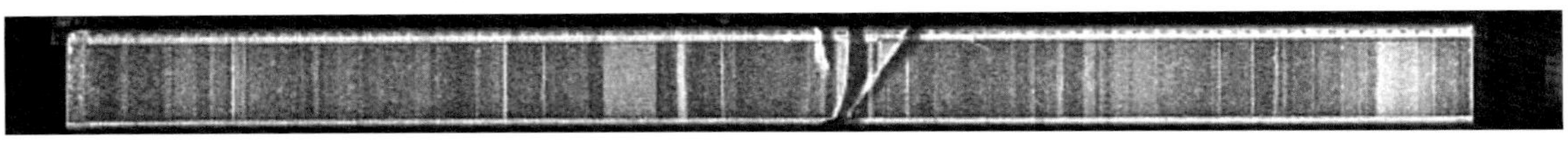

964
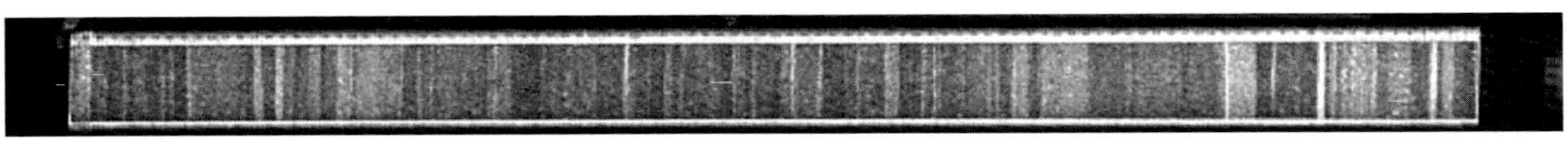

965
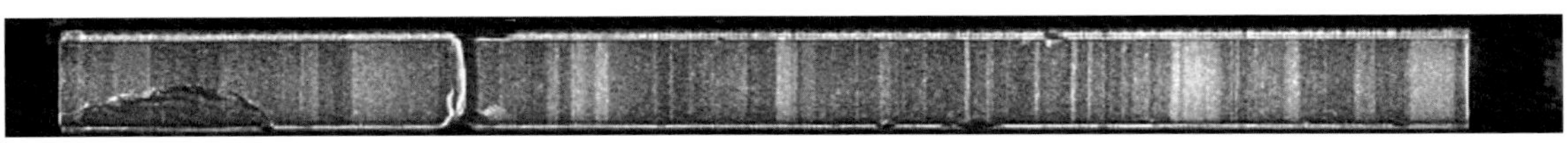

966
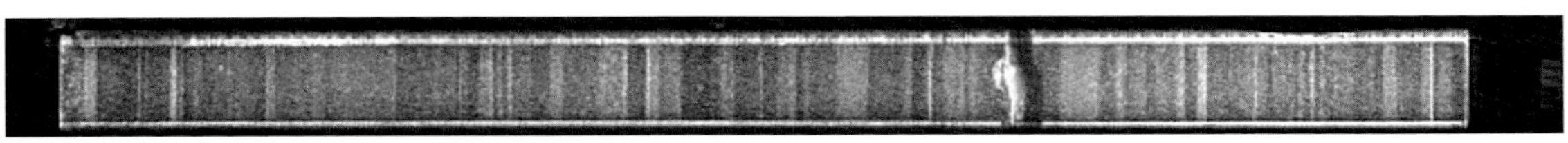

967
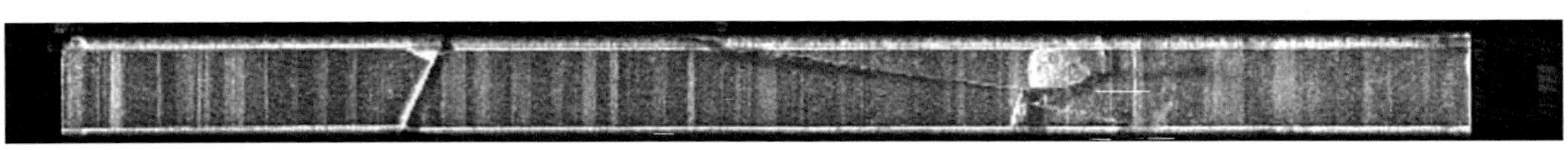

968
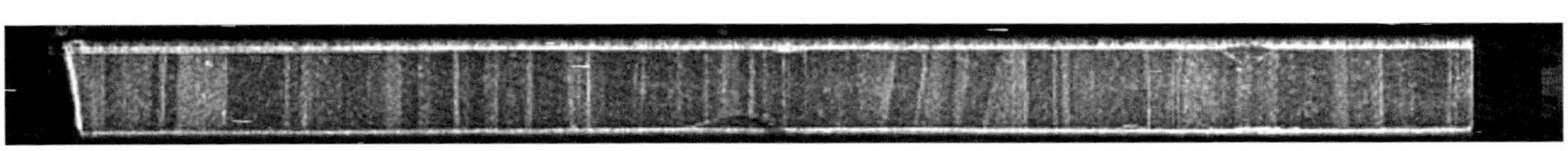

969
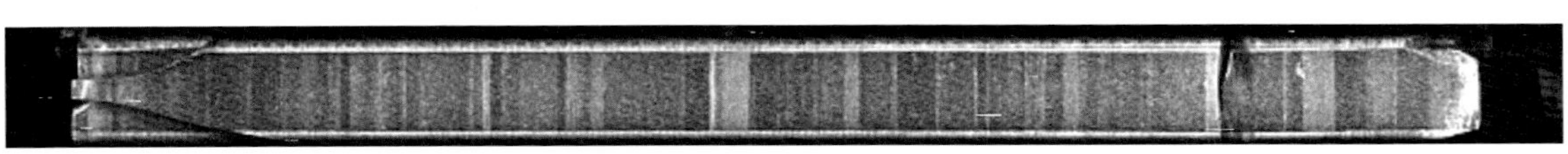

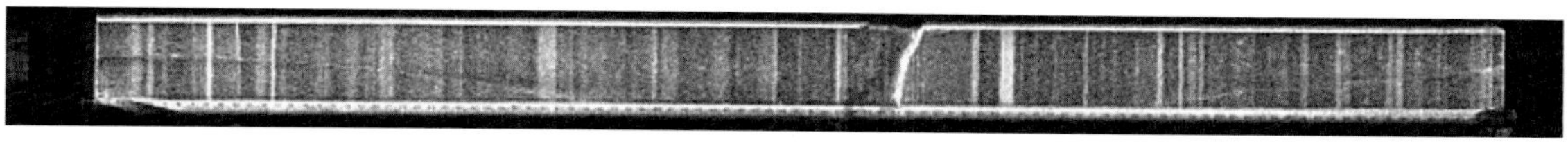

979

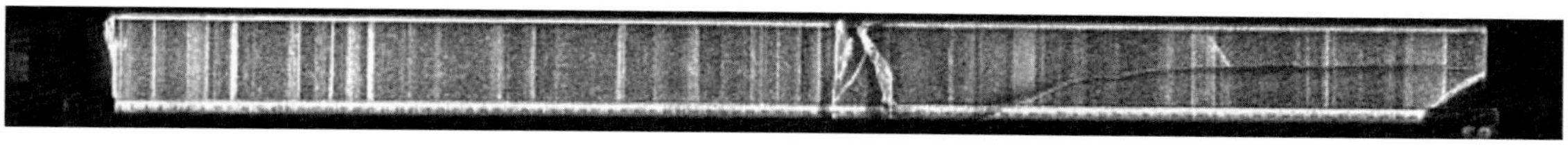

978

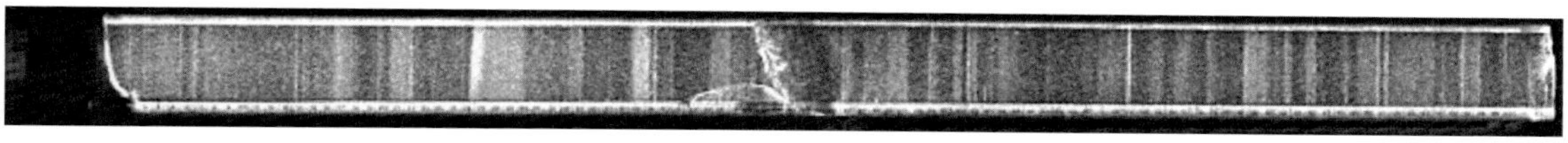

977

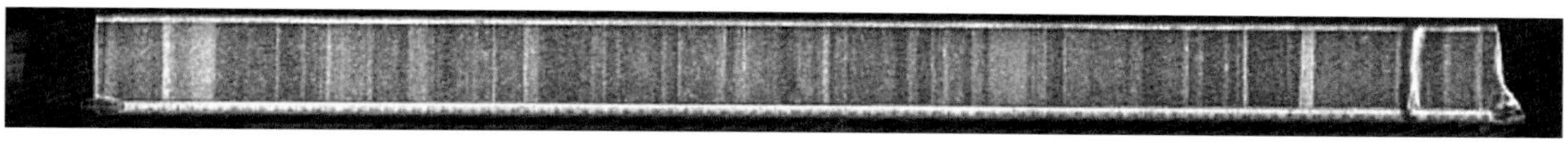

976

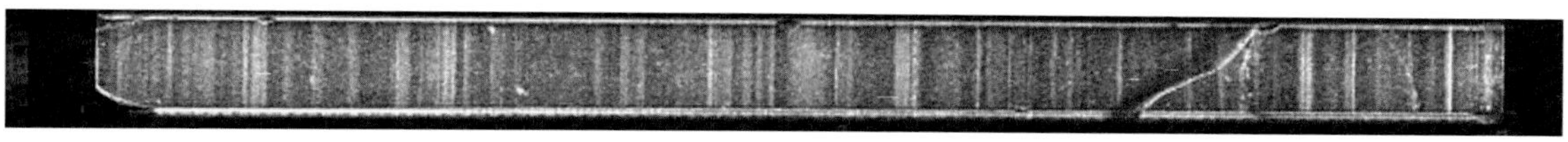

975

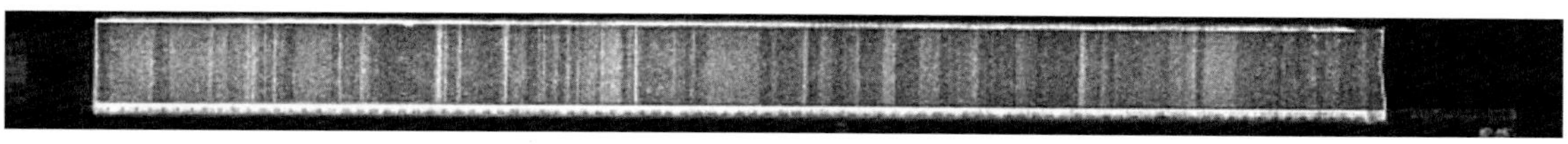

974

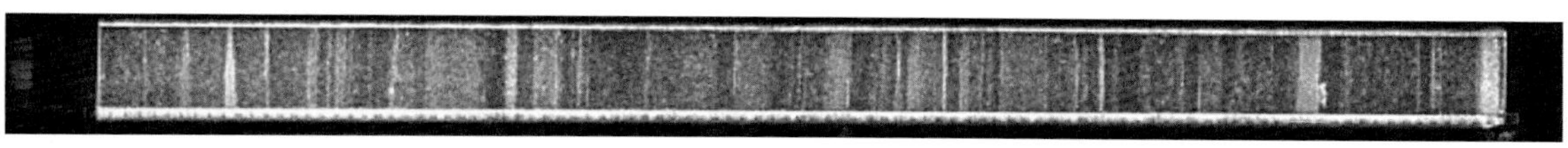

973

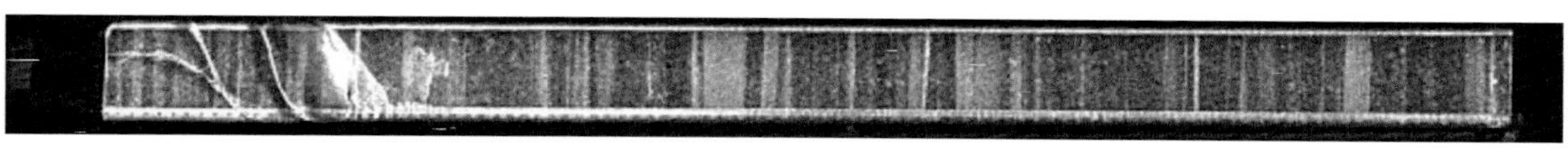

972

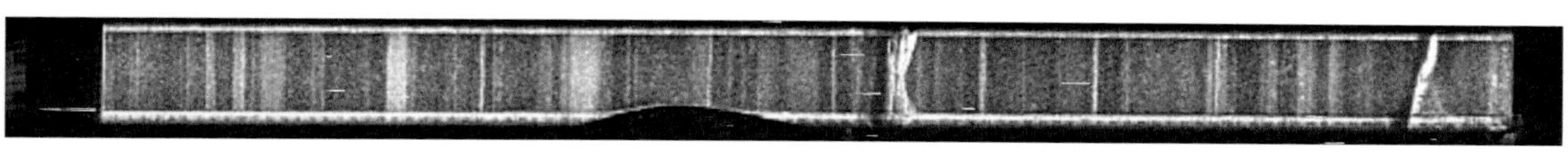

971

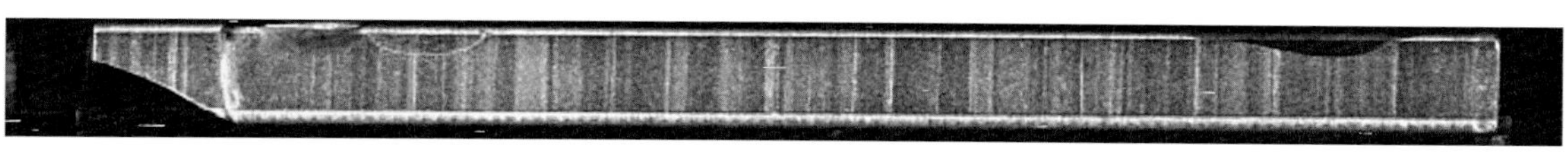

970

980

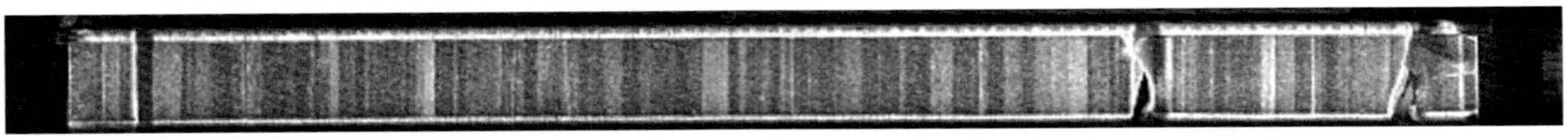

981

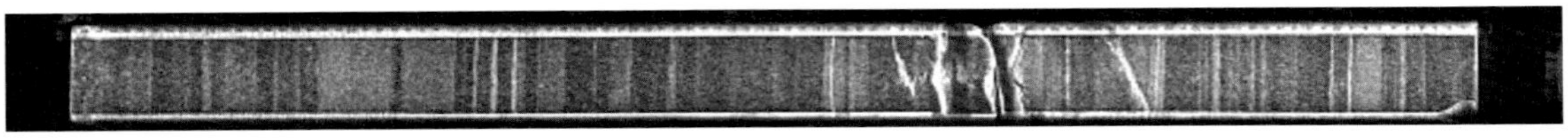

982

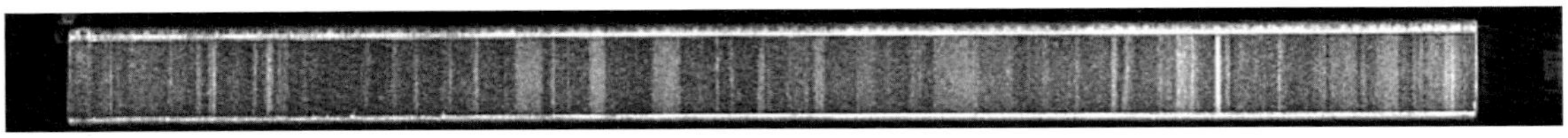

983

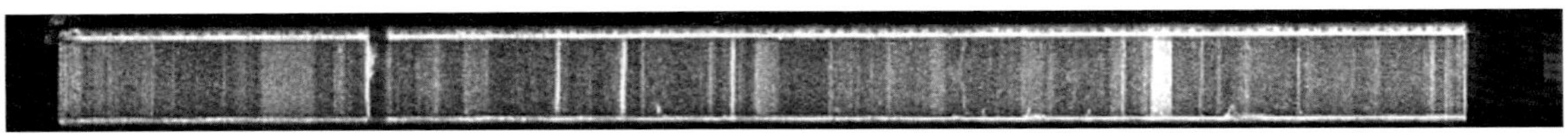

984

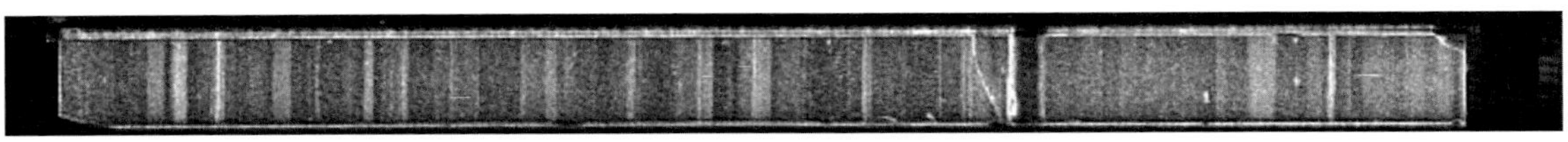

985

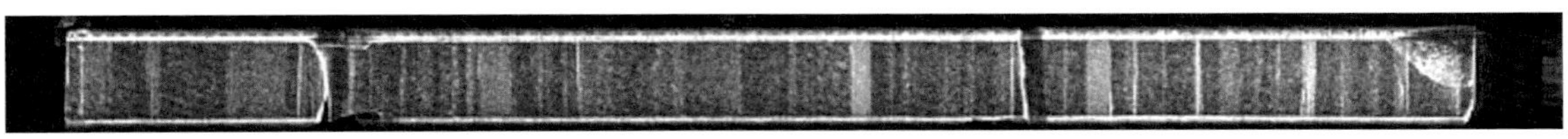

986

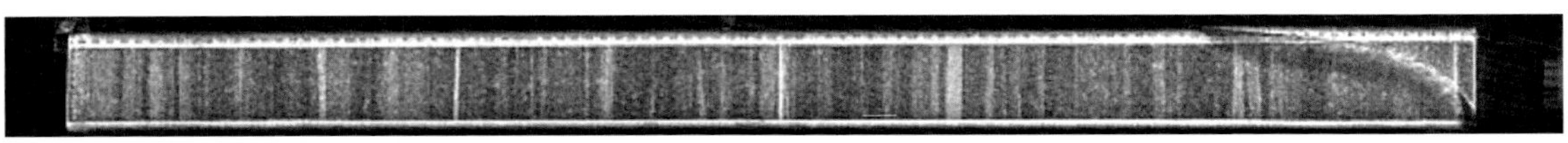

987

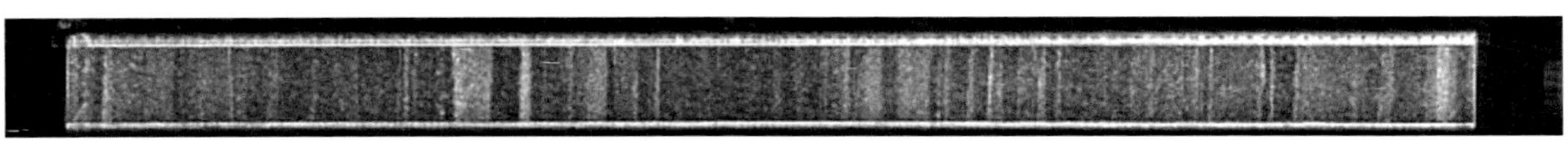

988

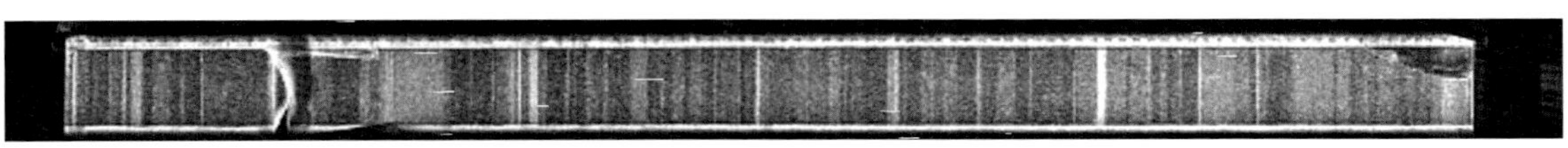

989

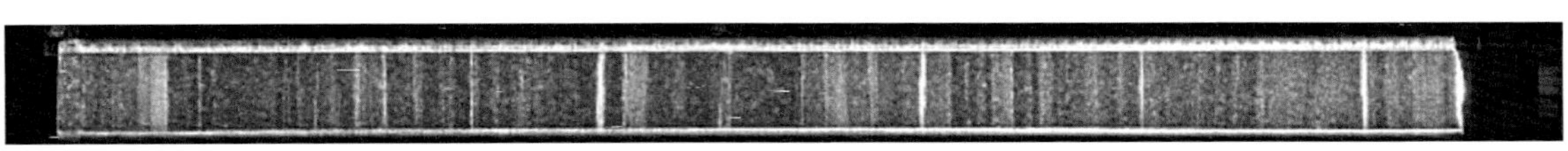

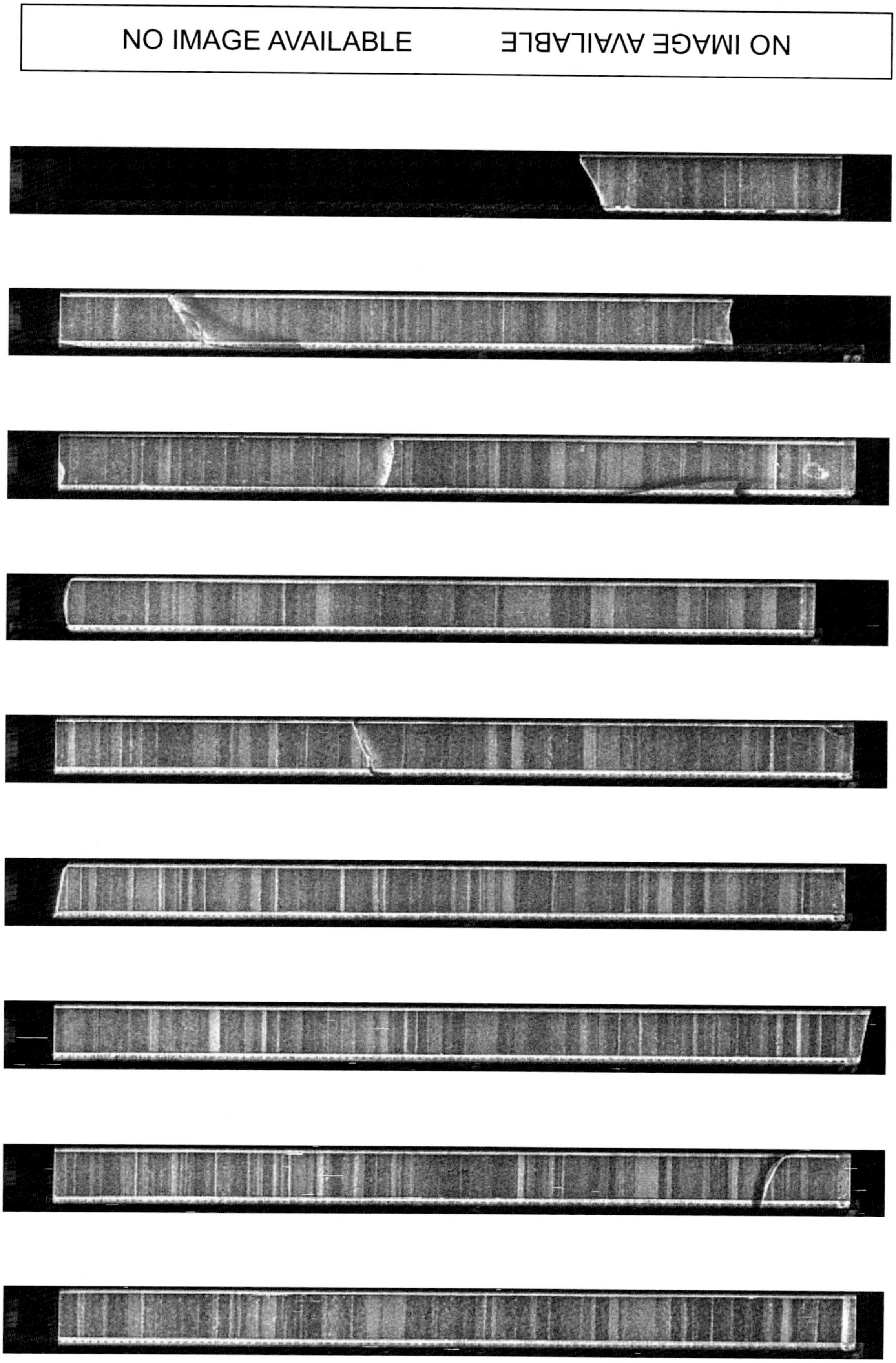

999

998

997

996

995

994

993

992

991

990

1000

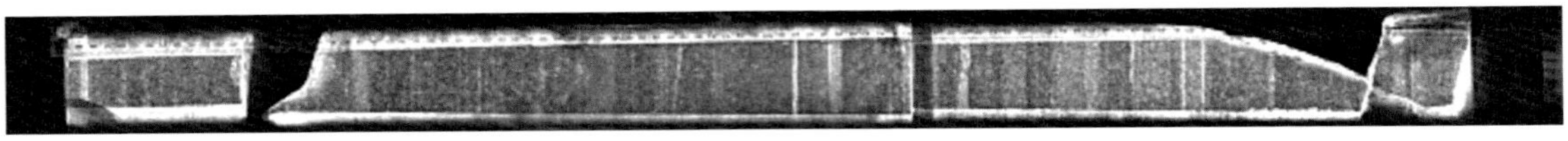

1001

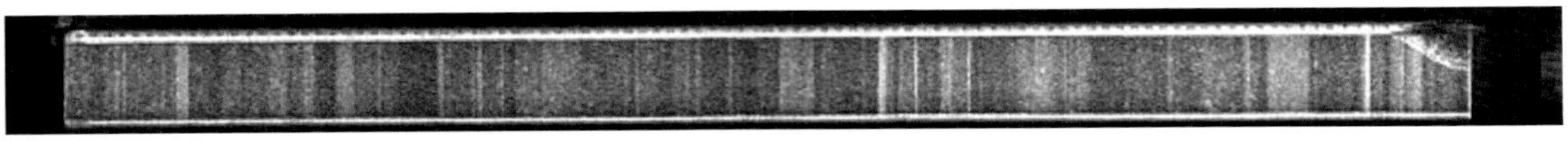

1002

1003

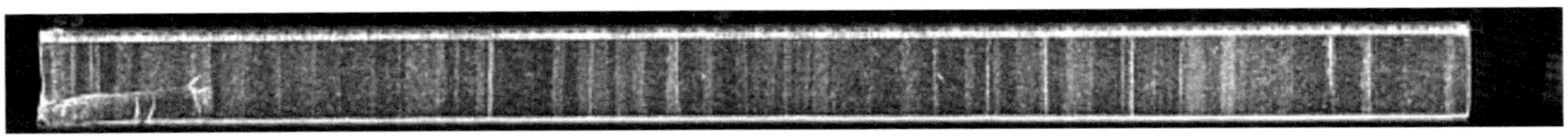

1004

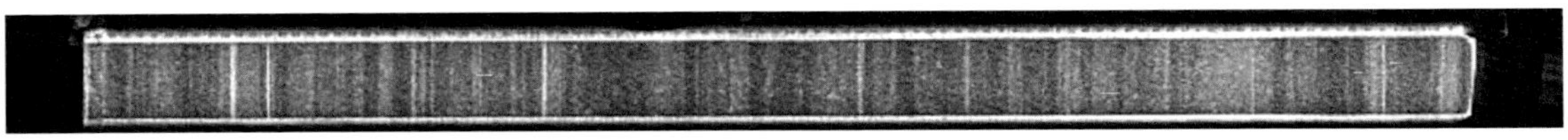

1005

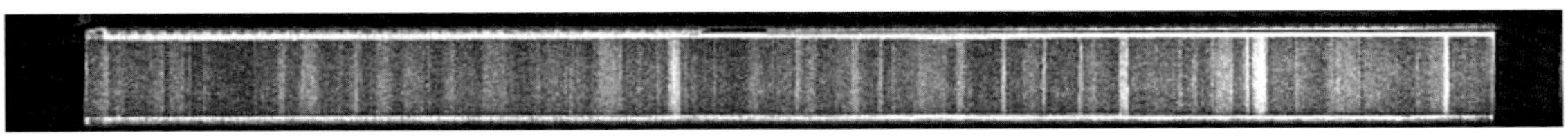

1006

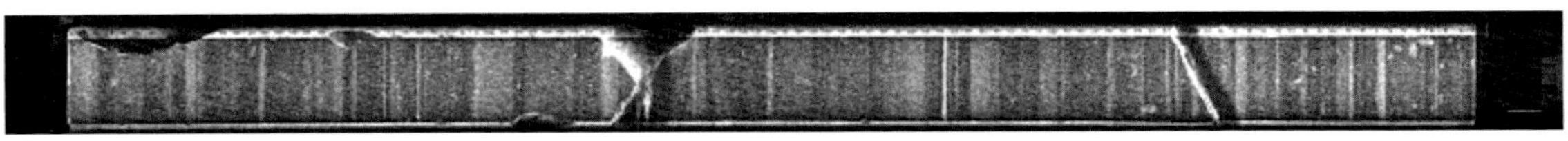

1007

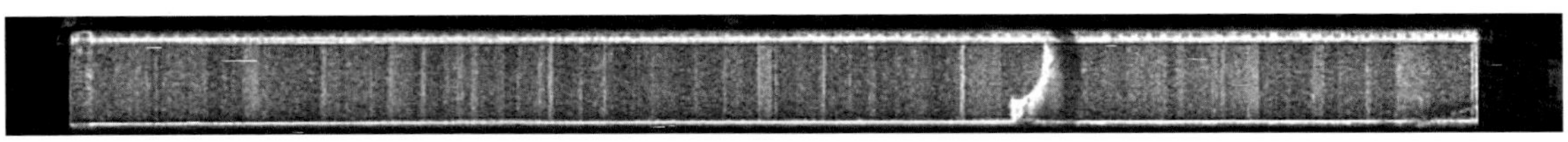

1008

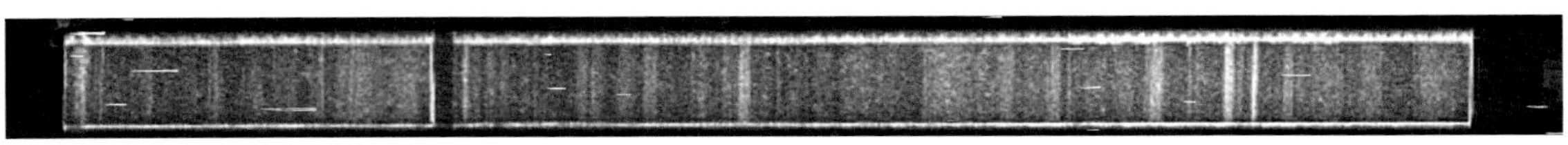

1009

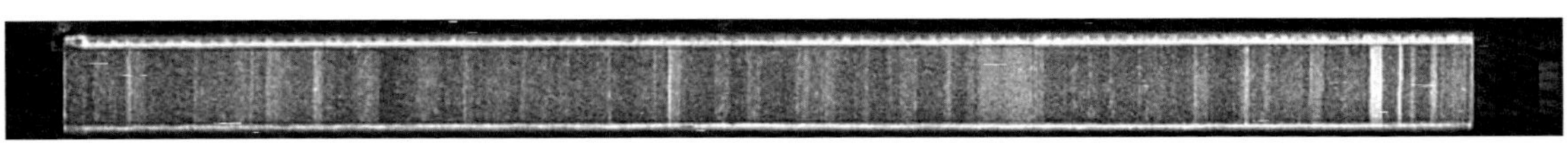

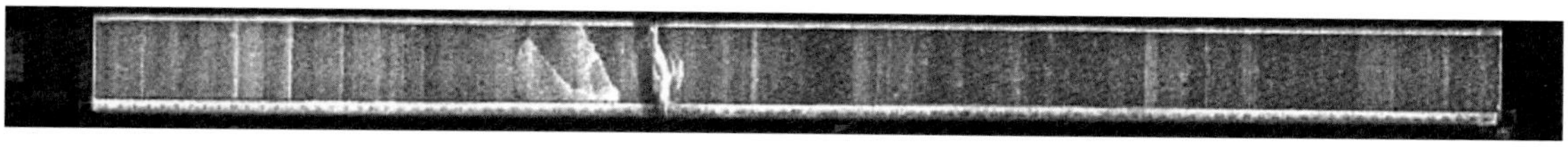
1019

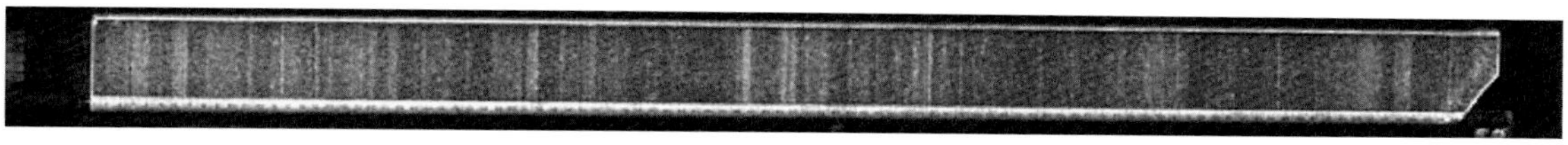
1018

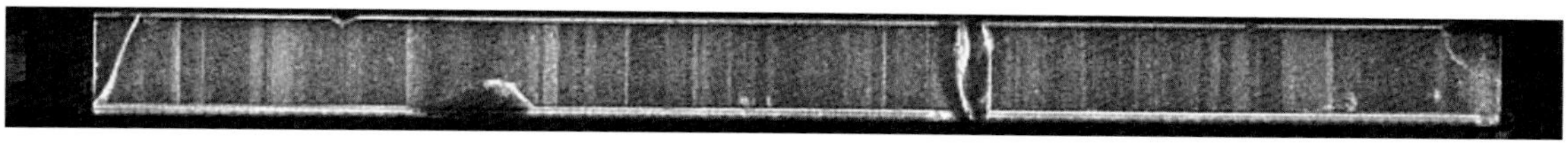
1017

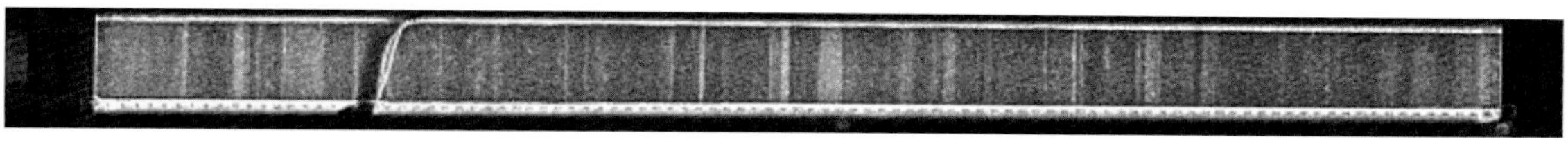
1016

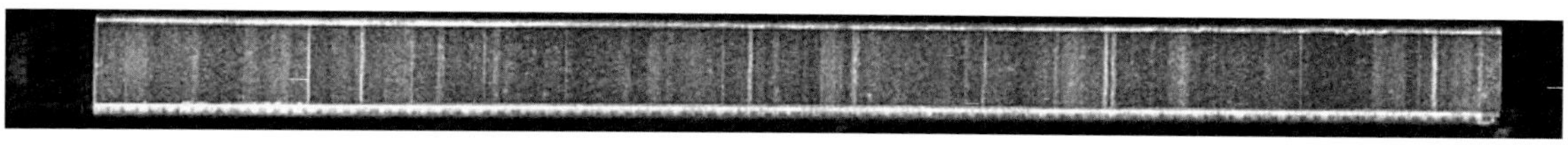
1015

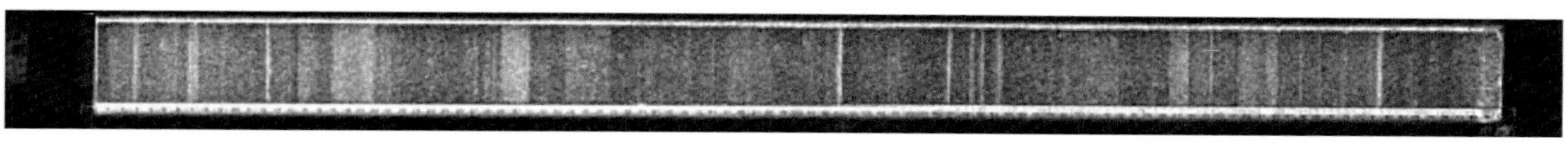
1014

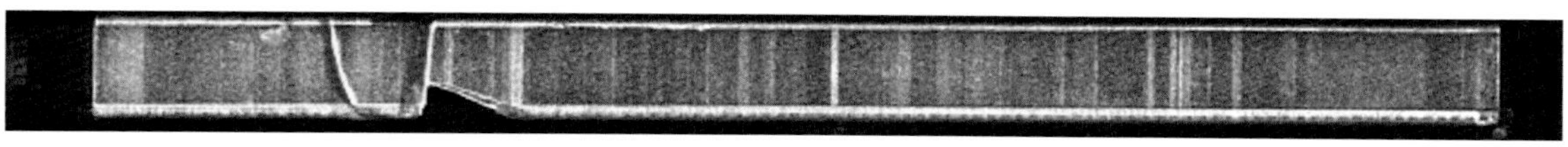
1013

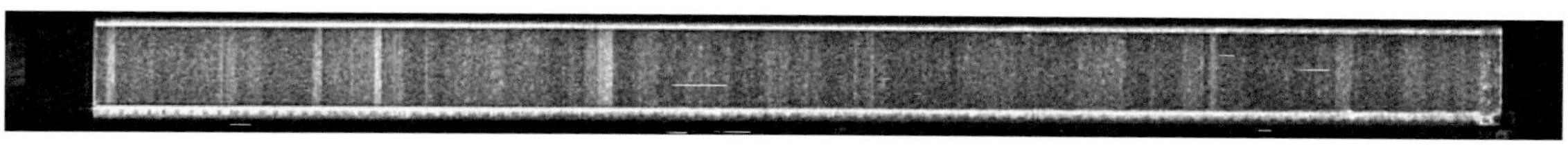
1012

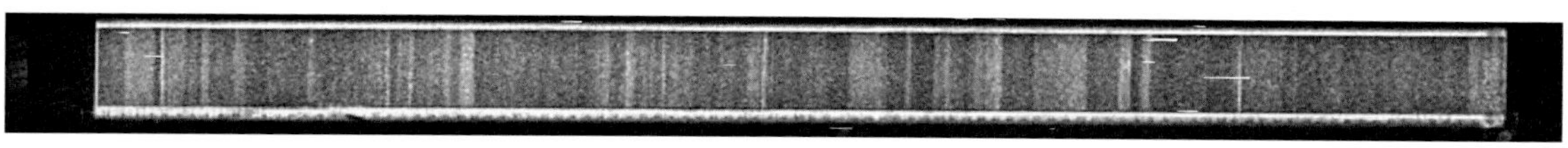
1011

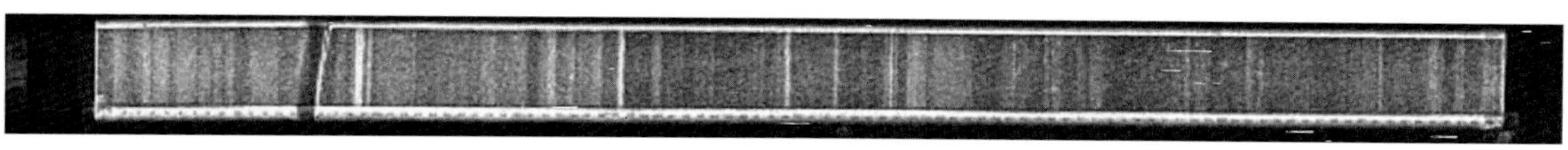
1010

1020

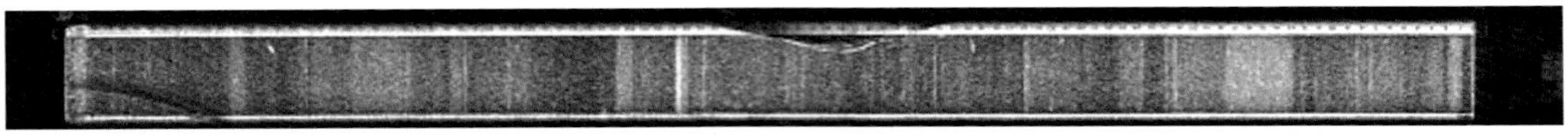

1021

1022

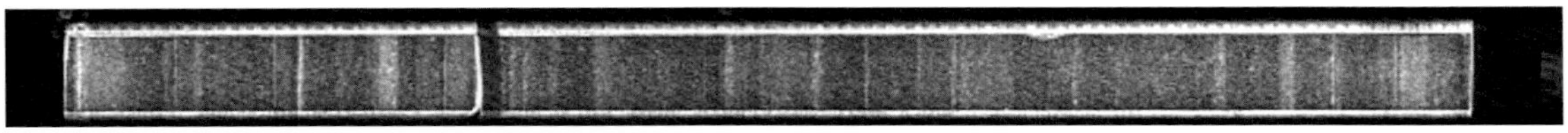

1023

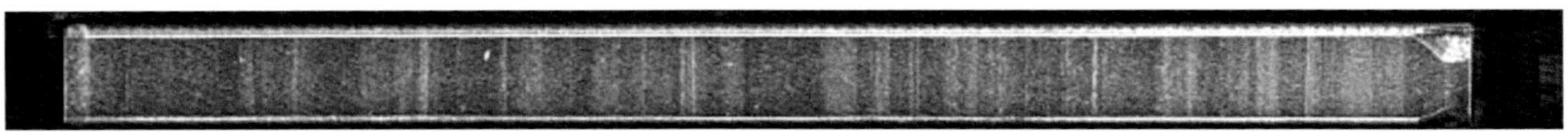

1024

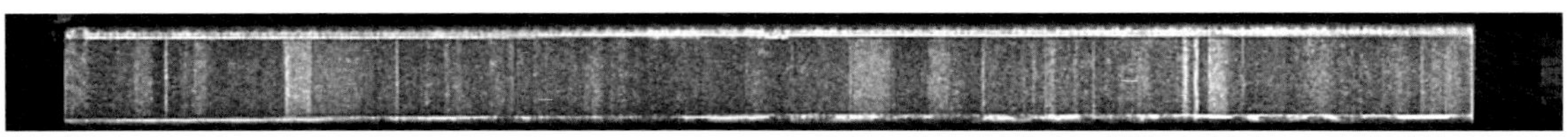

1025

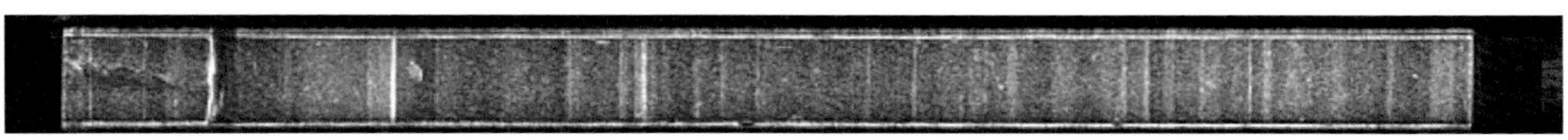

1026

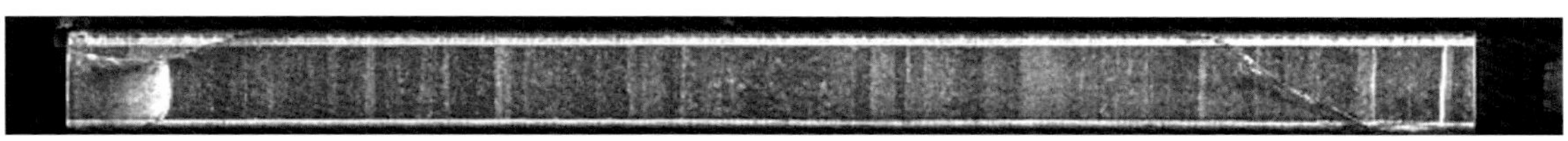

1027

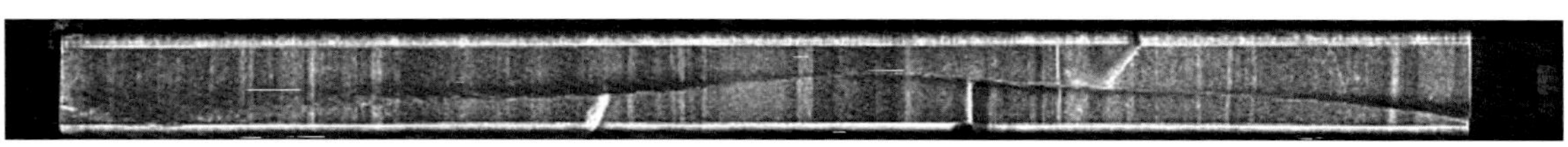

1028

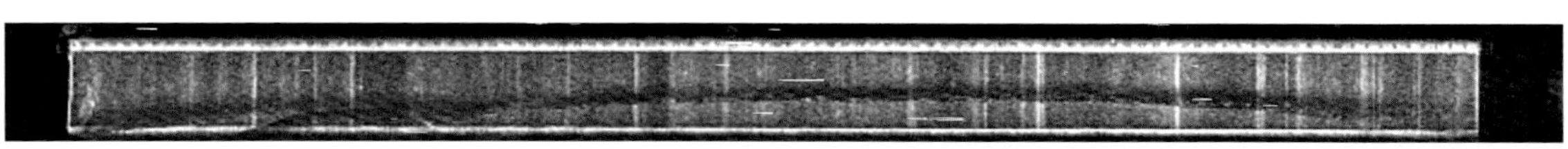

1029

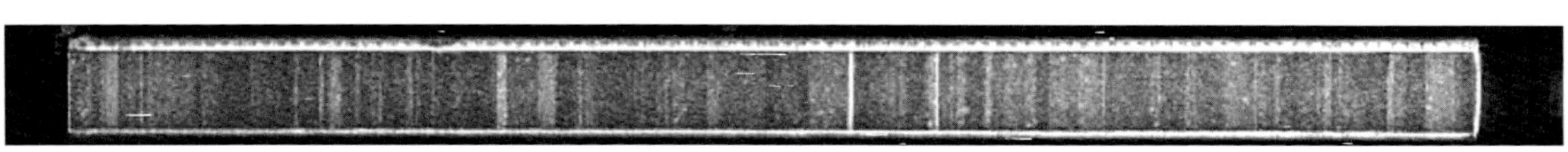

1039

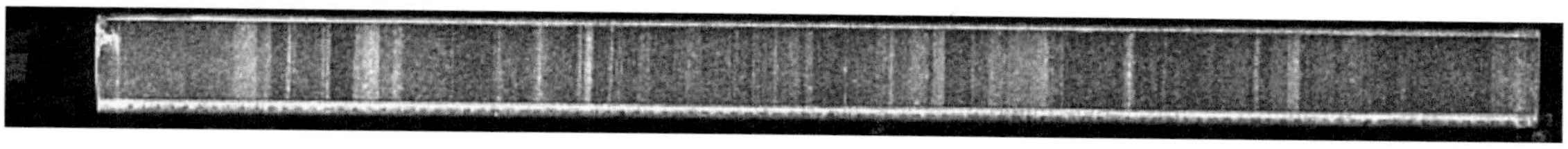

1038

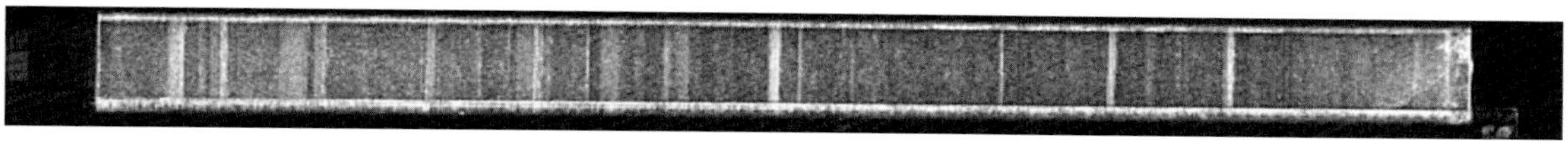

1037

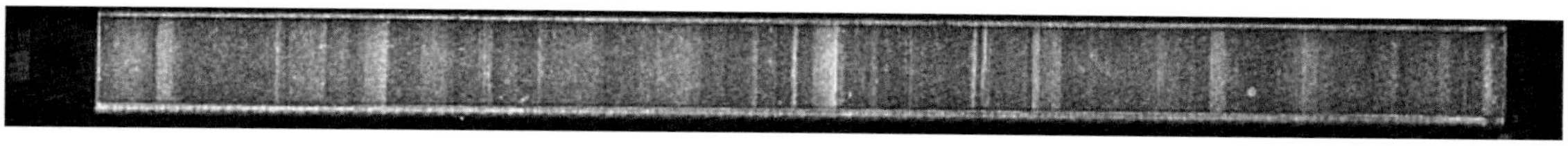

1036

1035

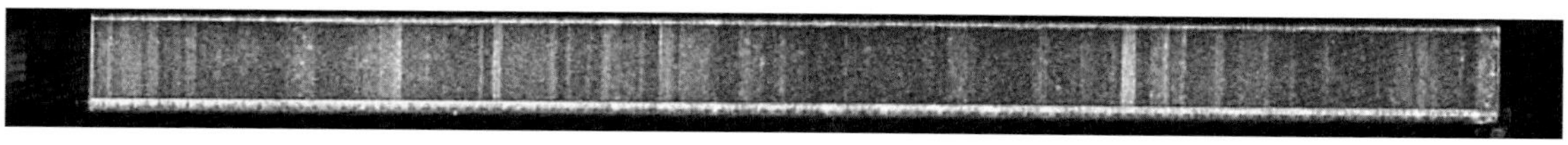

1034

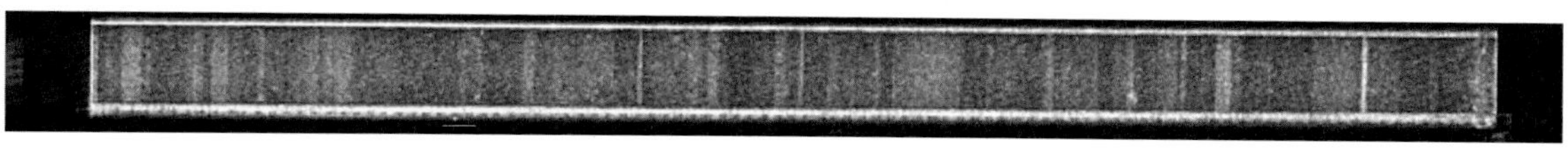

1033

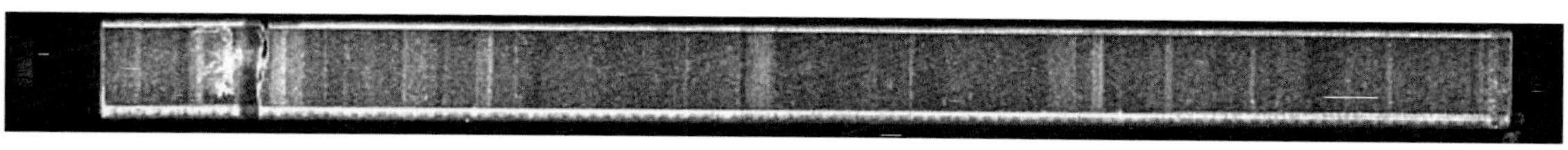

1032

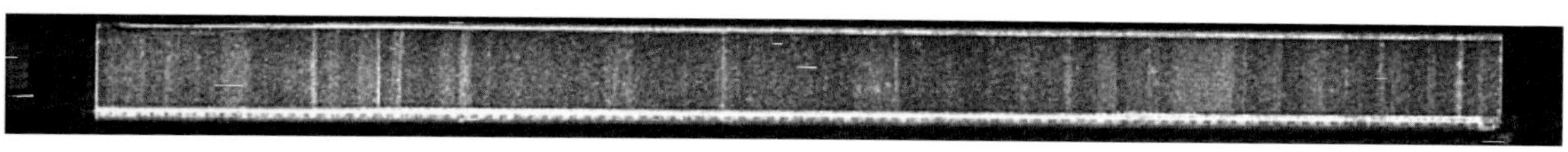

1031

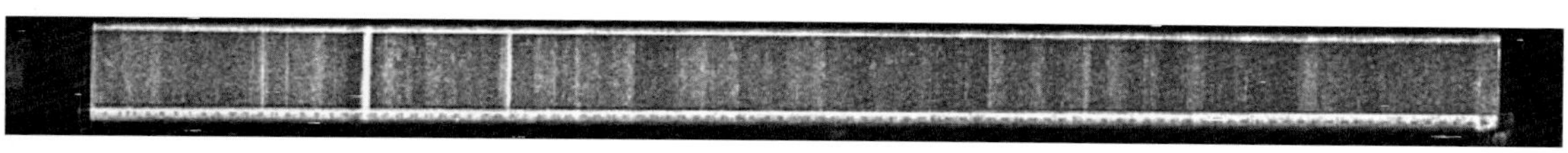

1030

1040

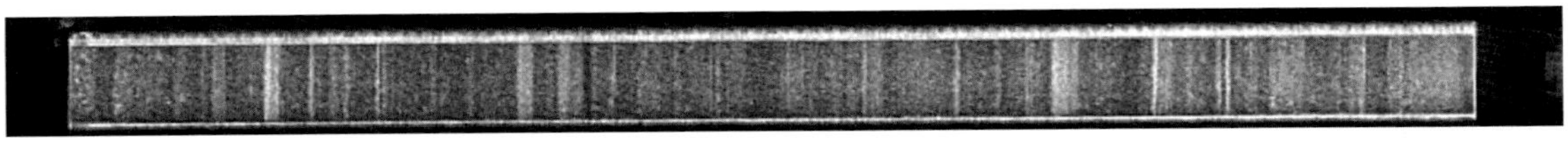

1041

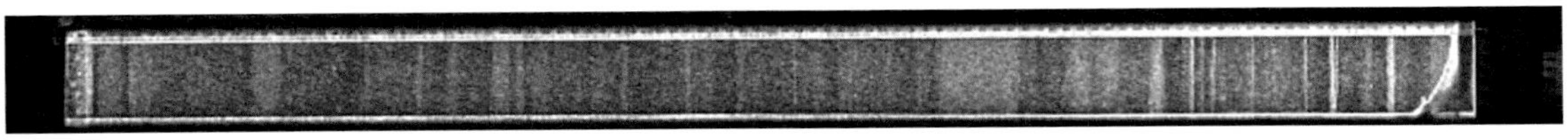

1042

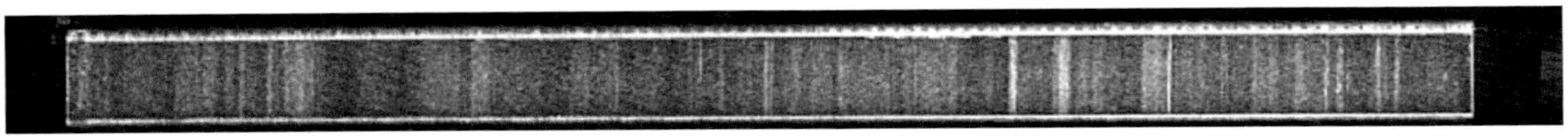

1043

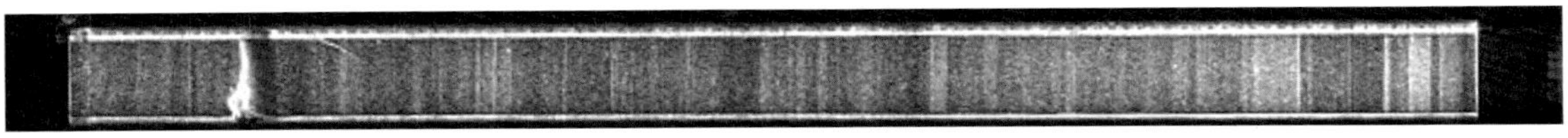

1044

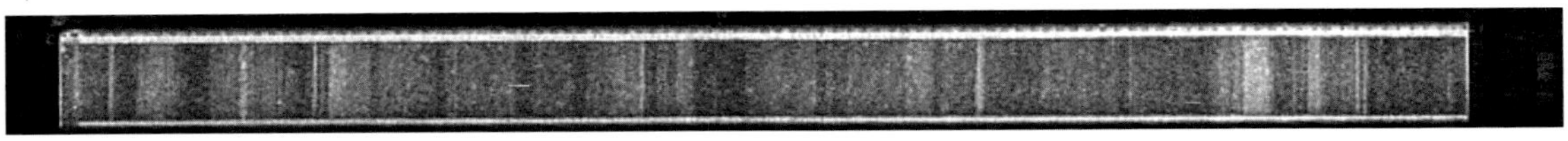

1045

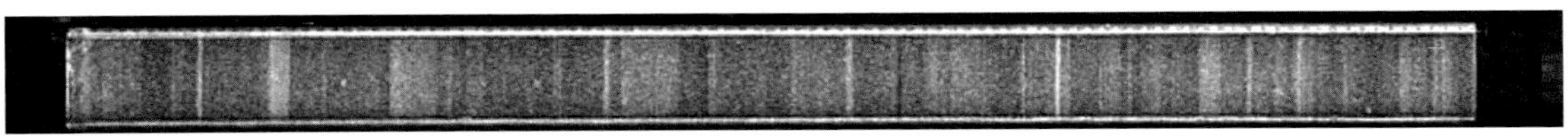

1046

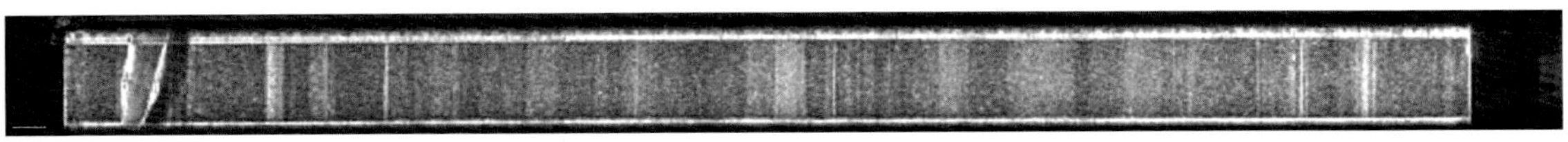

1047

1048

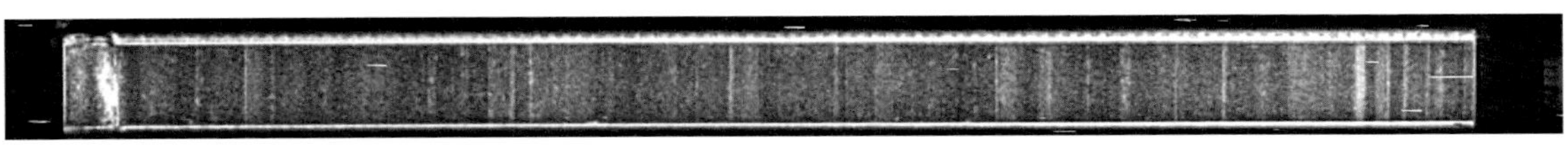

1049

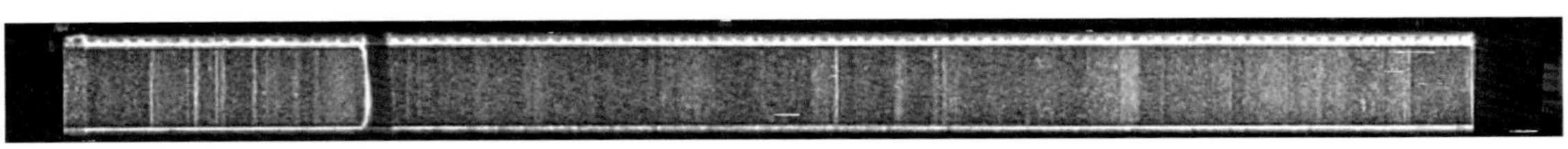

1059

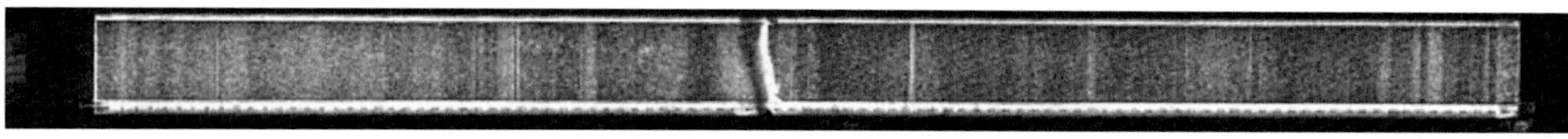

1058

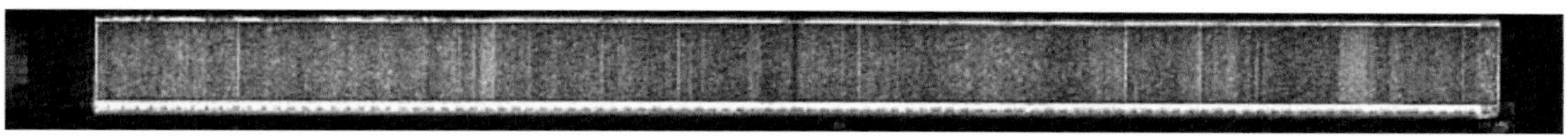

1057

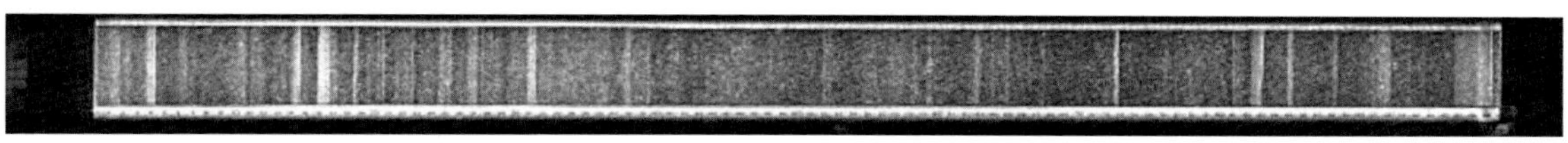

1056

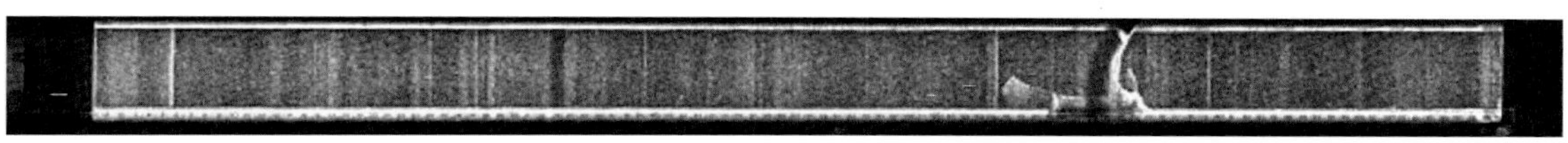

1055

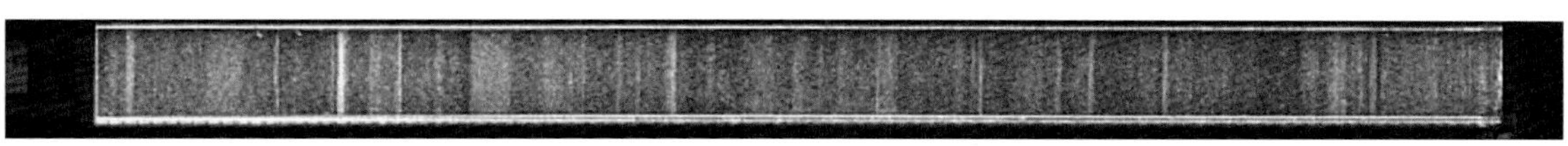

1054

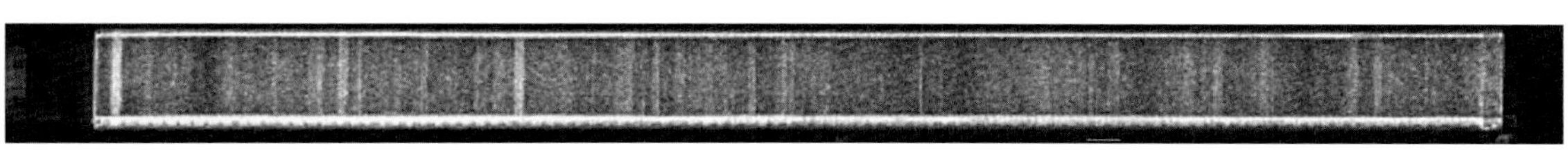

1053

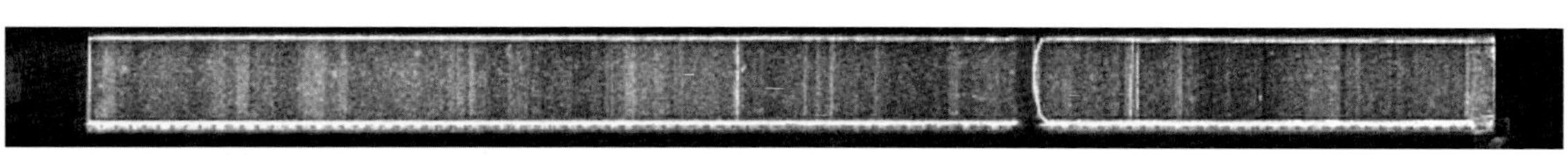

1052

1051

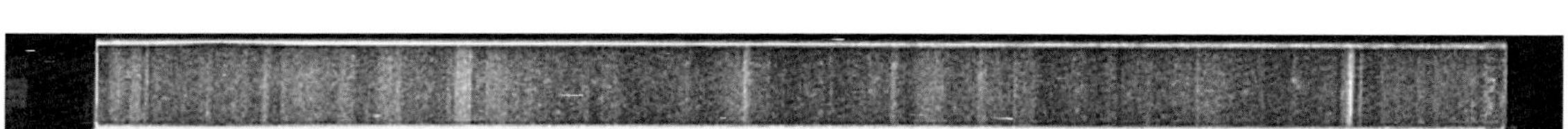

1050

1060

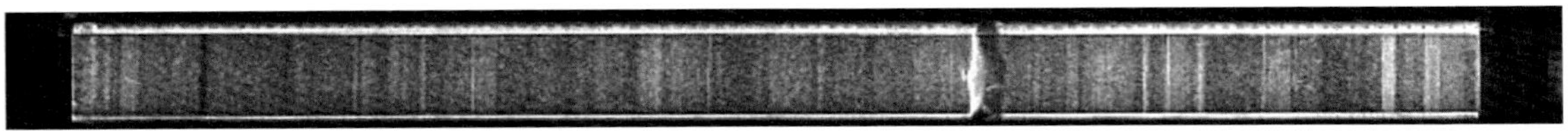

1061

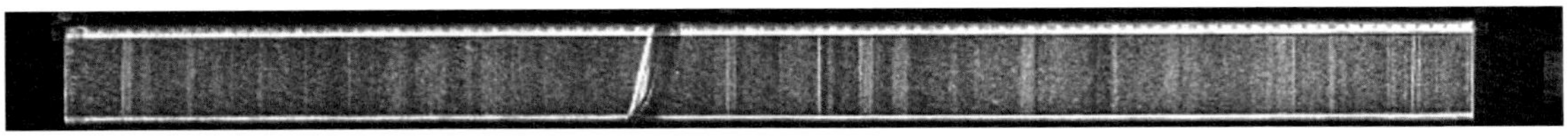

1062

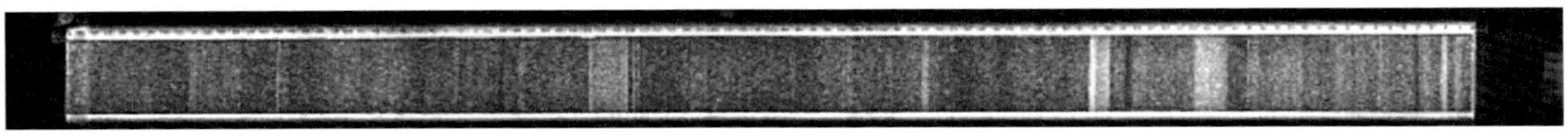

1063

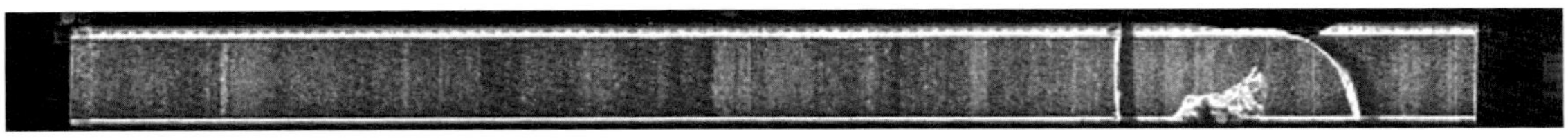

1064

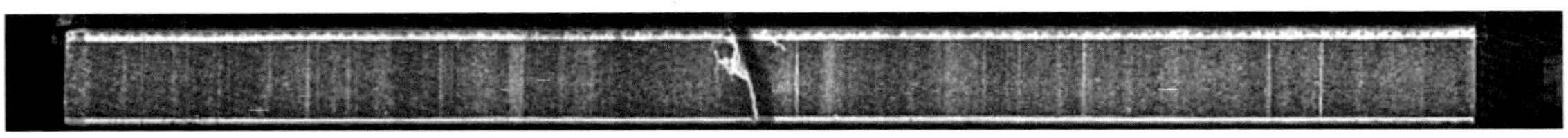

1065

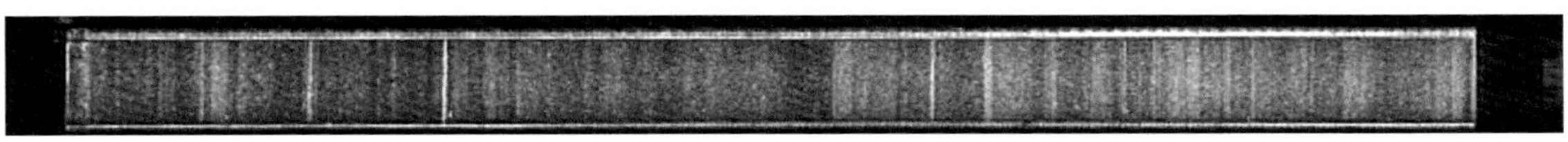

1066

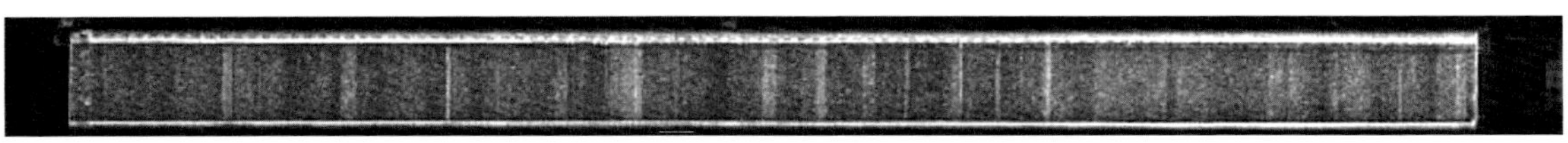

1067

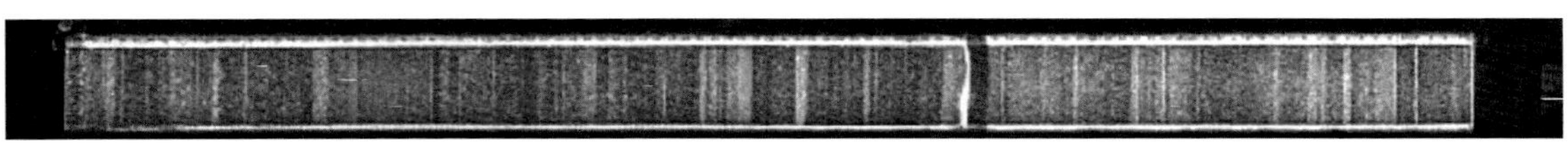

1068

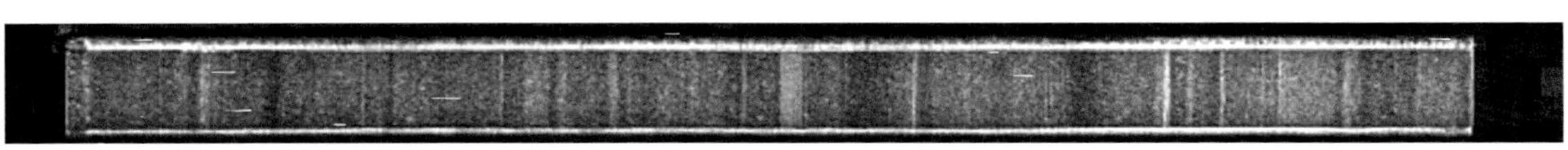

1069

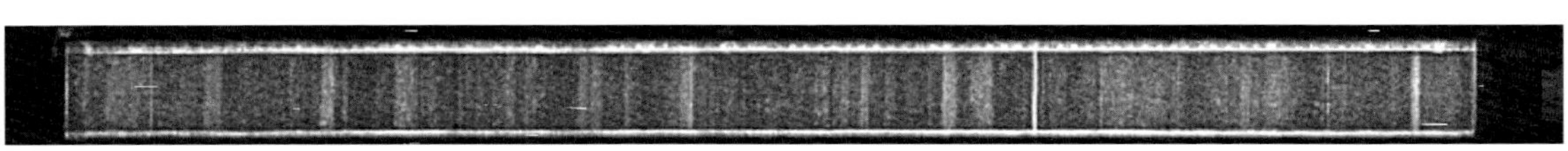

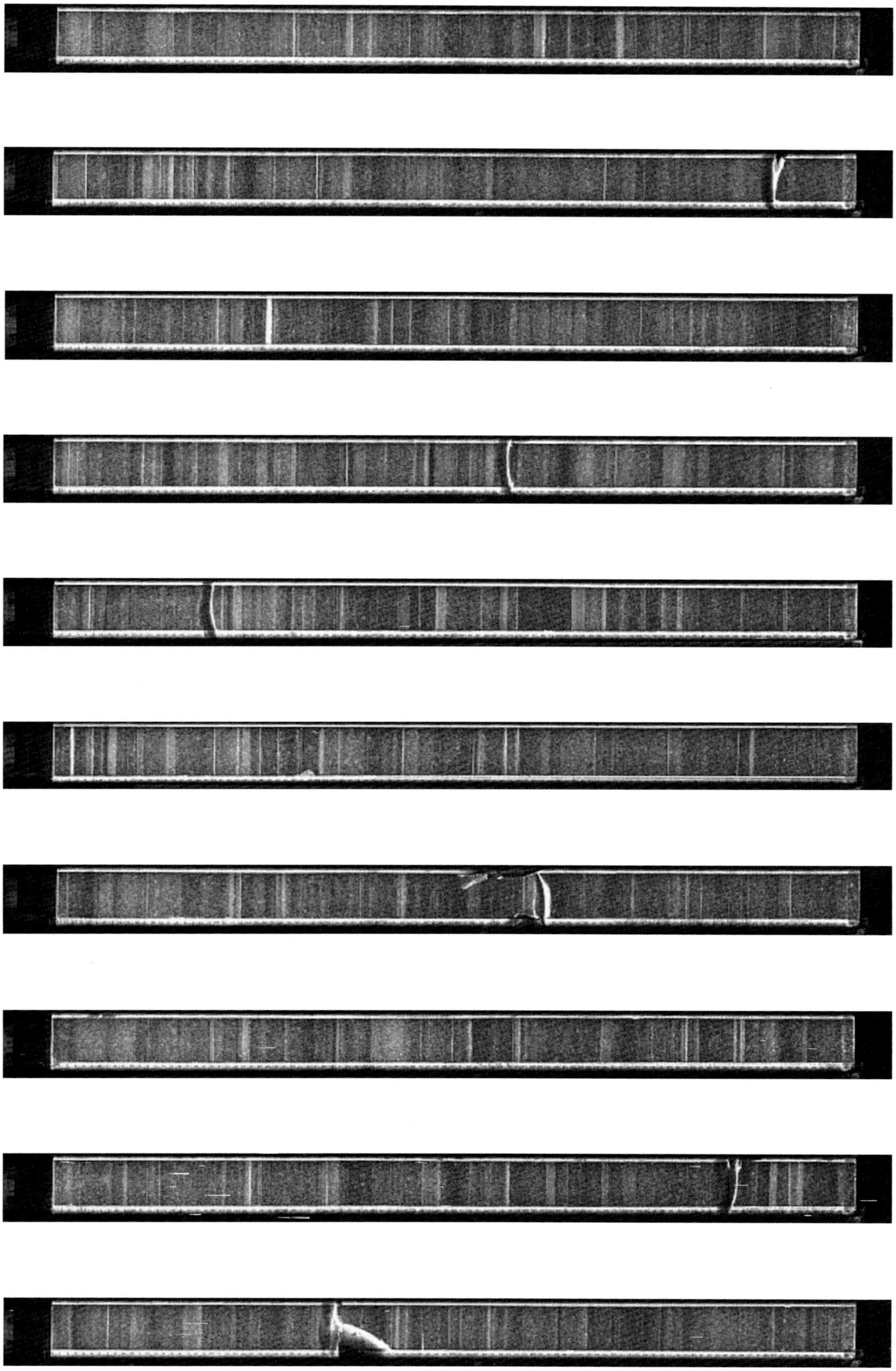

1079

1078

1077

1076

1075

1074

1073

1072

1071

1070

1080

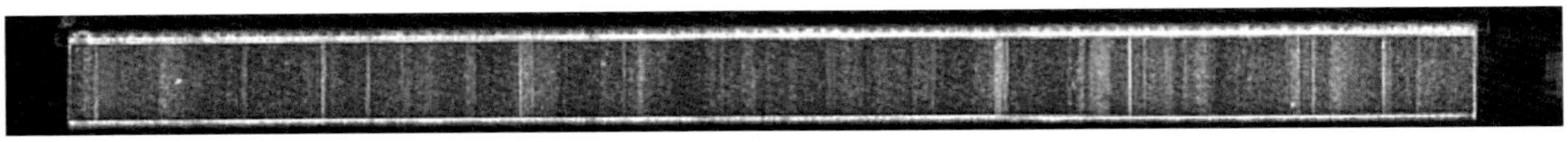

1081

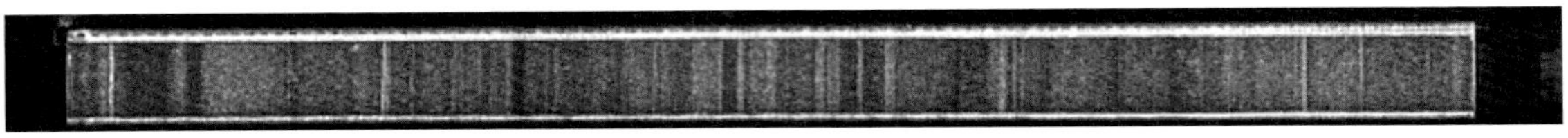

1082

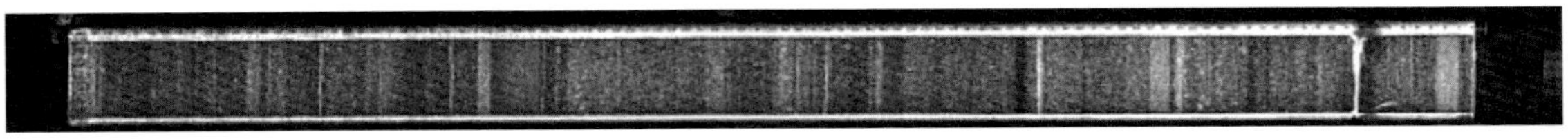

1083

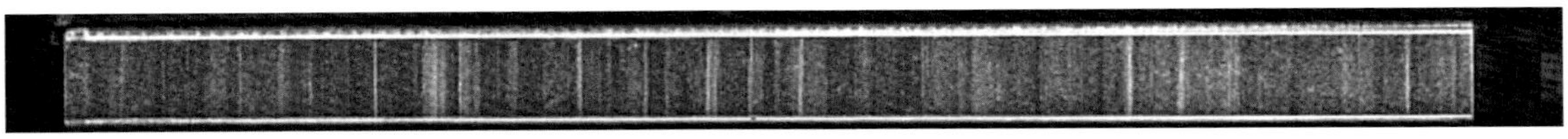

1084

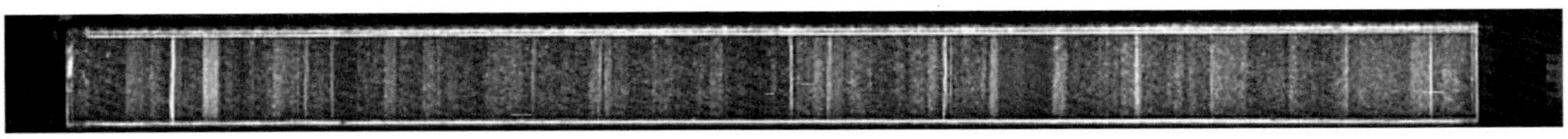

1085

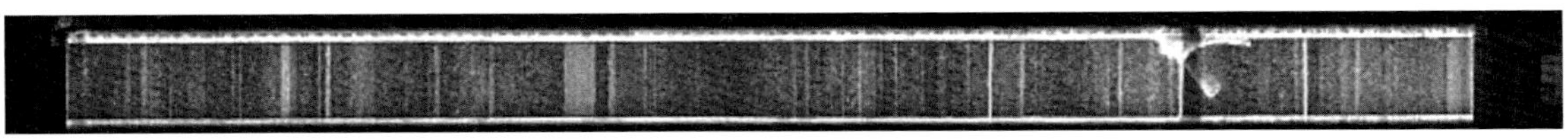

1086

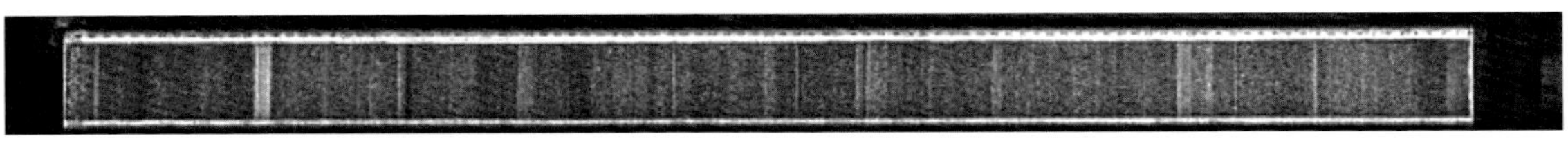

1087

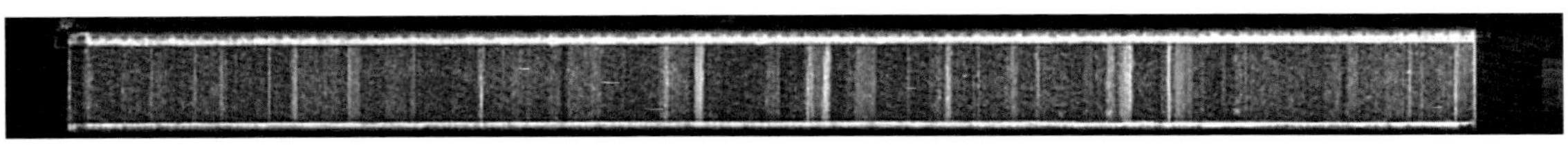

1088

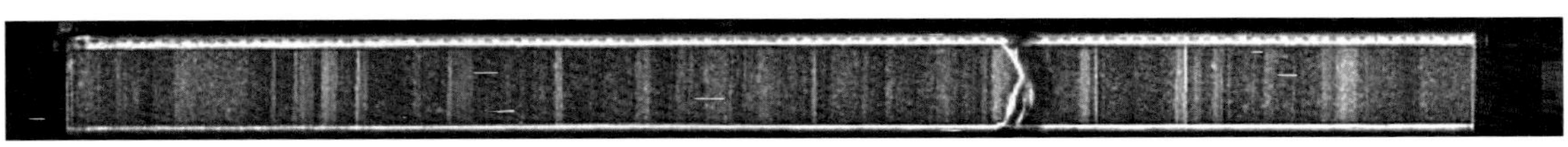

1089

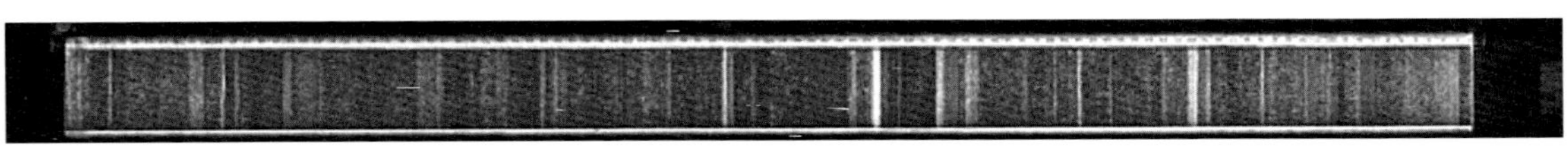

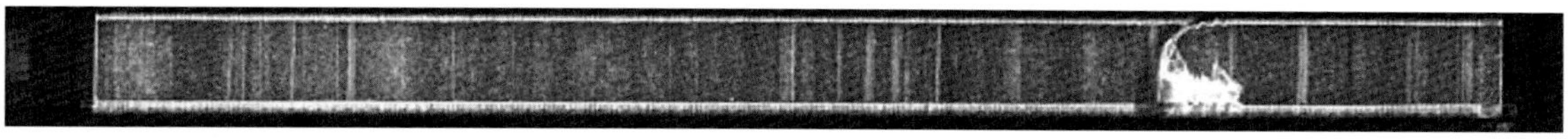
1099

1098

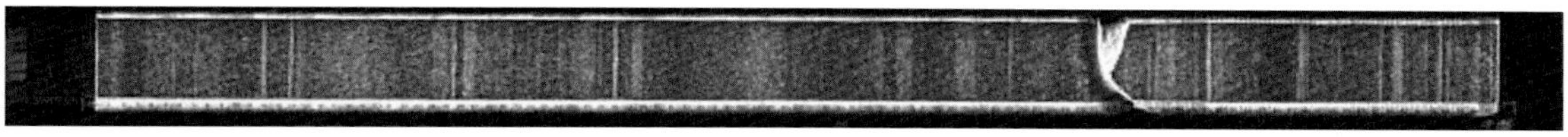
1097

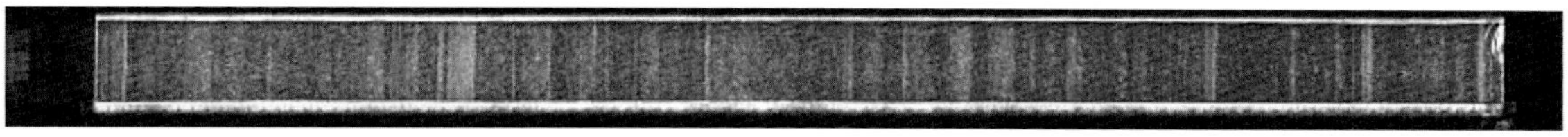
1096

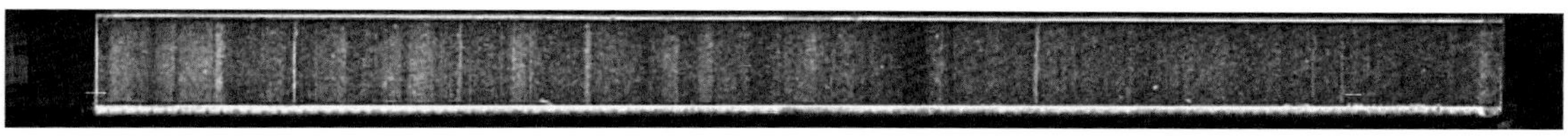
1095

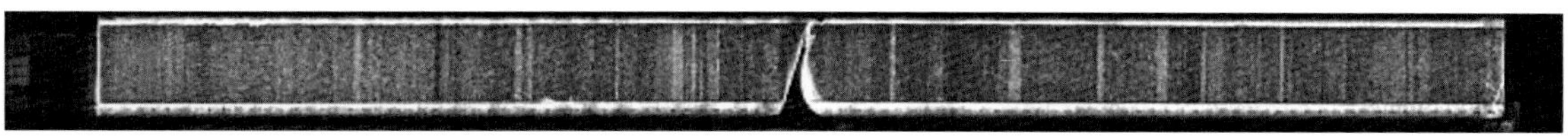
1094

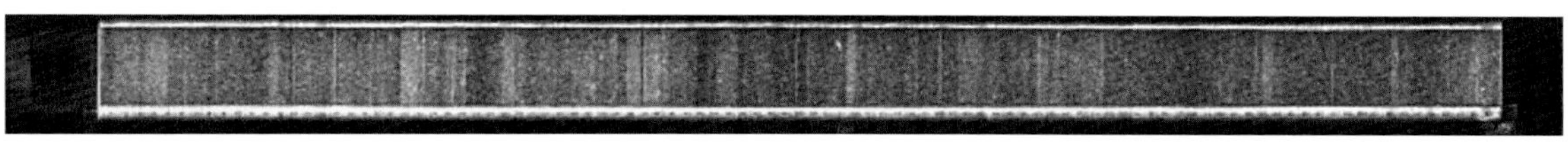
1093

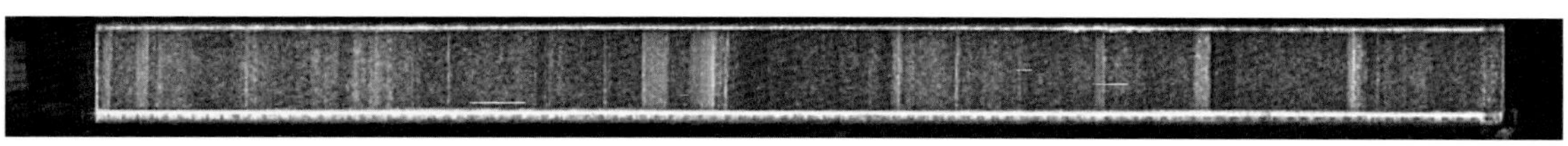
1092

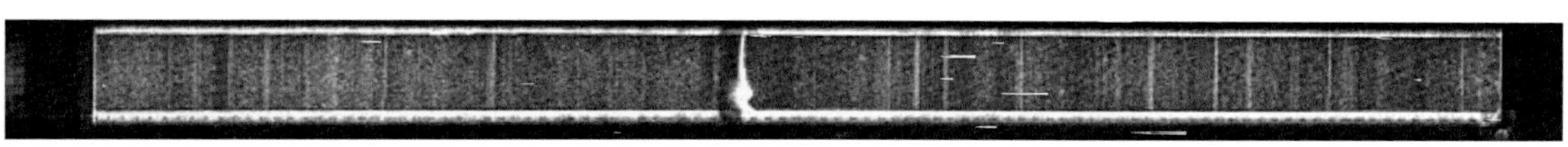
1091

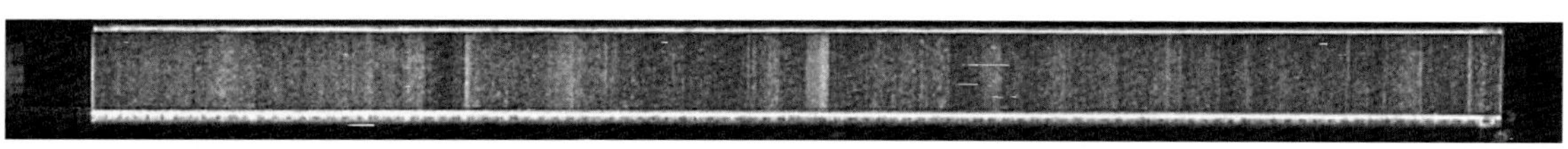
1090

1100

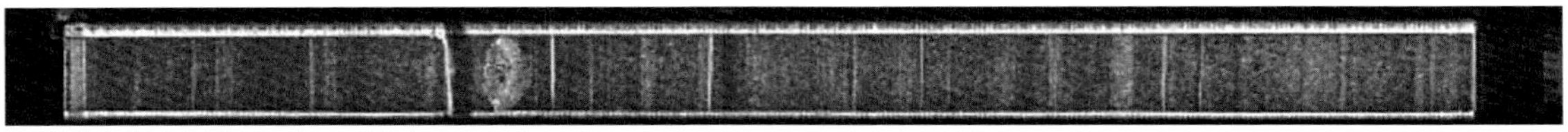

1101

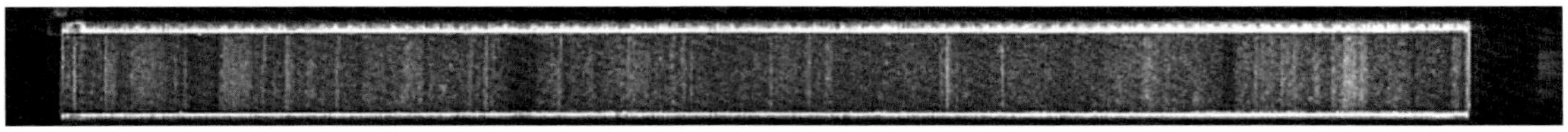

1102

1103

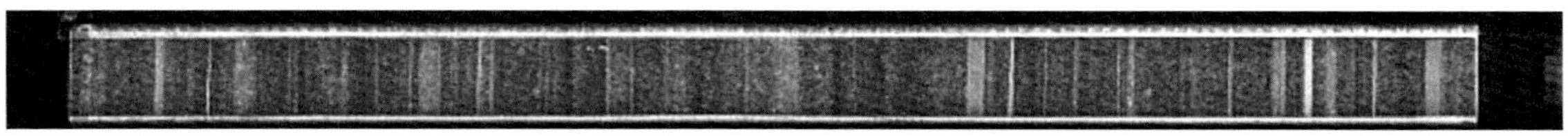

1104

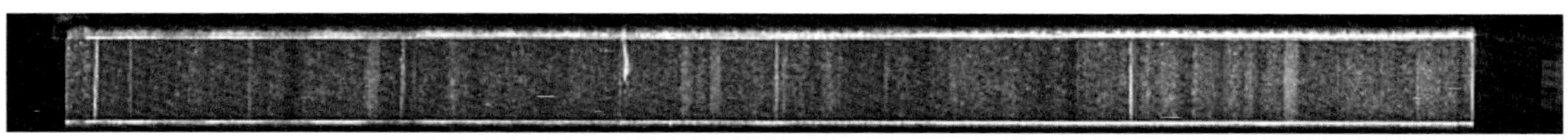

1105

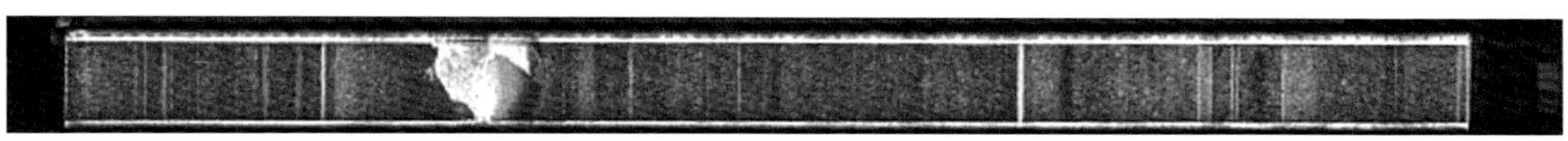

1106

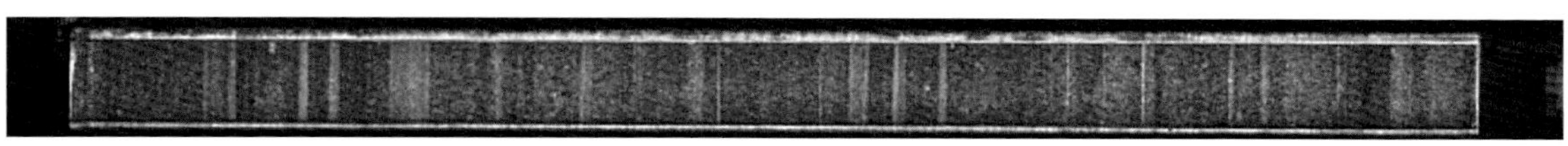

1107

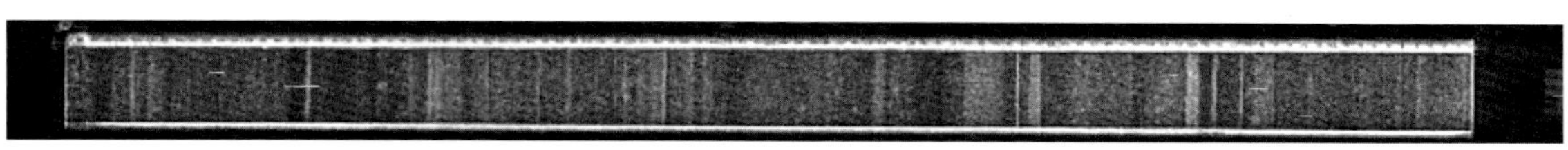

1108

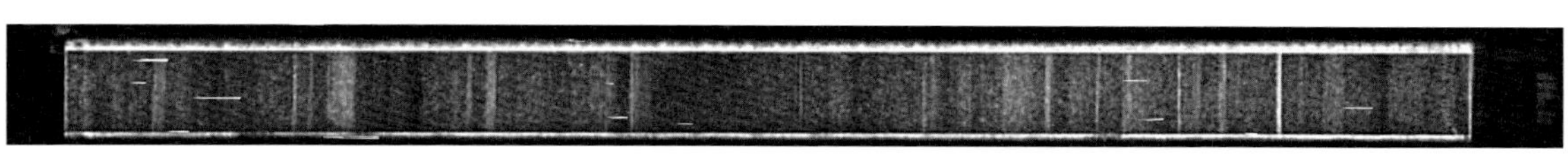

1109

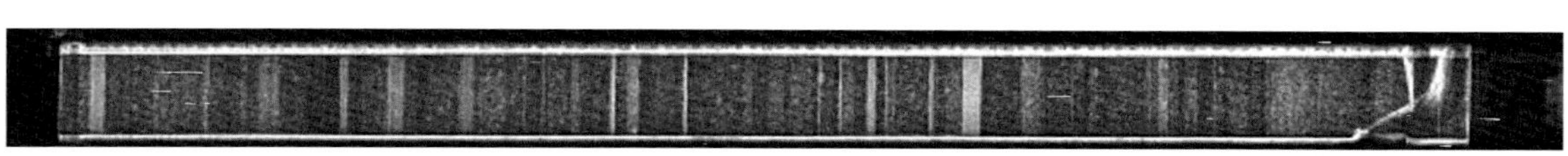

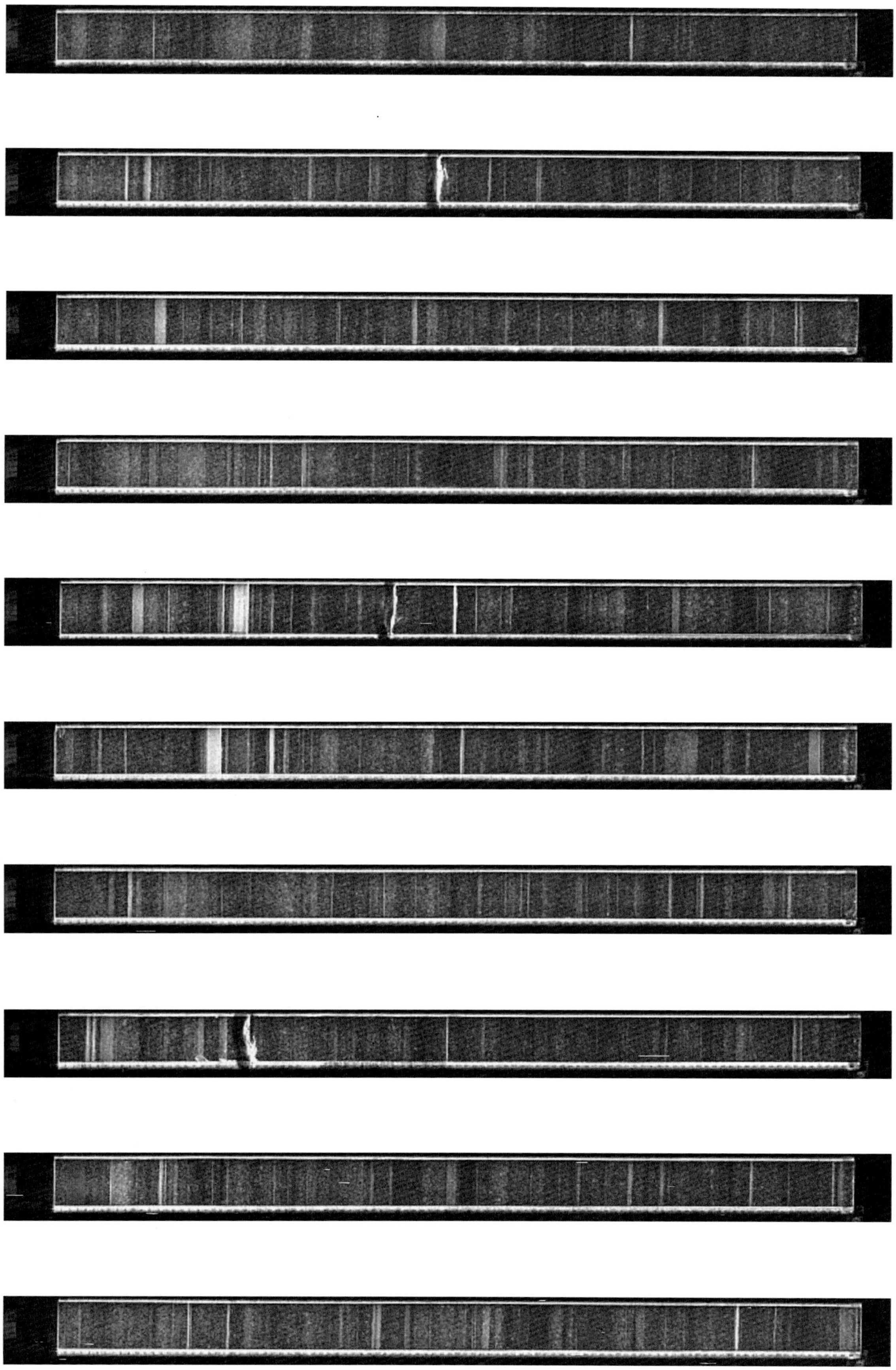

1119

1118

1117

1116

1115

1114

1113

1112

1111

1110

1120
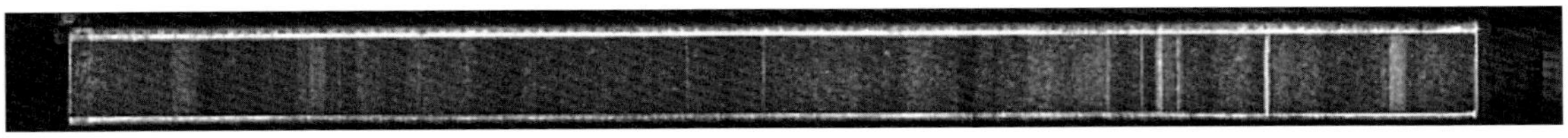

1121
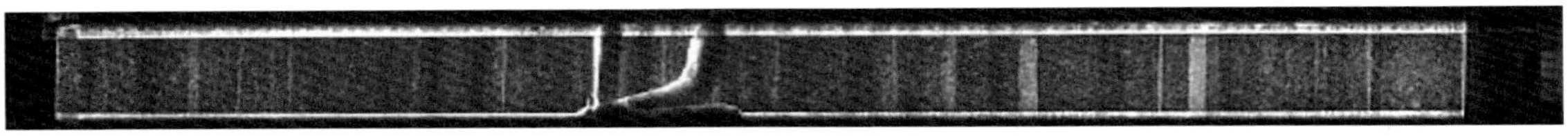

1122
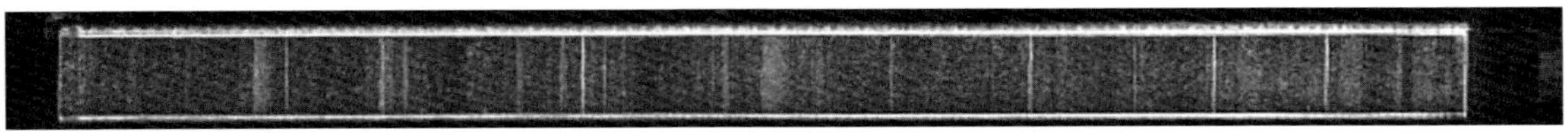

1123

1124

1125
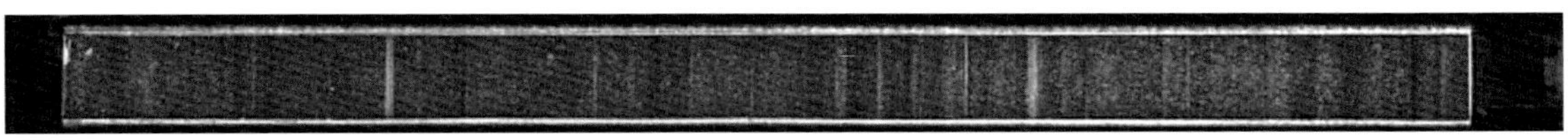

1126
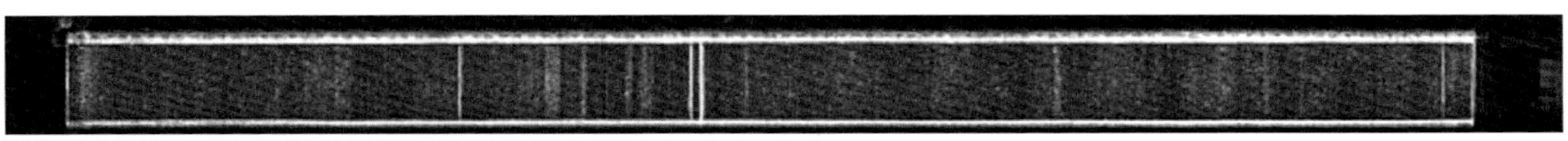

1127
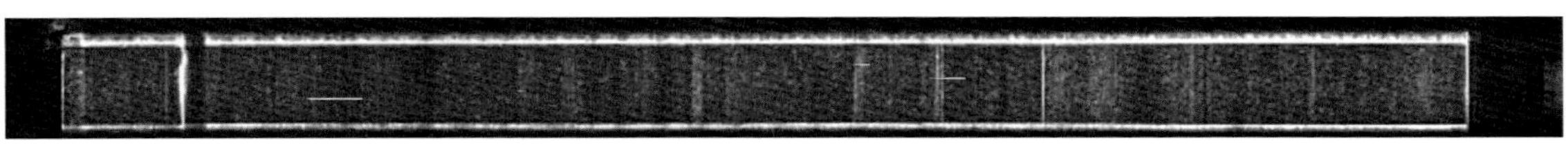

1128
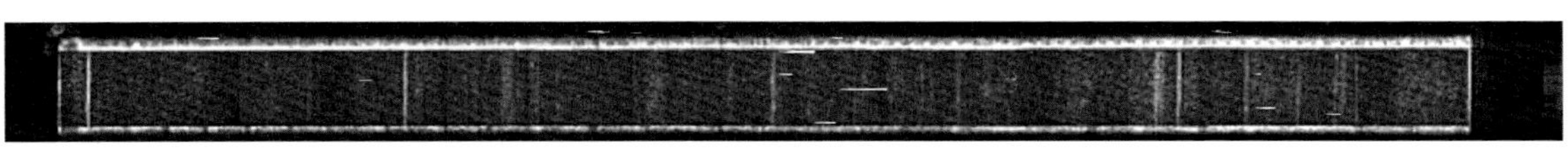

1129
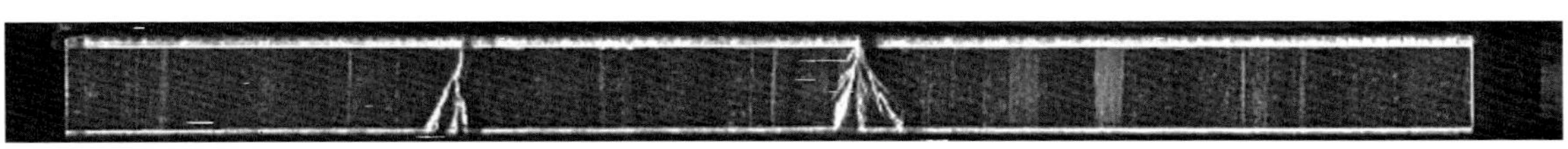

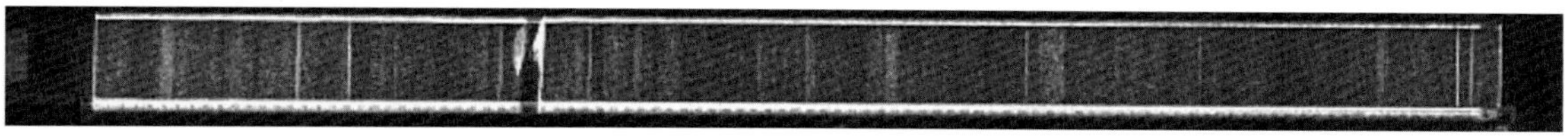
1139

1138

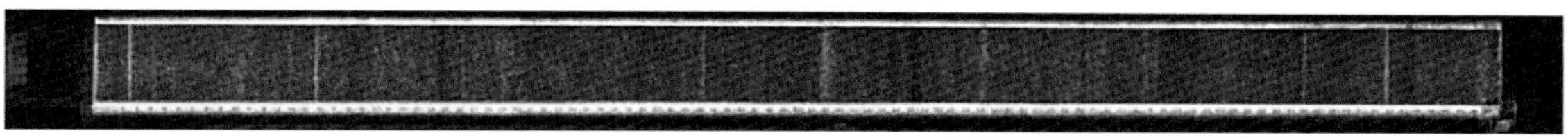
1137

1136

1135

1134

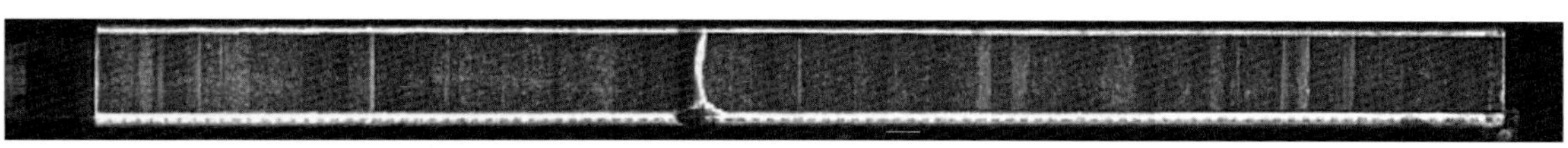
1133

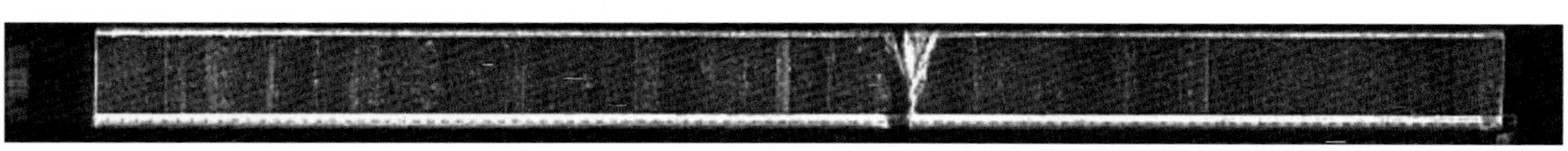
1132

1131

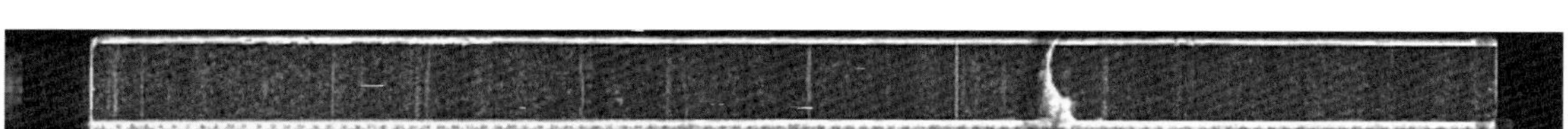
1130

1140
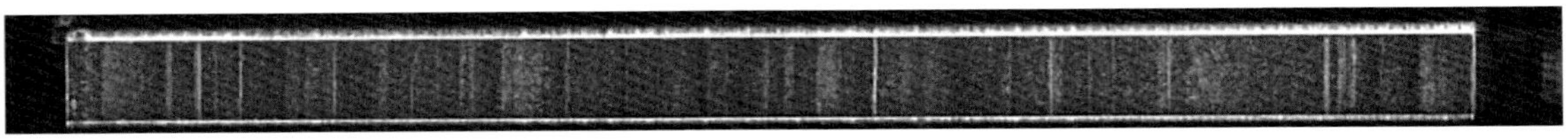

1141
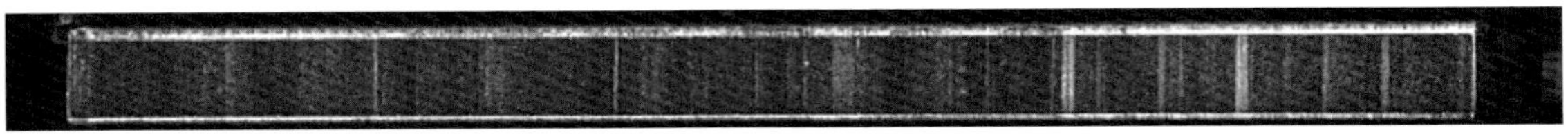

1142
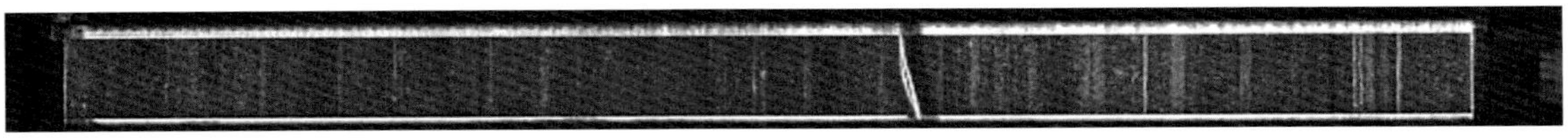

1143
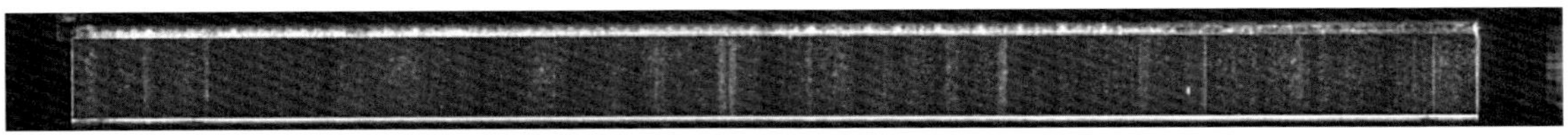

1144
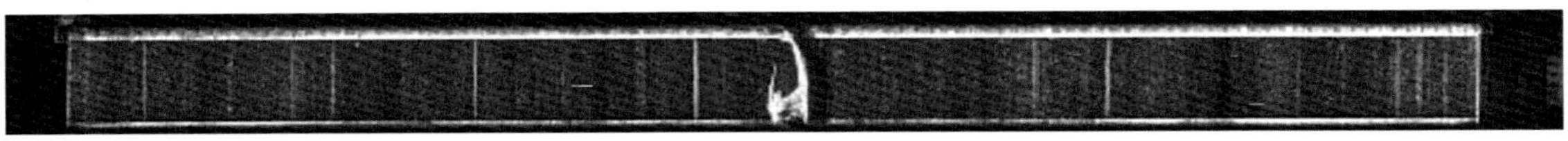

1145
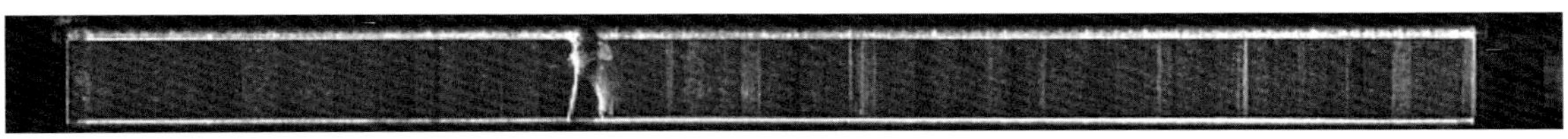

1146
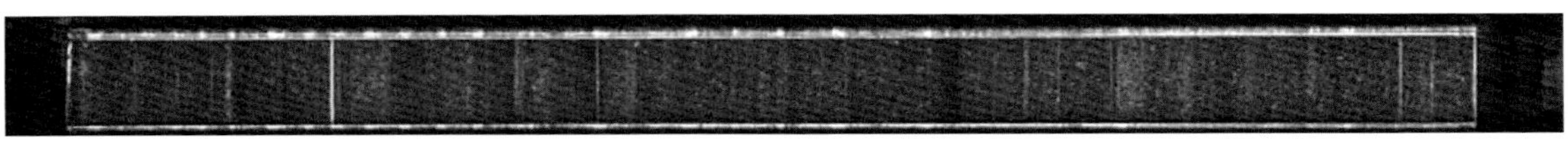

1147
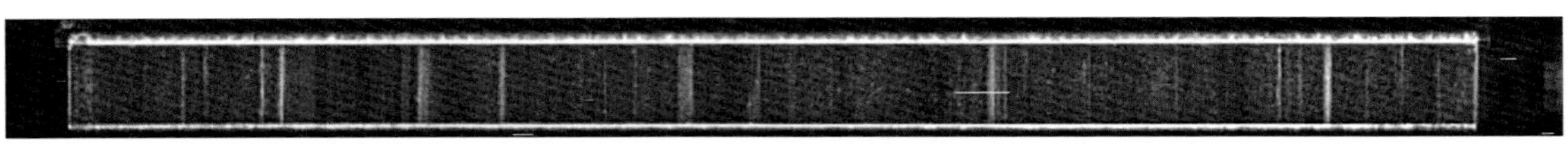

1148
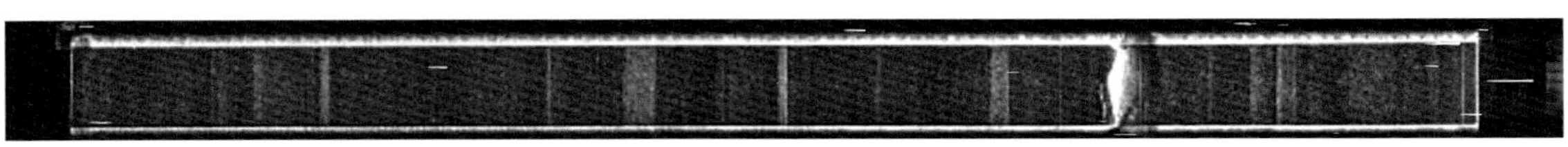

1149
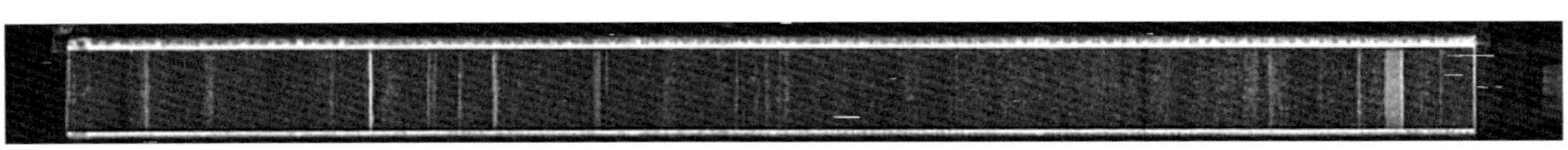

1159

1158

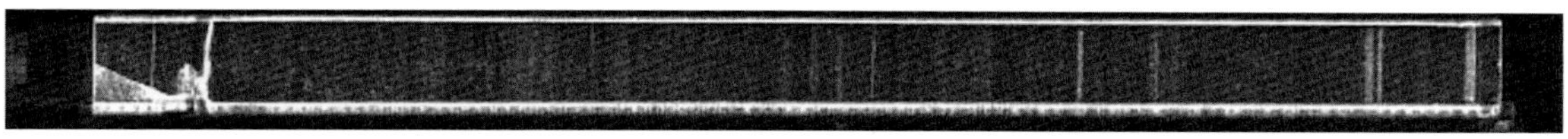
1157

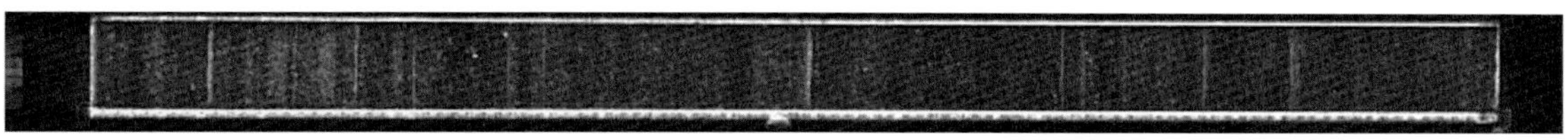
1156

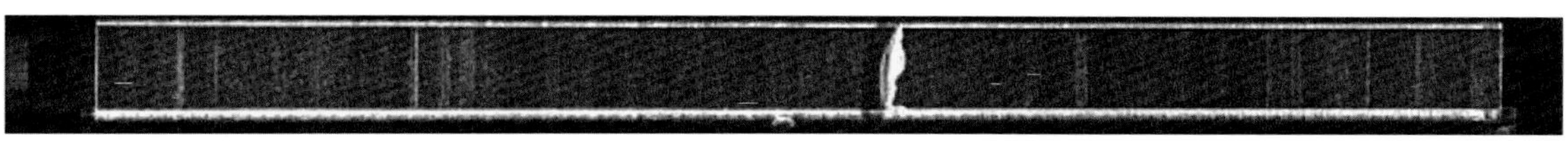
1155

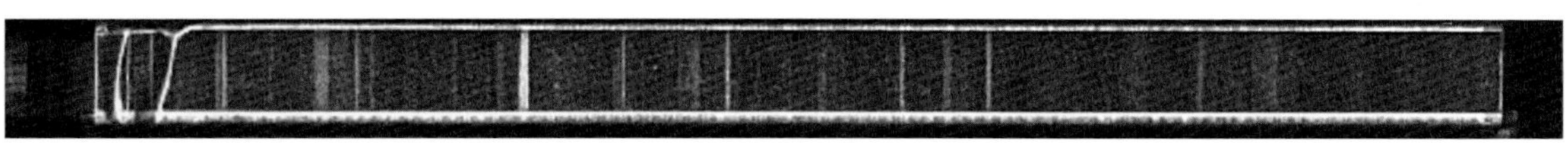
1154

1153

1152

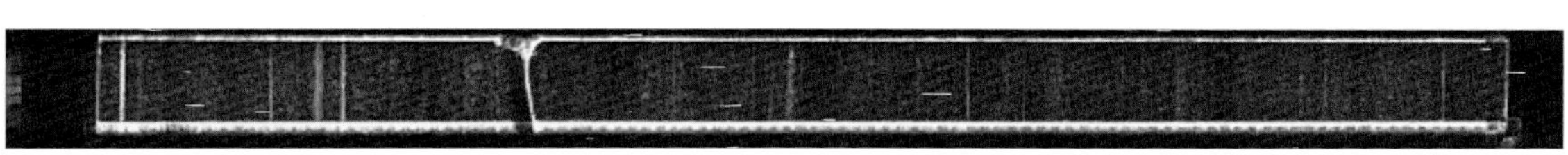
1151

1150

1160

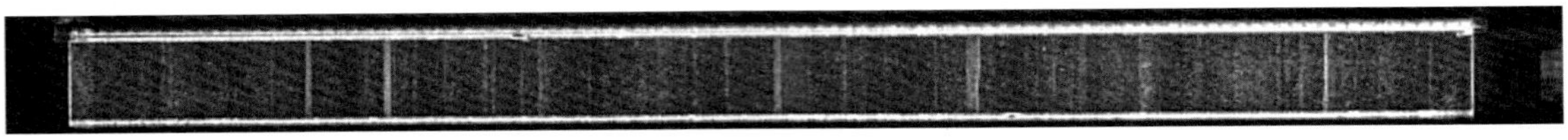

1161

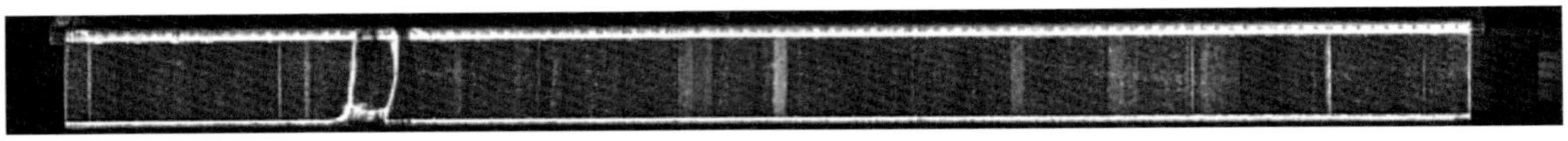

1162

1163

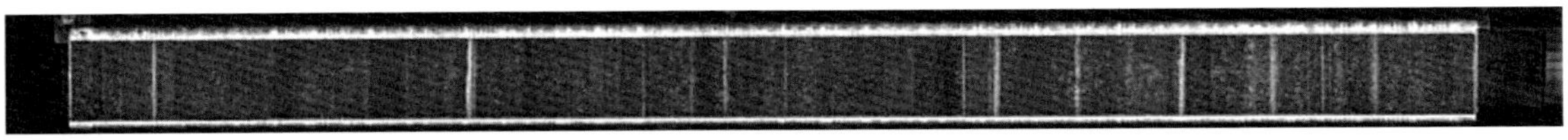

1164

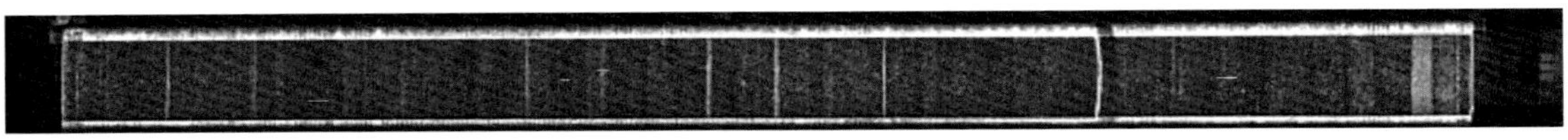

1165

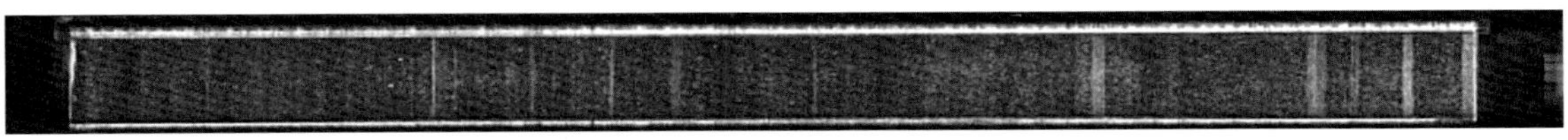

1166

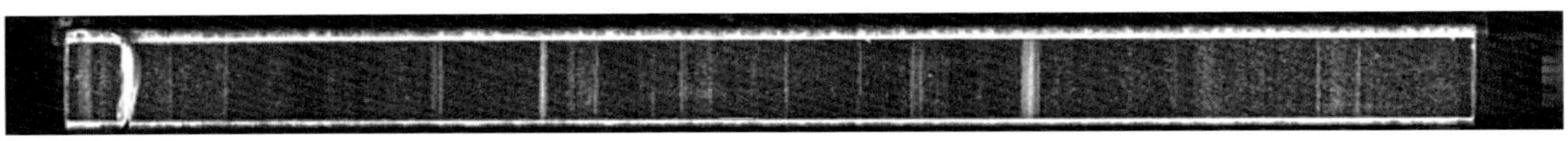

1167

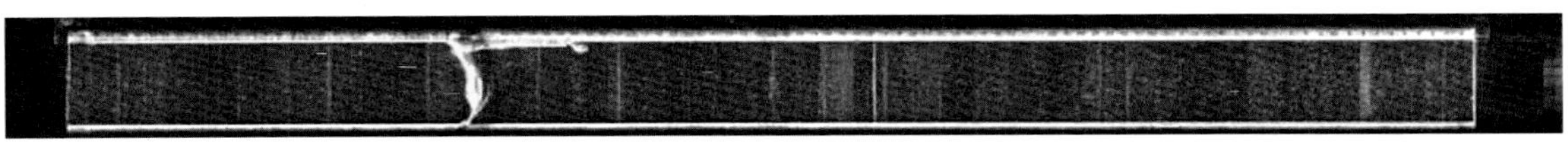

1168

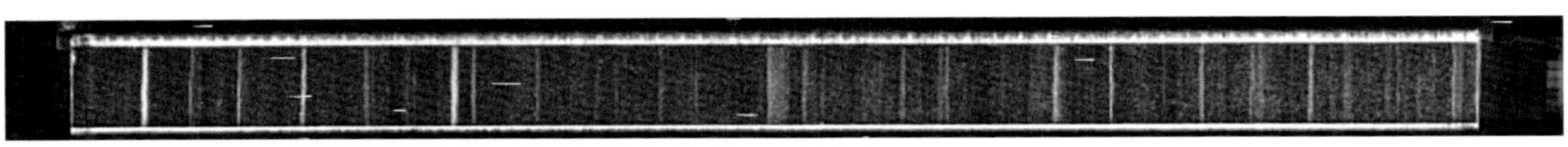

1169

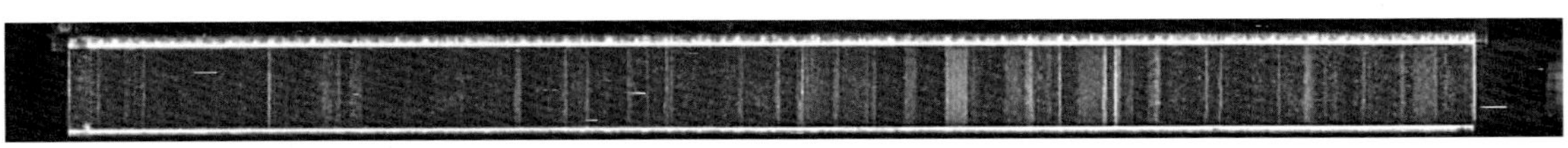

1179

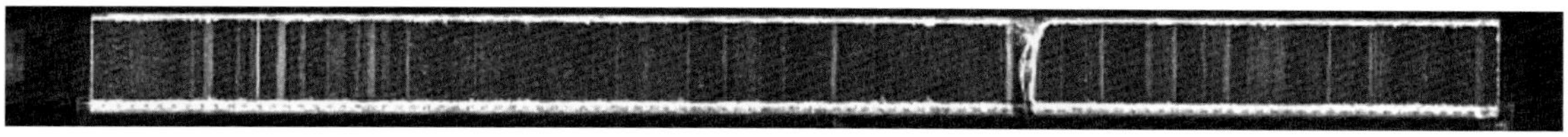
1178

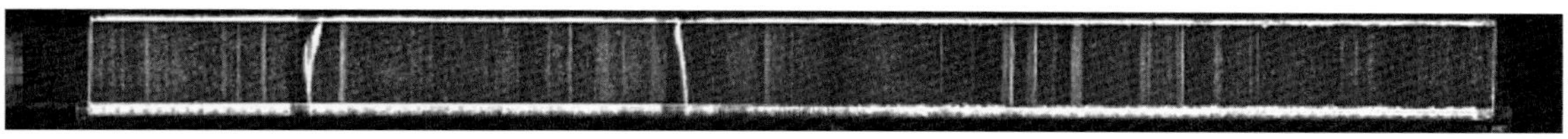
1177

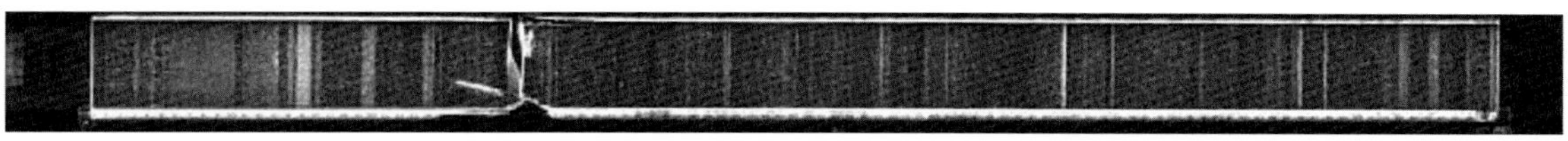
1176

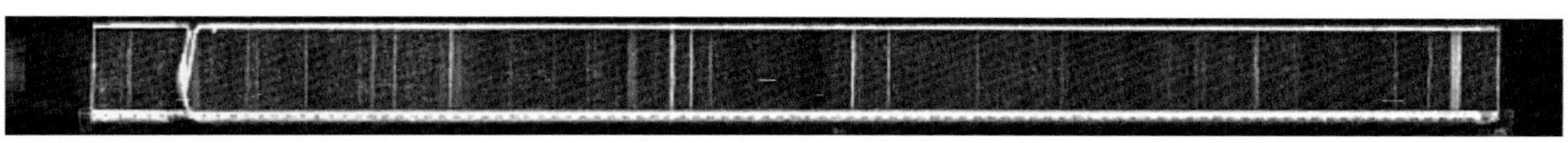
1175

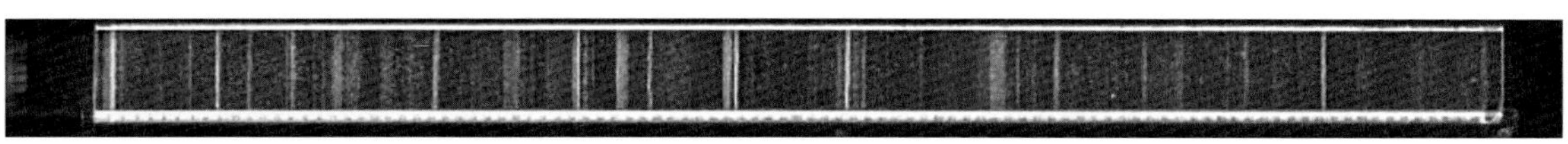
1174

1173

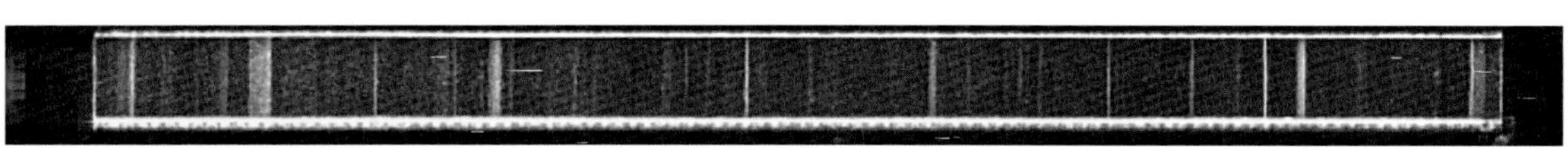
1172

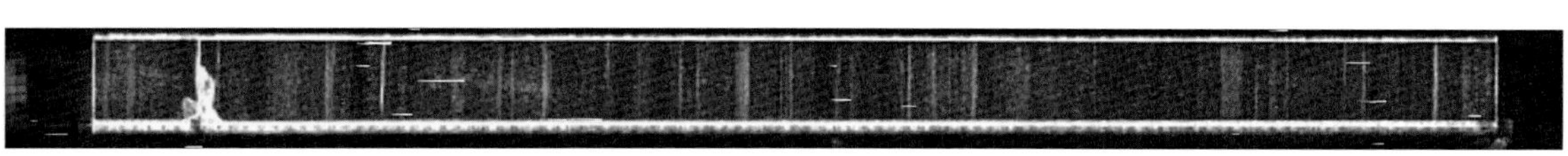
1171

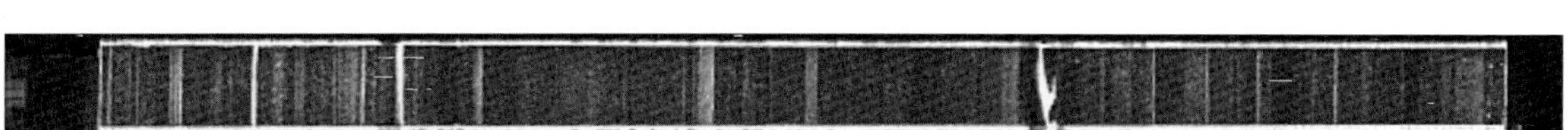
1170

1180

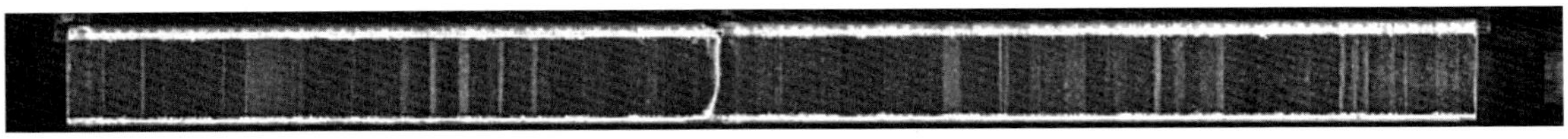

1181

1182

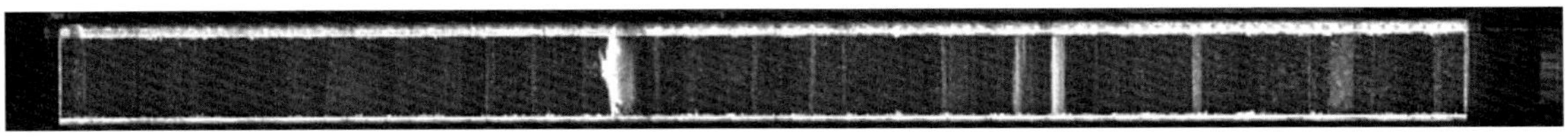

1183

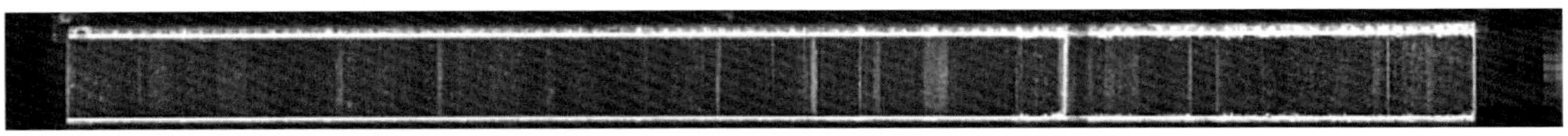

1184

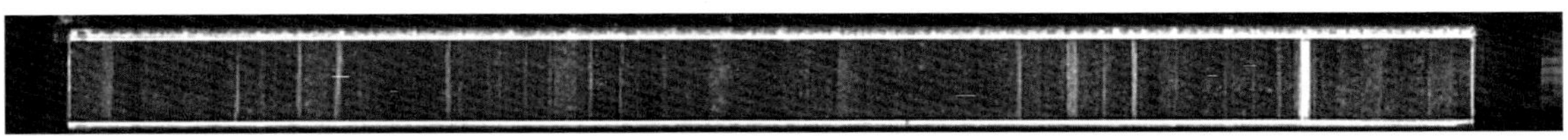

1185

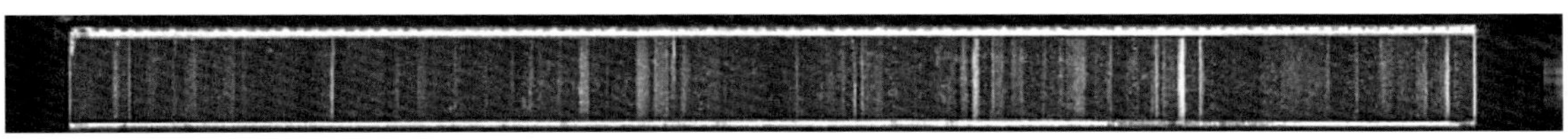

1186

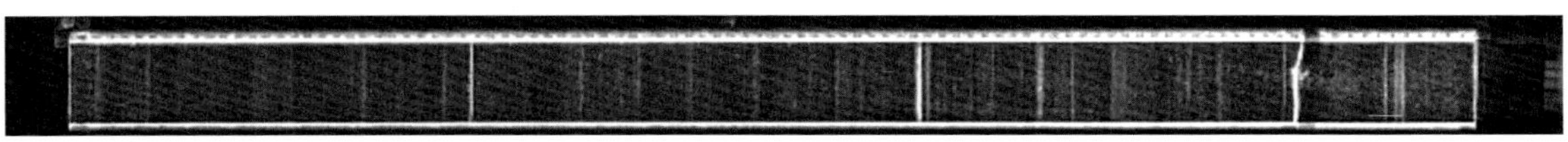

1187

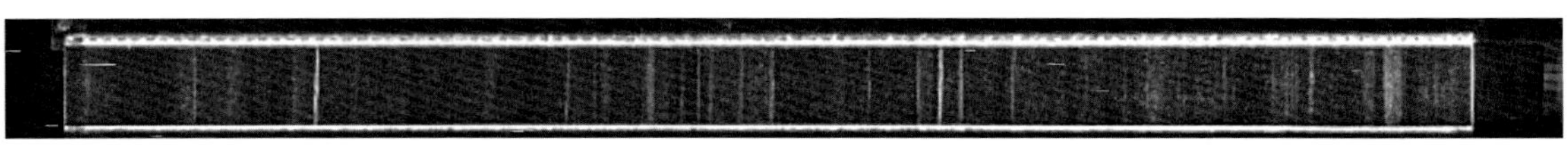

1188

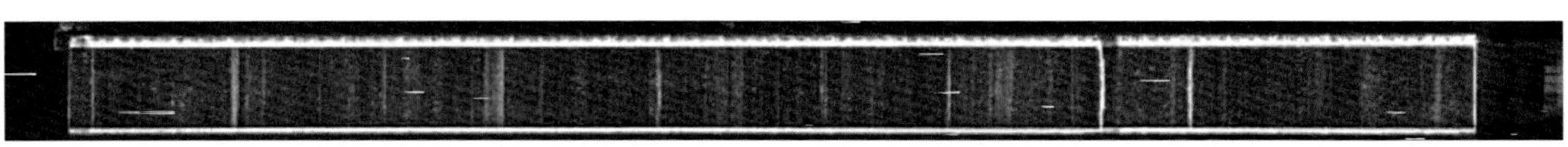

1189

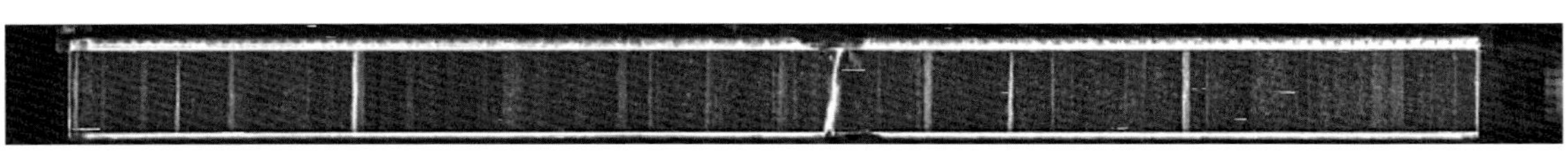

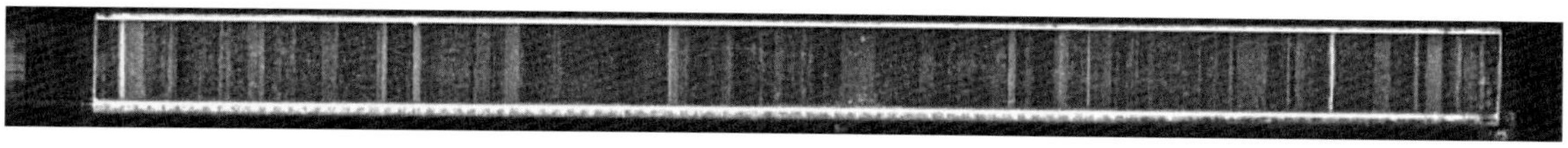
1199

1198

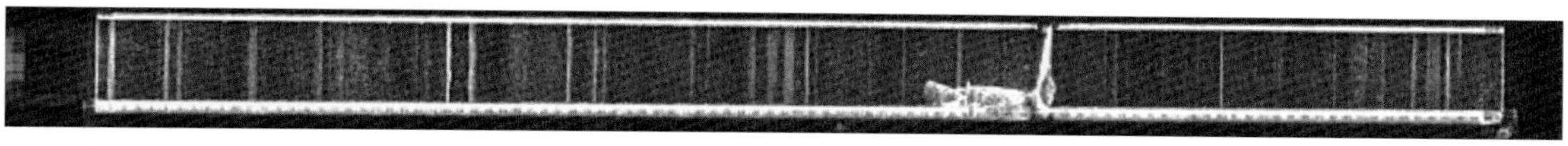
1197

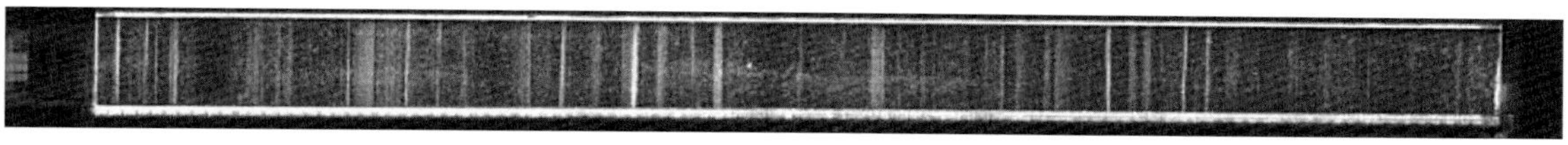
1196

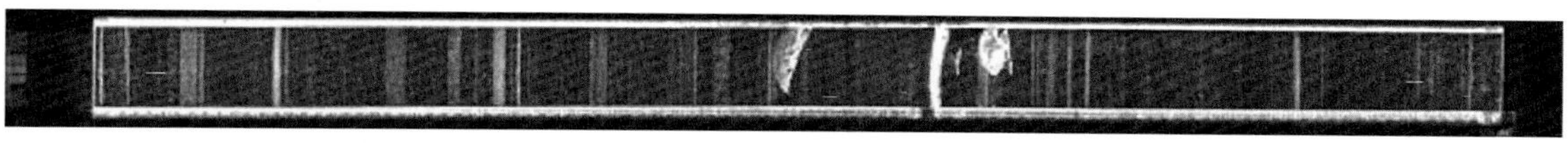
1195

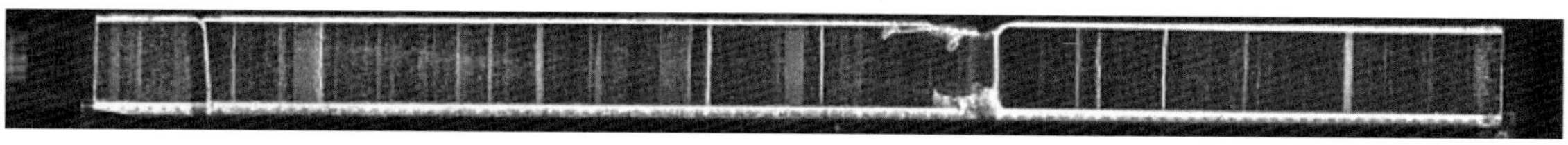
1194

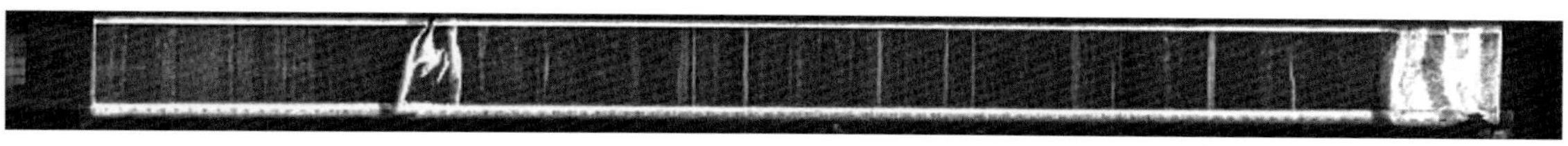
1193

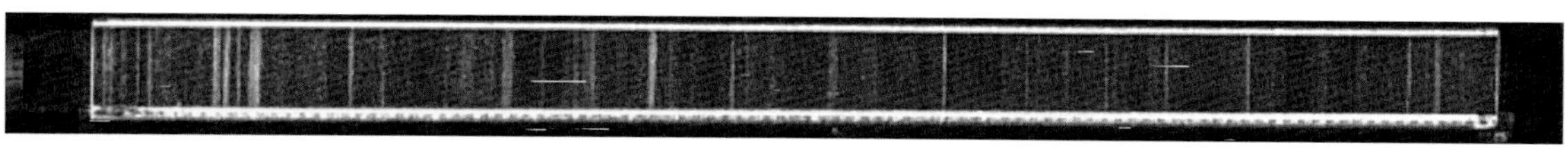
1192

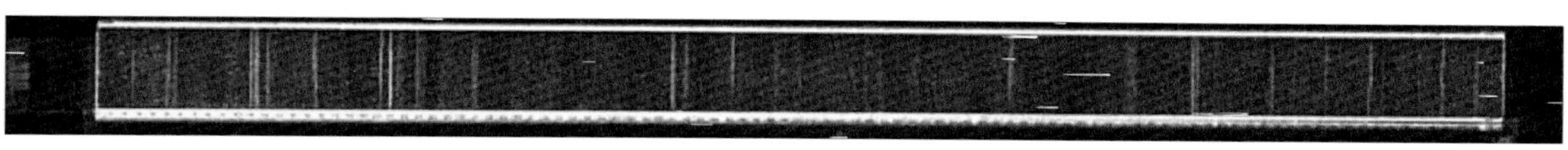
1191

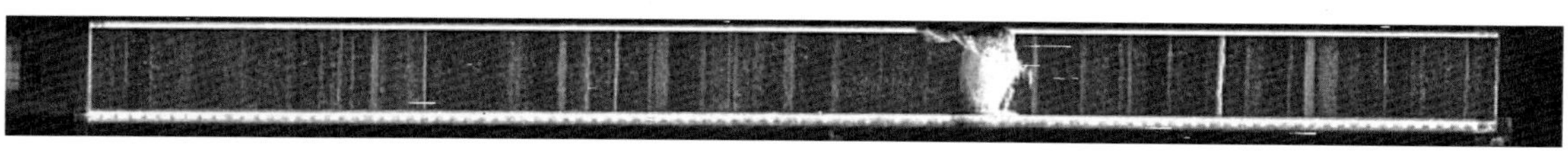
1190

1200

1201
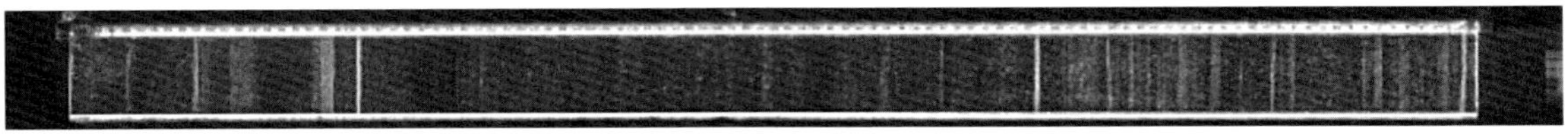

1202

1203
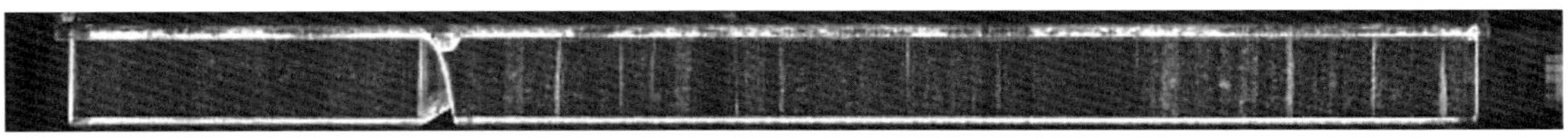

1204
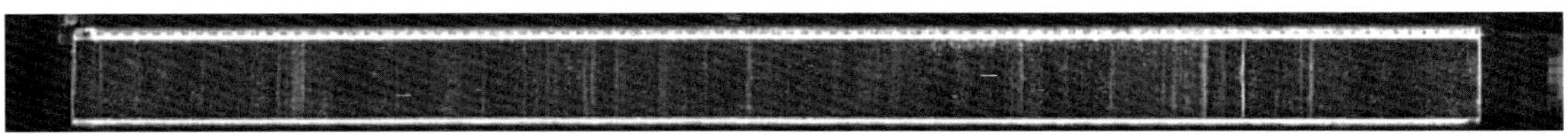

1205

1206
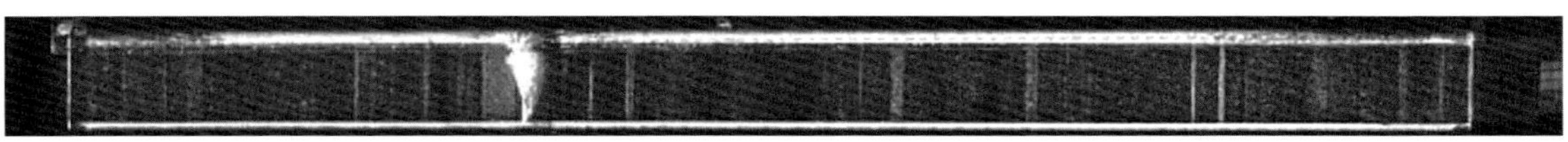

1207
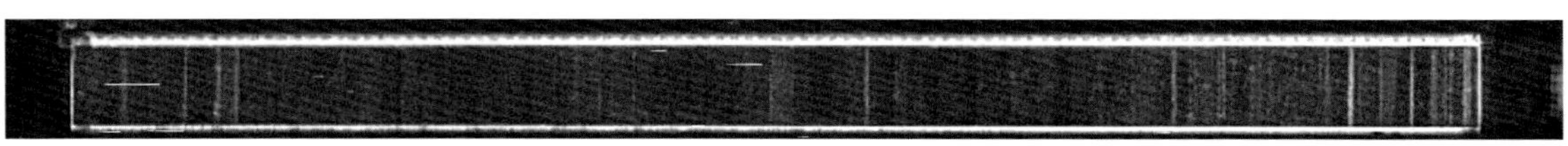

1208
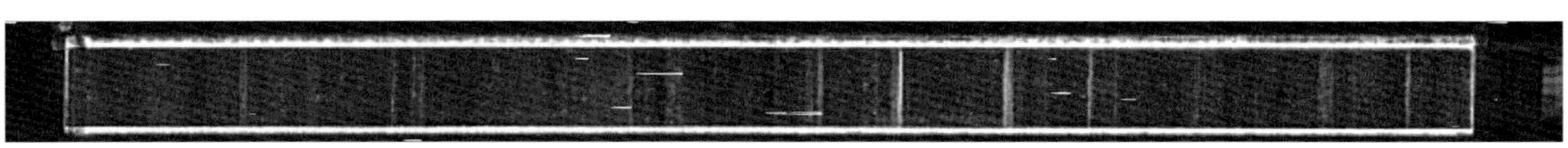

1209
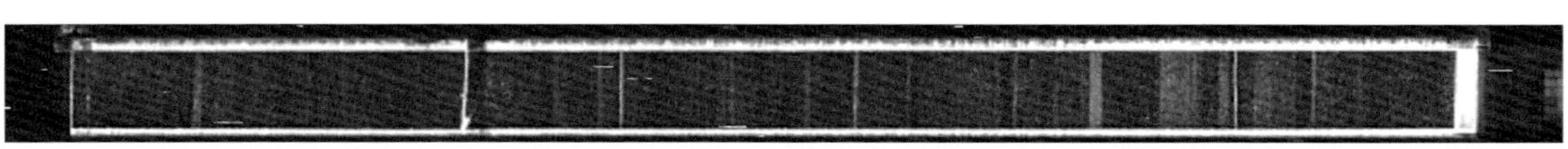

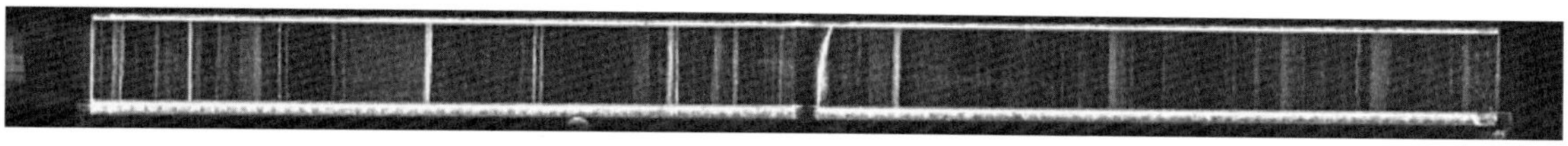
1219

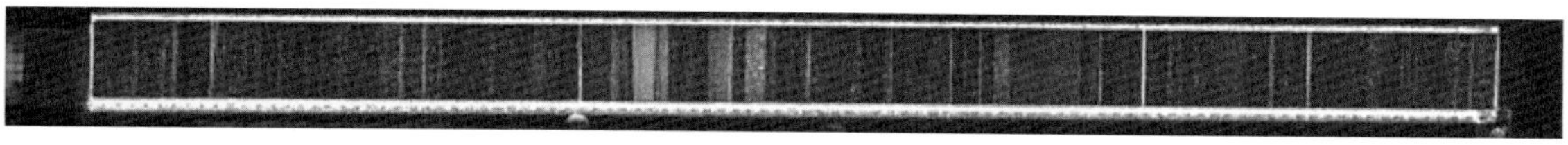
1218

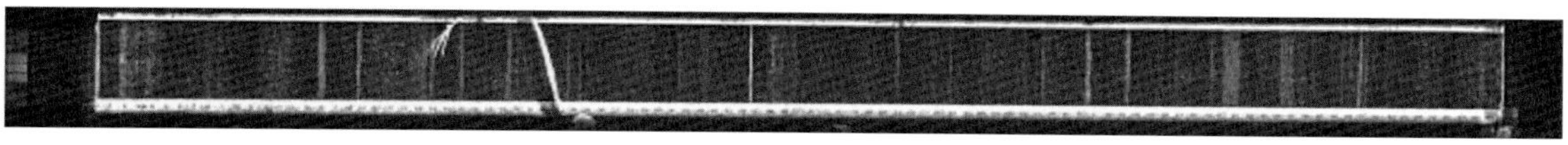
1217

1216

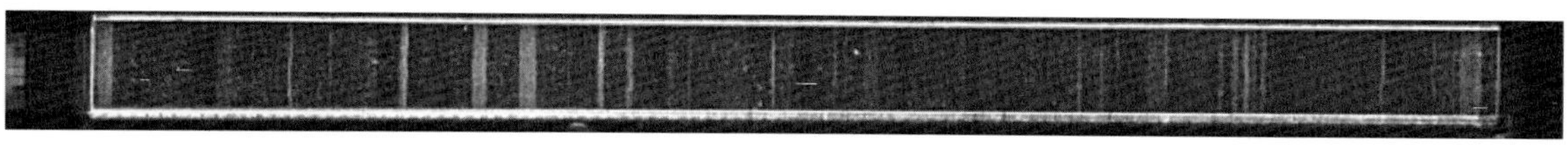
1215

1214

1213

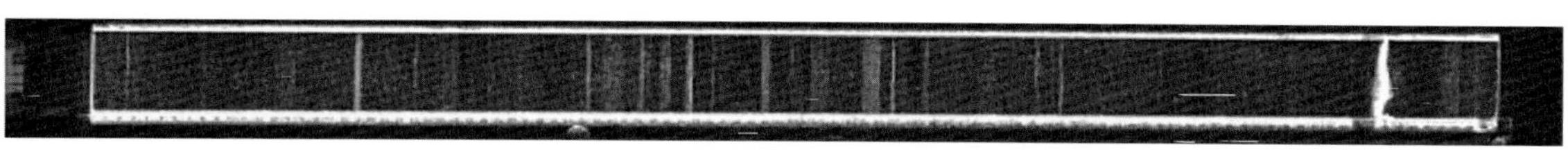
1212

1211

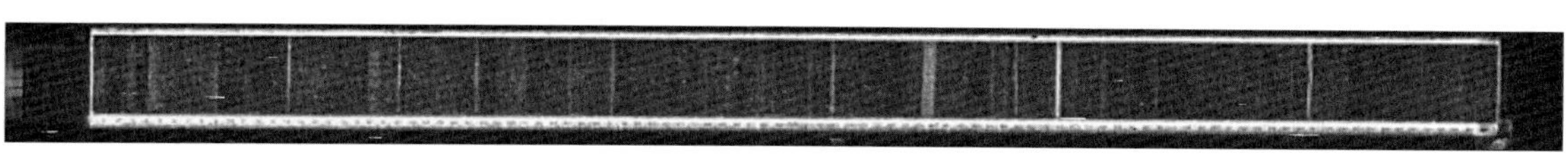
1210

1220

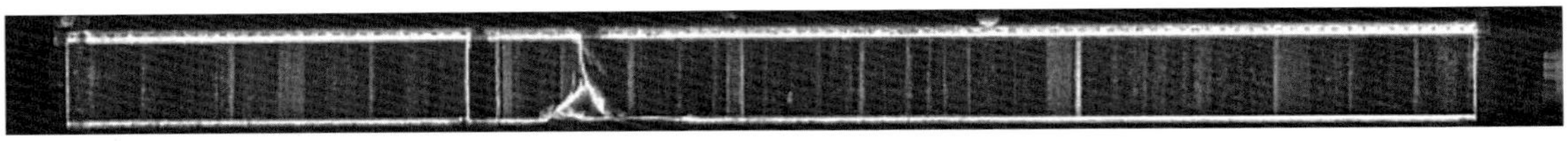

1221

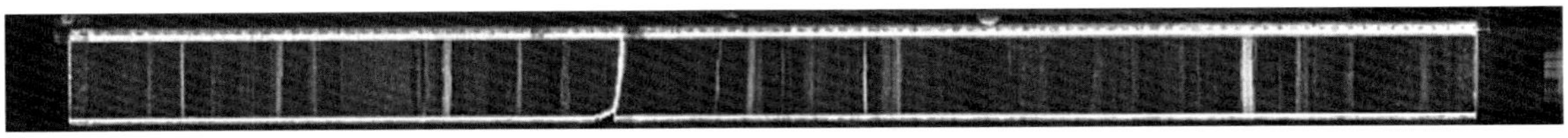

1222

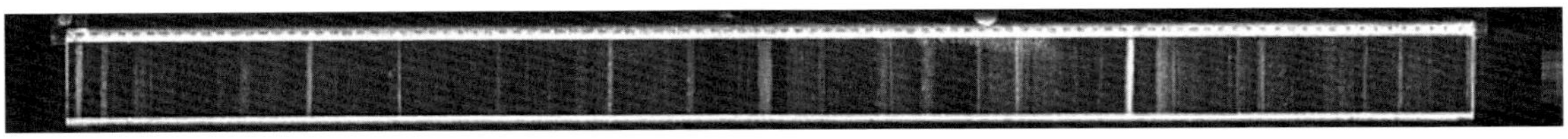

1223

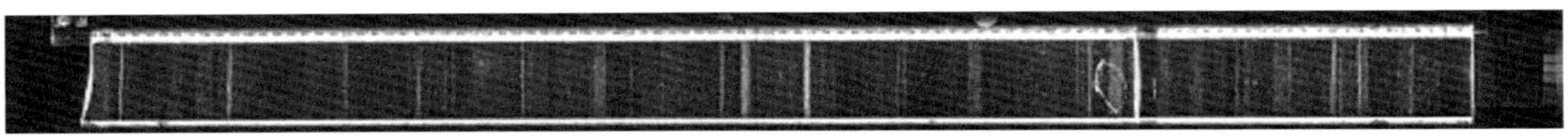

1224

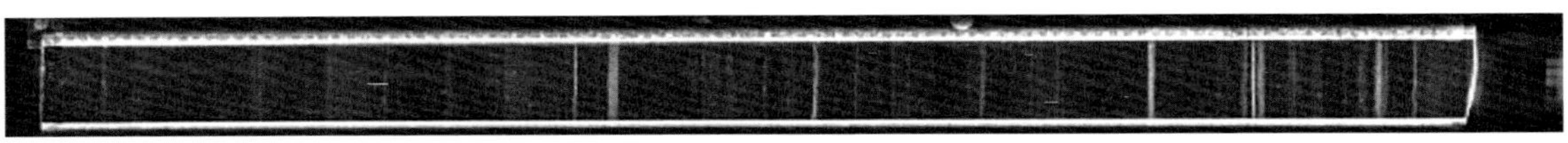

1225

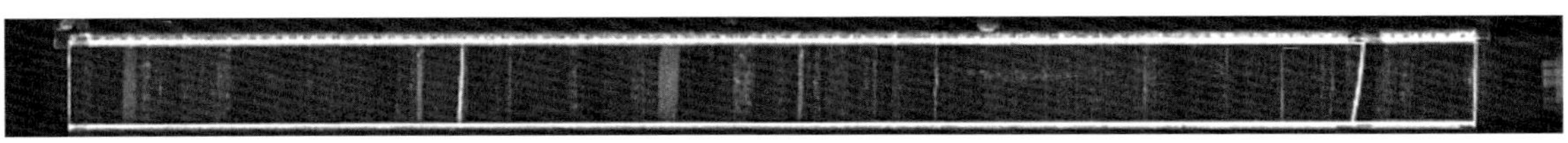

1226

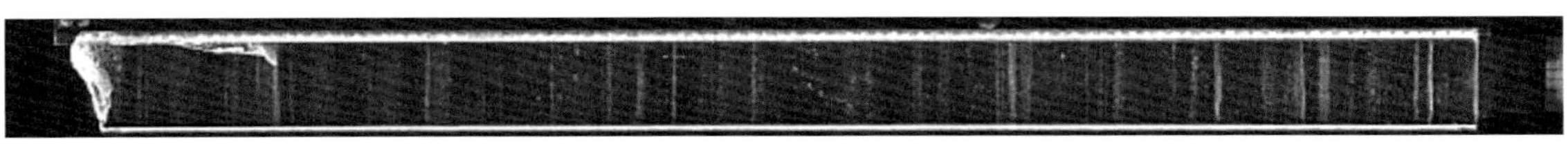

1227

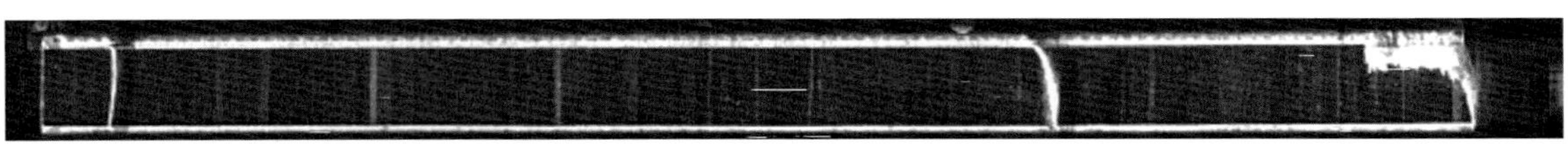

1228

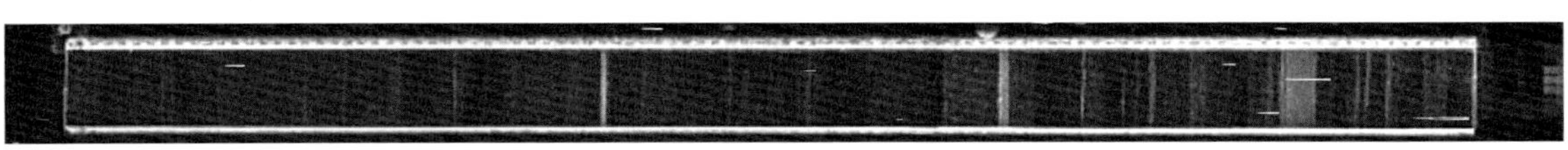

1229

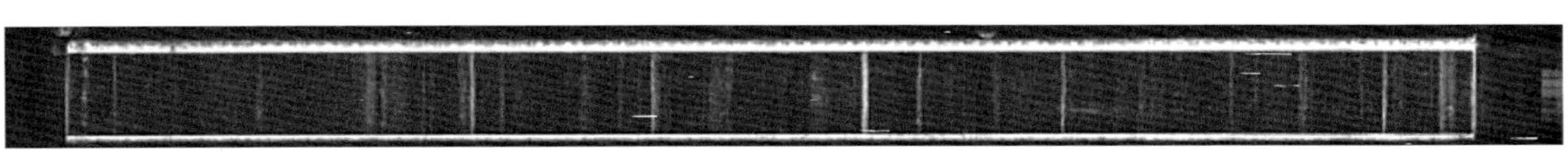

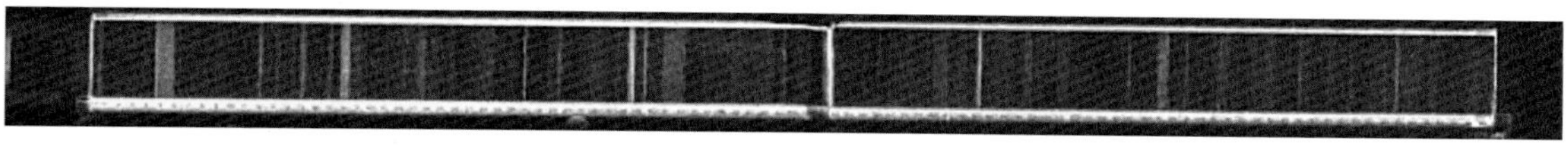
1239

1238

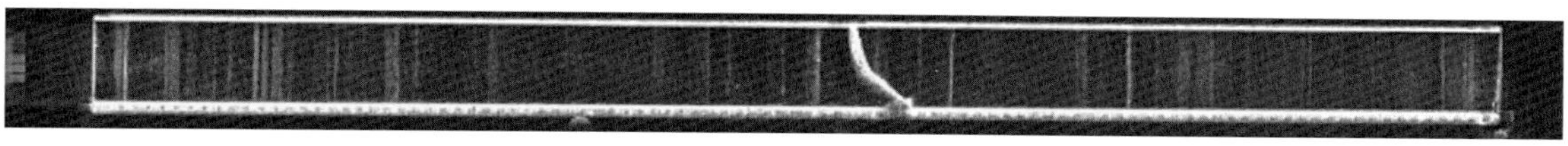
1237

1236

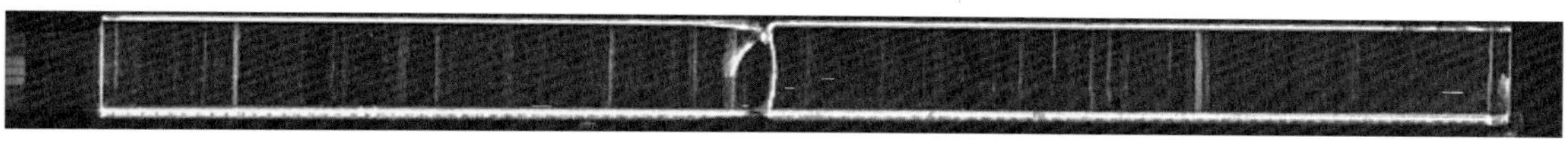
1235

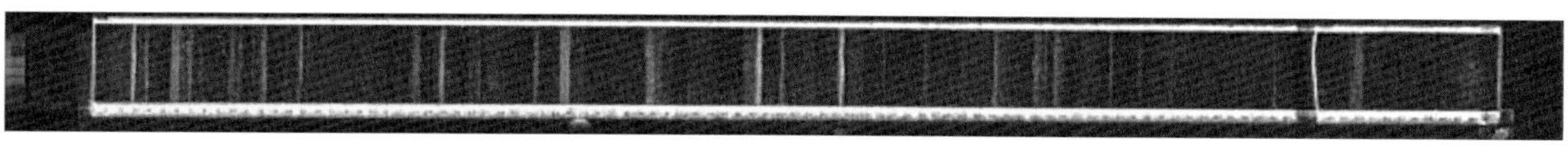
1234

1233

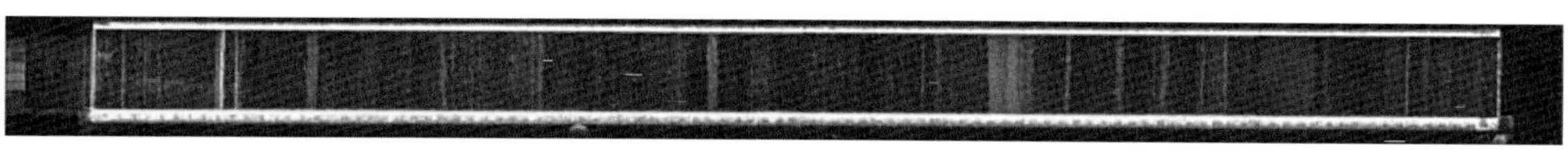
1232

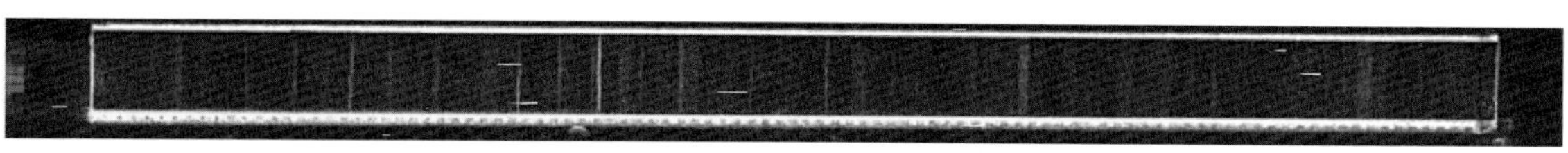
1231

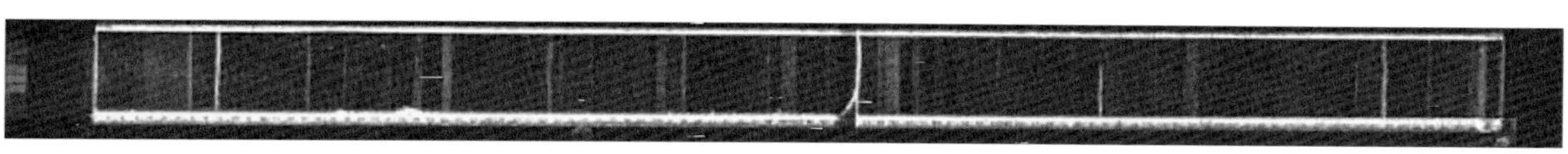
1230

1240

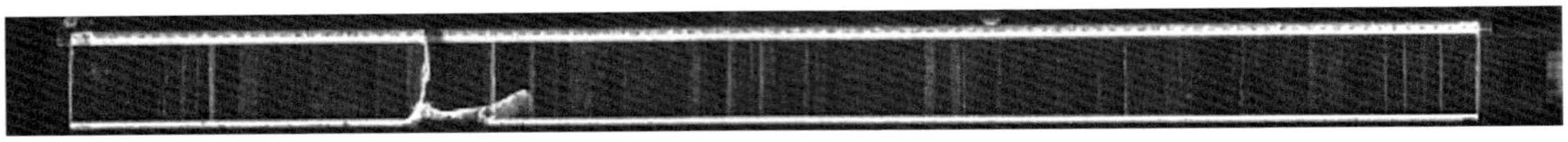

1241

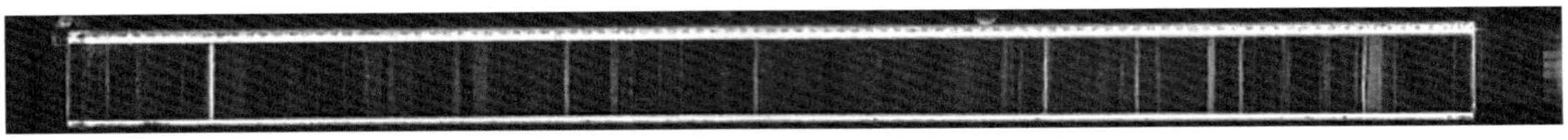

1242

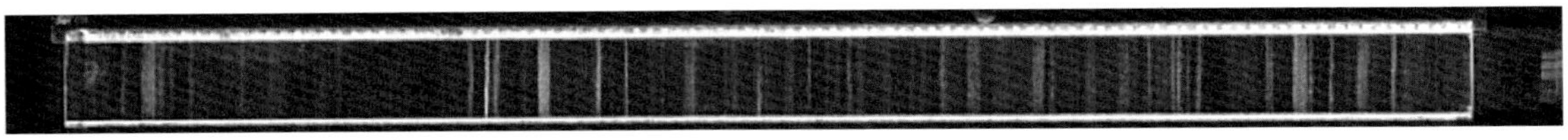

1243

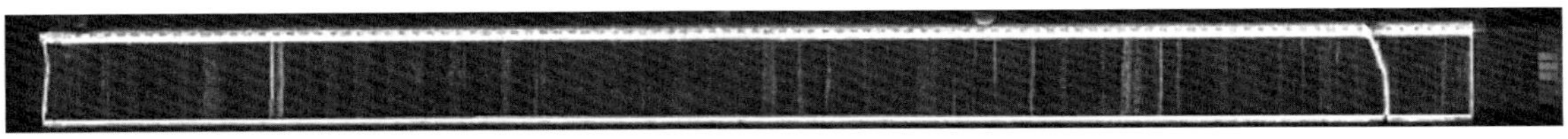

1244

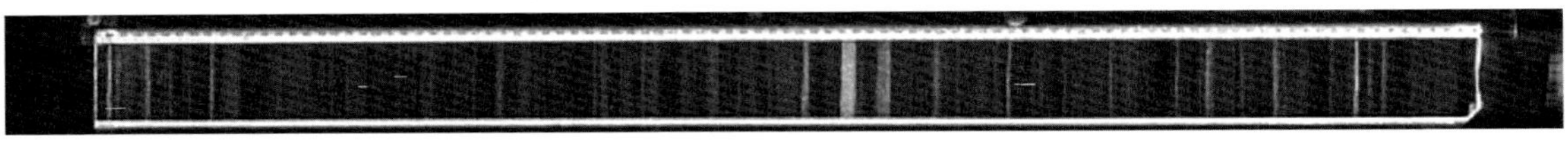

1245

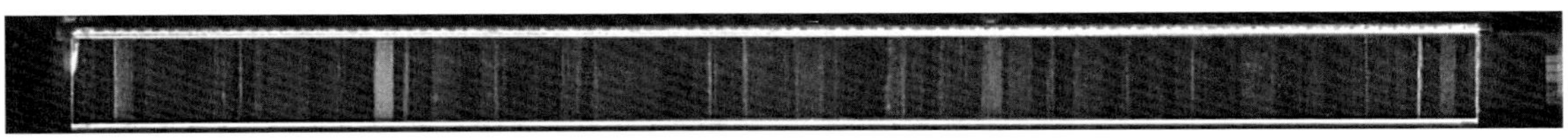

1246

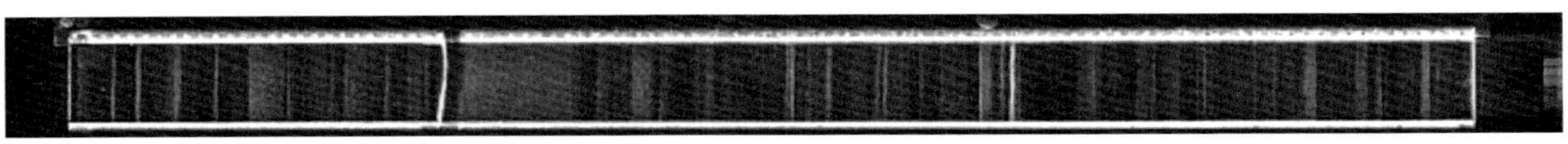

1247

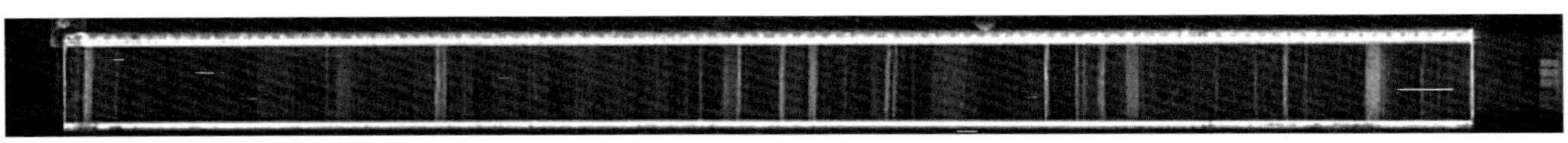

1248

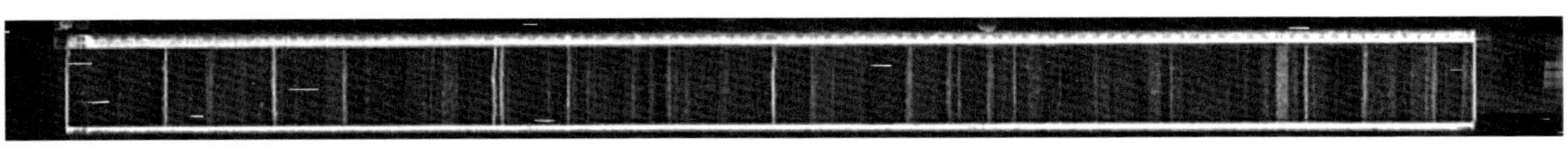

1249

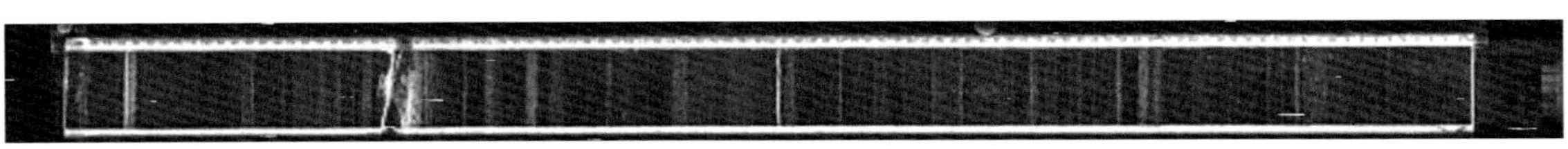

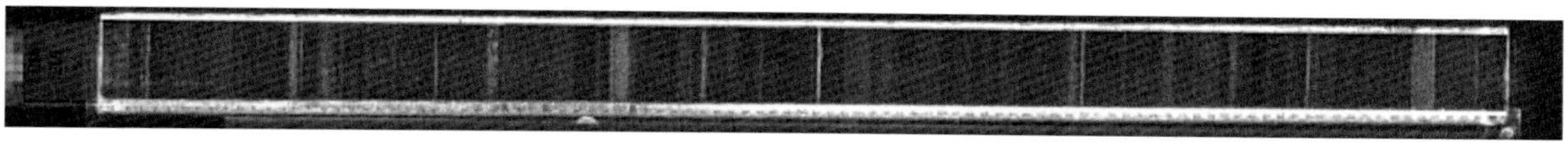
1259

1258

1257

1256

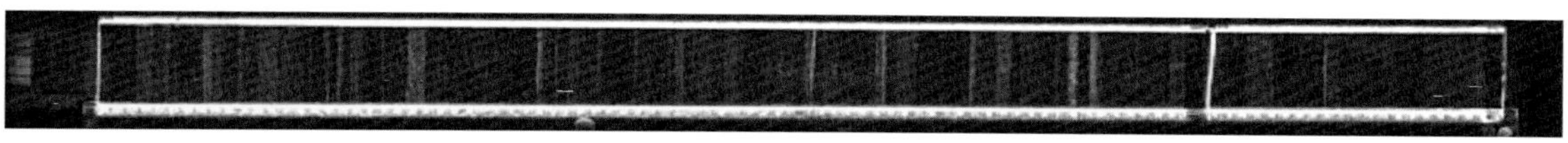
1255

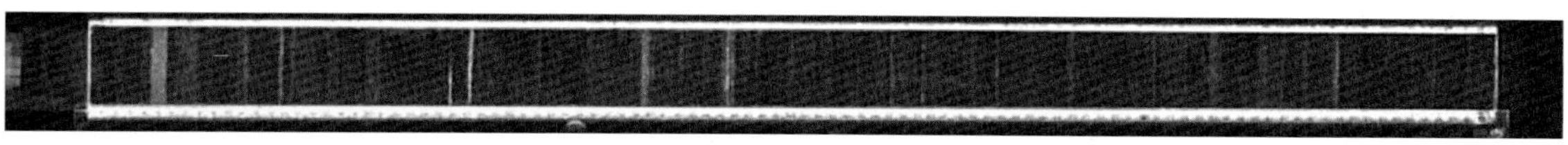
1254

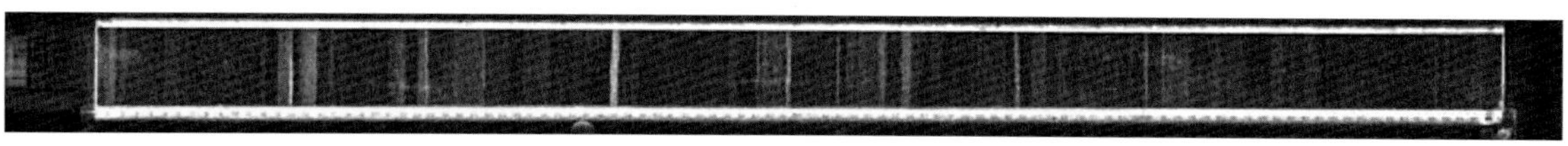
1253

1252

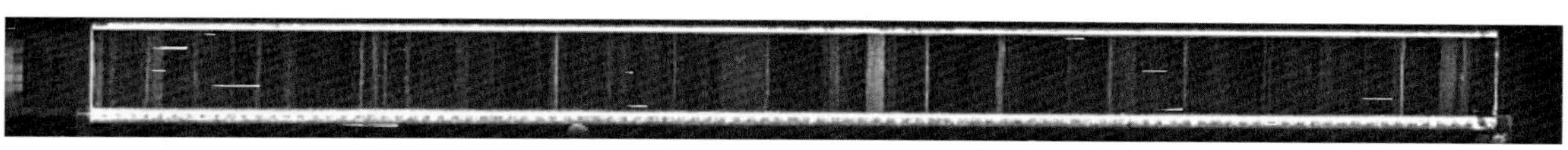
1251

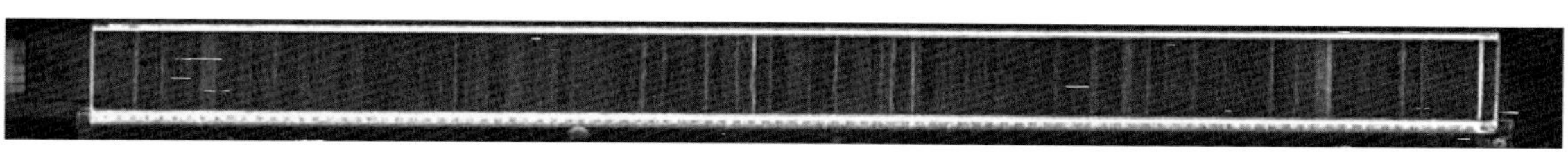
1250

1260

1261

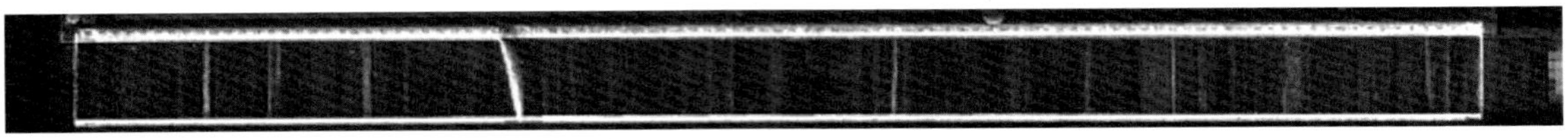

1262

1263

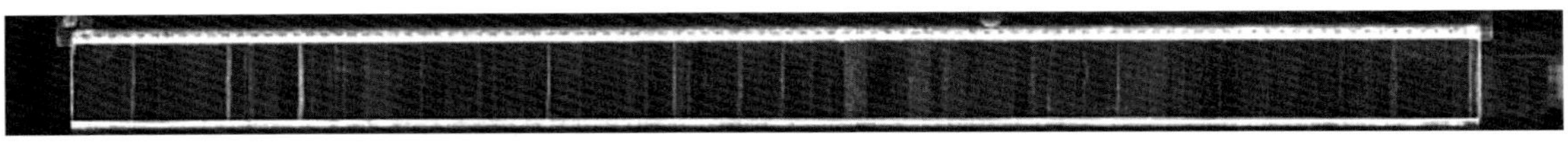

1264

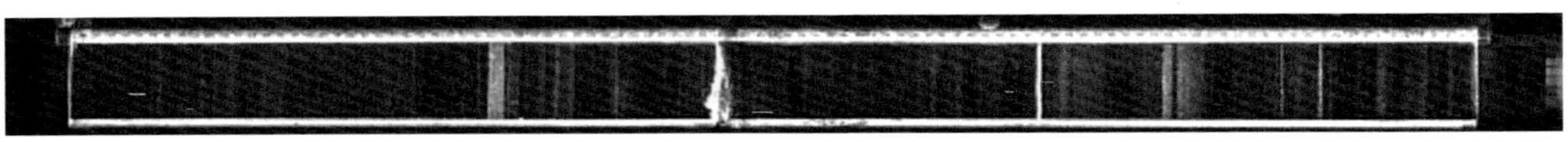

1265

1266

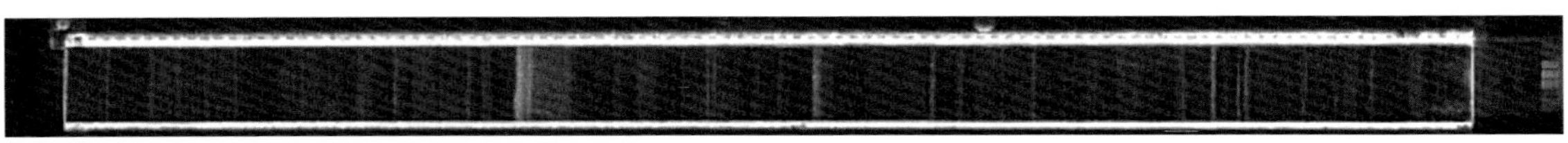

1267

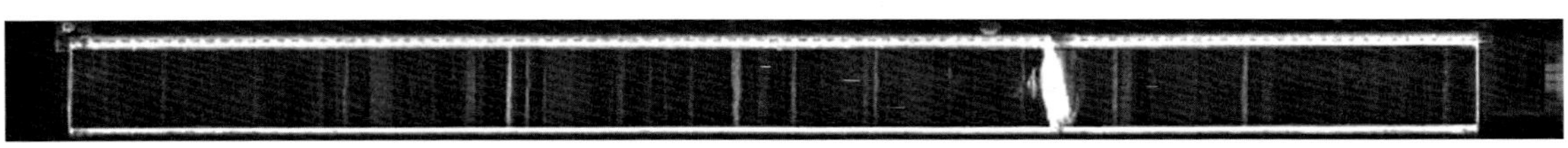

1268

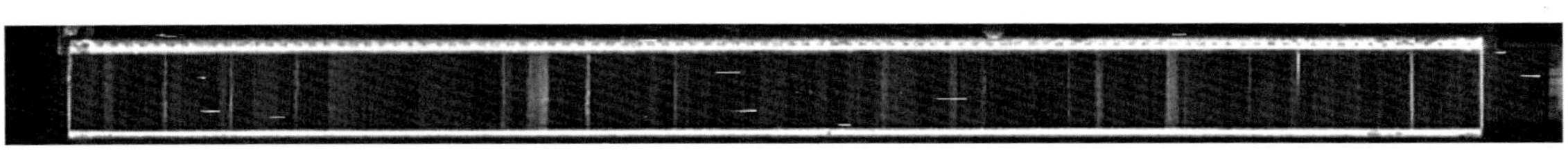

1269

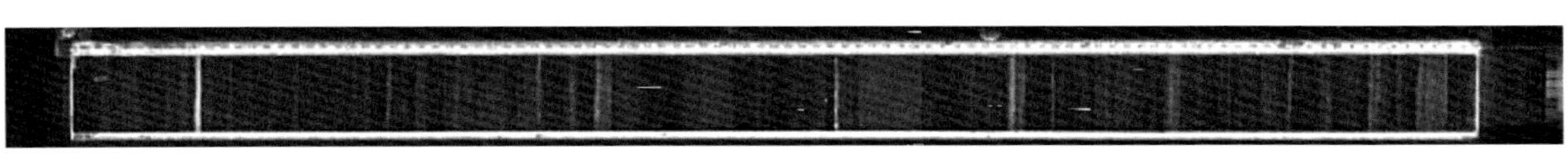

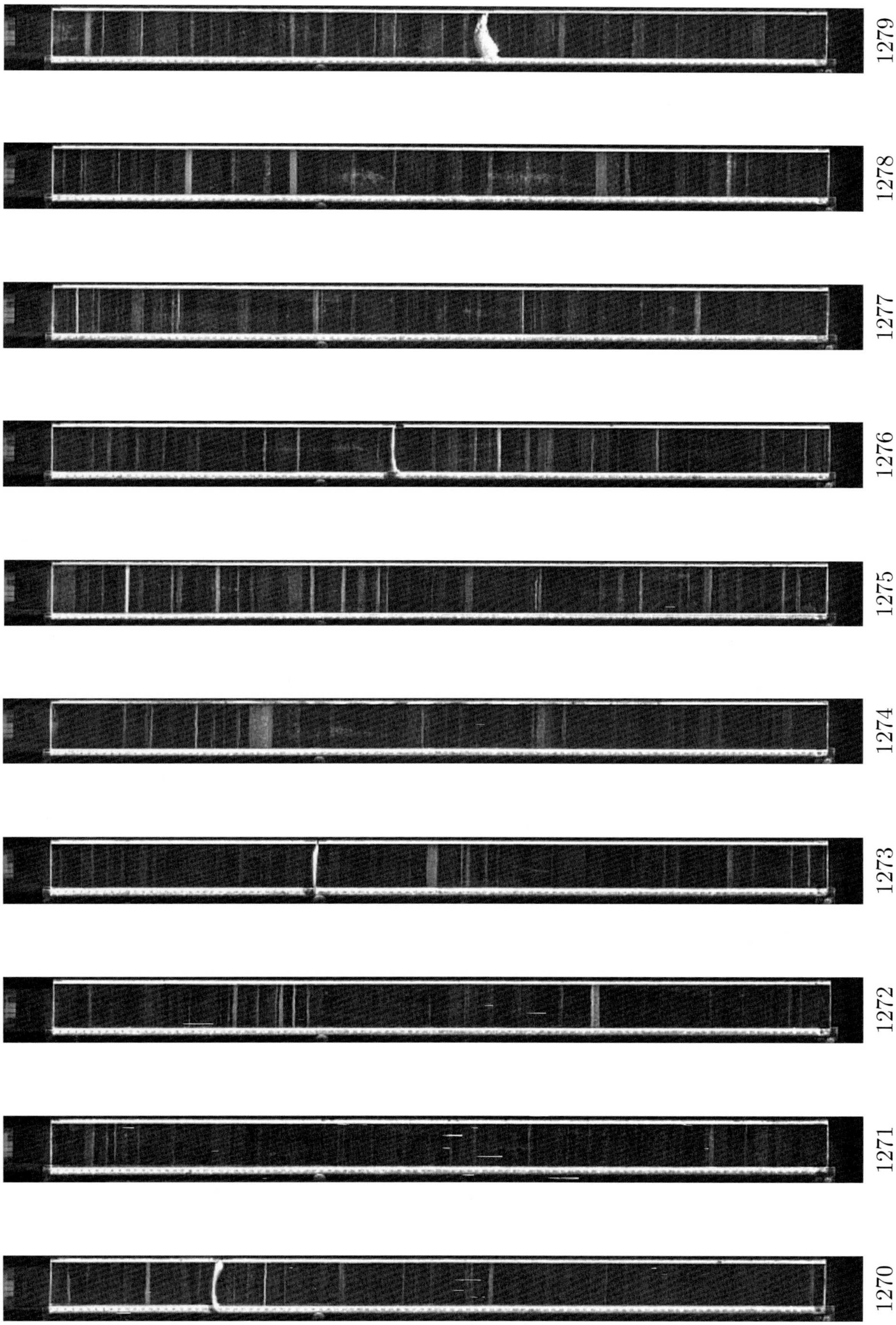
1279
1278
1277
1276
1275
1274
1273
1272
1271
1270

1280

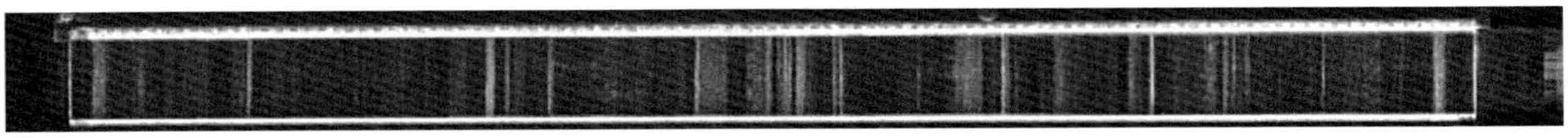

1281

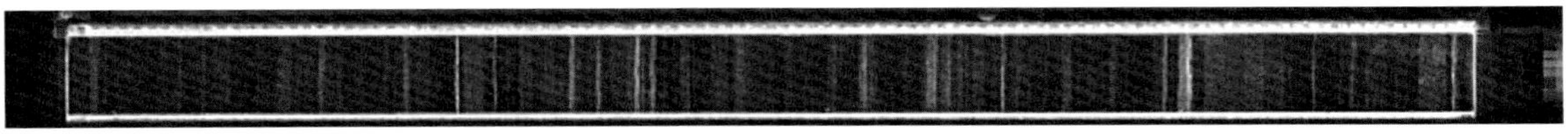

1282

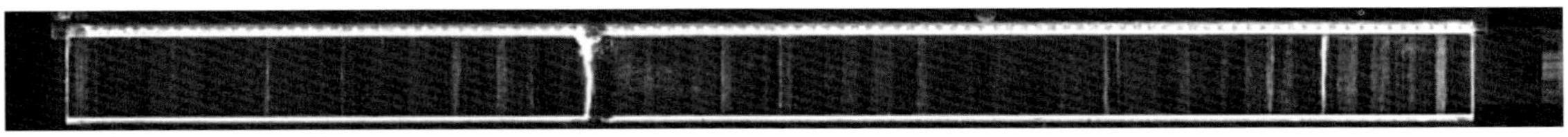

1283

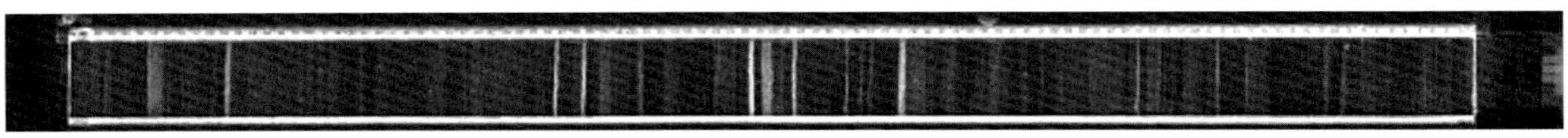

1284

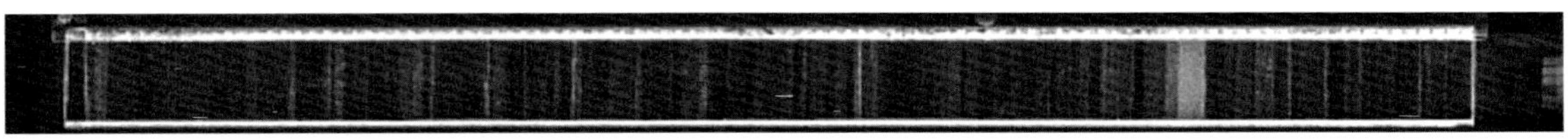

1285

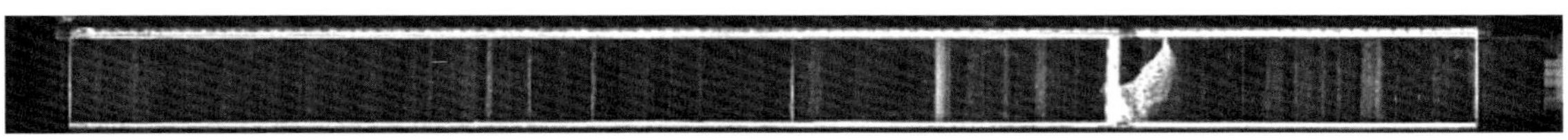

1286

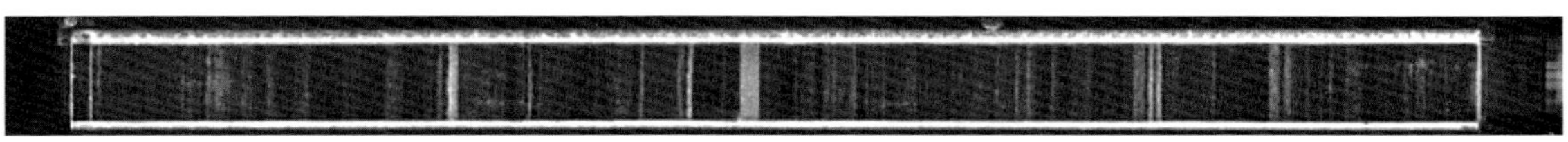

1287

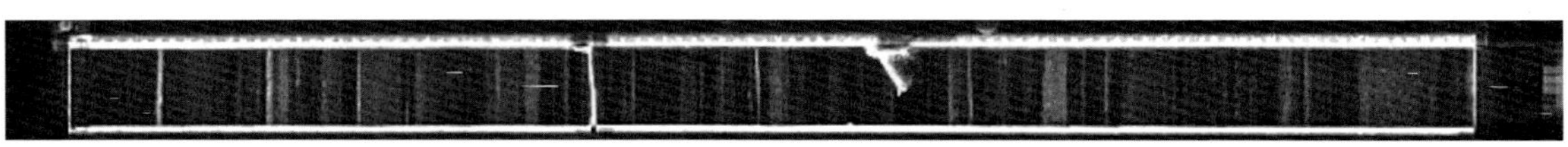

1288

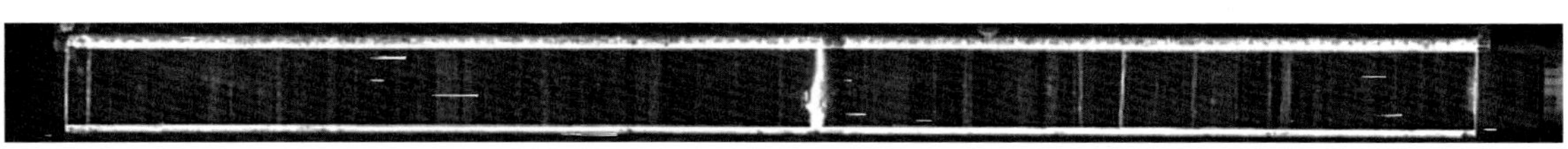

1289

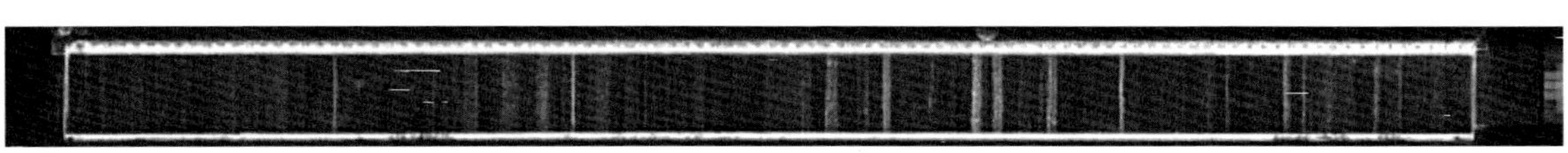

1299

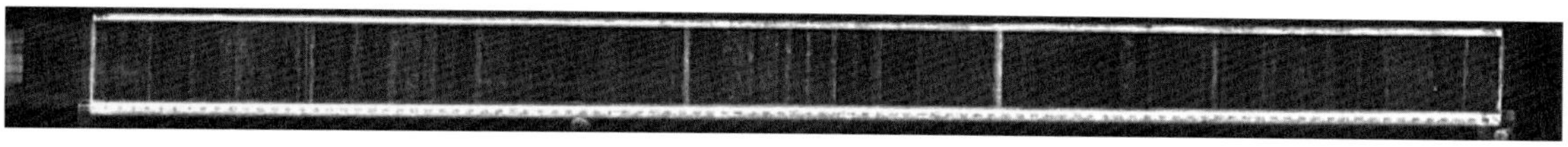
1298

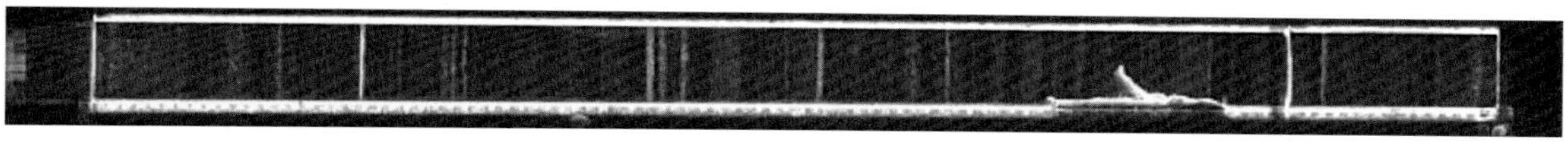
1297

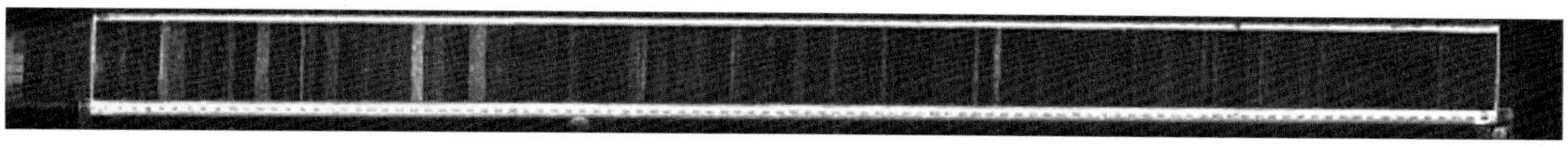
1296

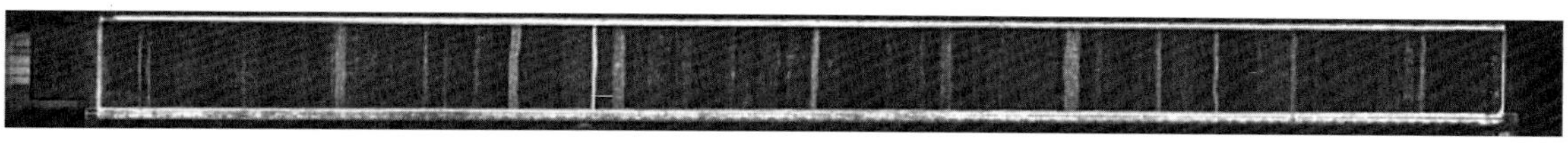
1295

1294

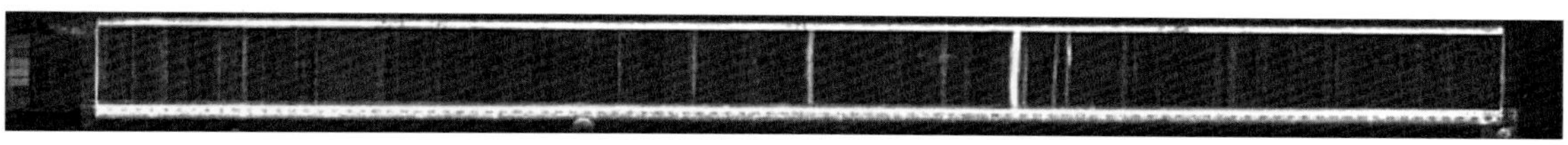
1293

1292

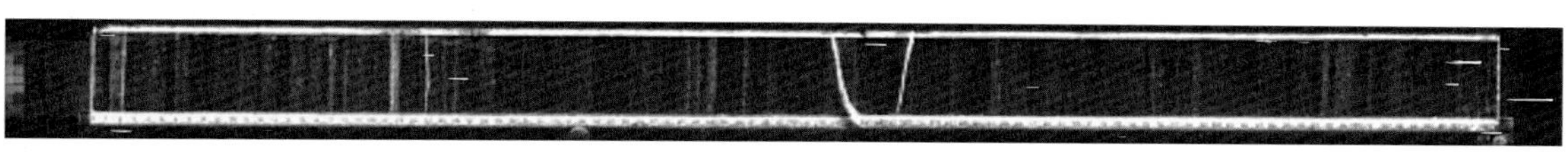
1291

1290

1300

1301

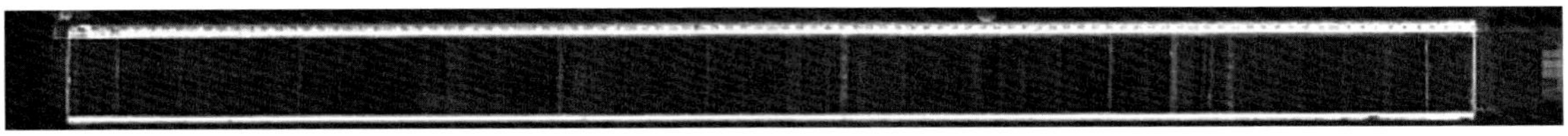

1302

1303

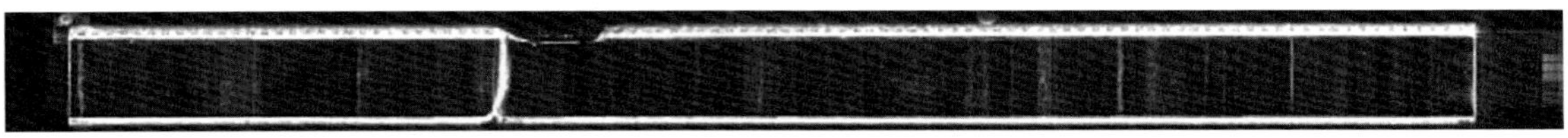

1304

1305

1306

1307

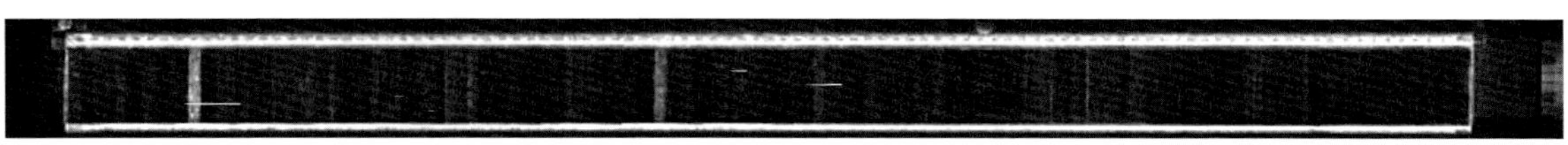

1308

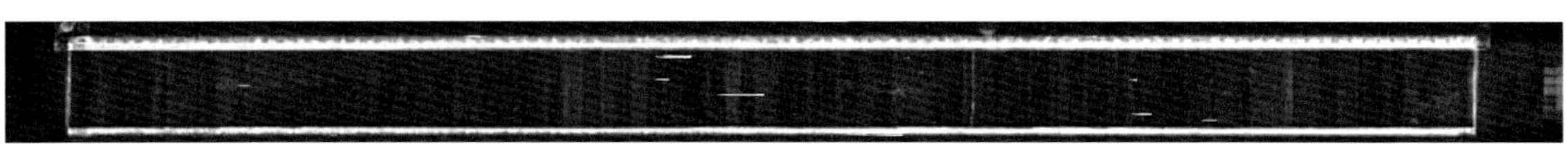

1309

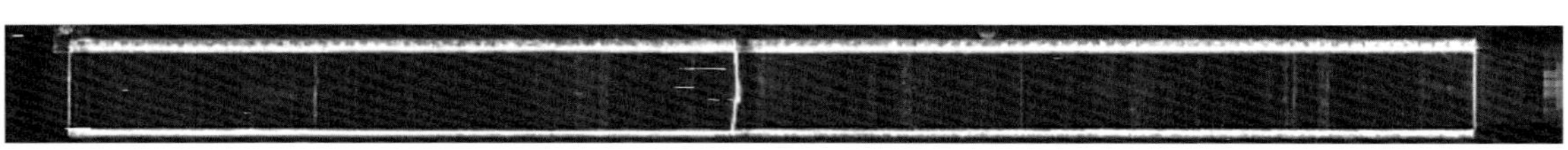

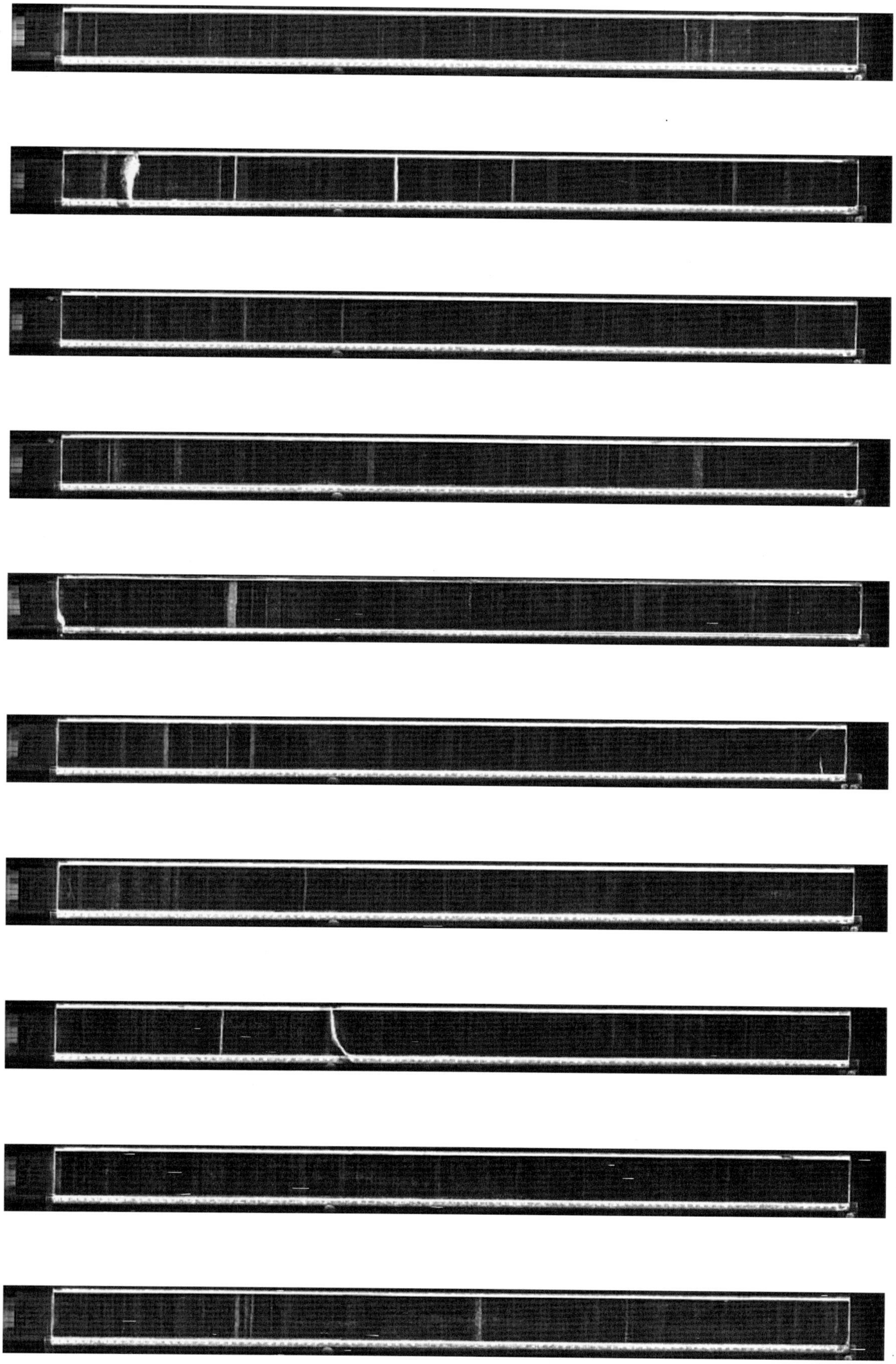

1319

1318

1317

1316

1315

1314

1313

1312

1311

1310

1320

1321

1322

1323

1324

1325

1326

1327
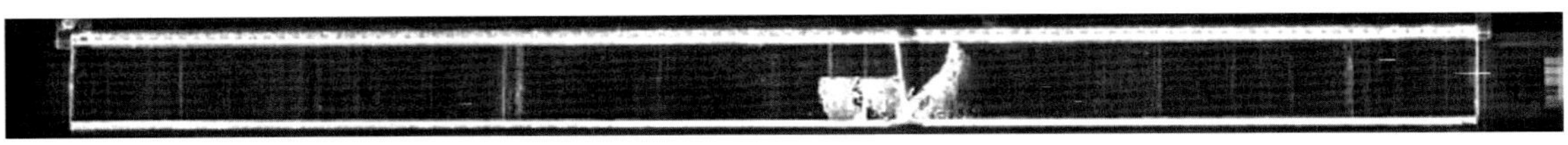

1328
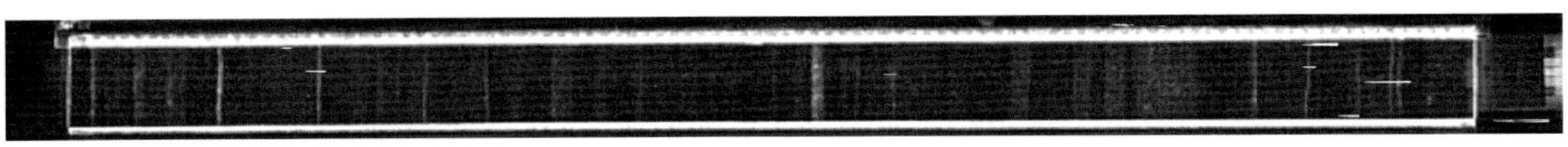

1329
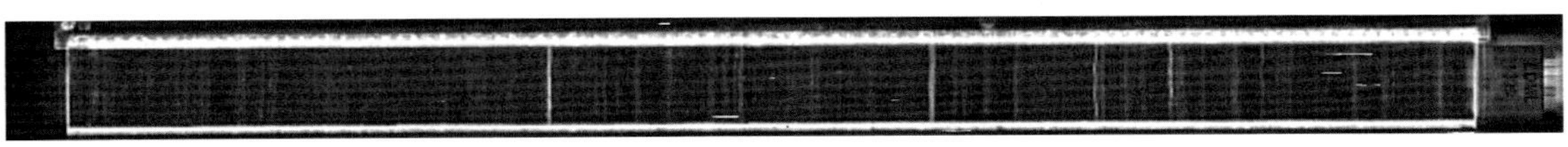

1339

1338

1337

1336

1335

1334

1333

1332

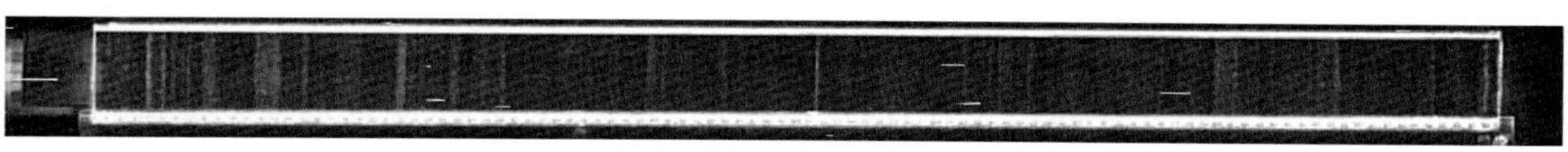
1331

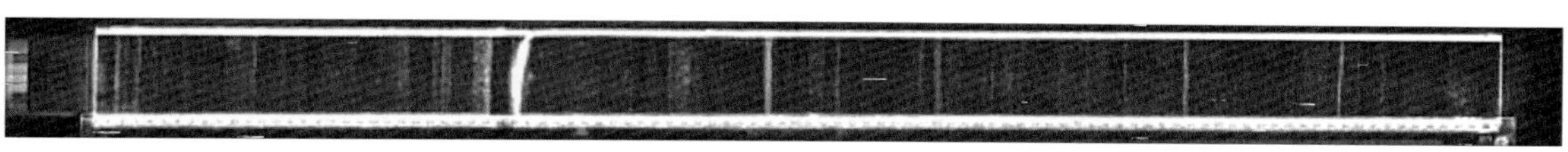
1330

1340
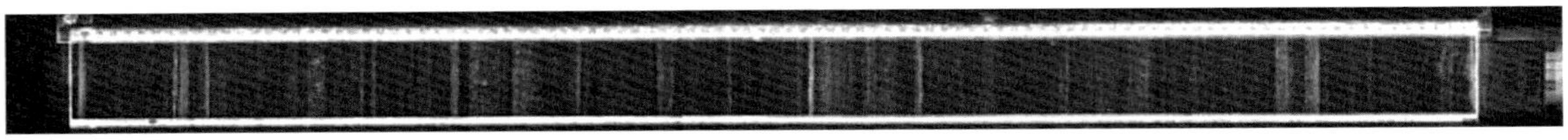

1341
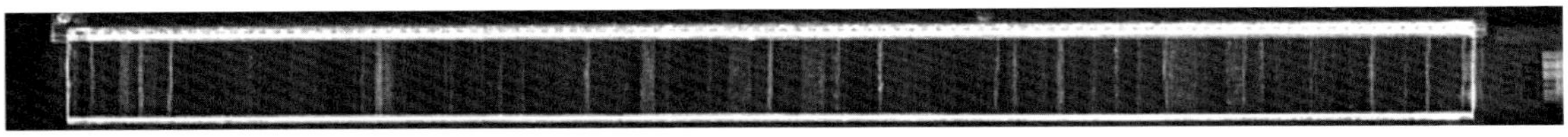

1342
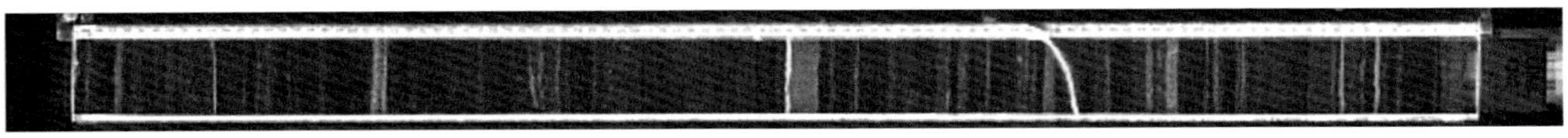

1343
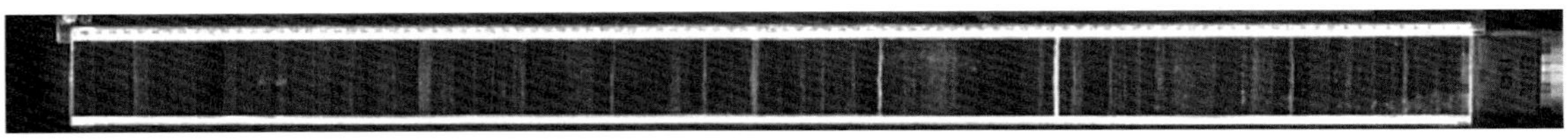

1344
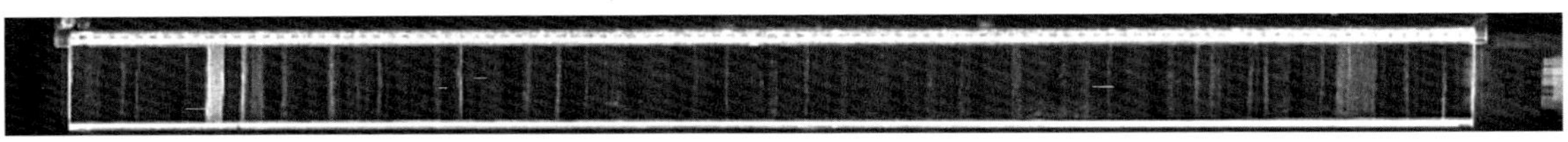

1345

1346
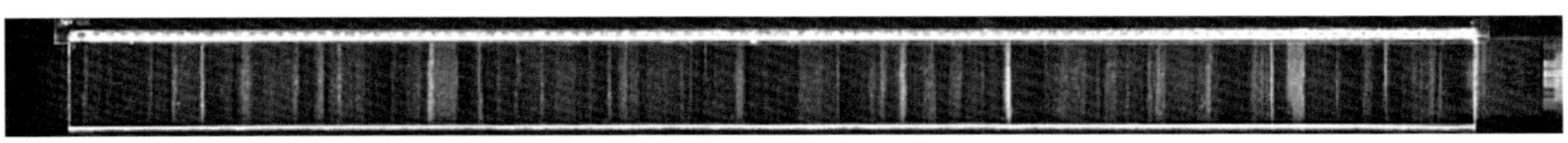

1347
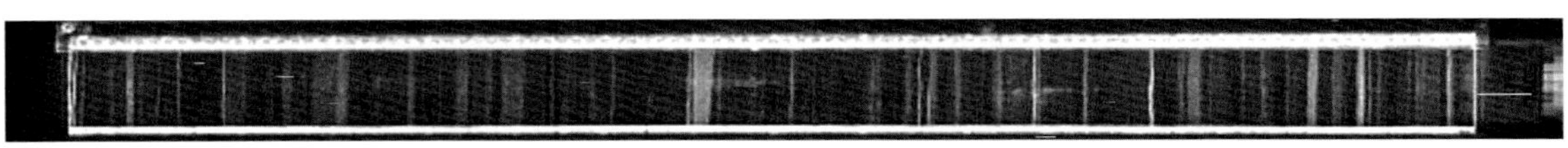

1348
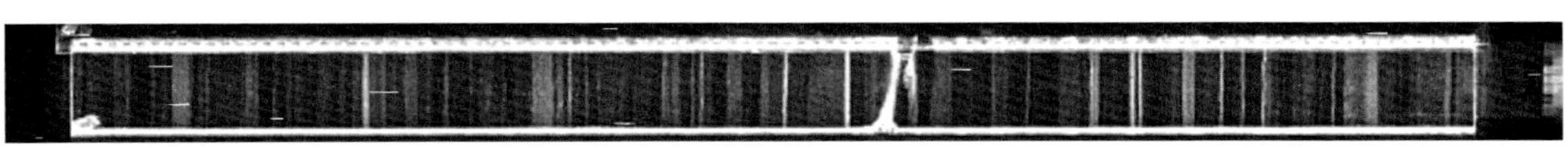

1349
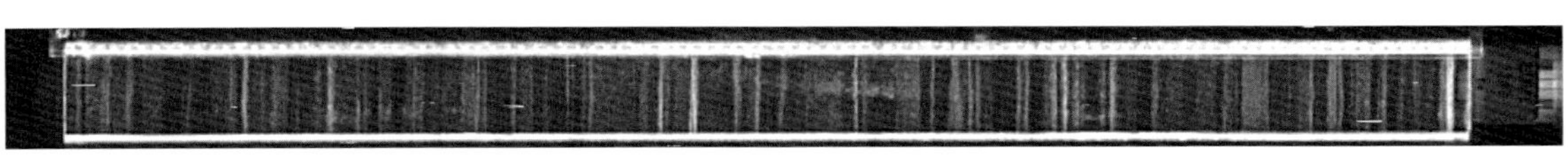

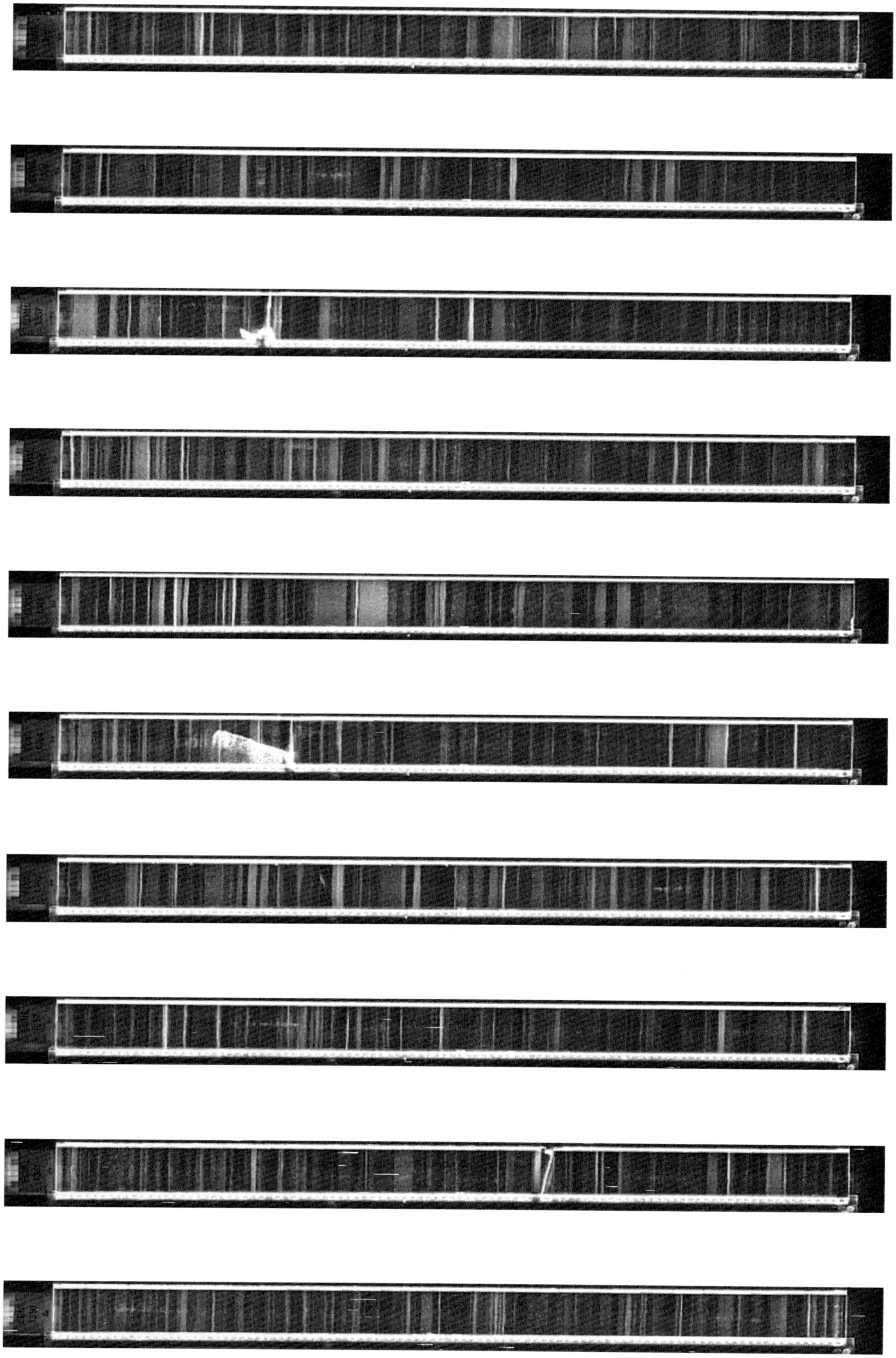

1359

1358

1357

1356

1355

1354

1353

1352

1351

1350

1360

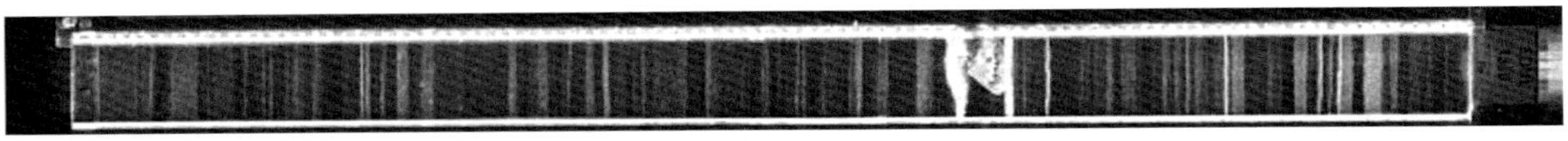

1361

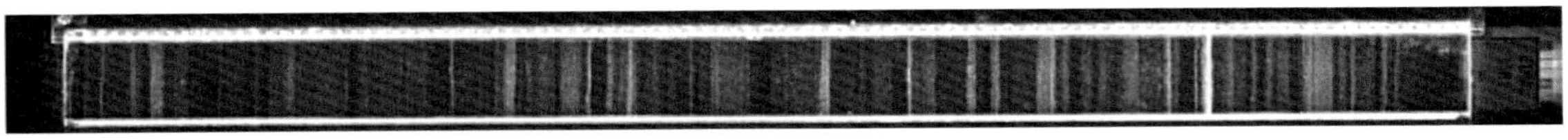

1362

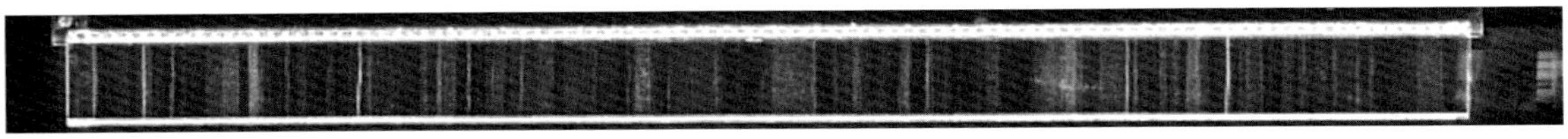

1363

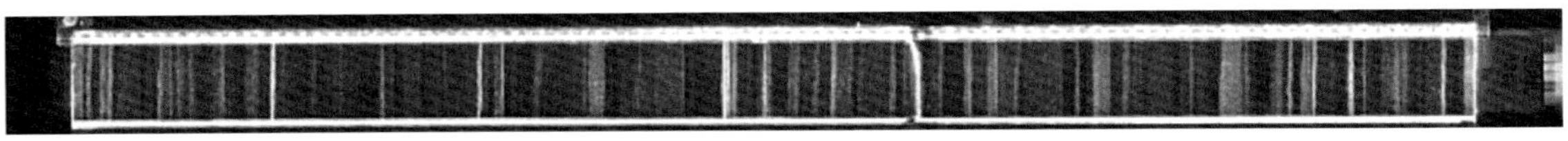

1364

1365

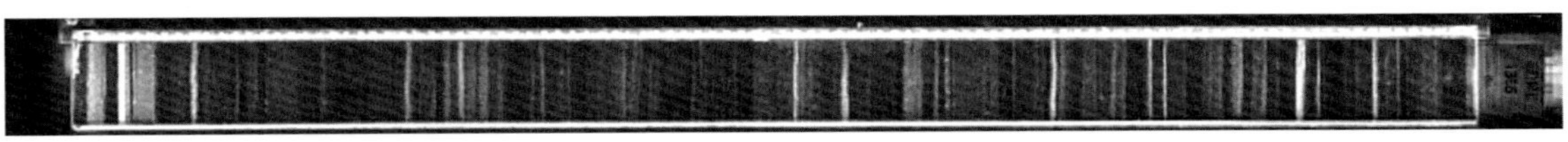

1366

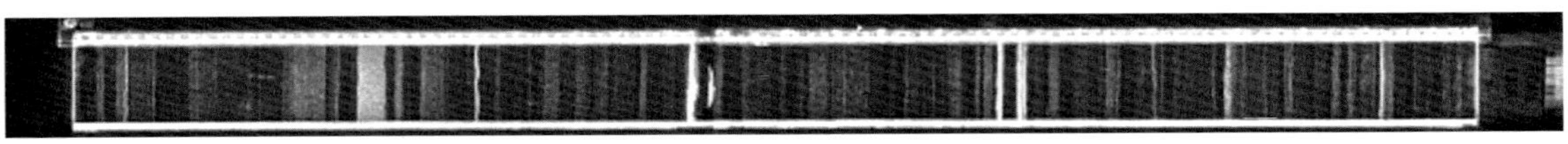

1367

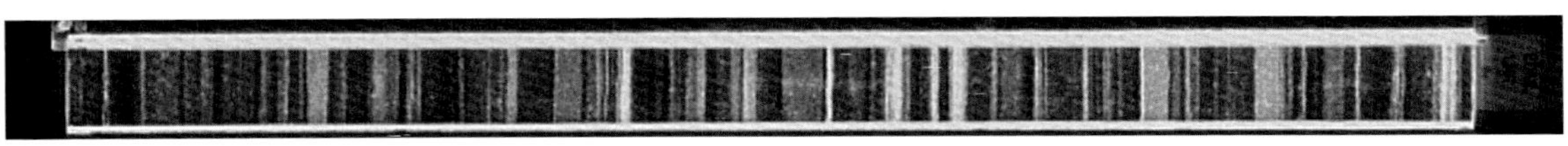

1368

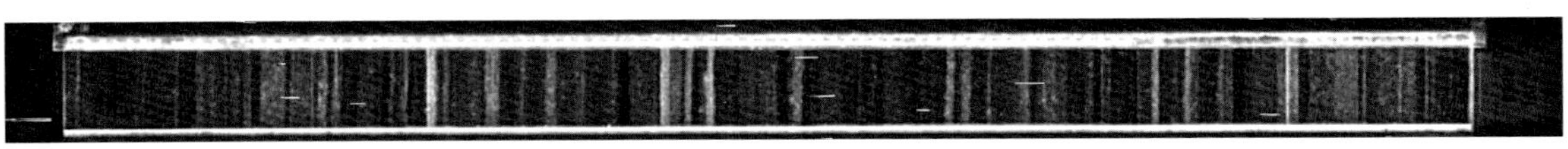

1369

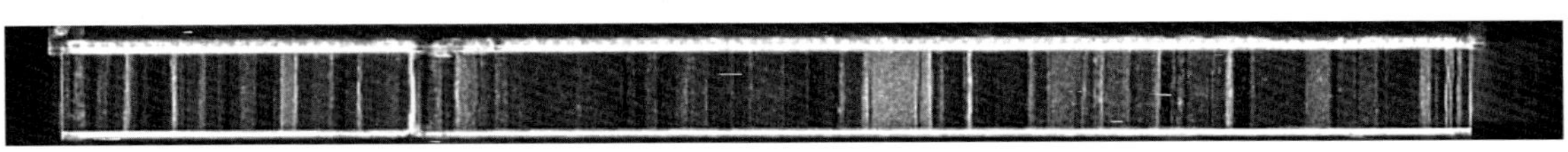

1379

1378

1377

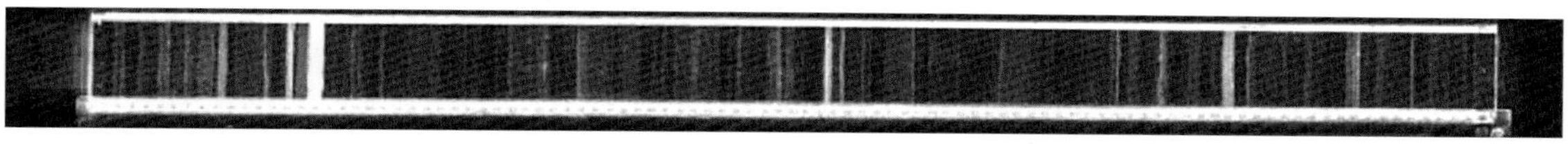
1376

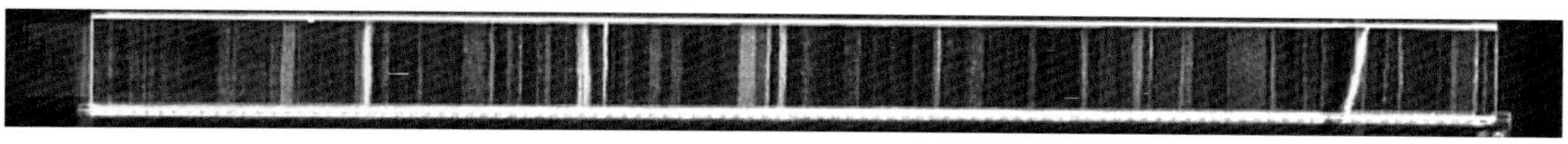
1375

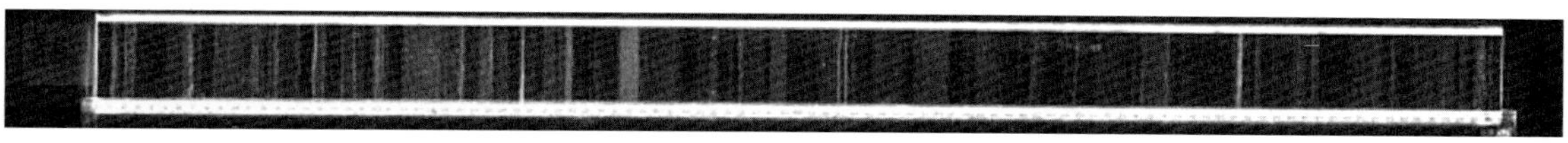
1374

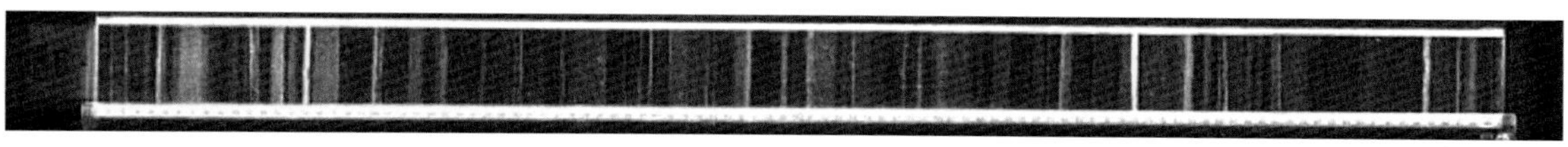
1373

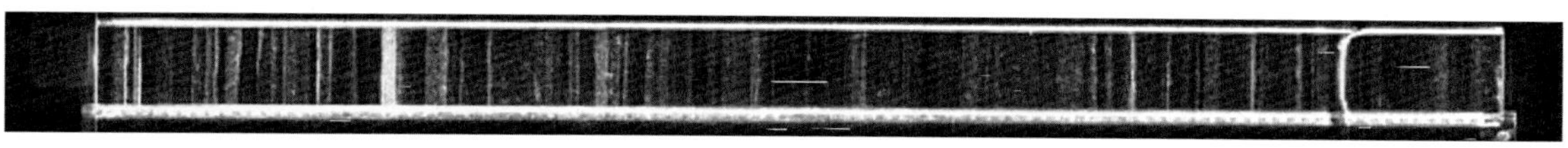
1372

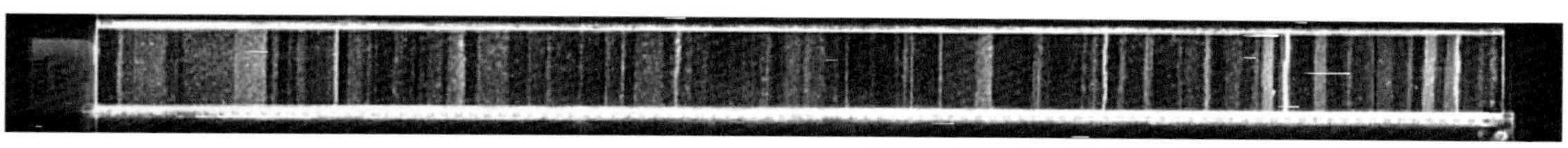
1371

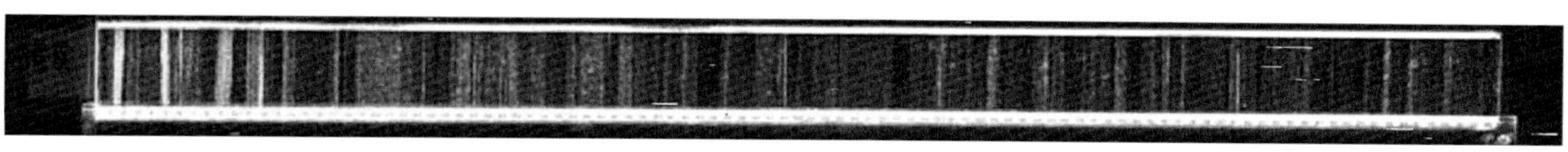
1370

1380
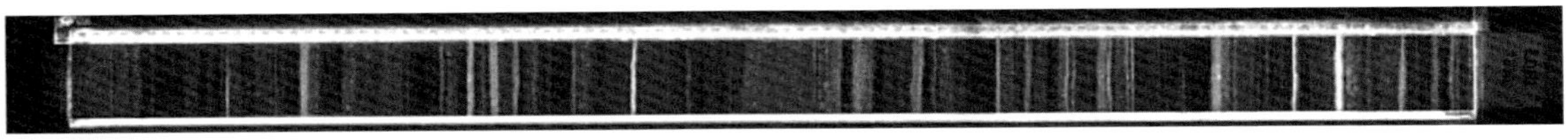

1381
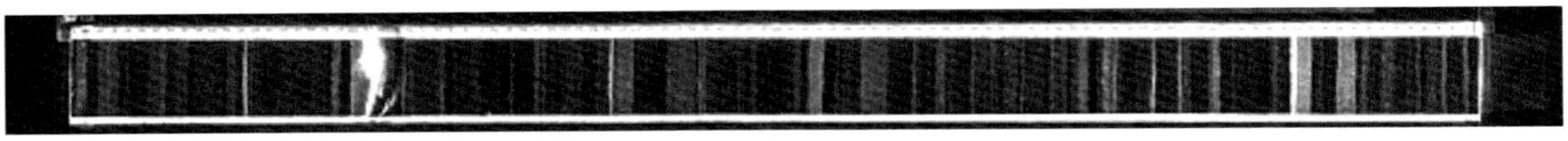

1382
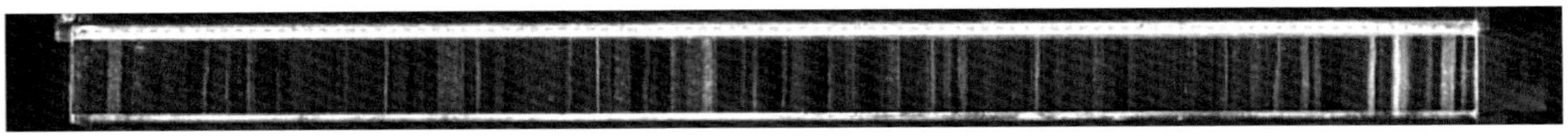

1383
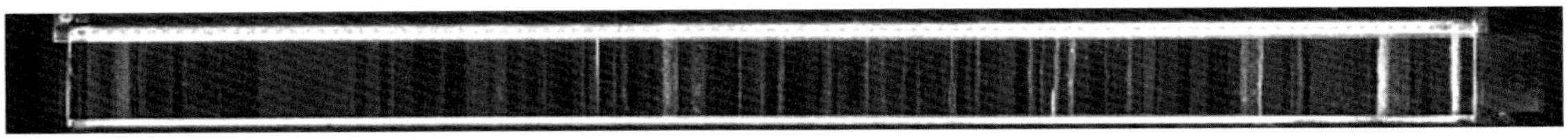

1384

1385

1386

1387
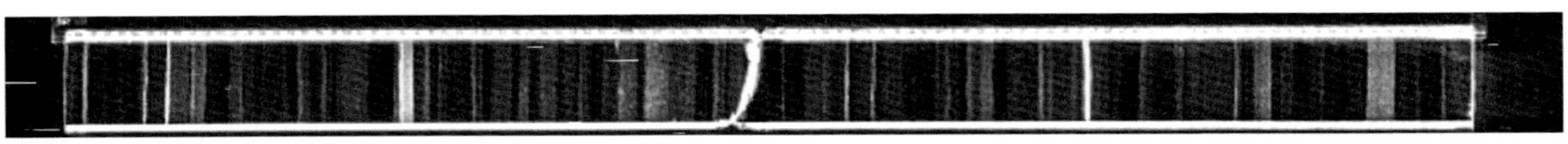

1388

1389
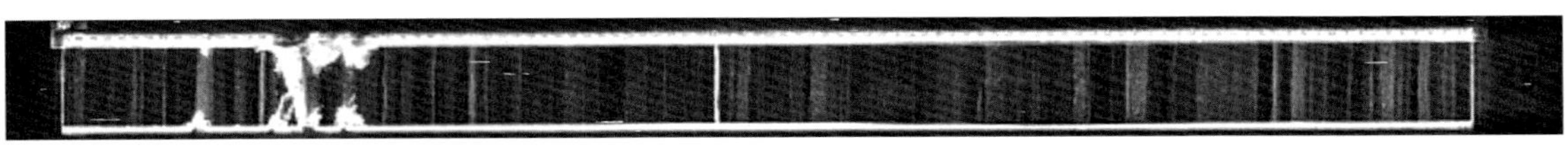

1399

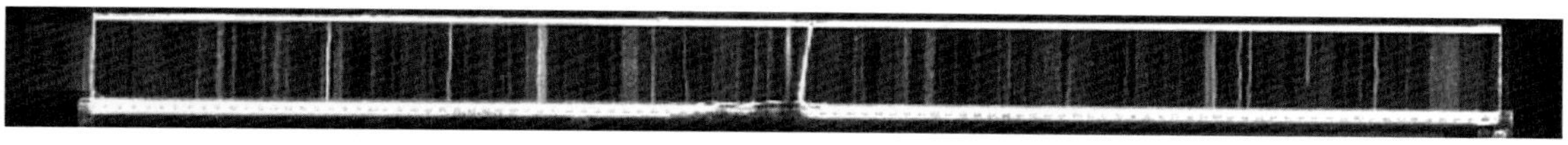

1398

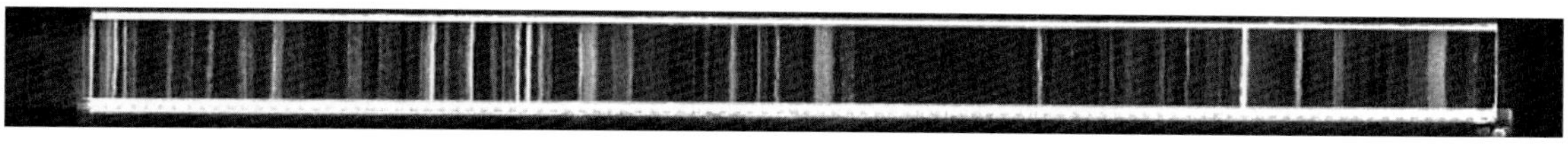

1397

1396

1395

1394

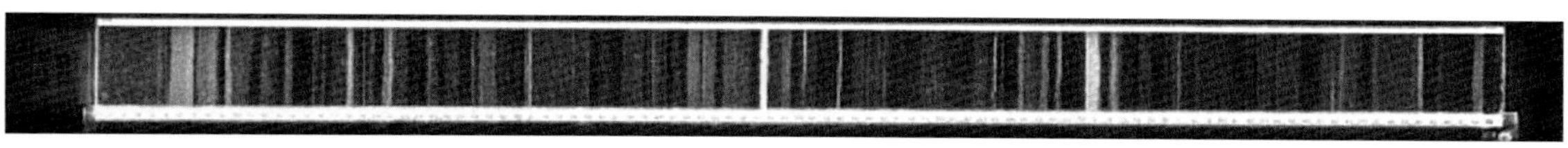

1393

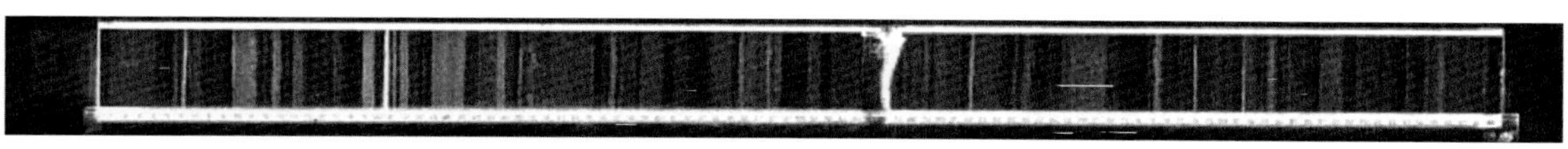

1392

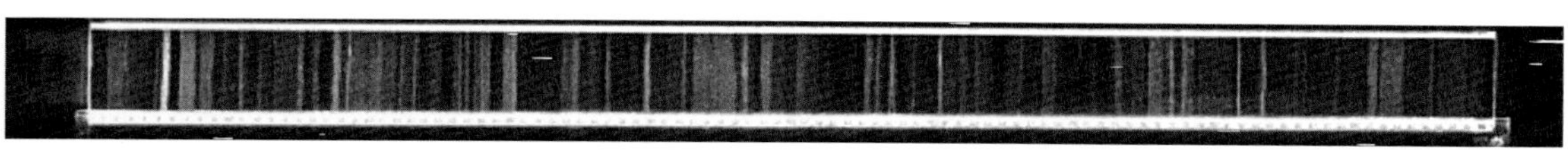

1391

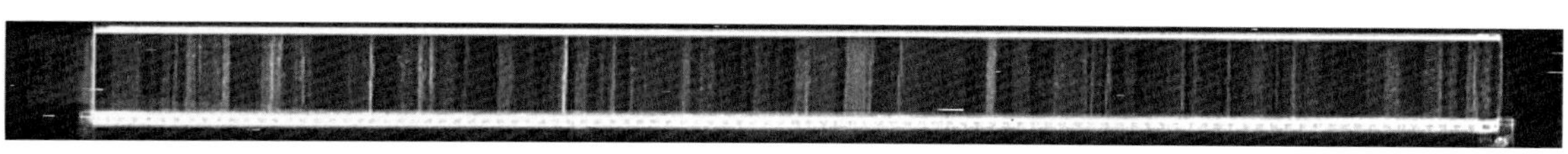

1390

1400

1401

1402

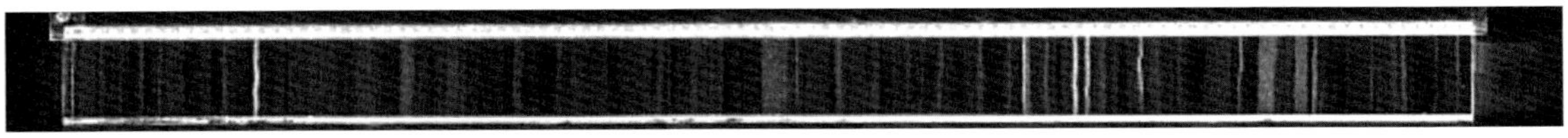

1403

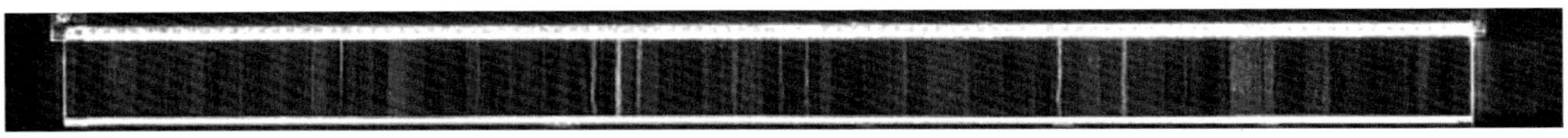

1404

1405

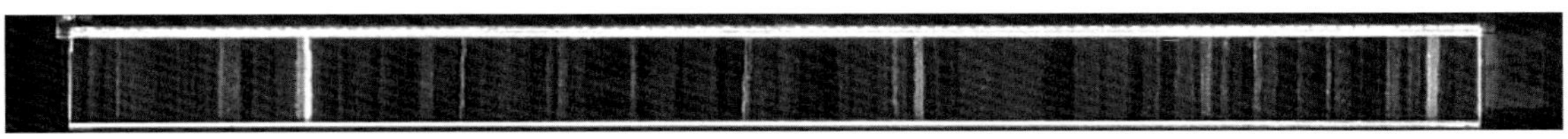

1406

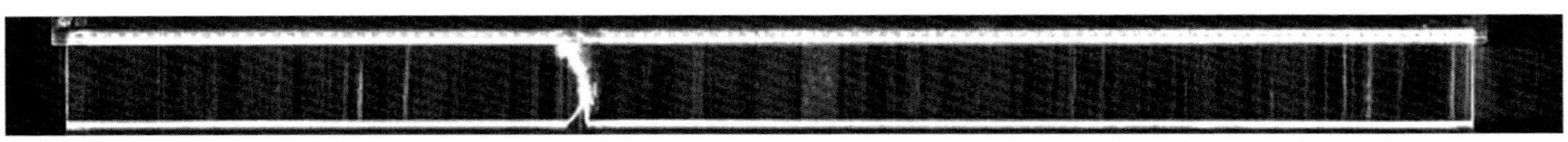

1407

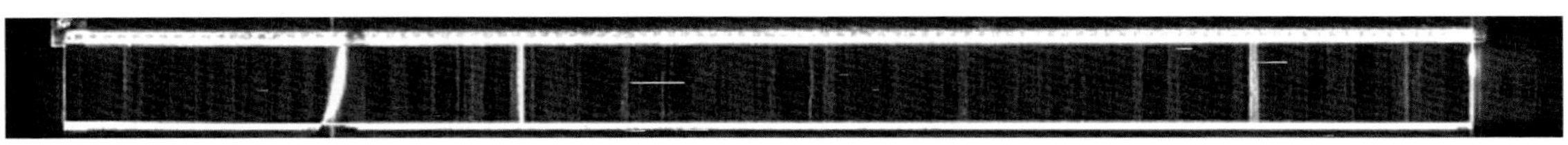

1408

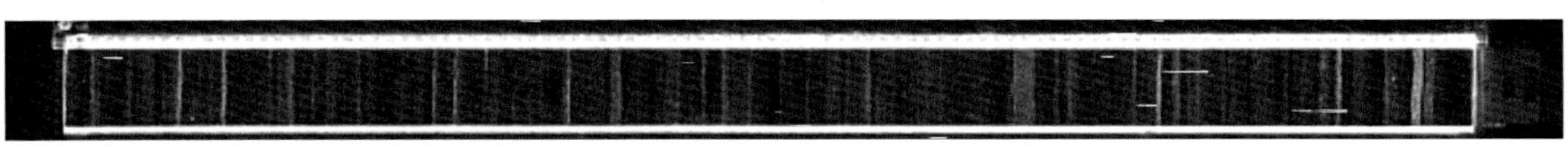

1409

1419

1418

1417

1416

1415

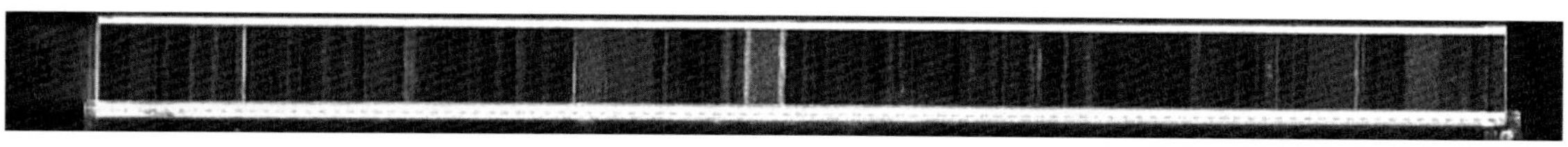

1414

1413

1412

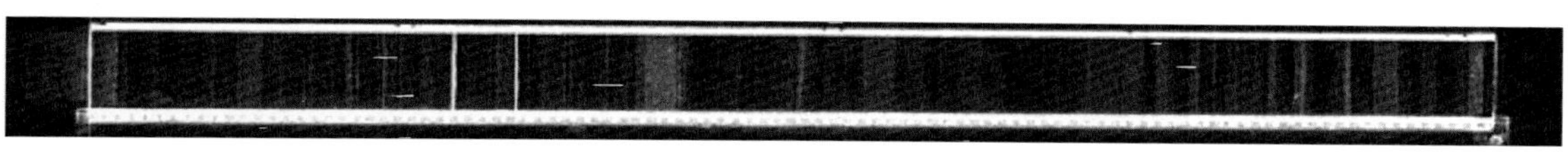

1411

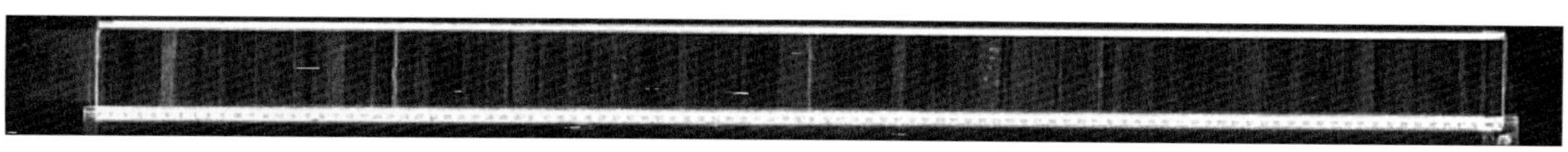

1410

1420
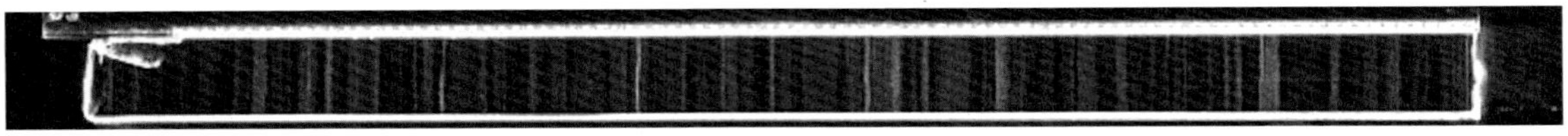

1421
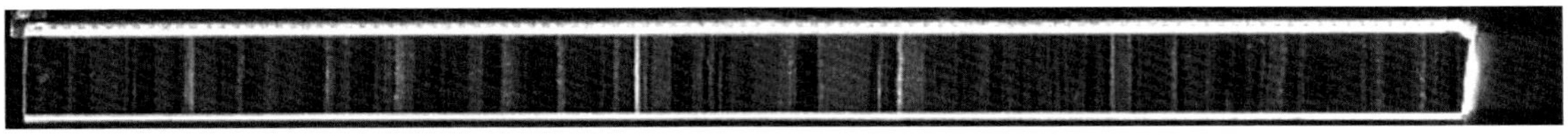

1422
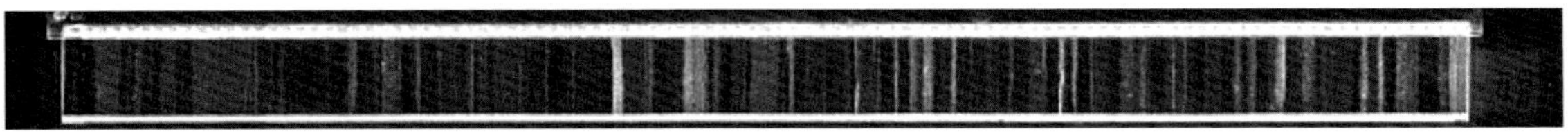

1423
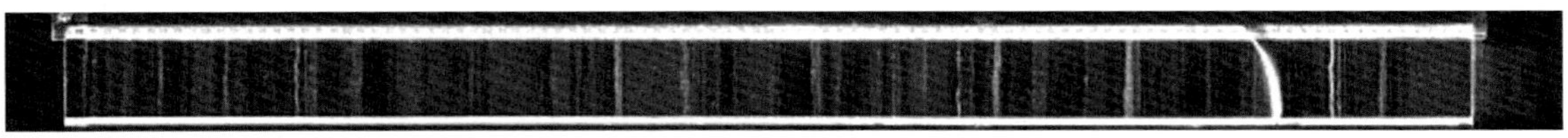

1424
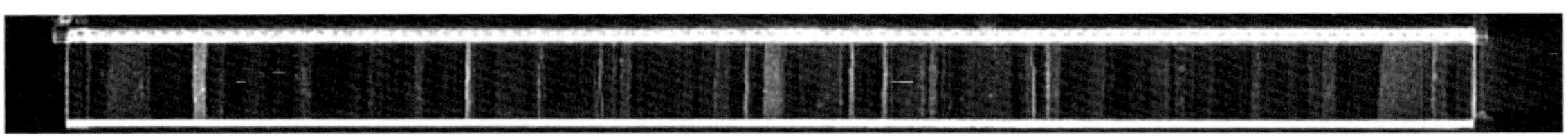

1425
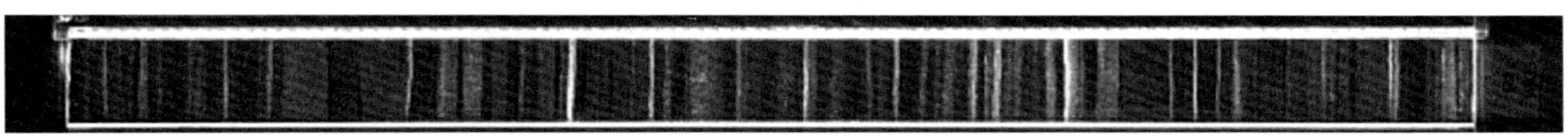

1426

1427
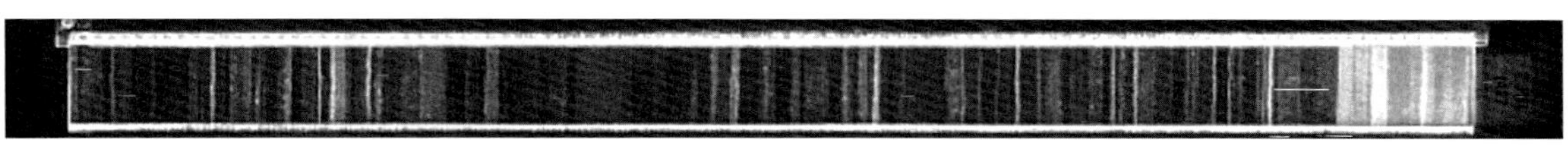

1428
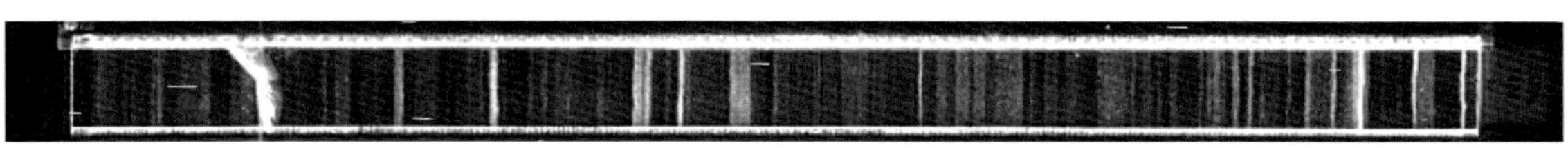

1429

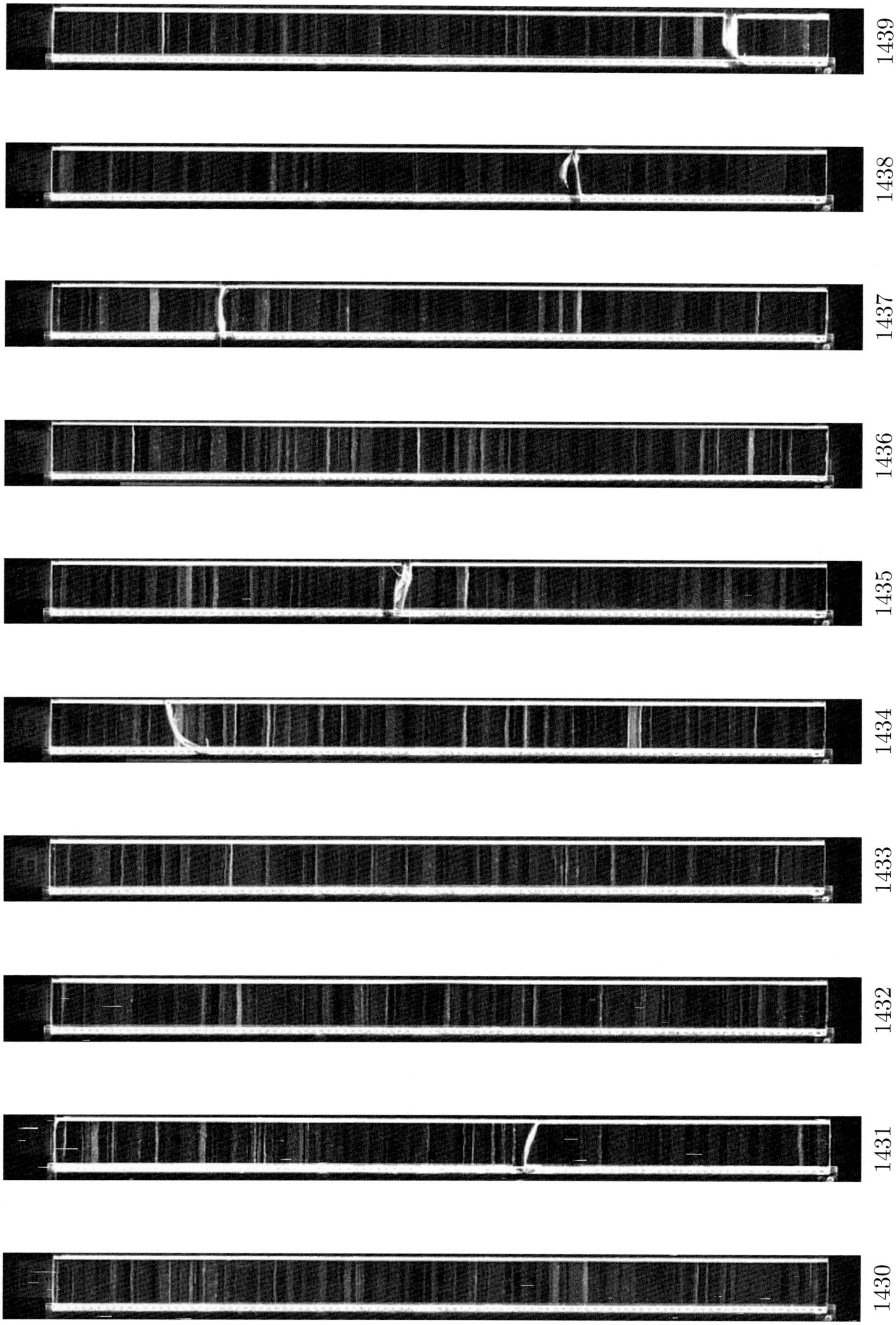
1439
1438
1437
1436
1435
1434
1433
1432
1431
1430

1440

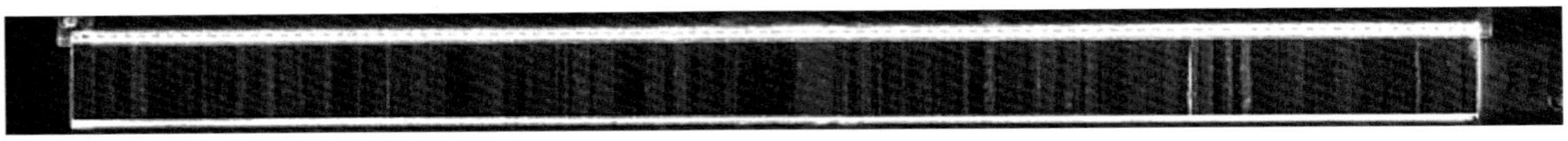

1441

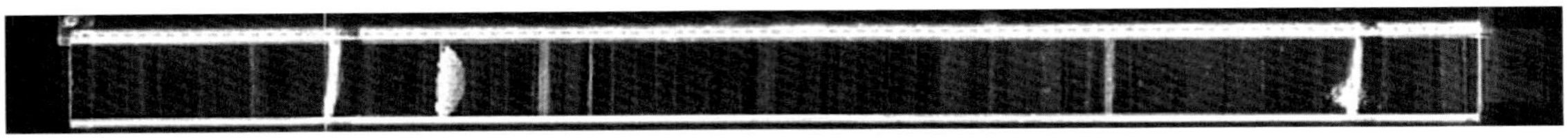

1442

1443

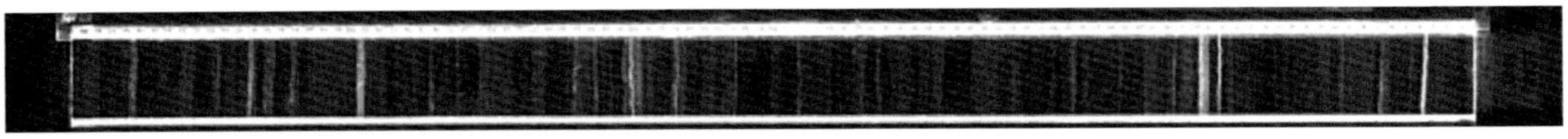

1444

1445

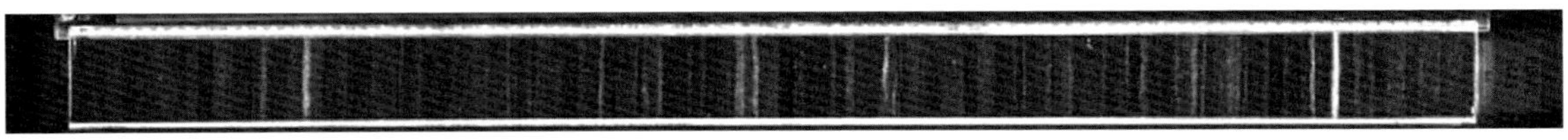

1446

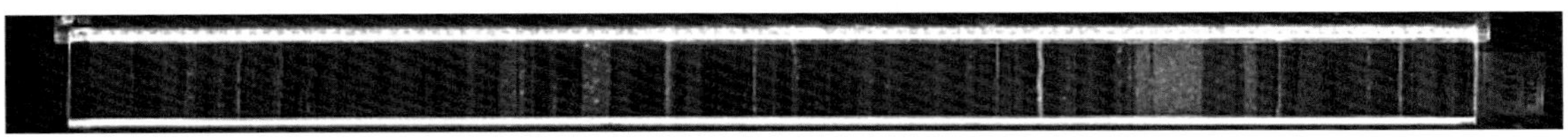

1447

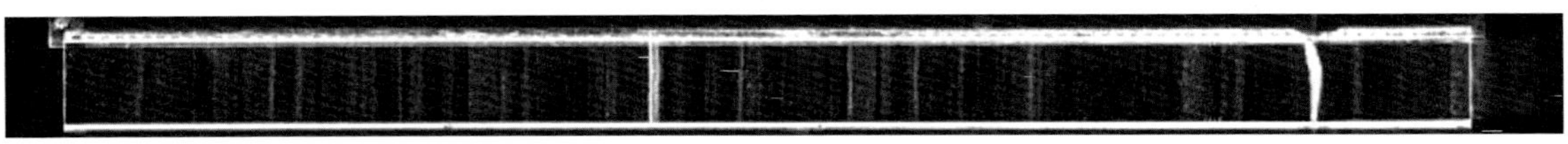

1448

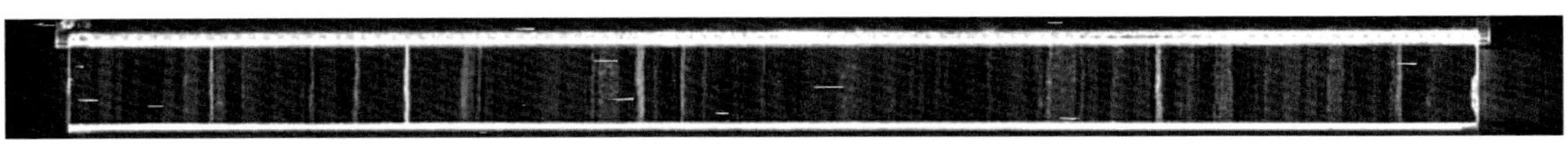

1449

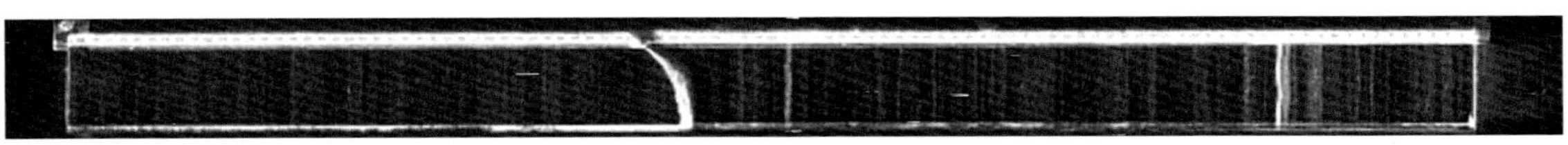

1459

1458

1457

1456

1455

1454

1453

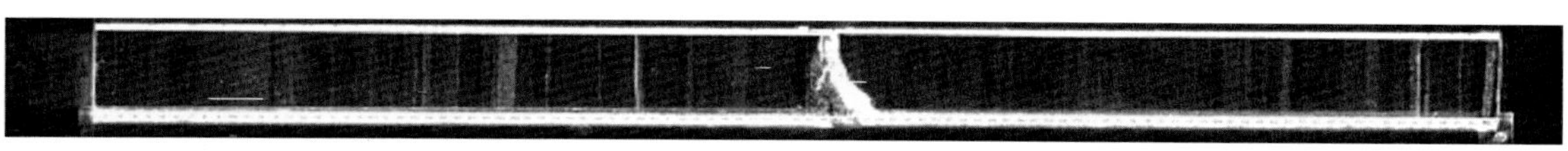
1452

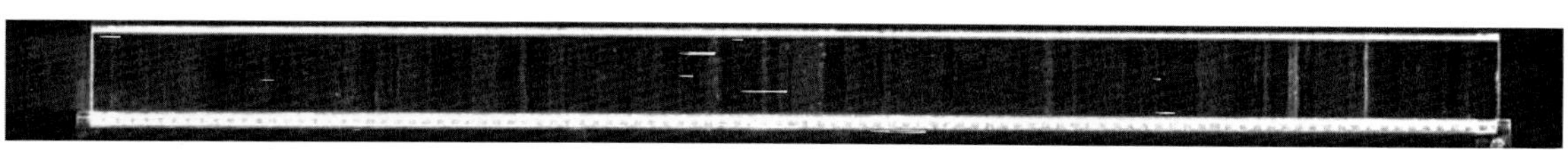
1451

1450

1460

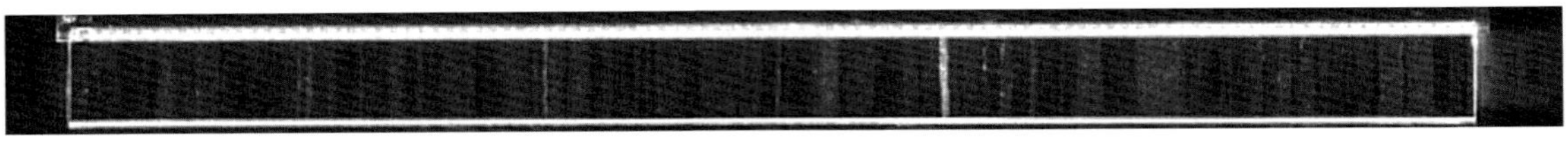

1461

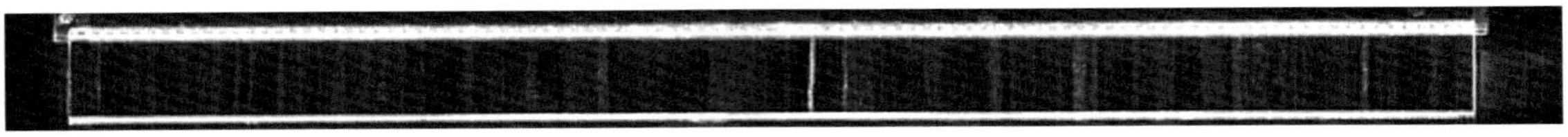

1462

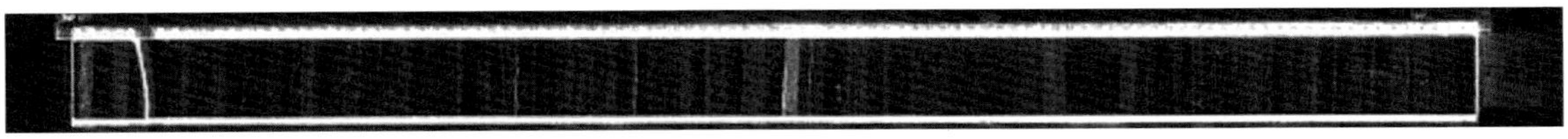

1463

1464

1465

1466

1467

1468

1469

1479

1478

1477

1476

1475

1474

1473

1472

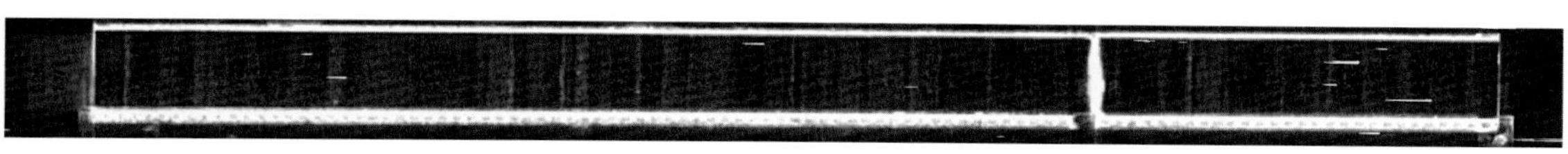
1471

1470

1480

1481

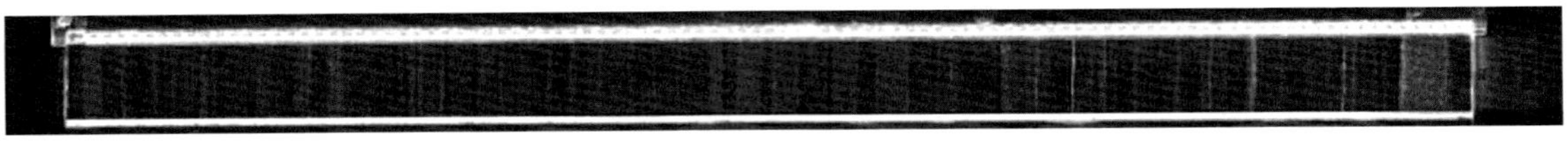

1482

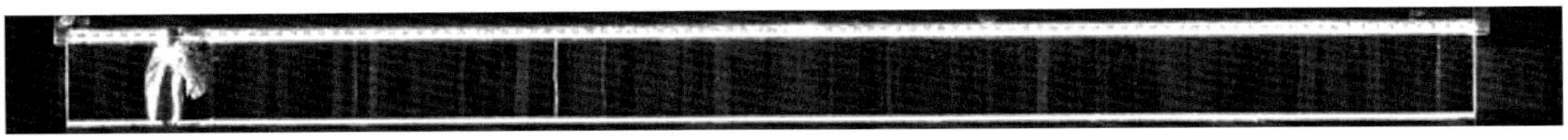

1483

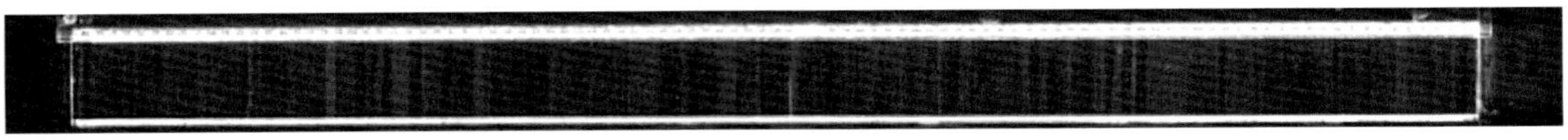

1484

1485

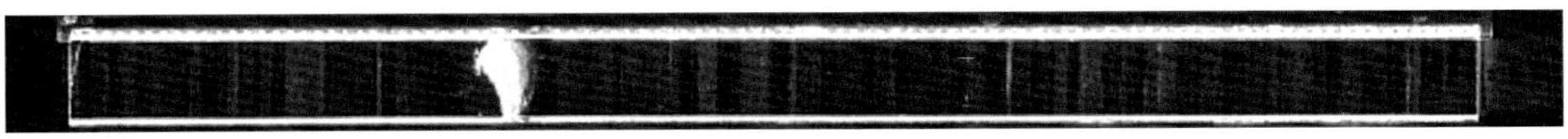

1486

1487

1488

1489

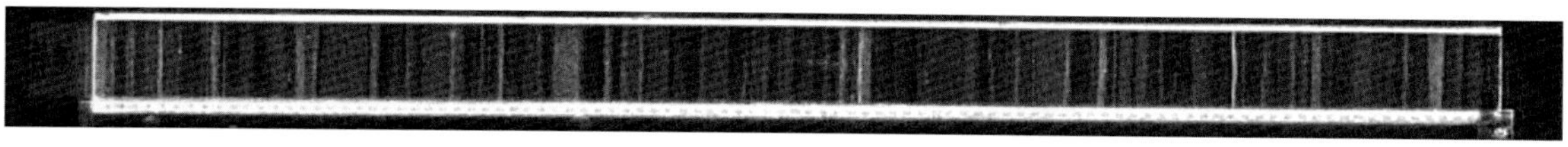

1499

1498

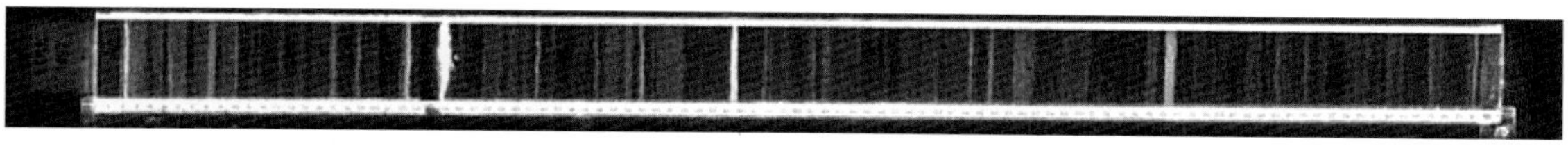

1497

1496

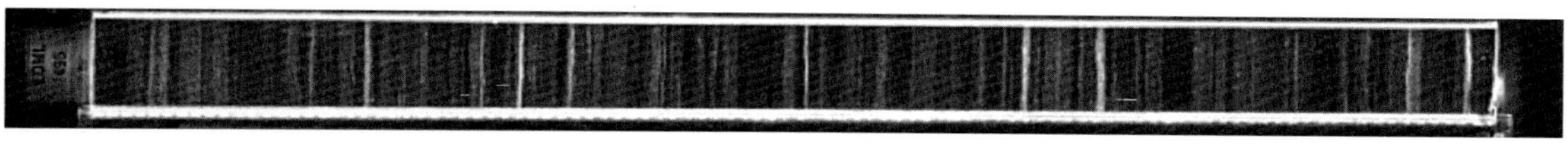

1495

1494

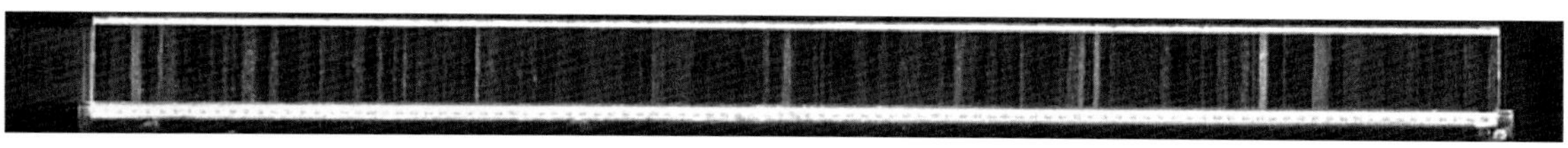

1493

1492

1491

1490

1500

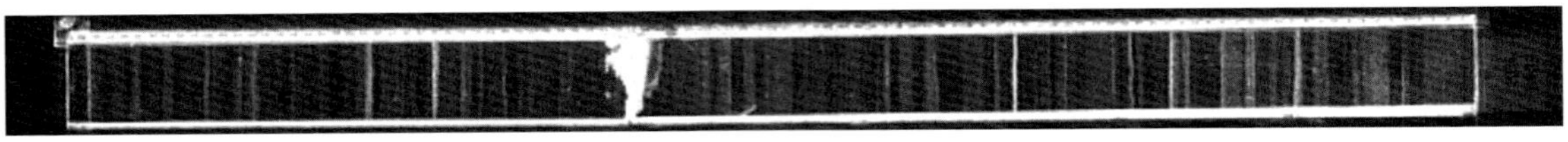

1501

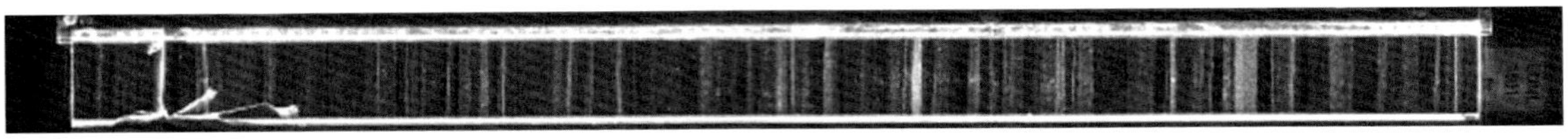

1502

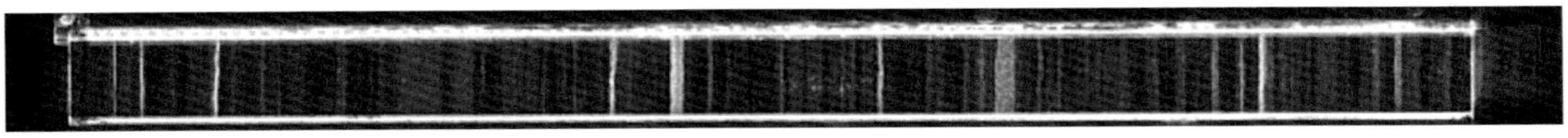

1503

1504

1505

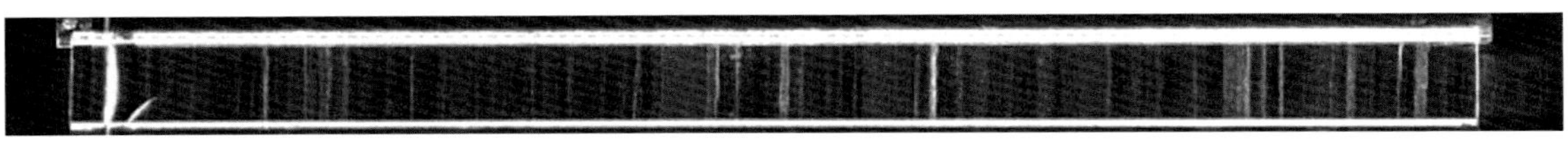

1506

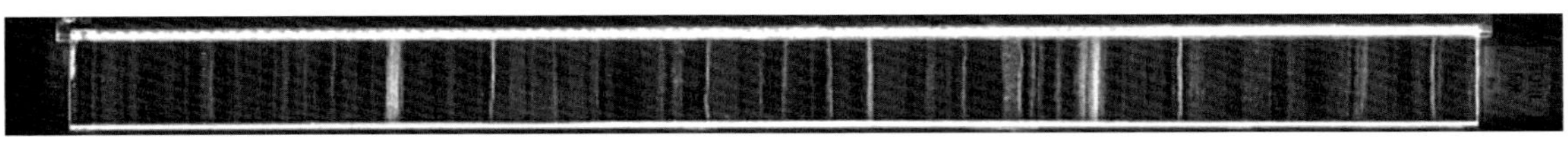

1507

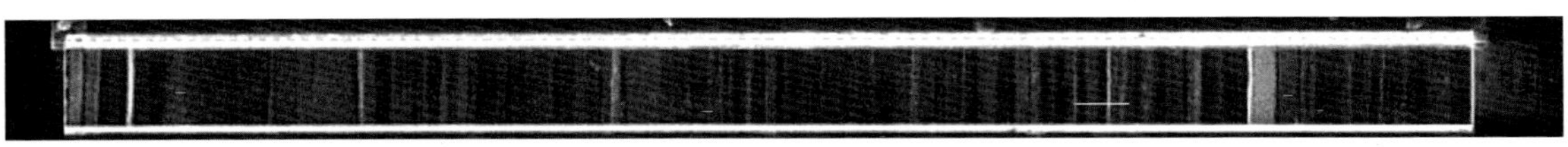

1508

1509

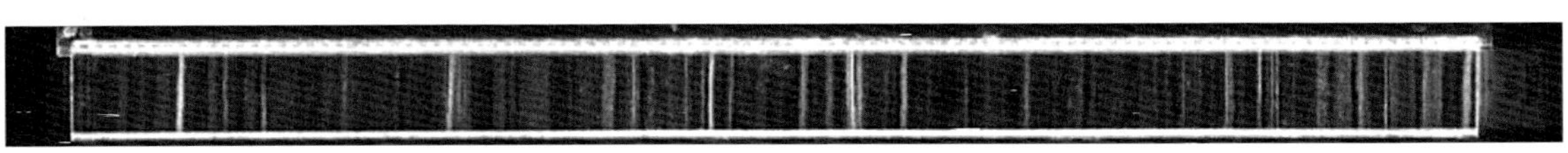

1519

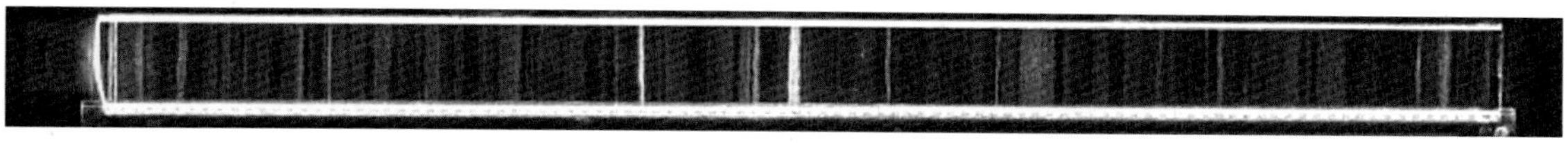
1518

1517

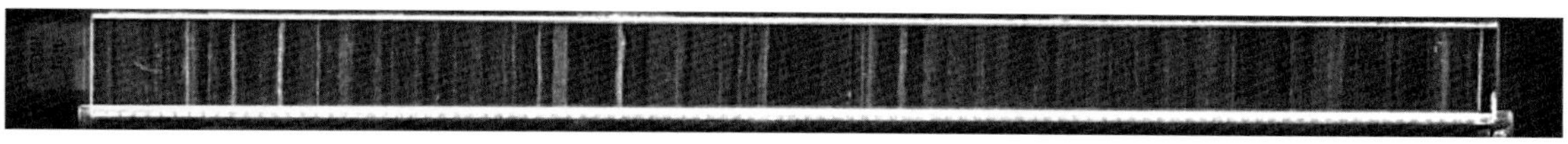
1516

1515

1514

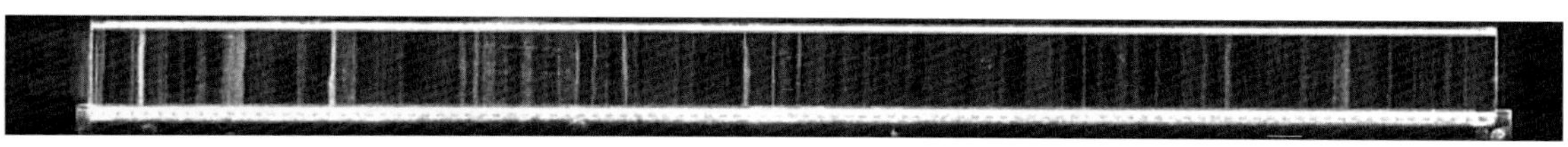
1513

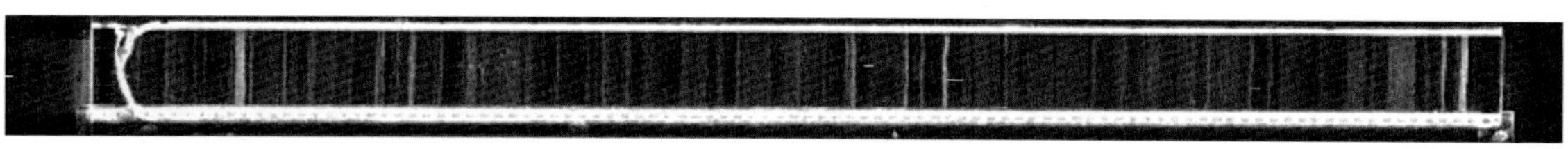
1512

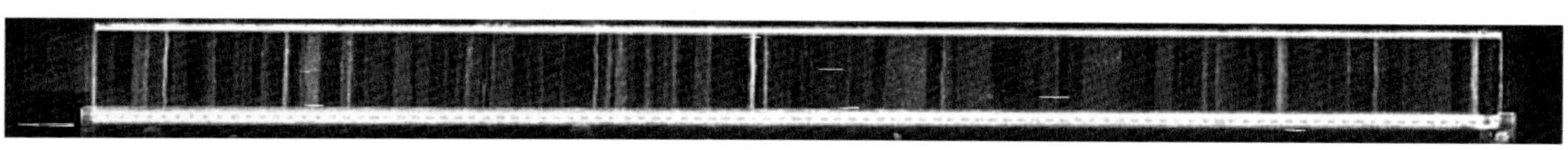
1511

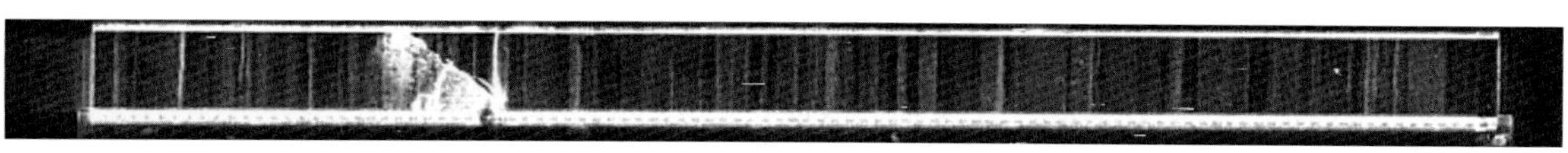
1510

1520

1521

1522

1523

1524
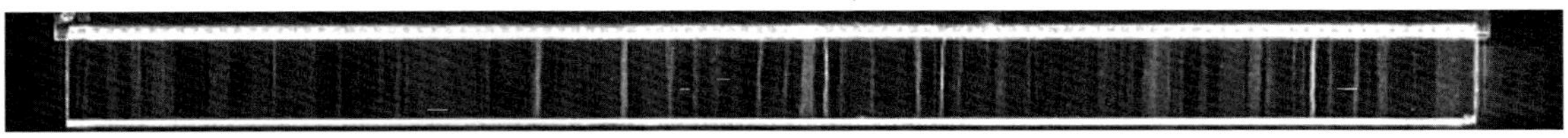

1525

1526
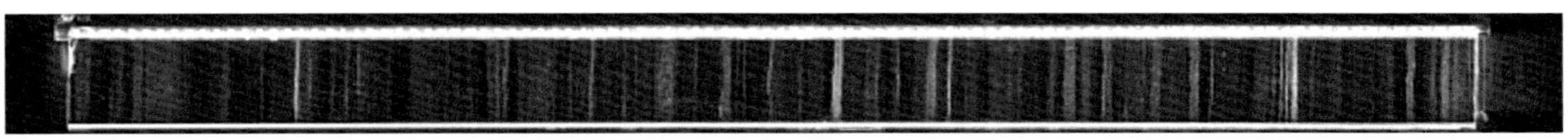

1527

1528
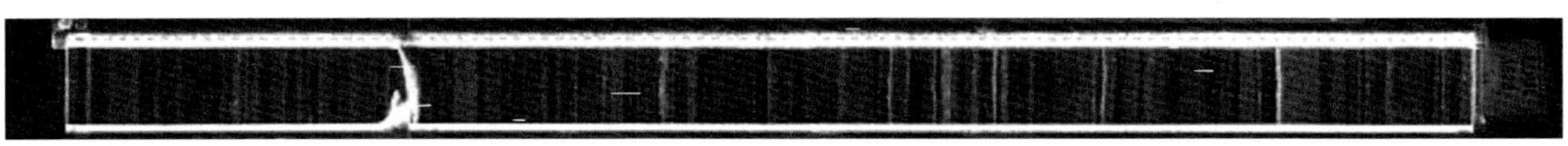

1529
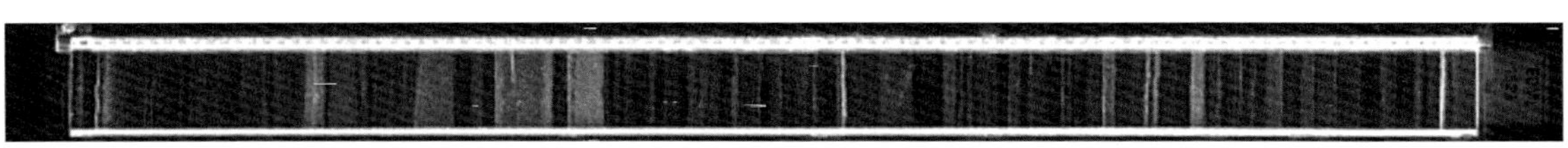

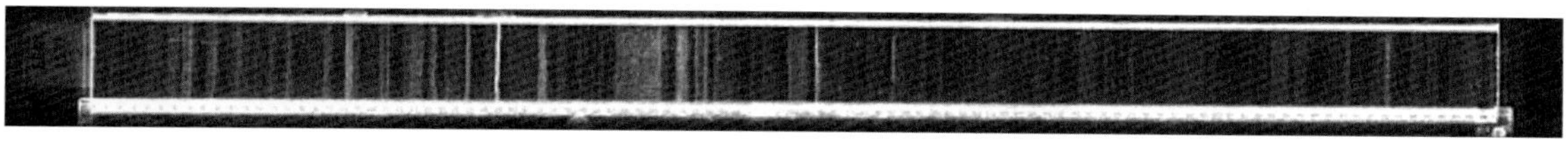

1539

1538

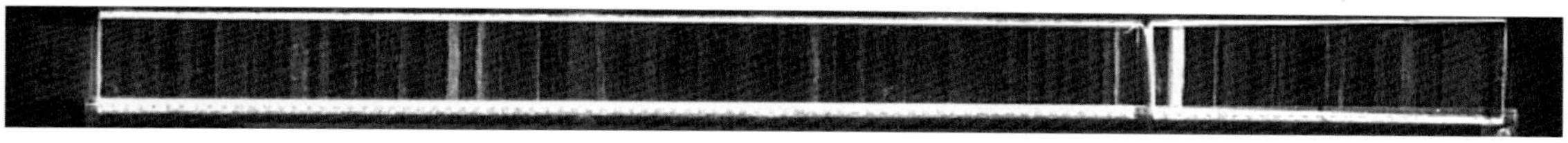

1537

1536

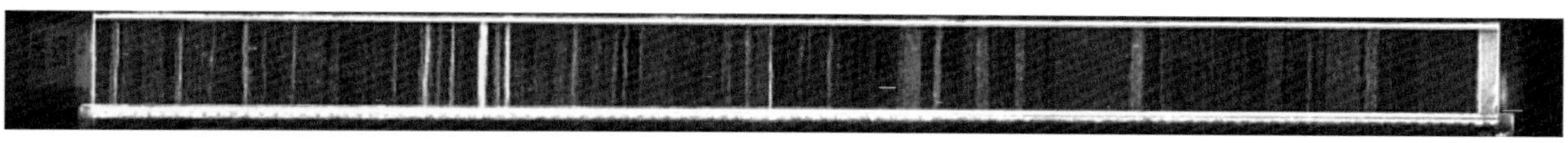

1535

1534

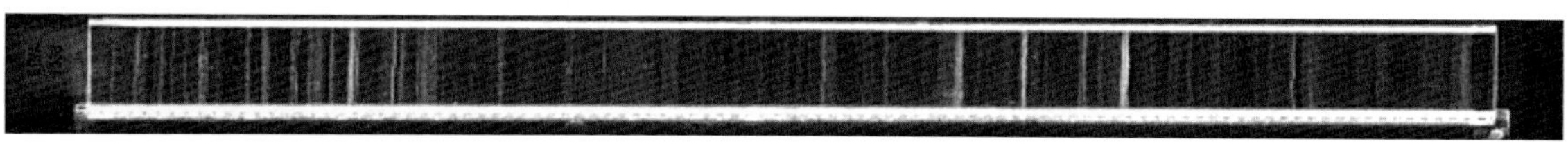

1533

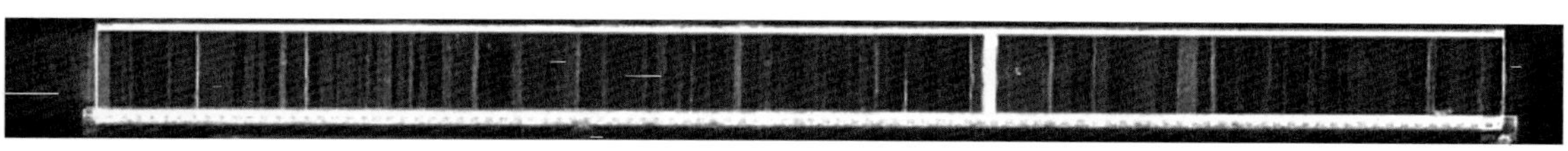

1532

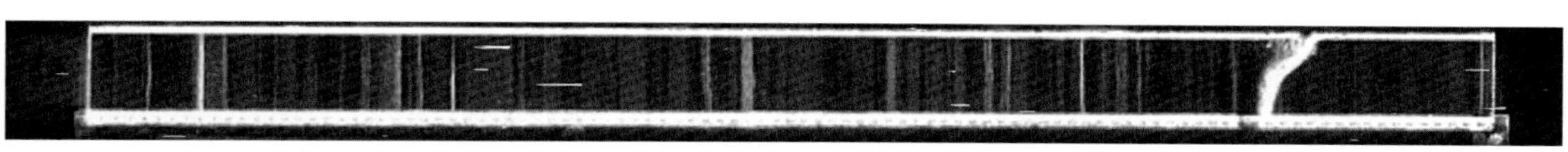

1531

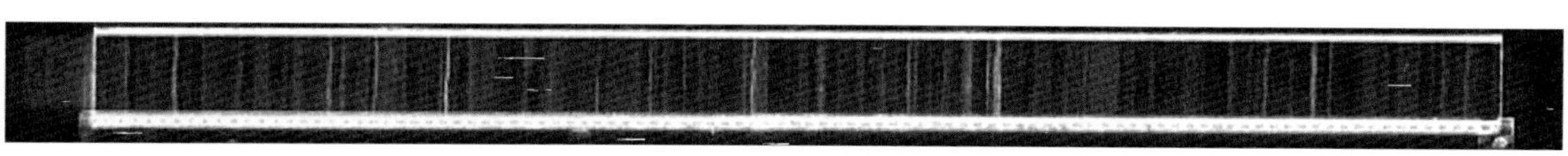

1530

1540

1541

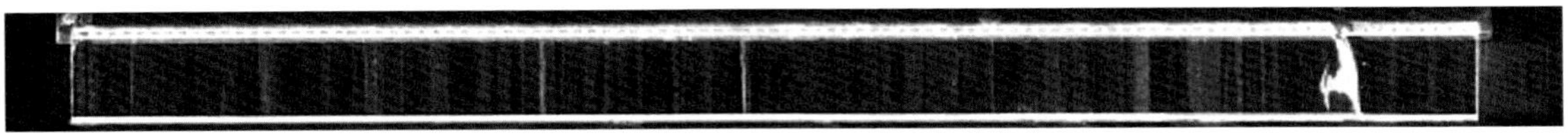

1542

1543

1544

1545

1546

1547

1548

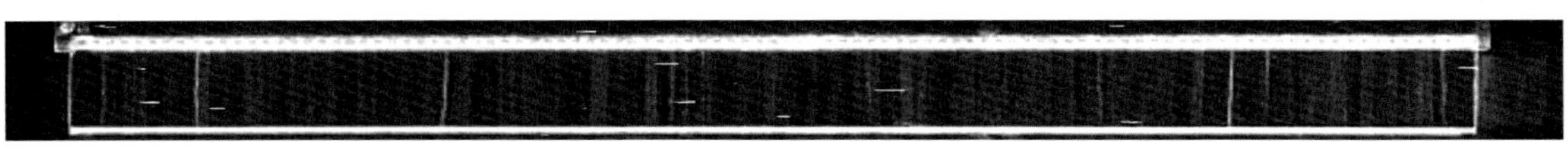

1549

1559

1558

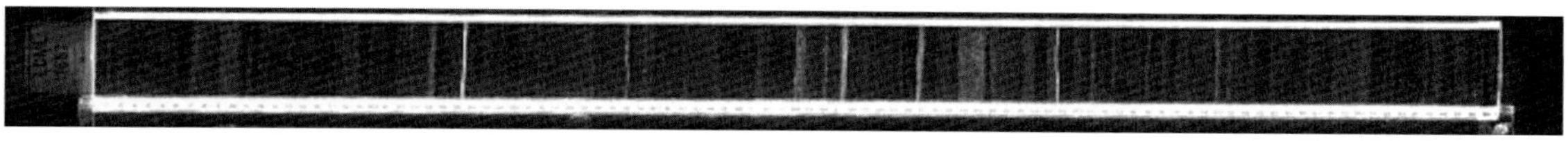
1557

1556

1555

1554

1553

1552

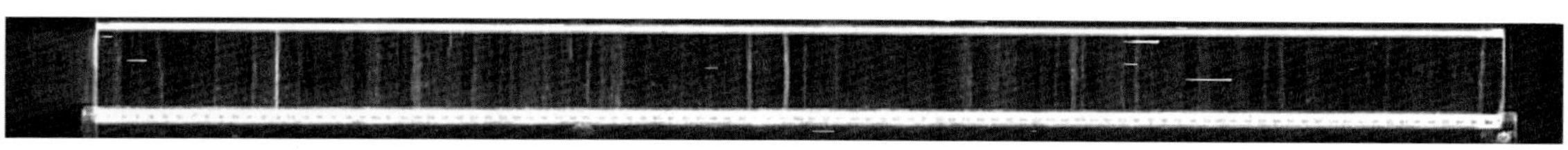
1551

1550

1560

1561

1562

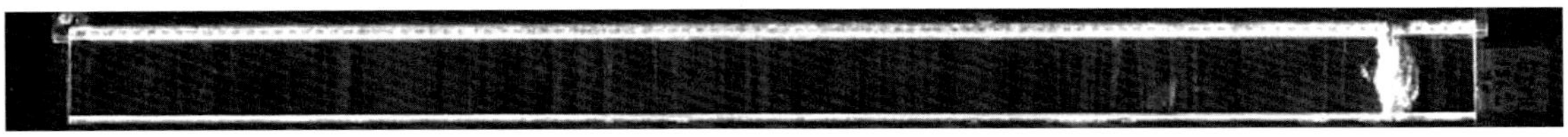

1563

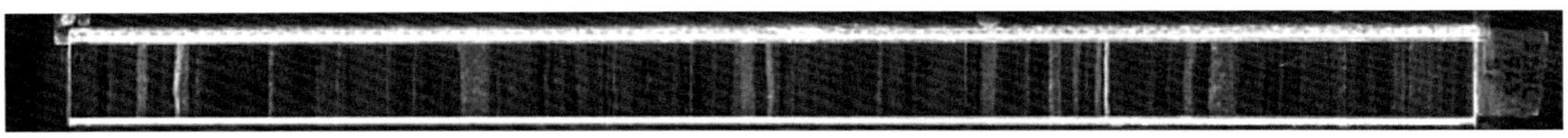

1564

1565

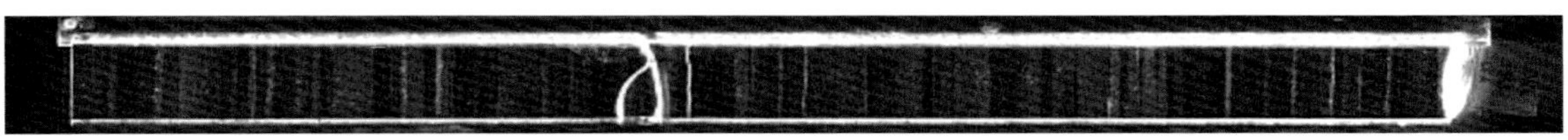

1566

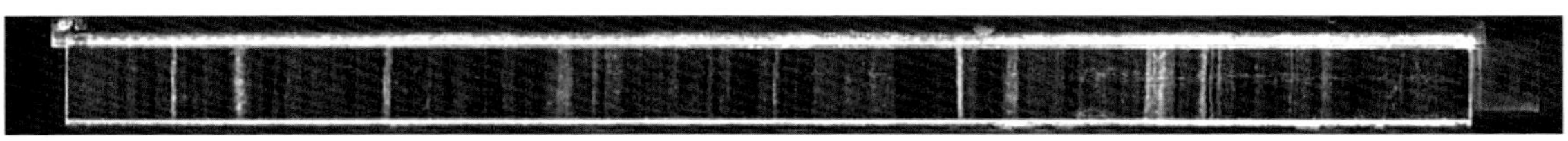

1567

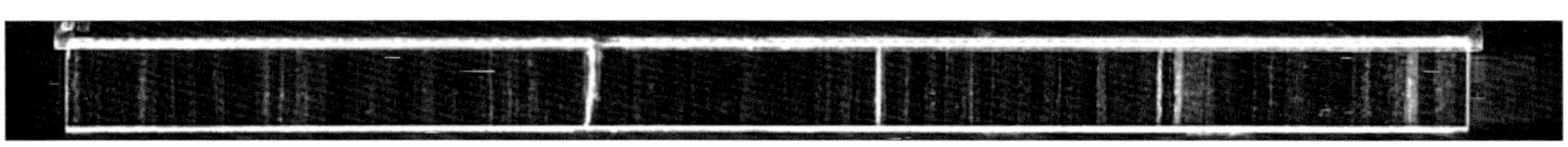

1568

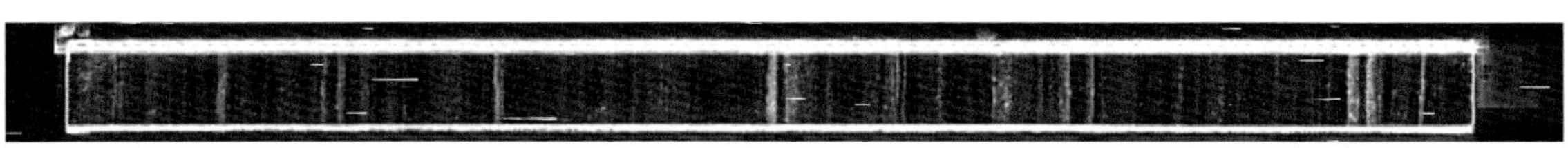

1569

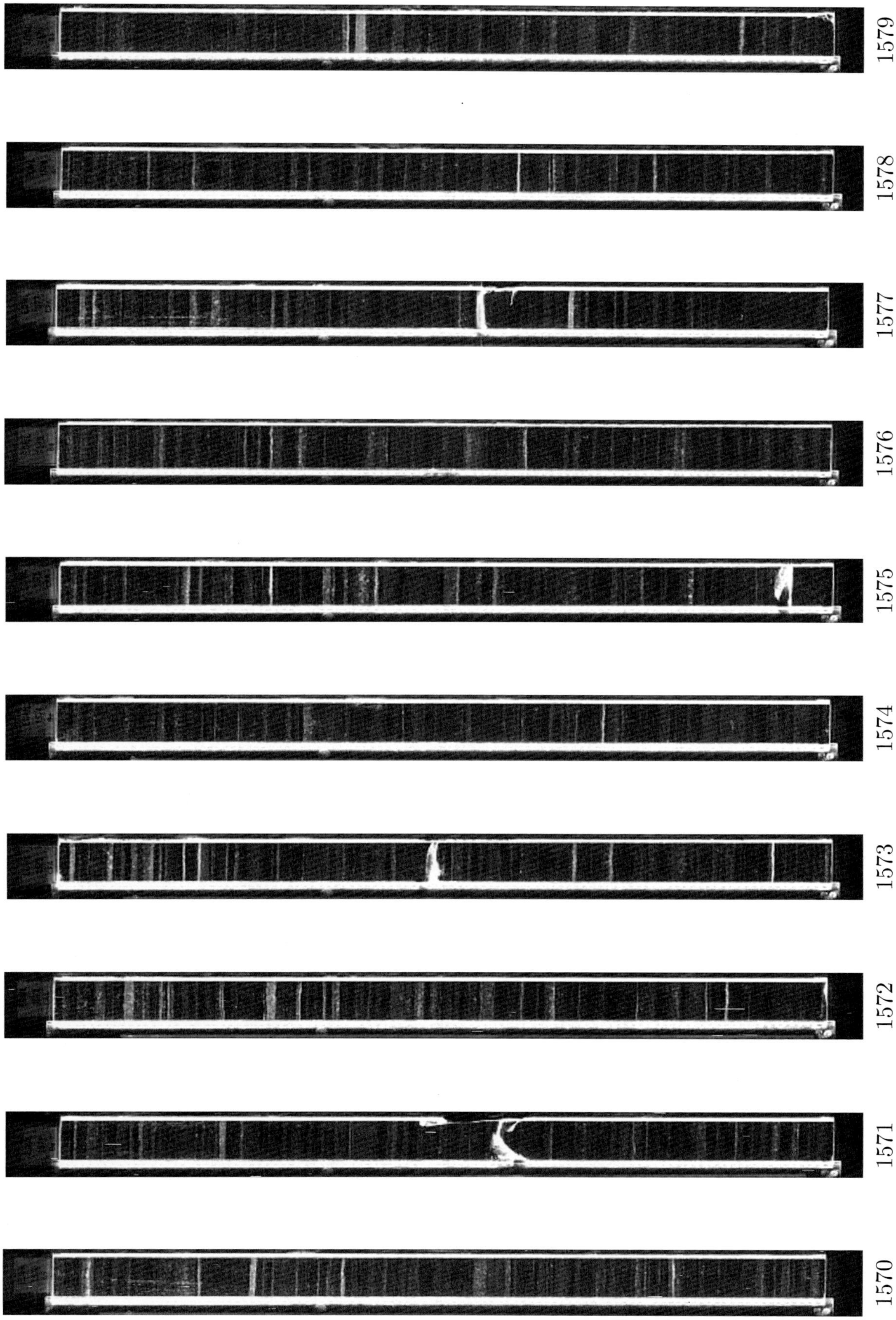
1579
1578
1577
1576
1575
1574
1573
1572
1571
1570

1580

1581

1582

1583

1584

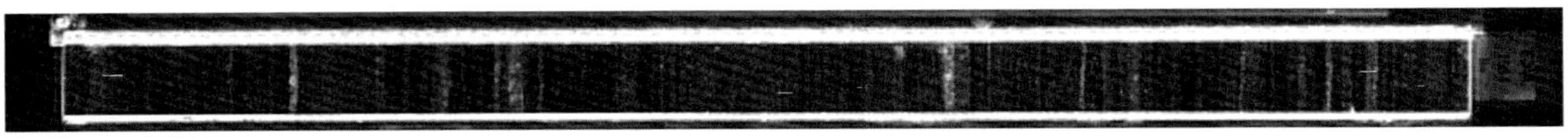

1585

1586

1587

1588

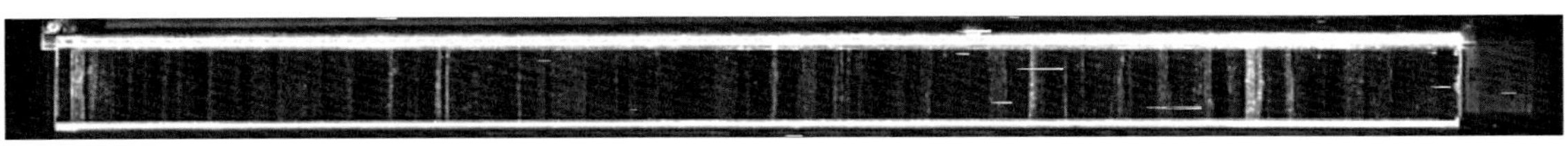

1589

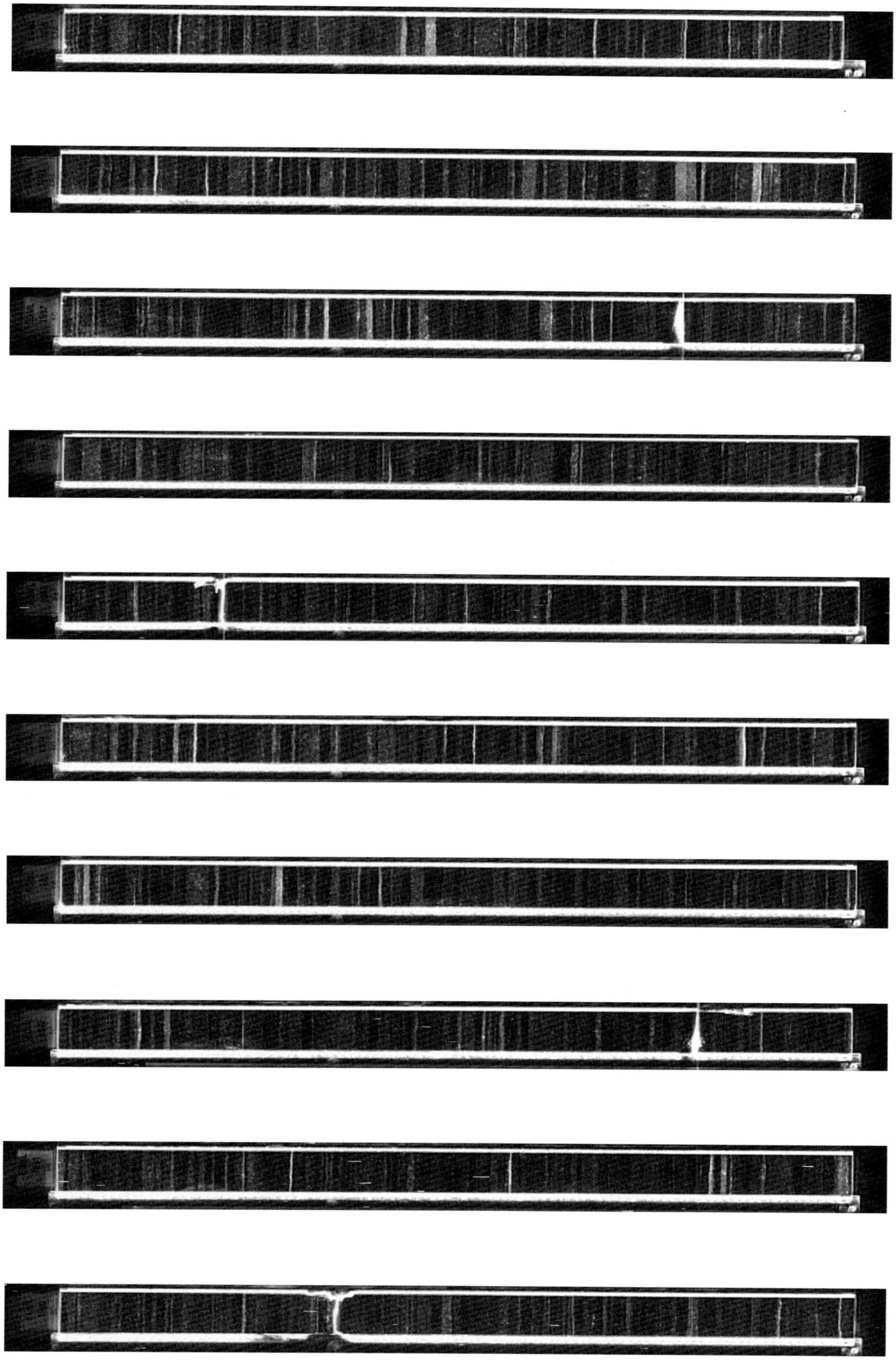

1599

1598

1597

1596

1595

1594

1593

1592

1591

1590

1600

1601

1602

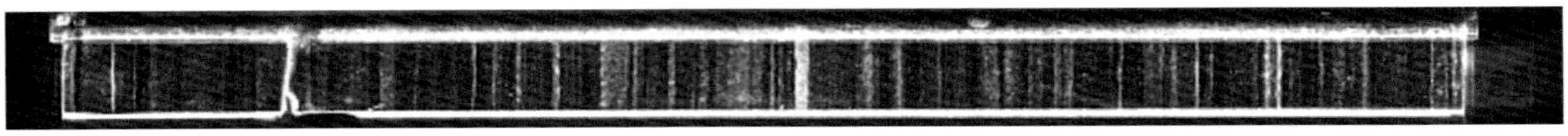

1603

1604

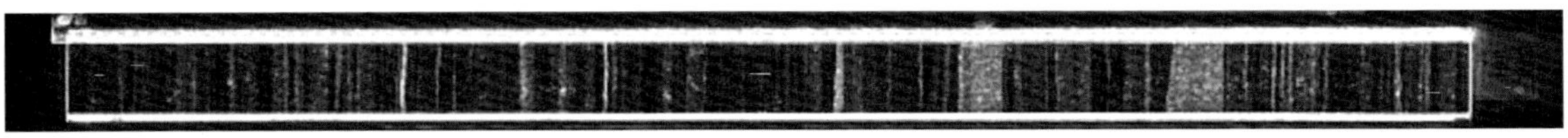

1605

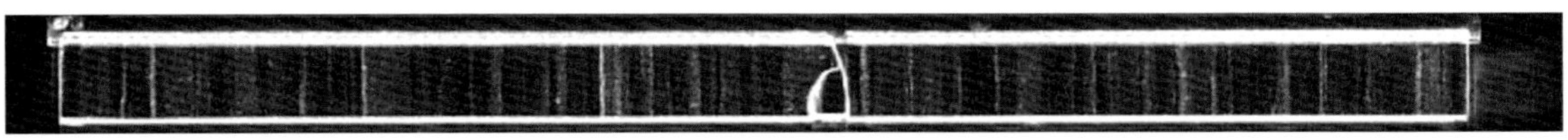

1606

1607

1608

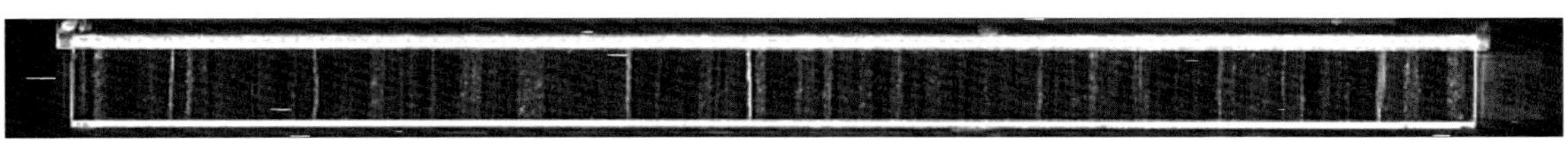

1609

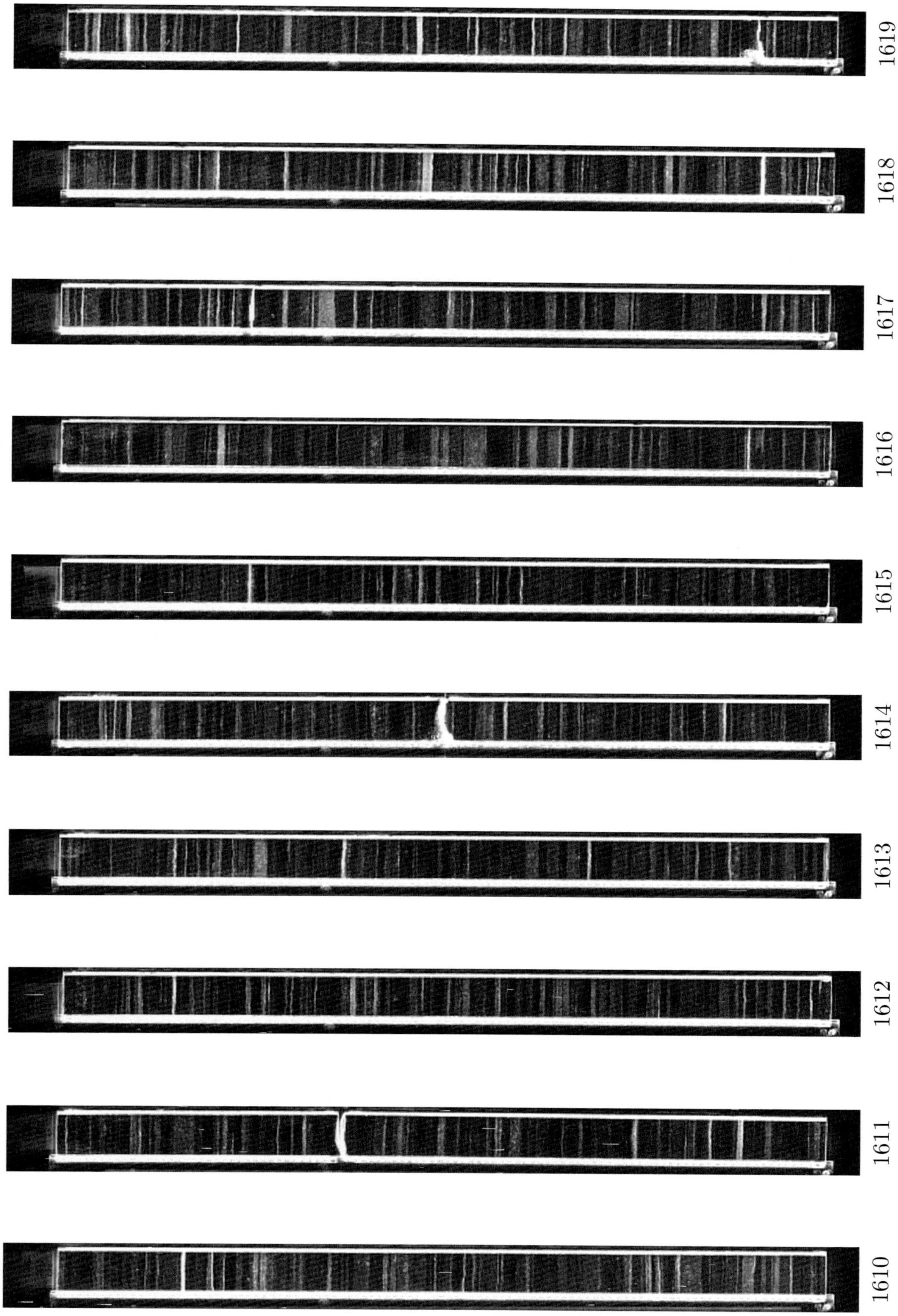
1619
1618
1617
1616
1615
1614
1613
1612
1611
1610

1620

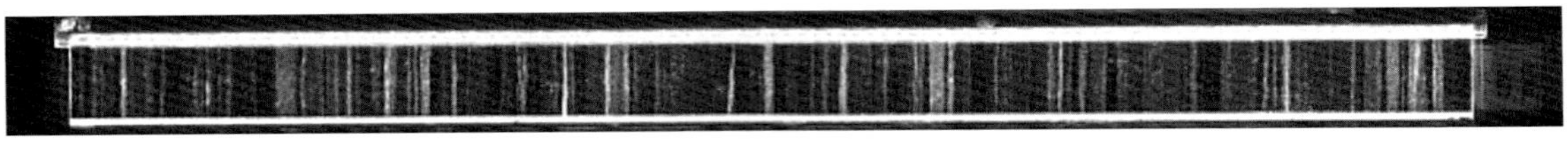

1621

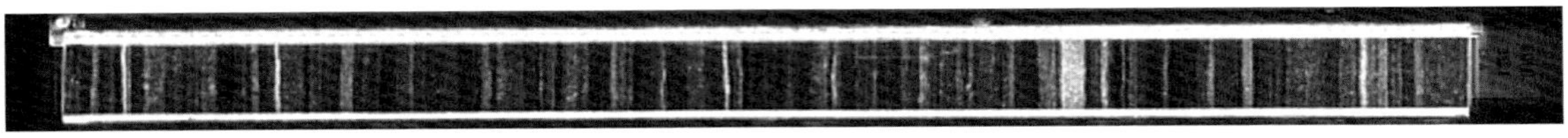

1622

1623

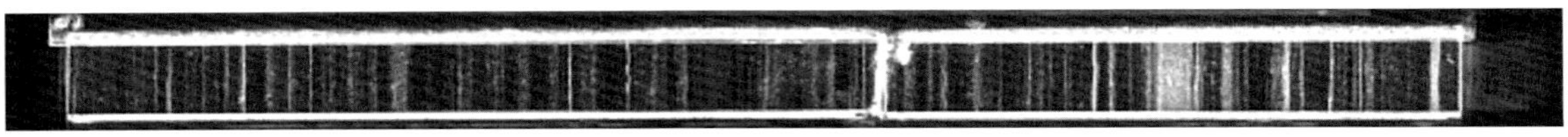

1624

1625

1626

1627

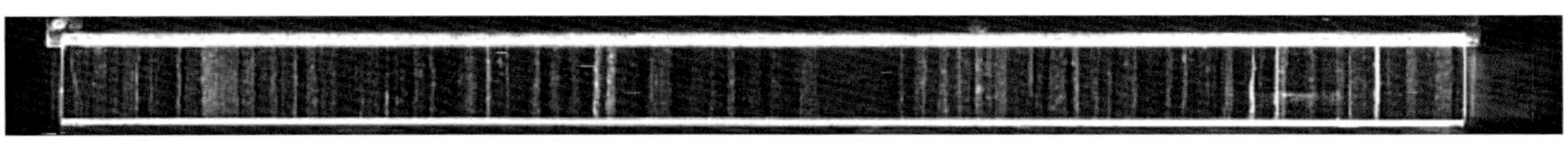

1628

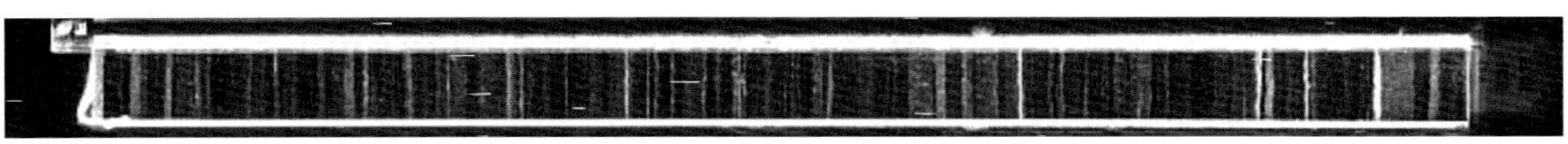

1629

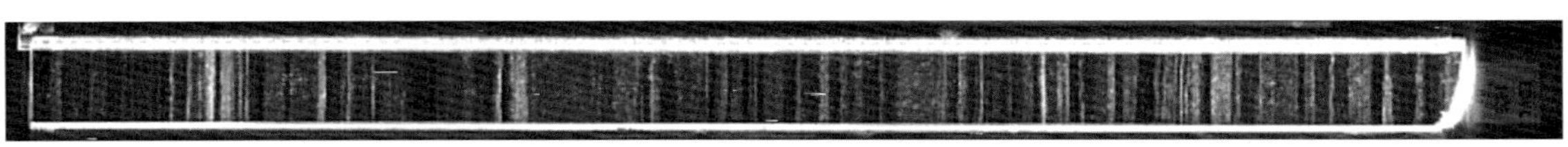

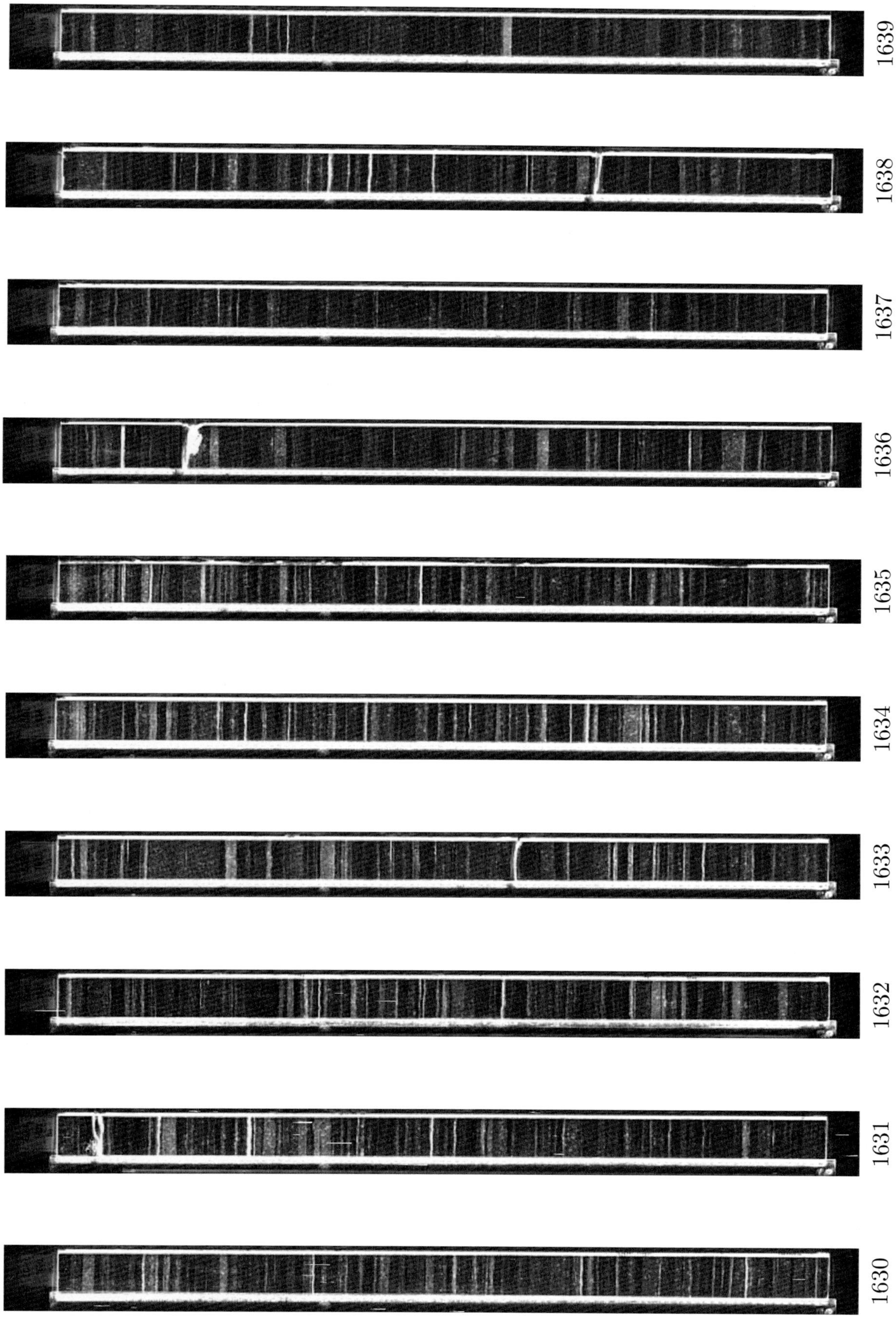
1639
1638
1637
1636
1635
1634
1633
1632
1631
1630

1640

1641

1642

1643

1644

1645

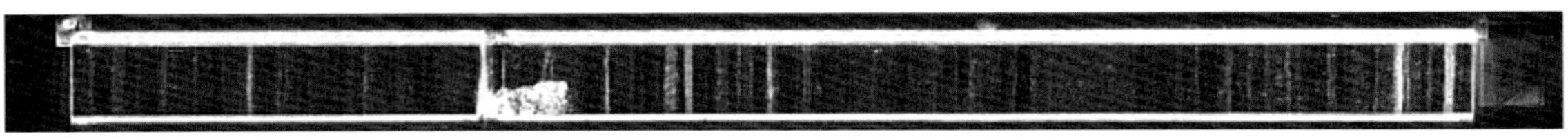

1646

1647

1648

1649

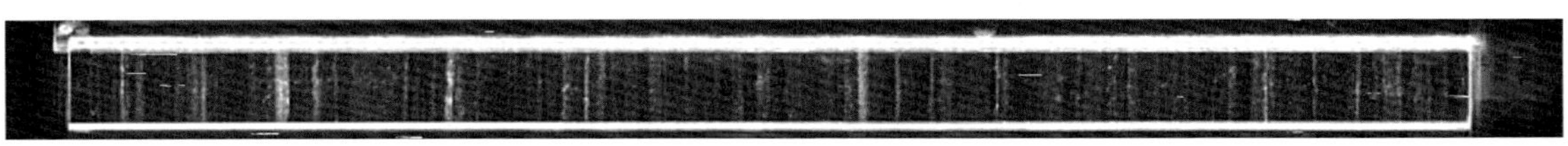

1659

1658

1657

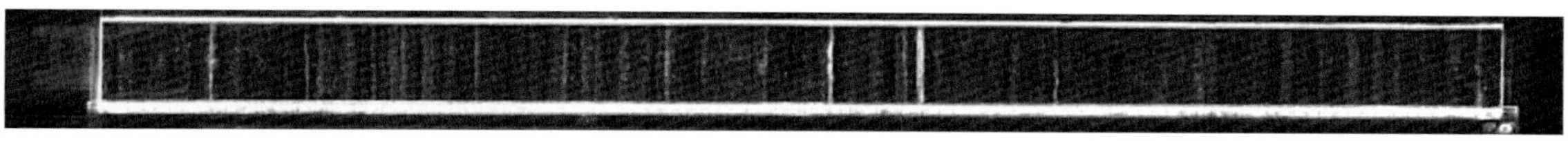

1656

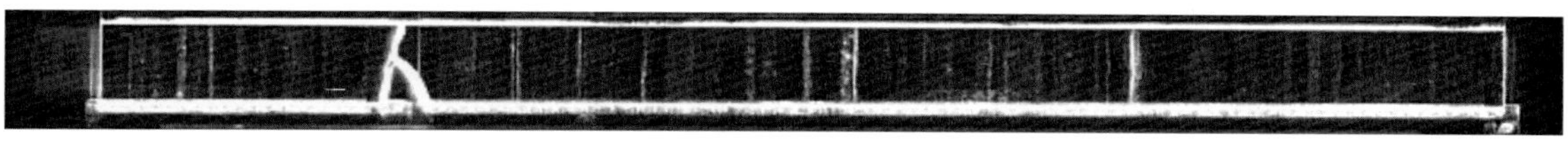

1655

1654

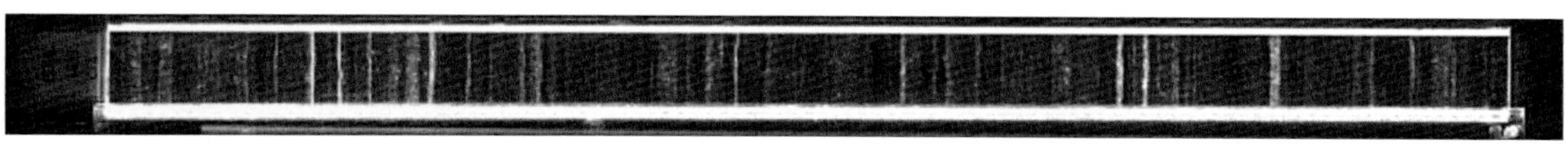

1653

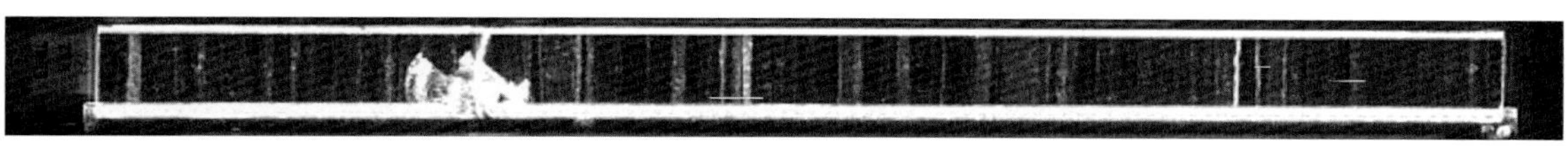

1652

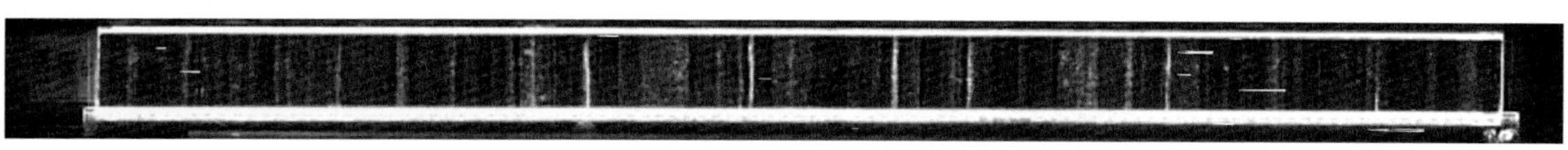

1651

1650

1660

1661

1662

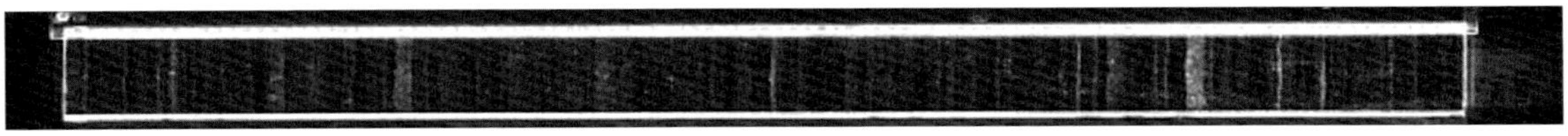

1663

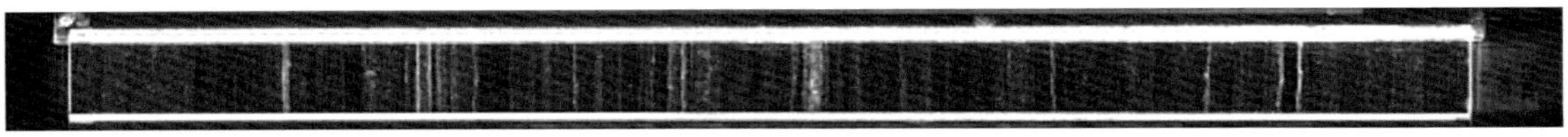

1664

1665

1666

1667

1668

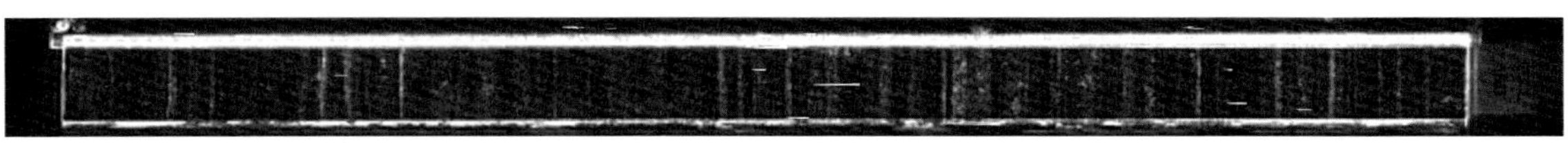

1669

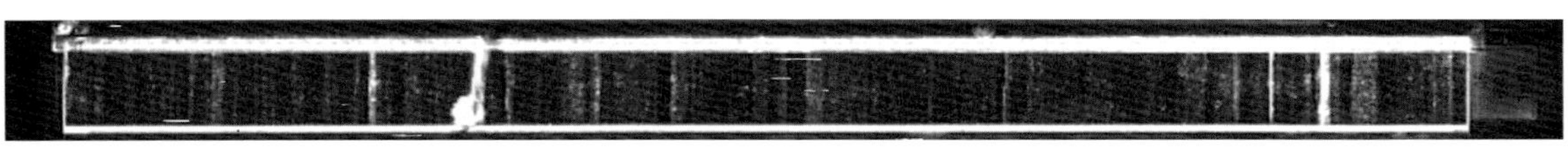

1679

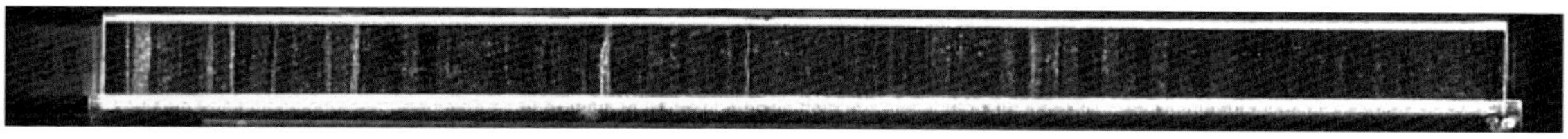
1678

1677

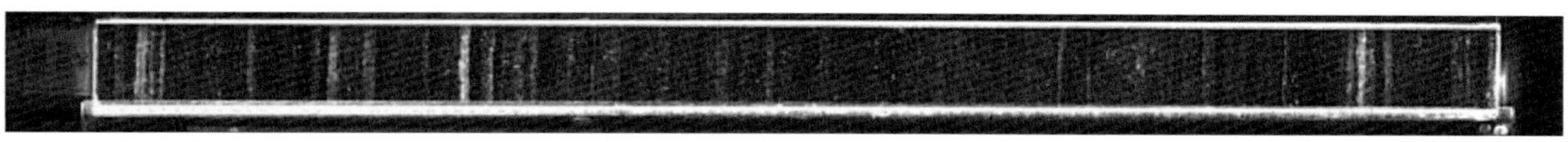
1676

1675

1674

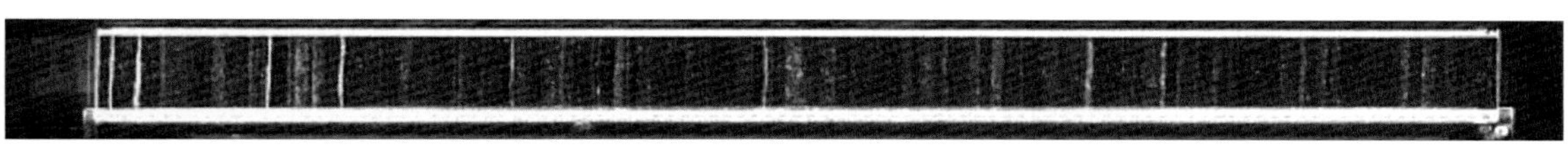
1673

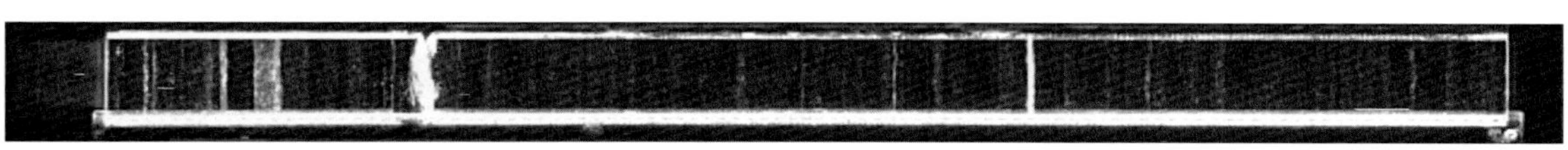
1672

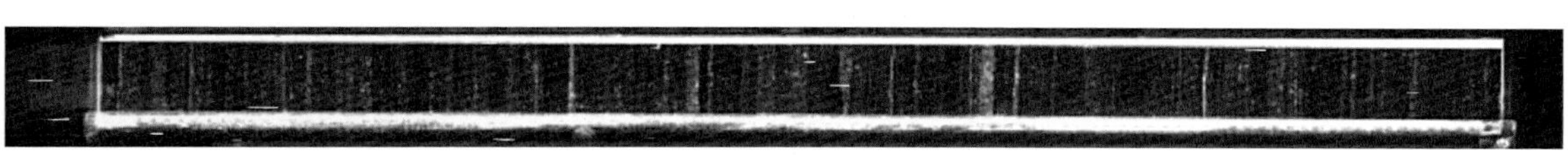
1671

1670

1680

1681

1682

1683

1684
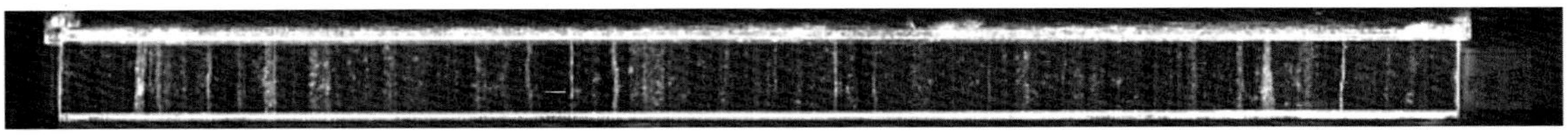

1685

1686

1687
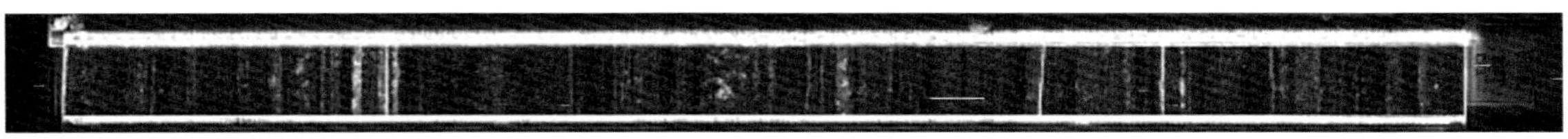

1688
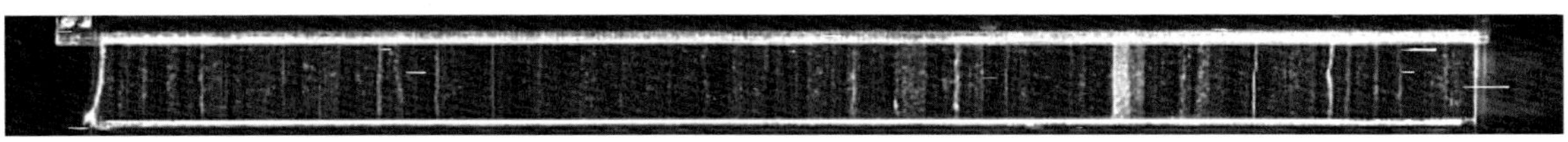

1689

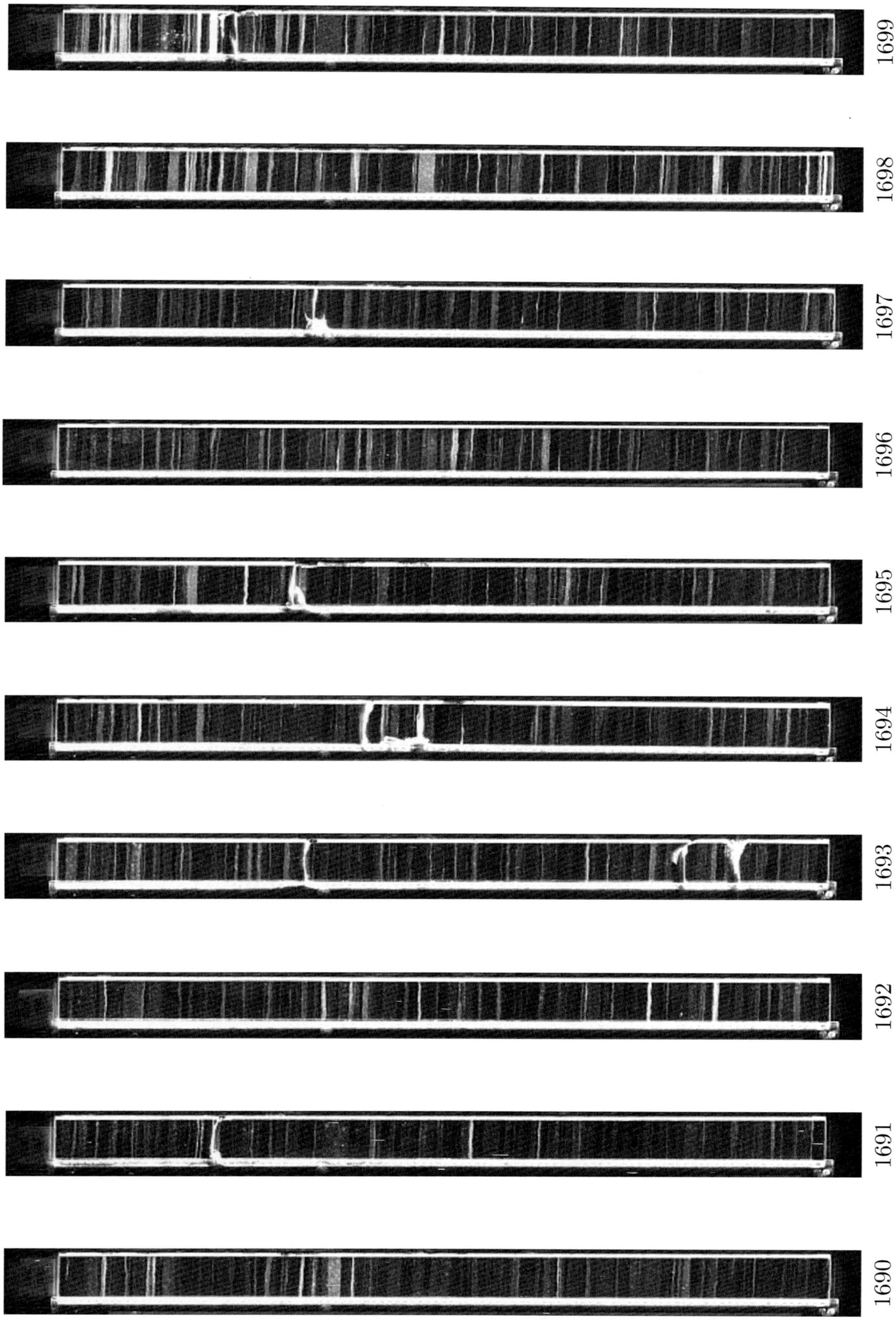
1699
1698
1697
1696
1695
1694
1693
1692
1691
1690

1700

1701

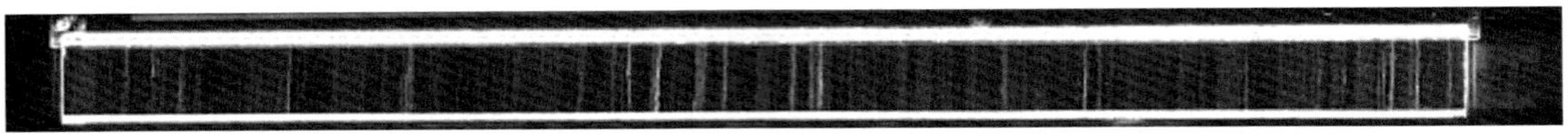

1702

1703

1704

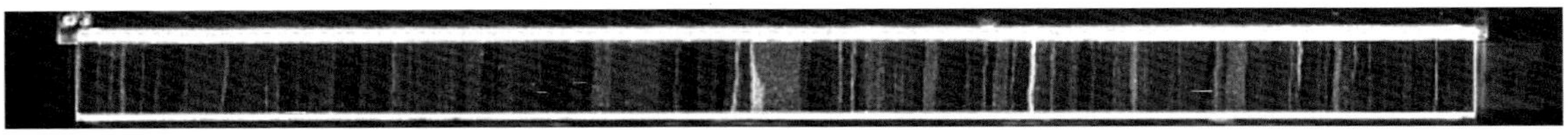

1705

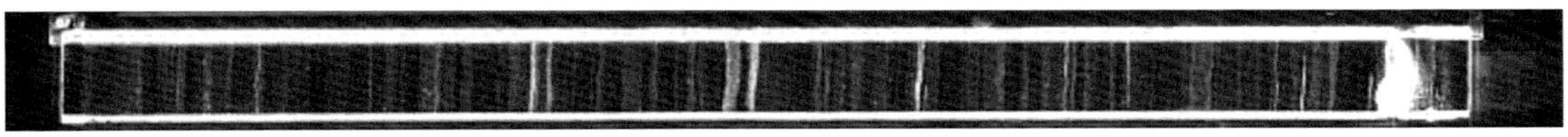

1706

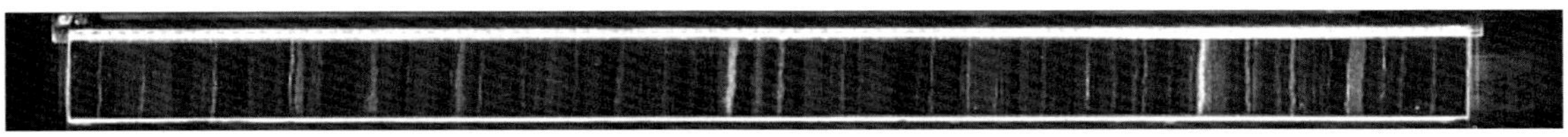

1707

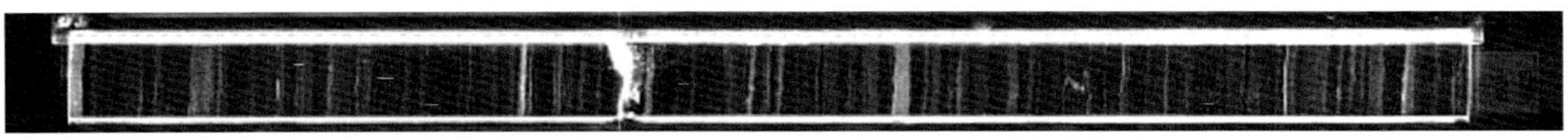

1708

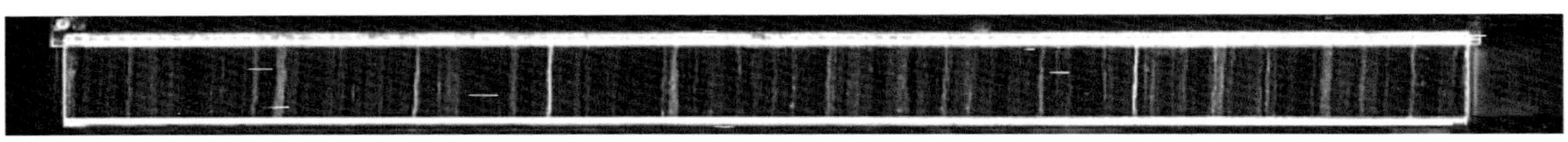

1709

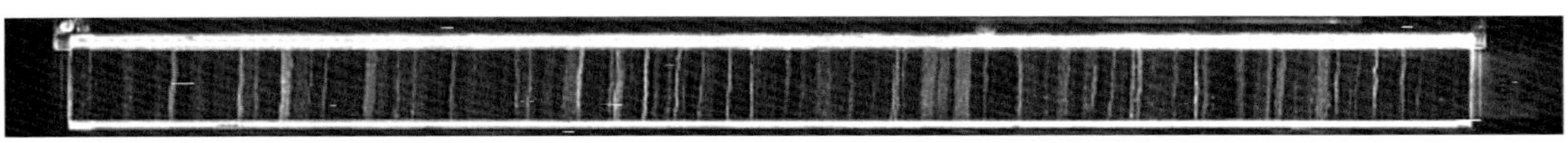

1719

1718

1717

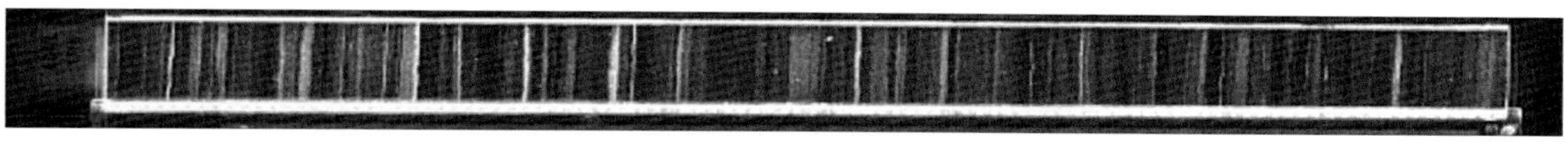
1716

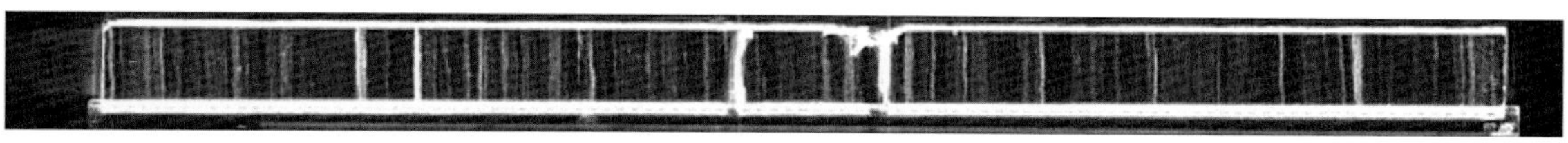
1715

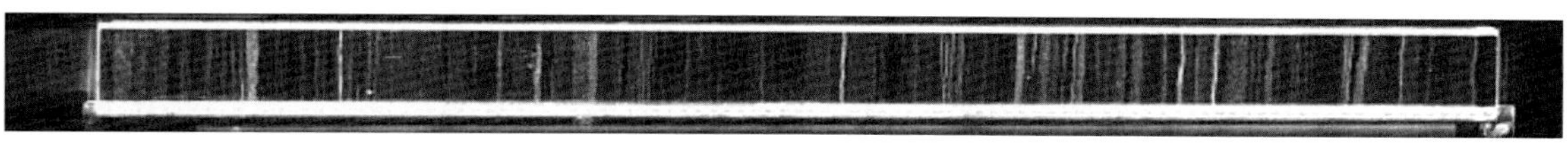
1714

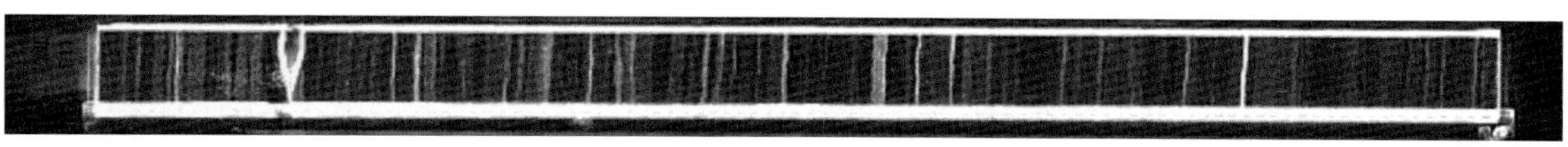
1713

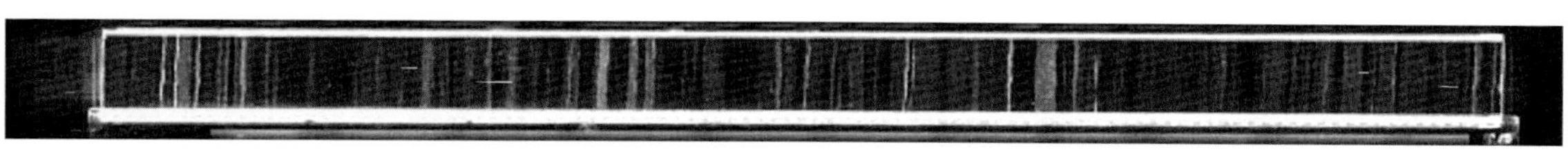
1712

1711

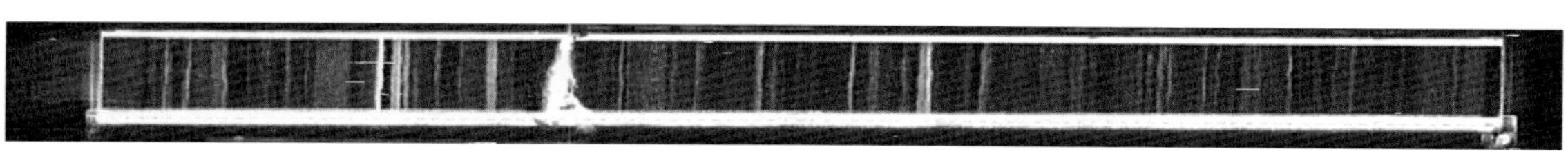
1710

1720

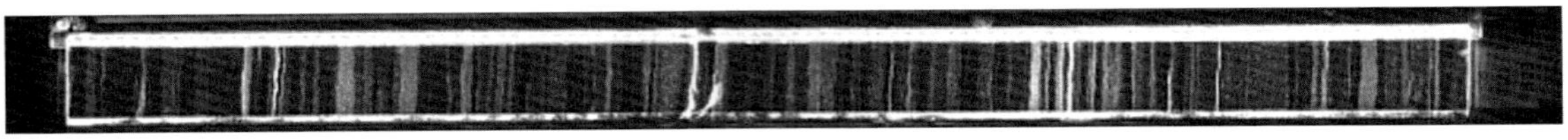

1721

1722

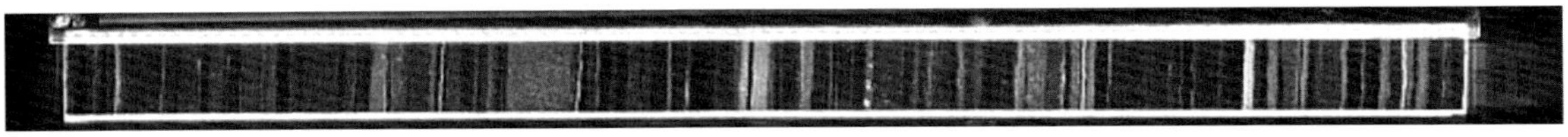

1723

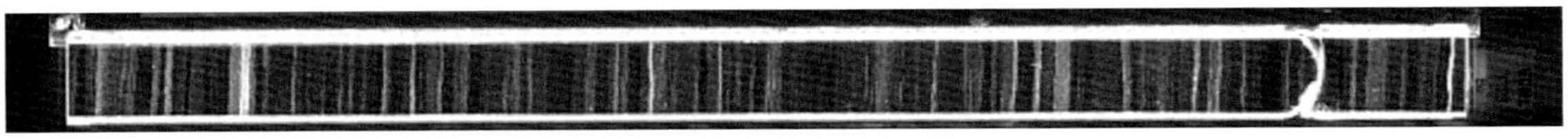

1724

1725

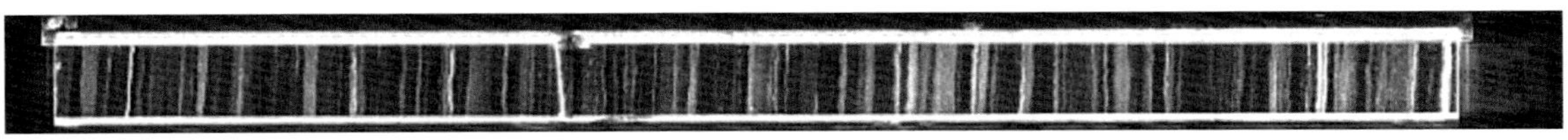

1726

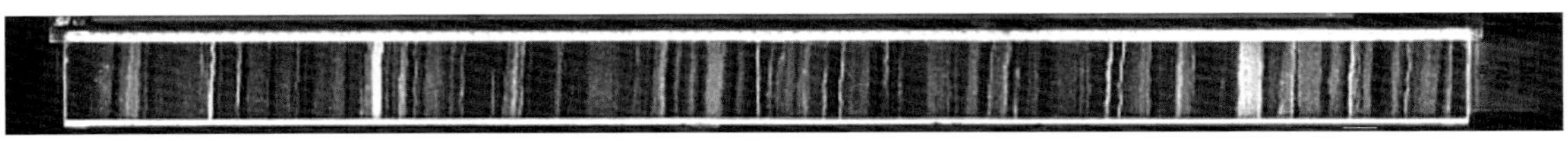

1727

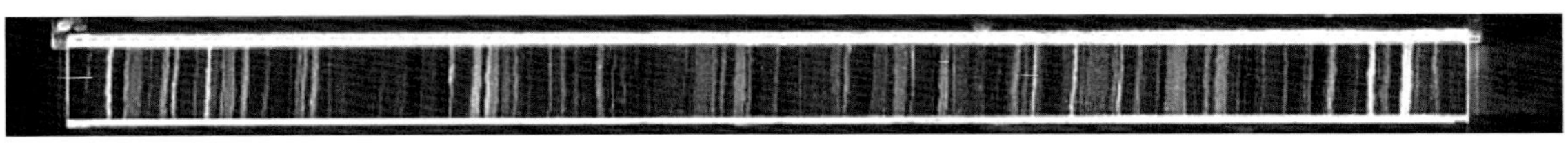

1728

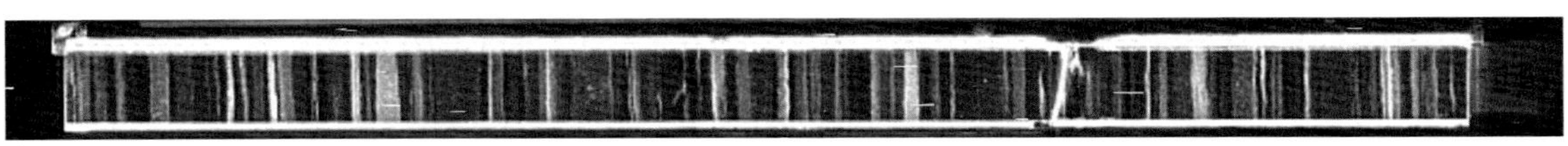

1729

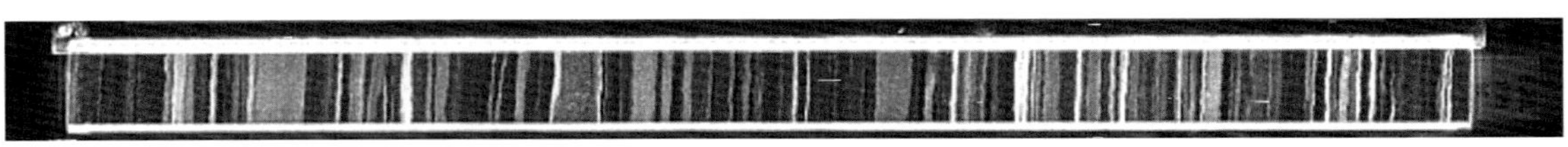

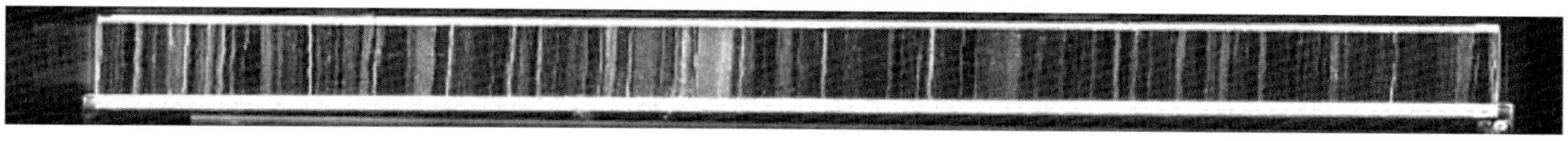

1739

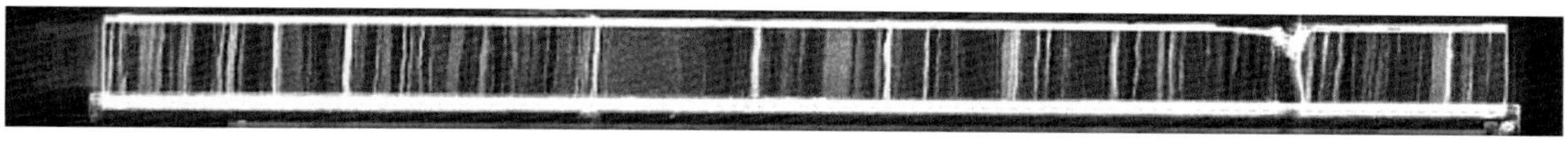

1738

1737

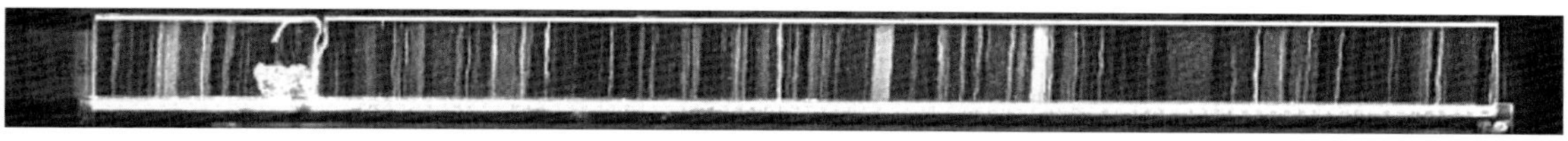

1736

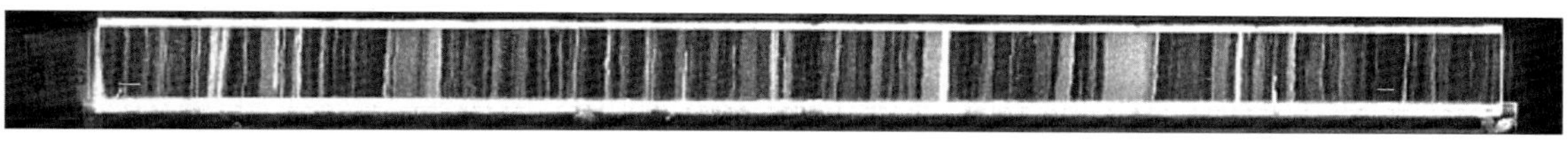

1735

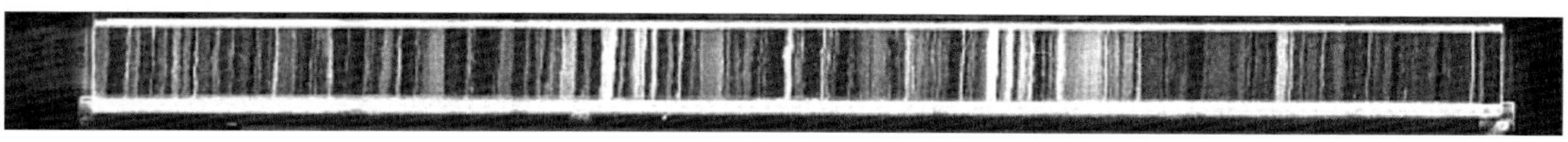

1734

1733

1732

1731

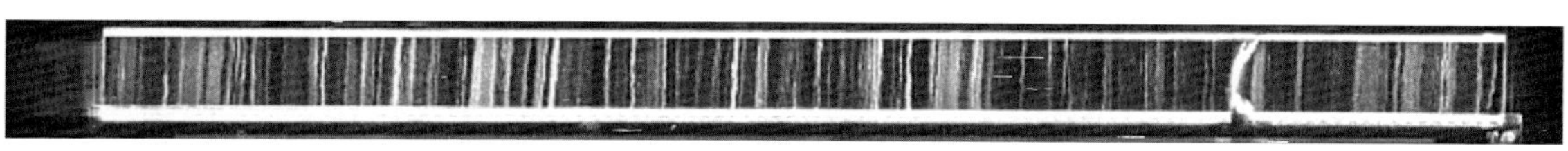

1730

1740

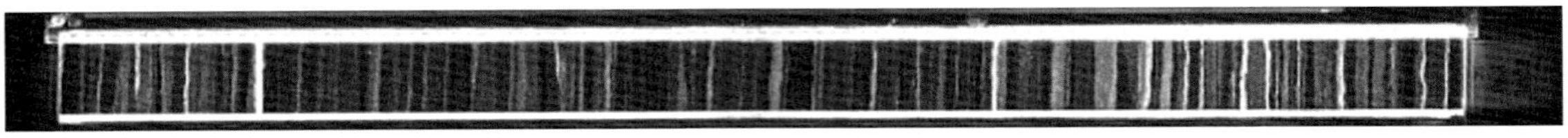

1741

1742

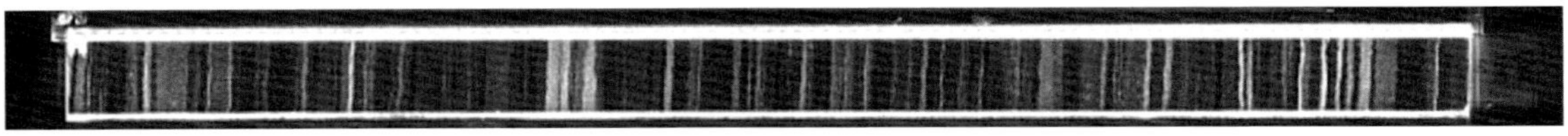

1743

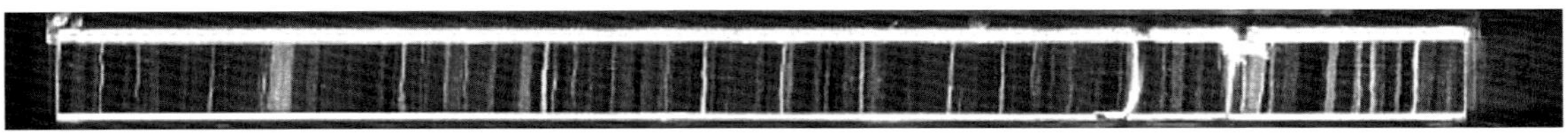

1744

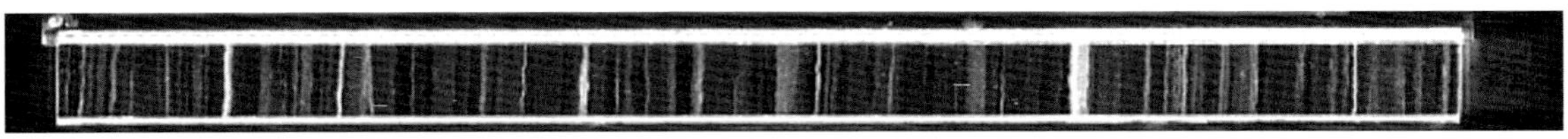

1745

1746

1747

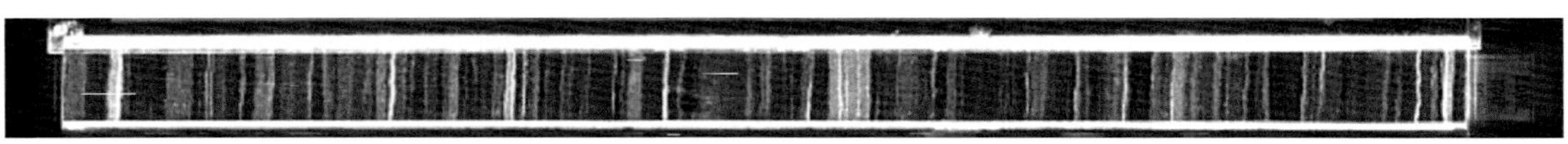

1748

1749

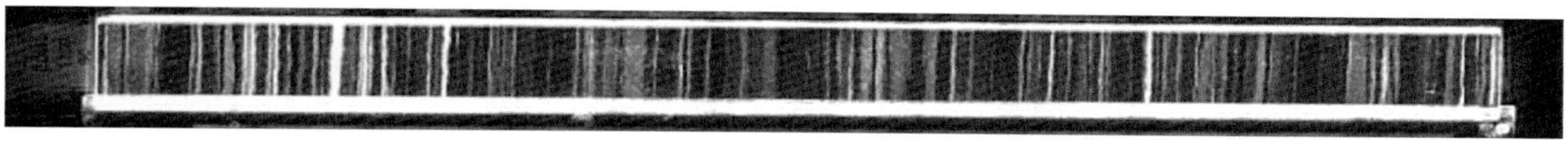

1759

1758

1757

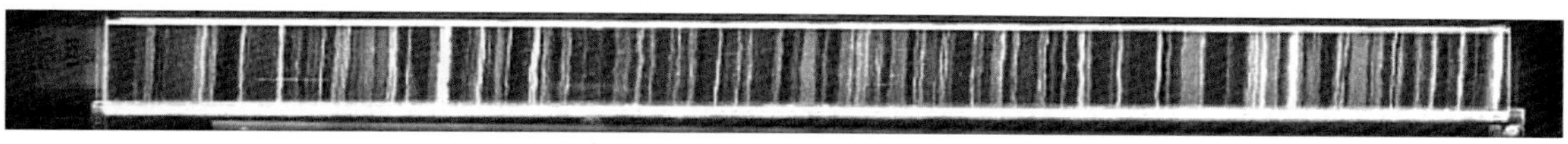

1756

1755

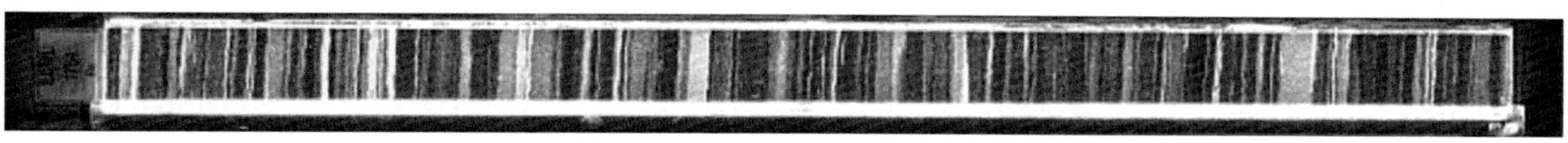

1754

1753

1752

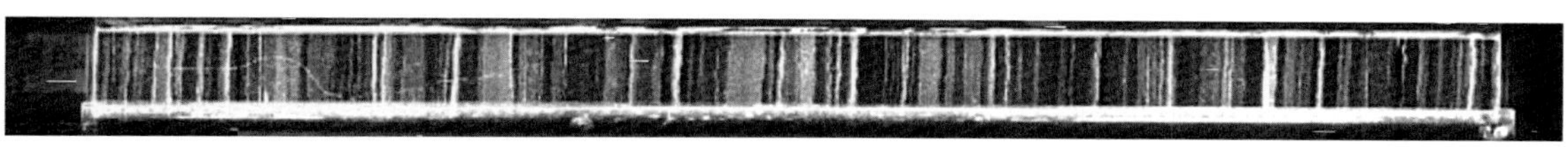

1751

1750

1760

1761

1762

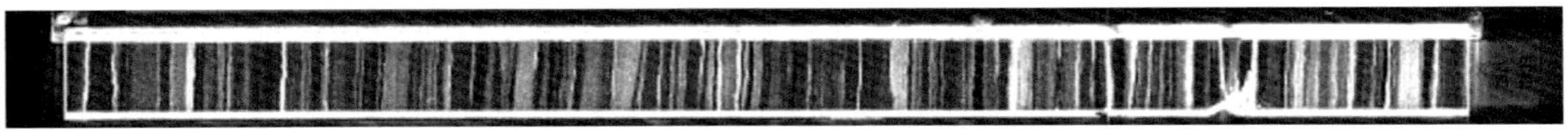

1763

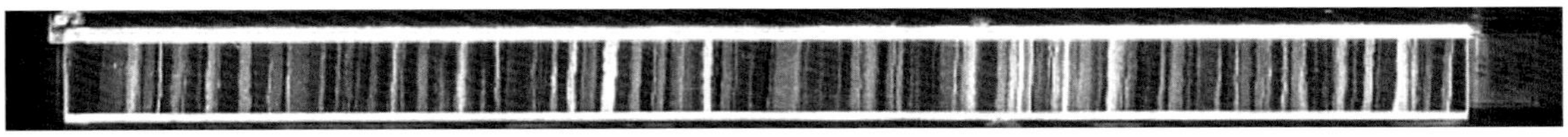

1764

1765

1766

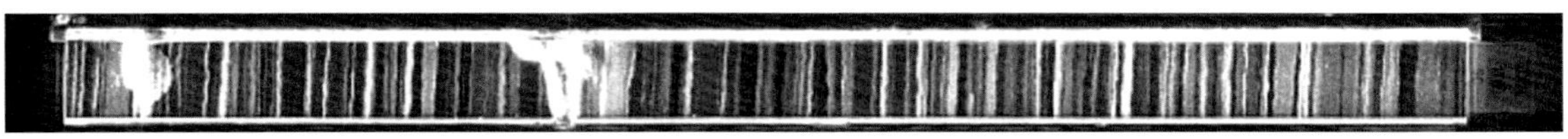

1767

1768

1769

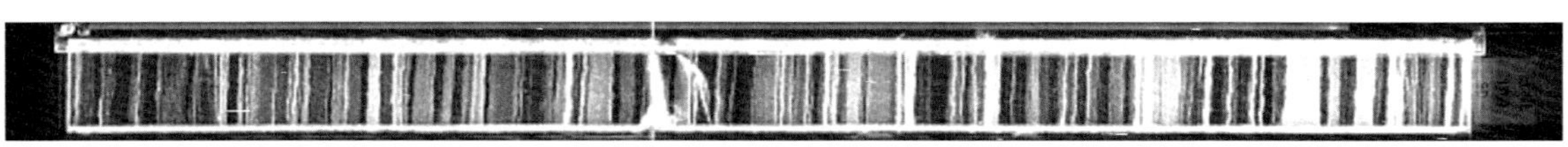

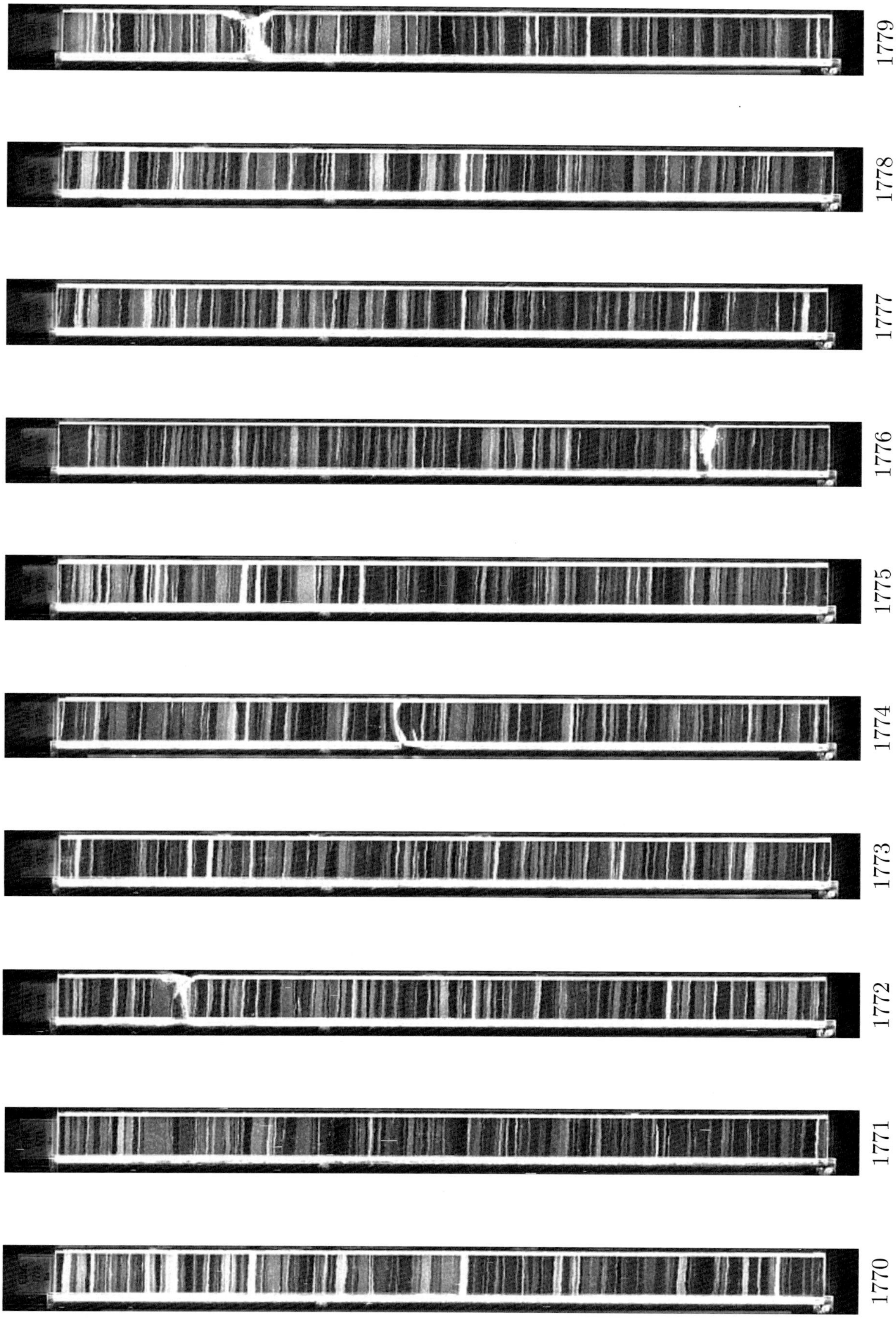
1779
1778
1777
1776
1775
1774
1773
1772
1771
1770

1780

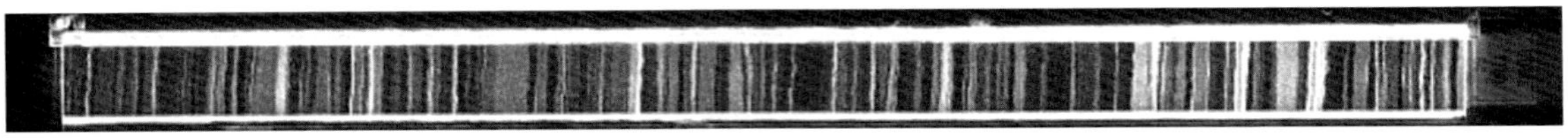

1781

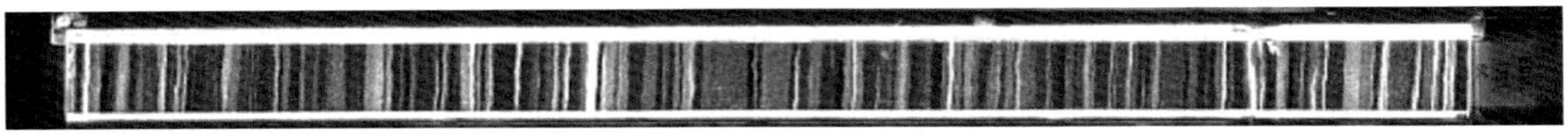

1782

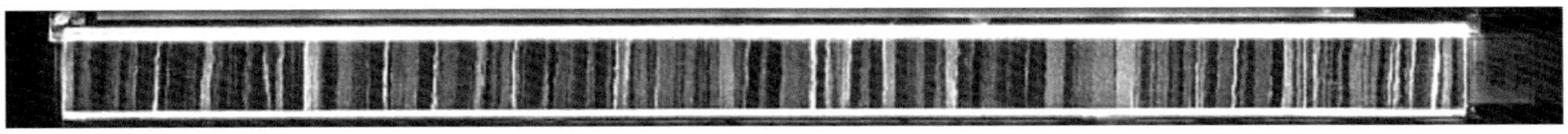

1783

1784

1785

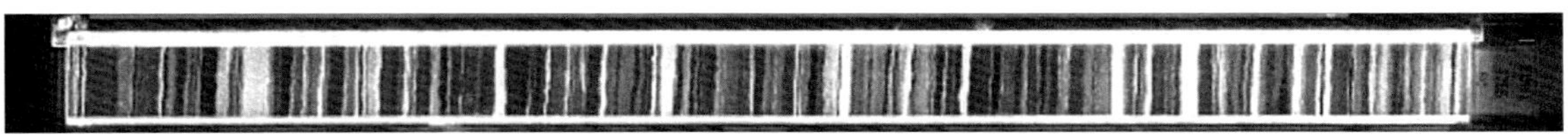

1786

1787

1788

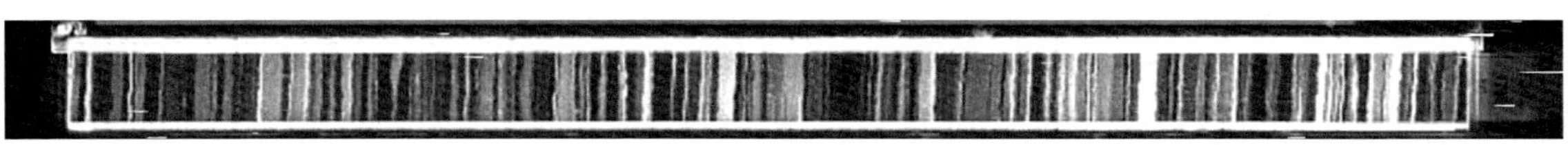

1789

1799

1798

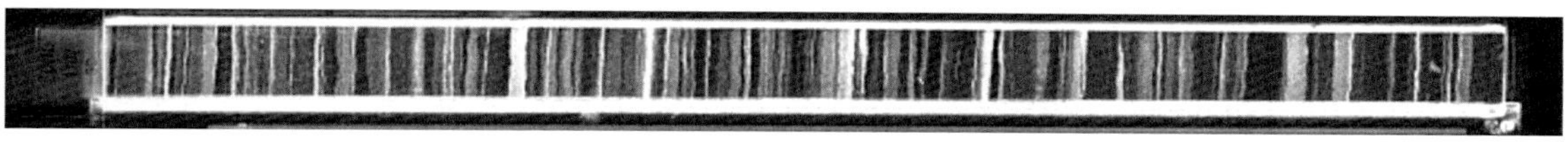

1797

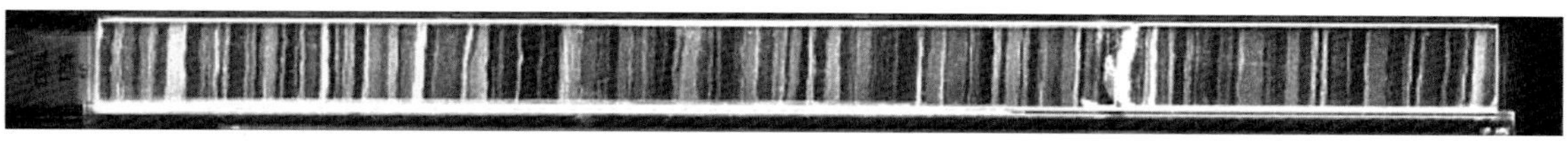

1796

1795

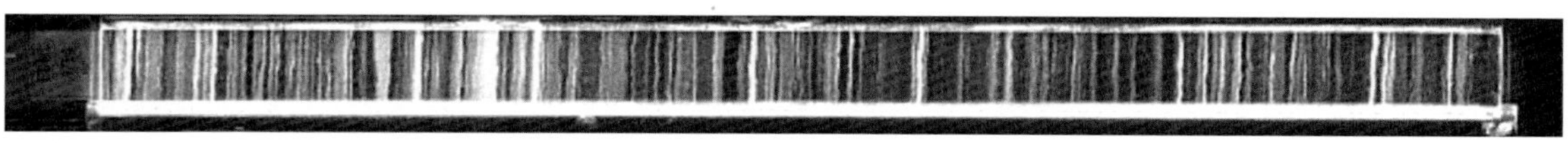

1794

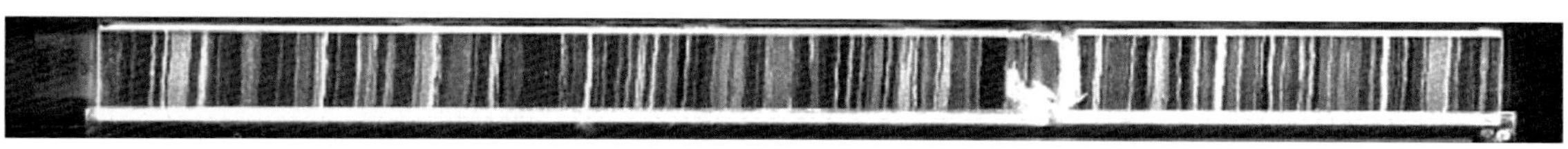

1793

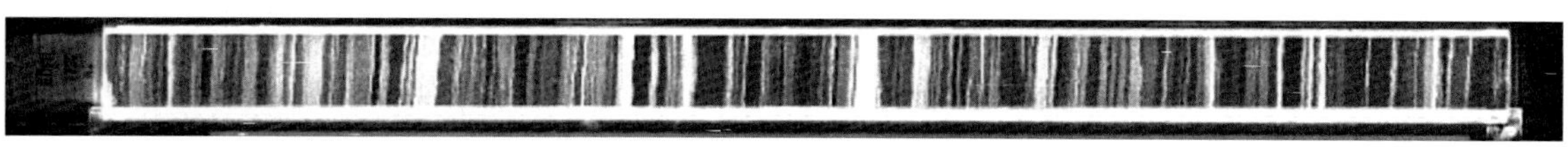

1792

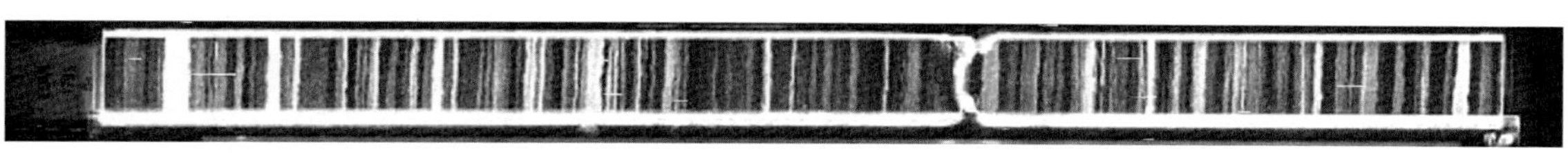

1791

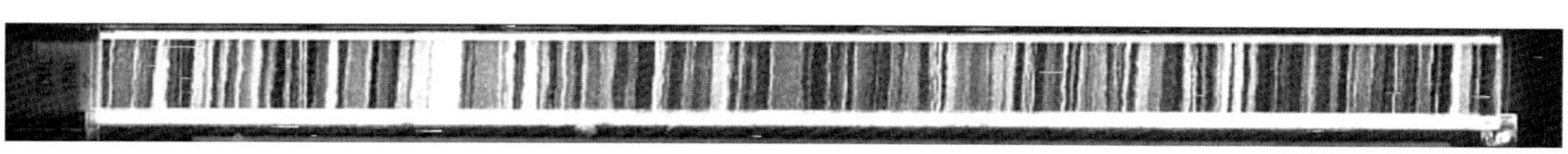

1790

1800

1801

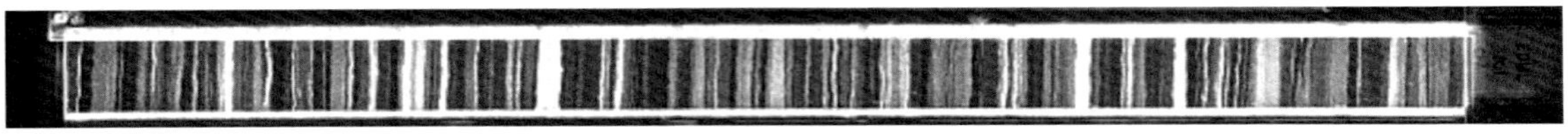

1802

1803

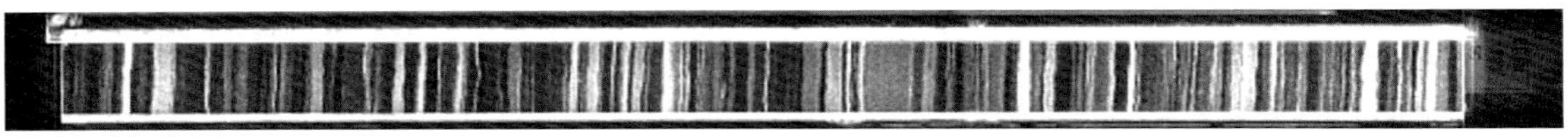

1804

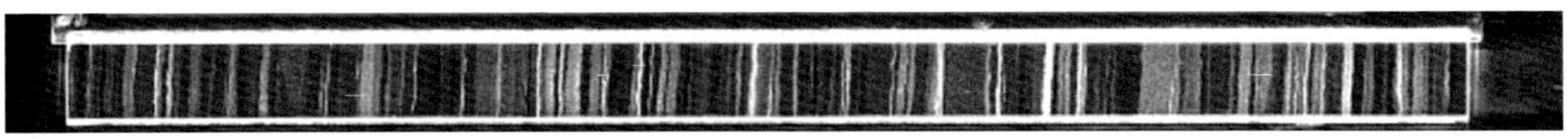

1805

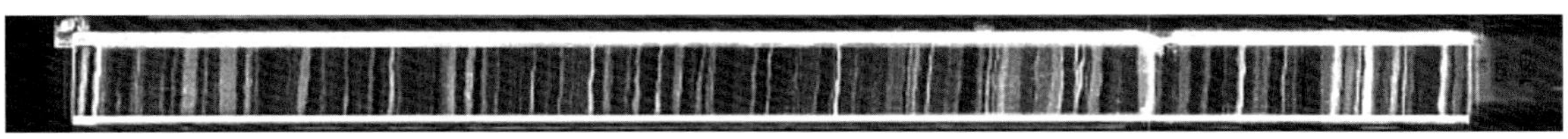

1806

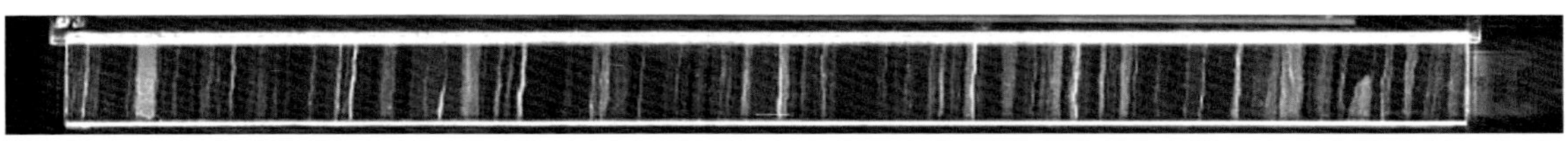

1807

1808

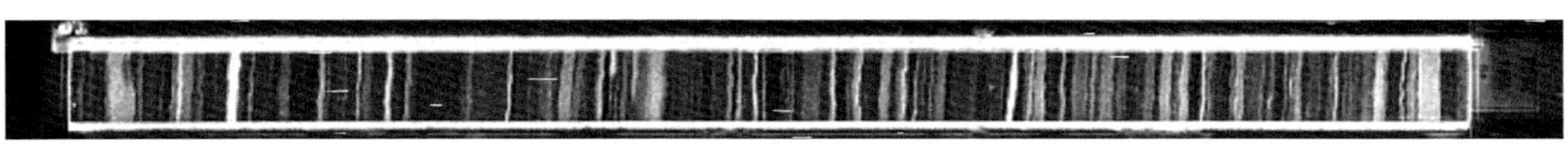

1809

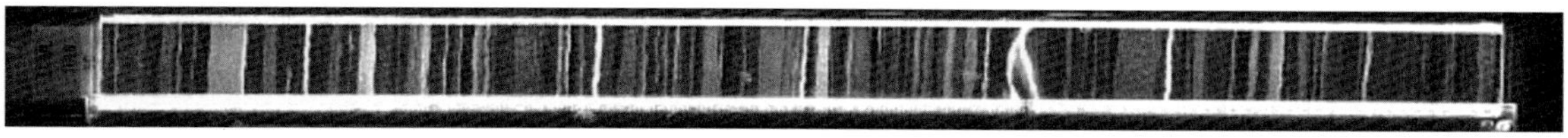

1819

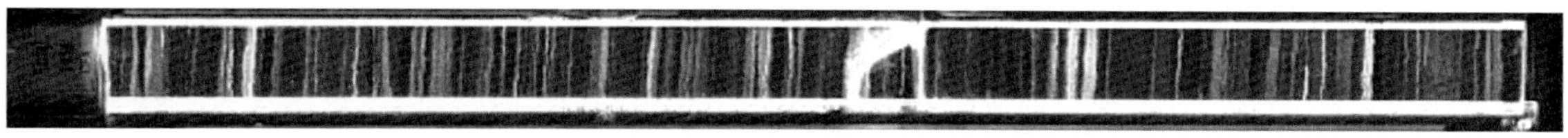

1818

1817

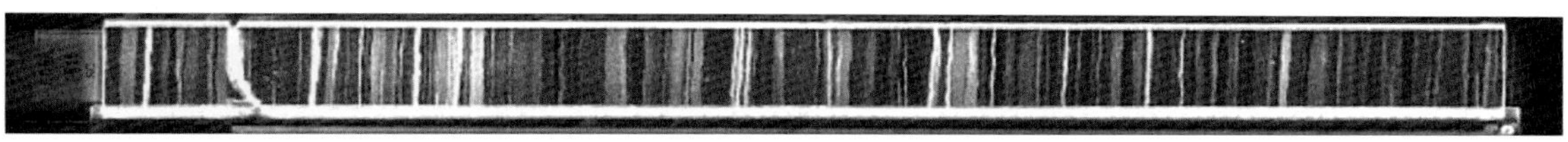

1816

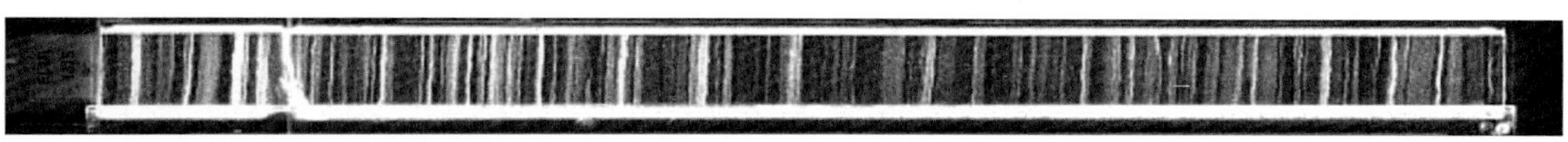

1815

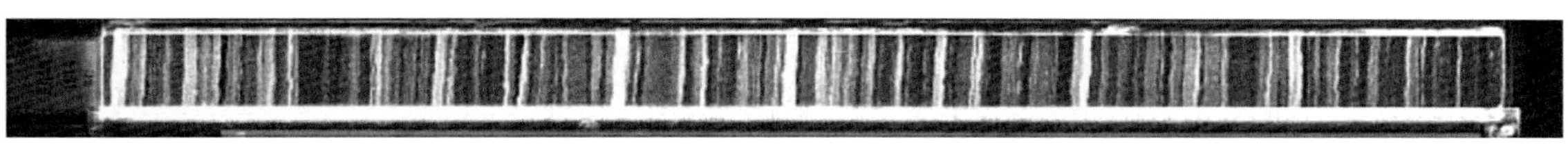

1814

1813

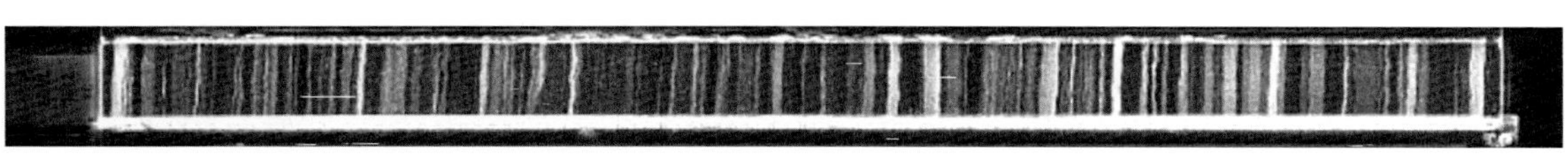

1812

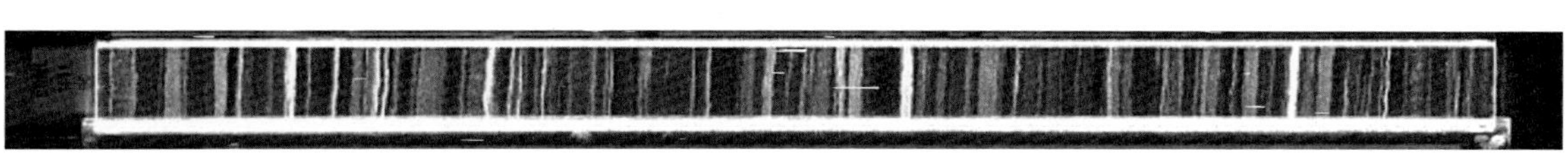

1811

1810

1820
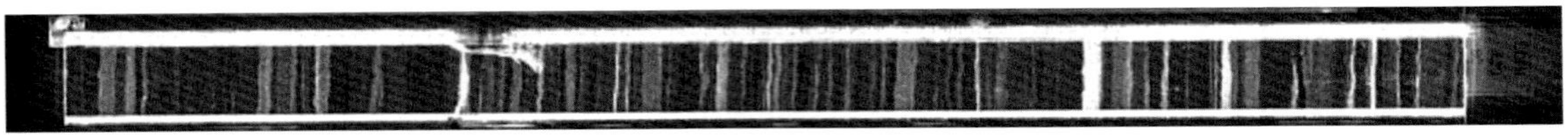

1821
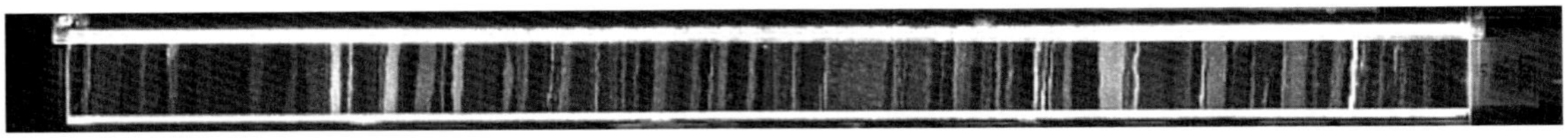

1822

1823

1824
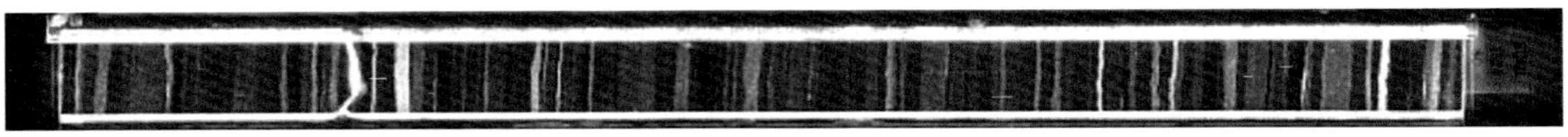

1825
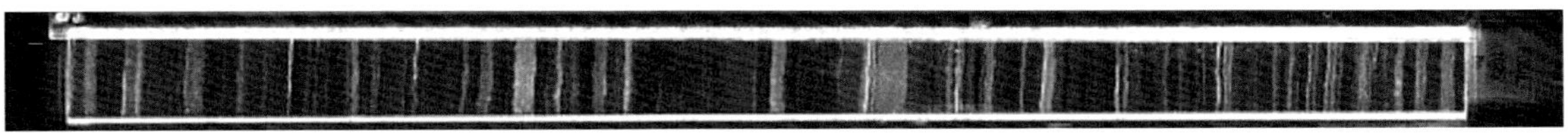

1826
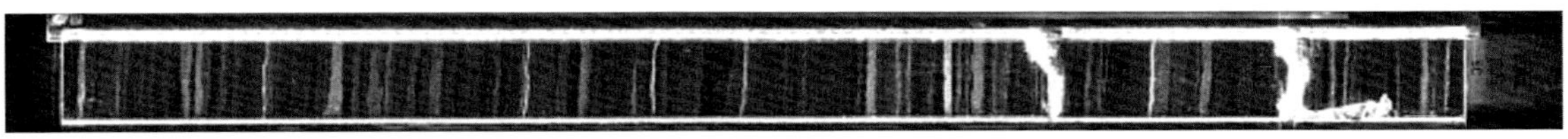

1827
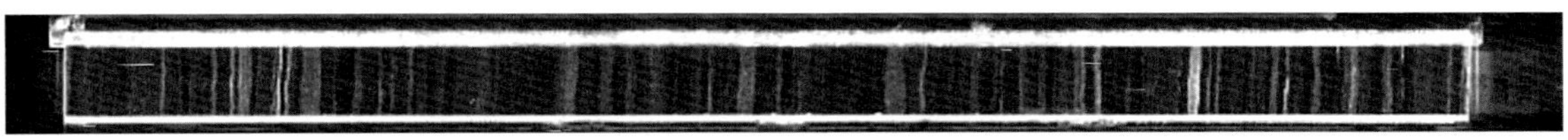

1828

1829
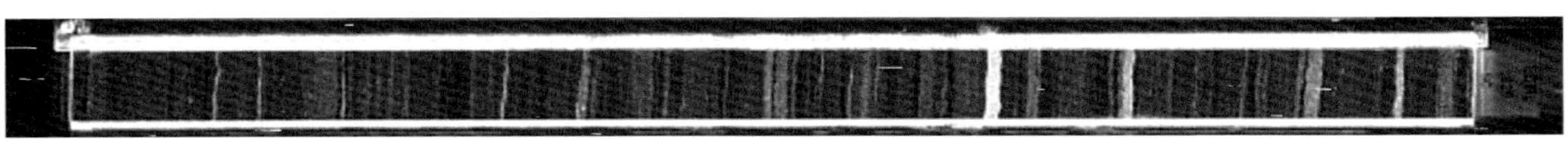

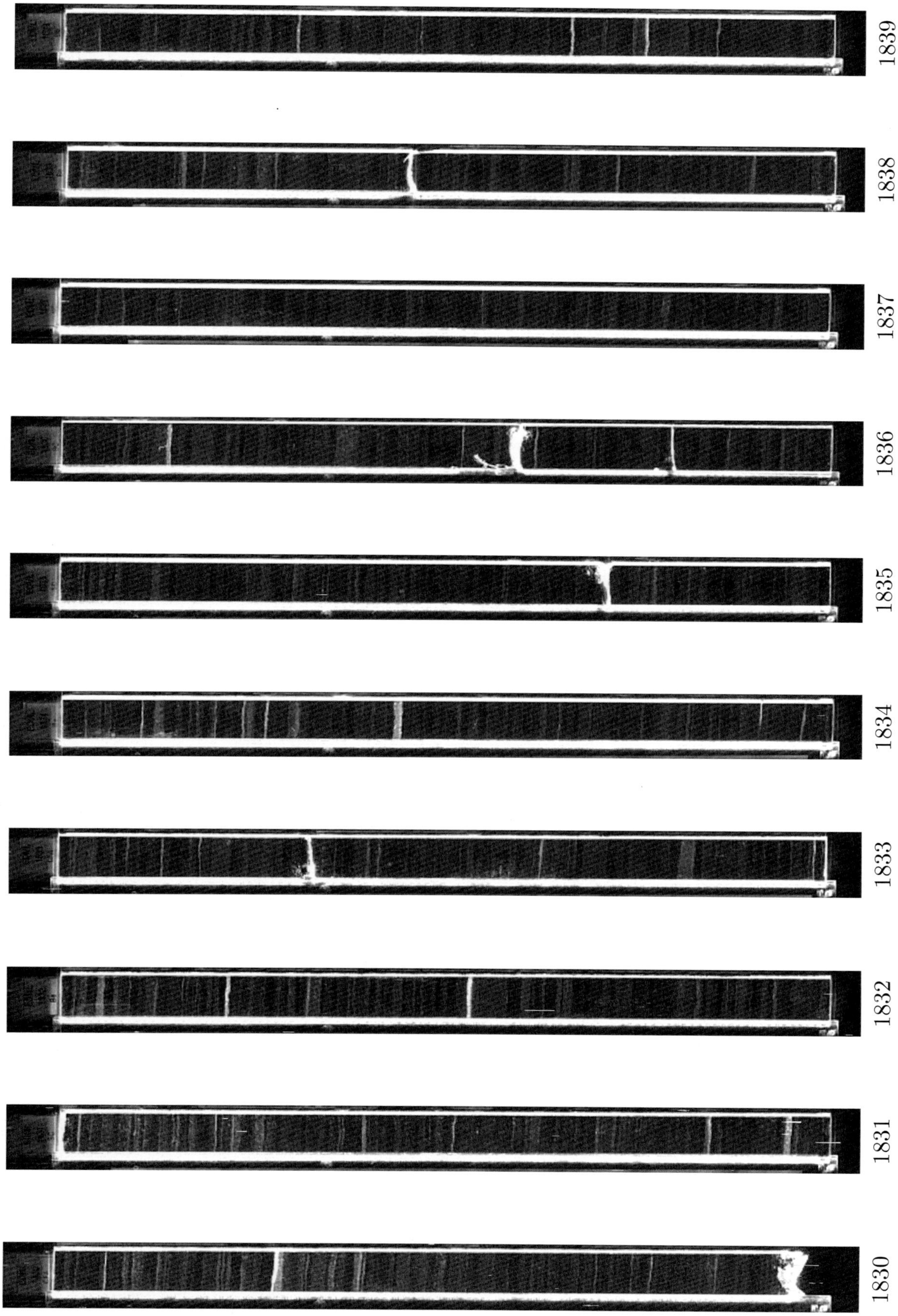
1839
1838
1837
1836
1835
1834
1833
1832
1831
1830

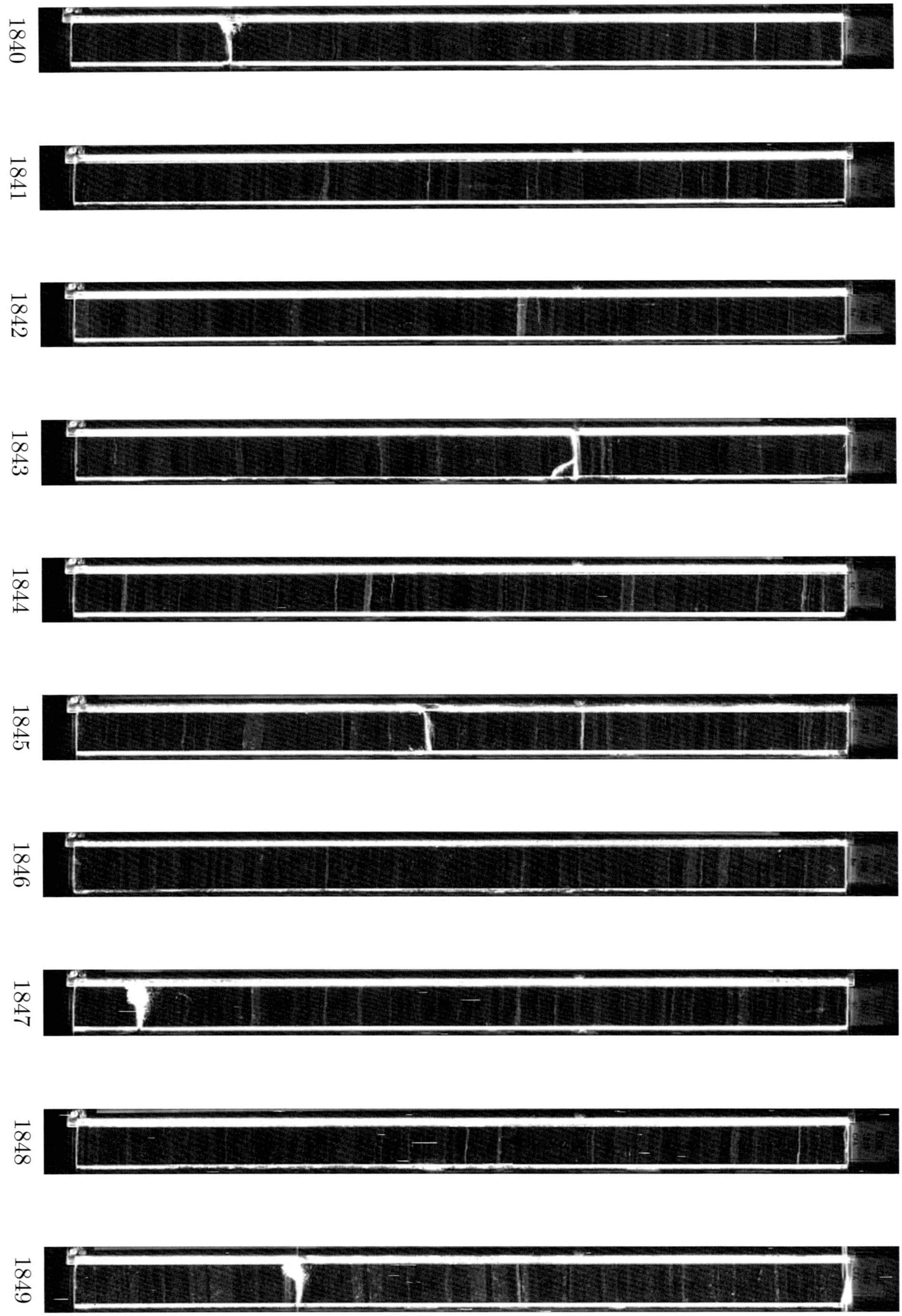
1840
1841
1842
1843
1844
1845
1846
1847
1848
1849

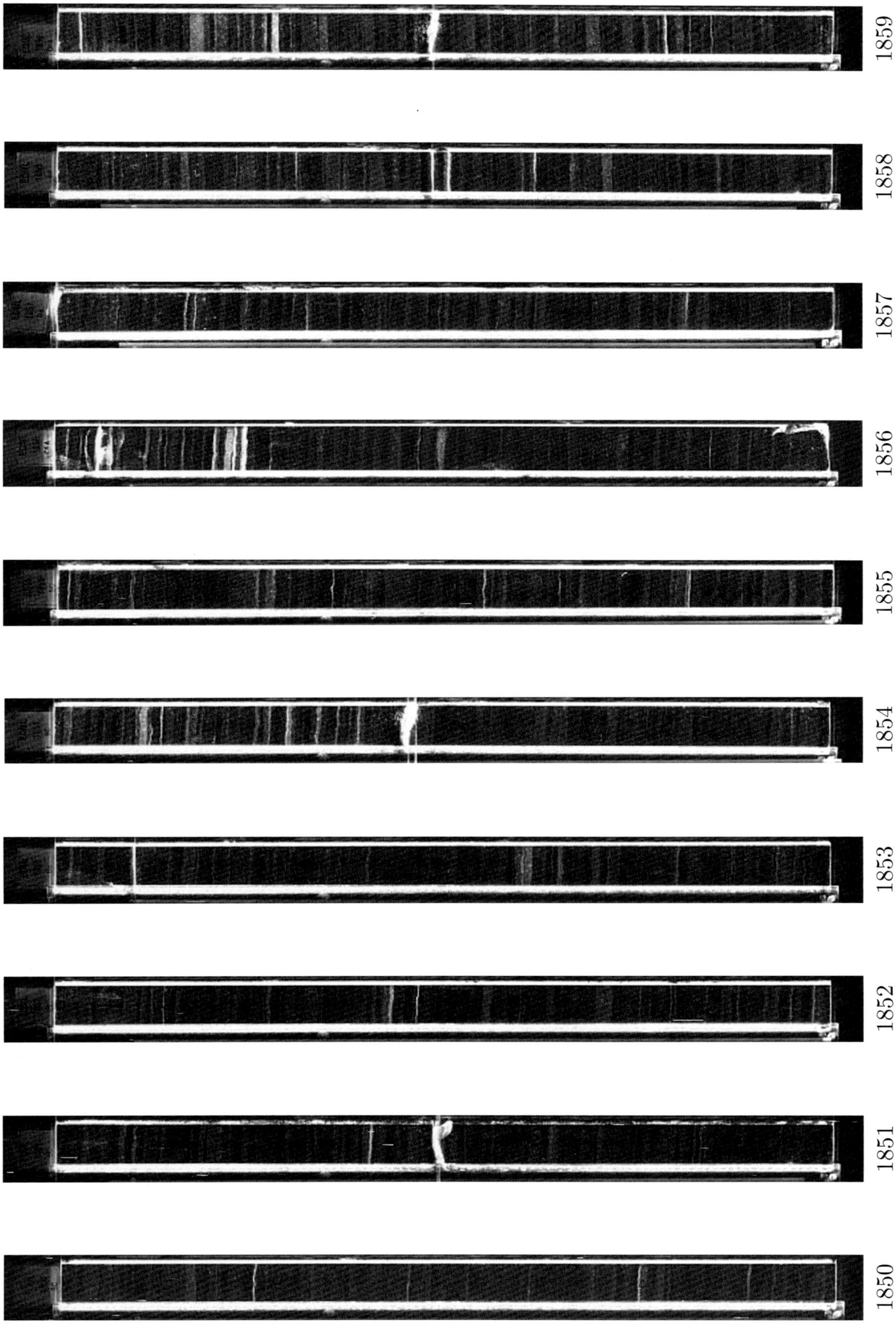
1859
1858
1857
1856
1855
1854
1853
1852
1851
1850

1860

1861

1862

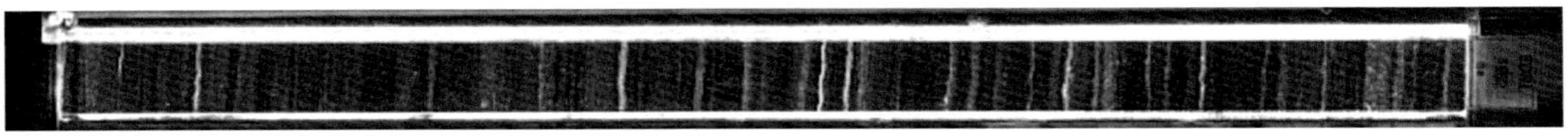

1863

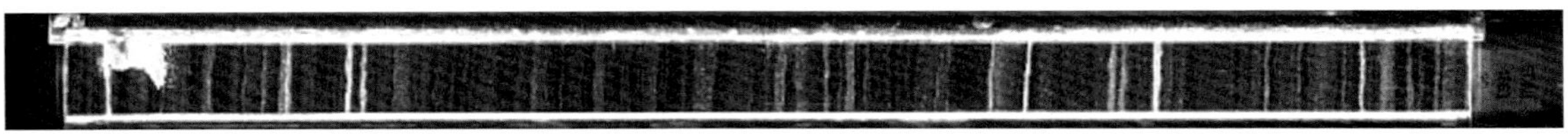

1864

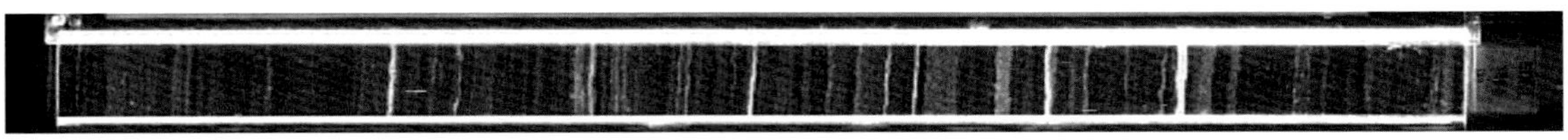

1865

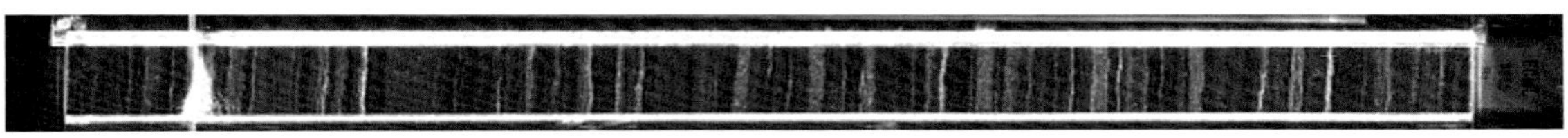

1866

1867

1868

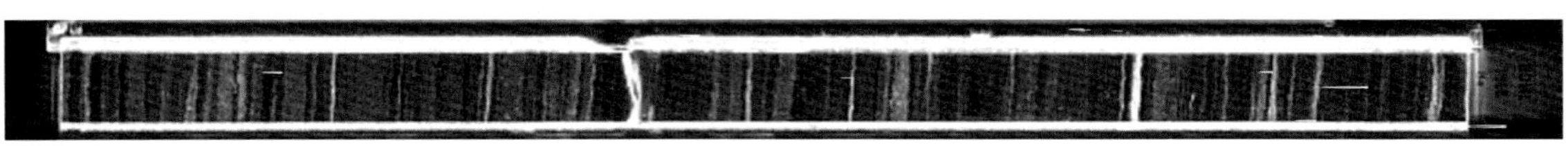

1869

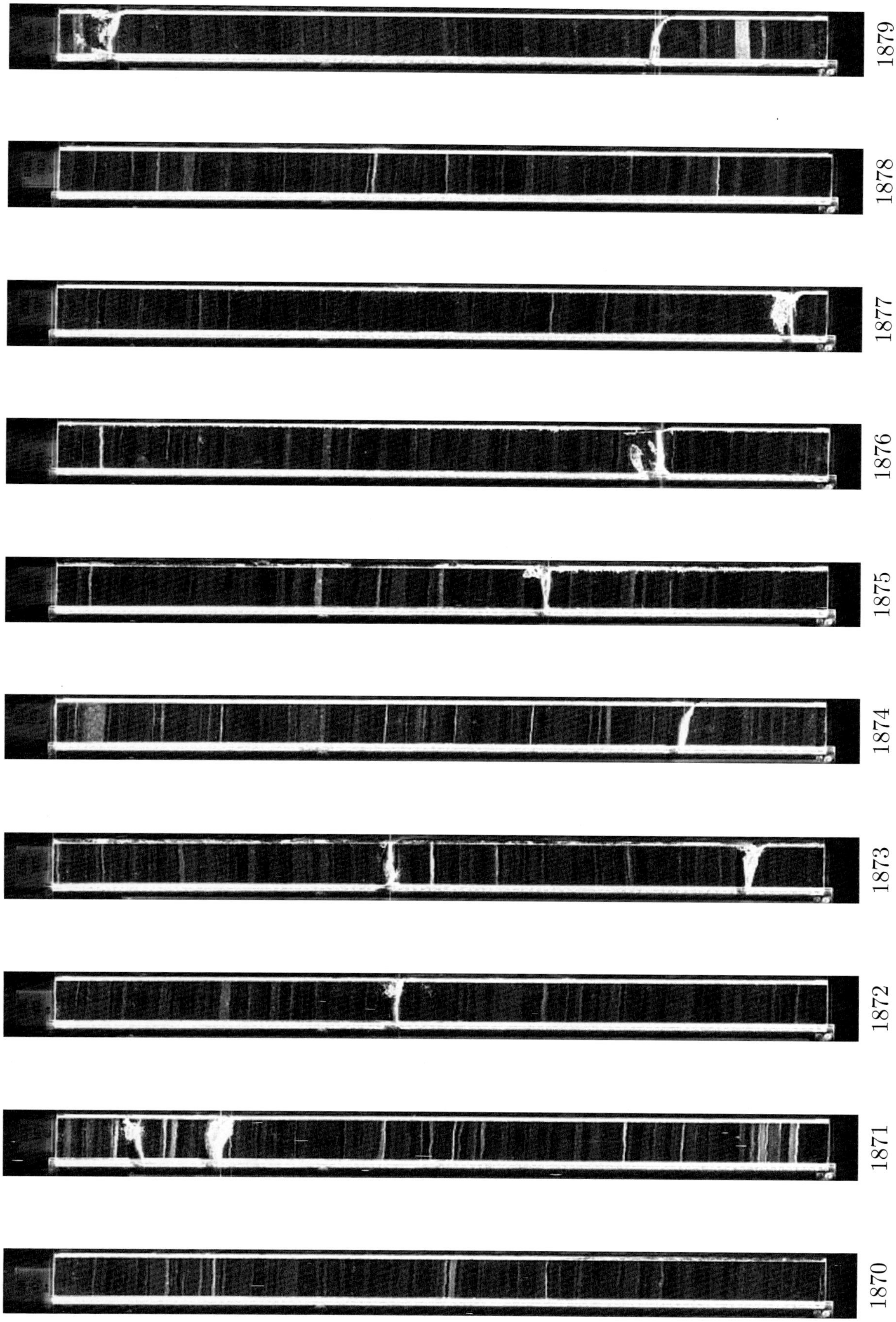
1879
1878
1877
1876
1875
1874
1873
1872
1871
1870

1880

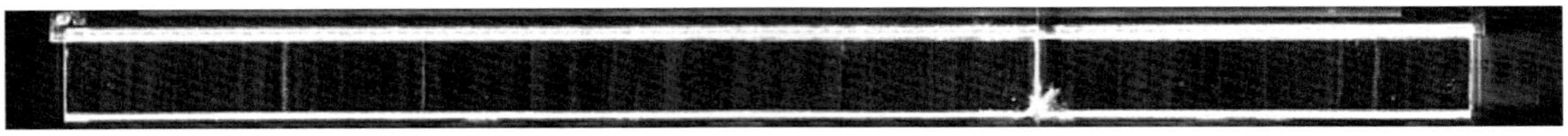

1881

1882

1883

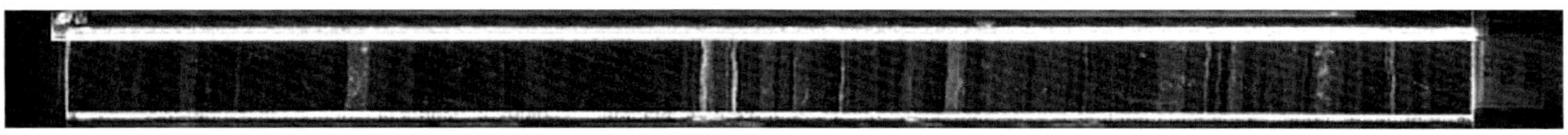

1884

1885

1886

1887

1888

1889

1899

1898

1897

1896

1895

1894

1893

1892

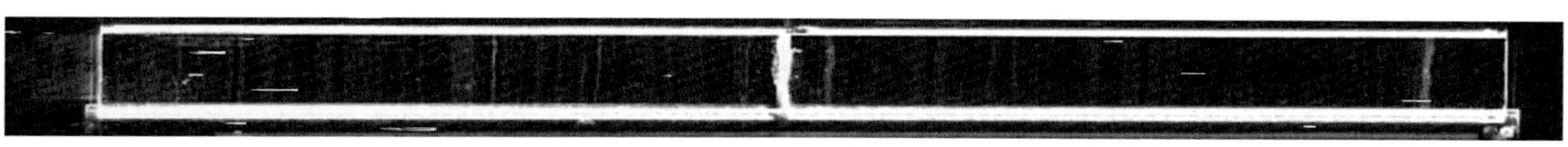

1891

1890

1900

1901

1902
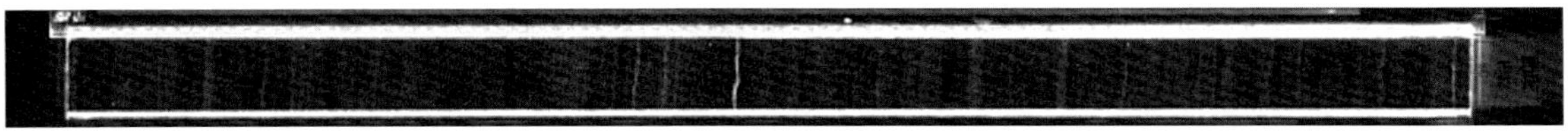

1903

1904

1905

1906
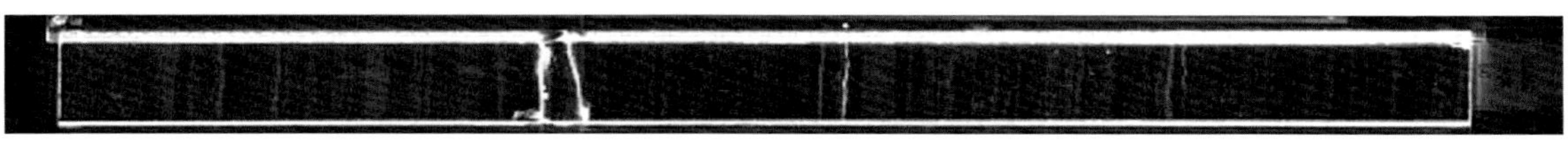

1907

1908
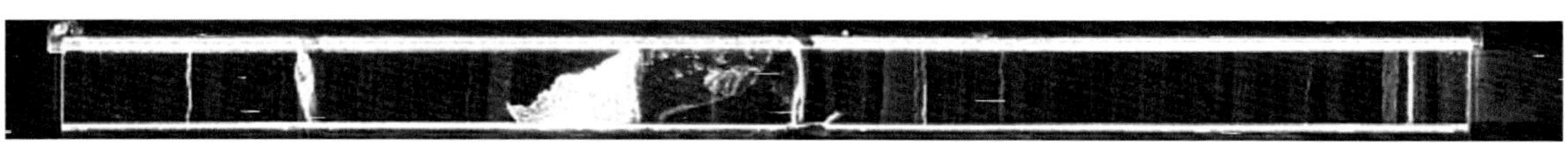

1909

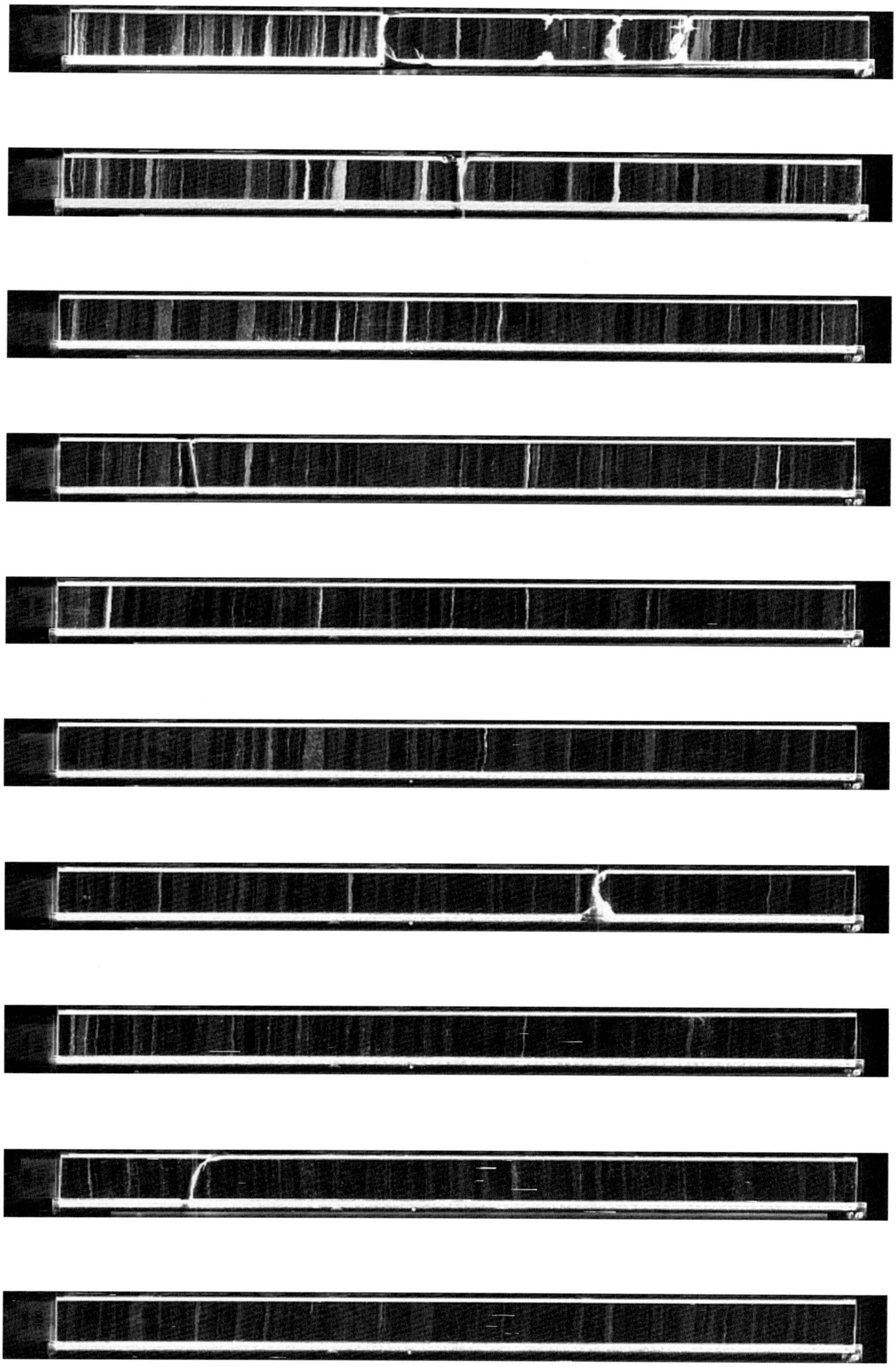

1919

1918

1917

1916

1915

1914

1913

1912

1911

1910

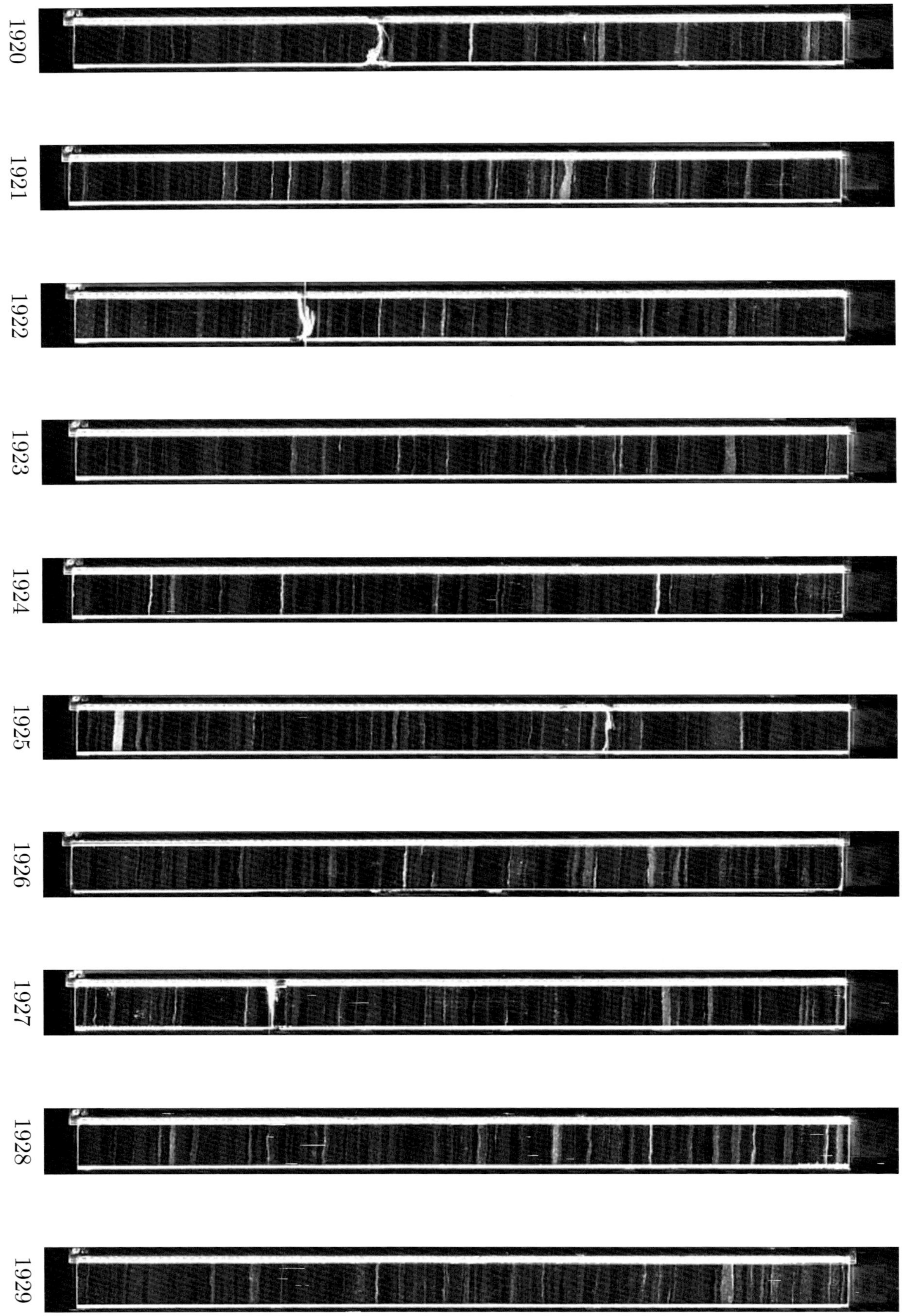
1920
1921
1922
1923
1924
1925
1926
1927
1928
1929

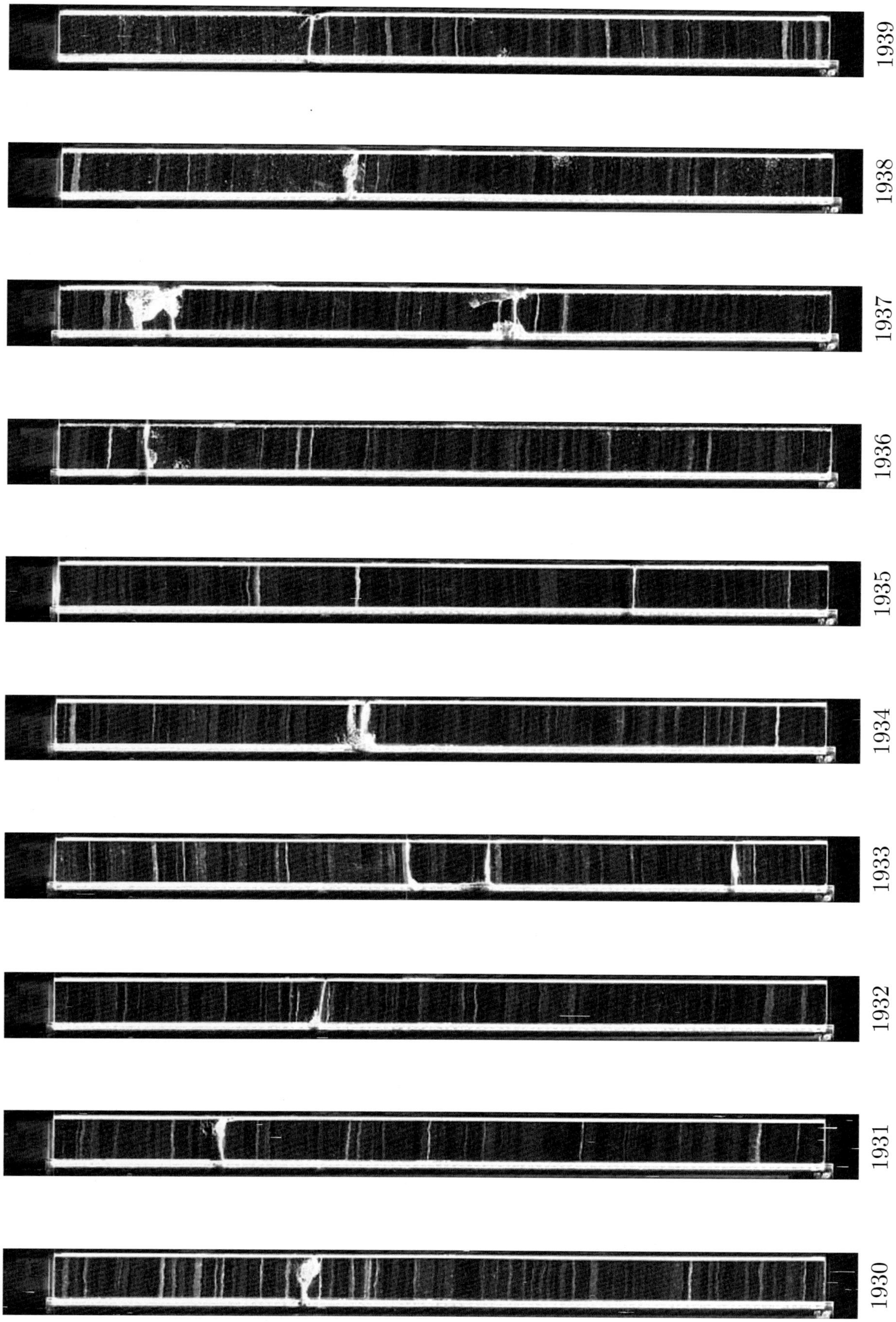
1939
1938
1937
1936
1935
1934
1933
1932
1931
1930

1940

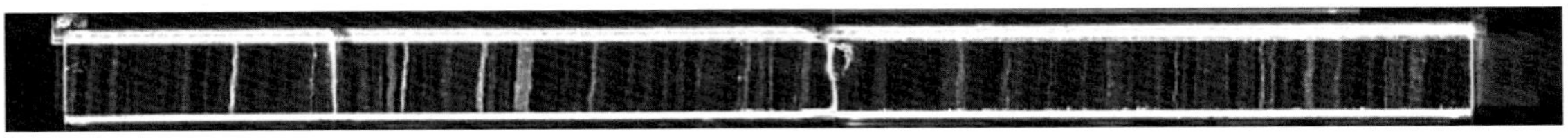

1941

1942

1943

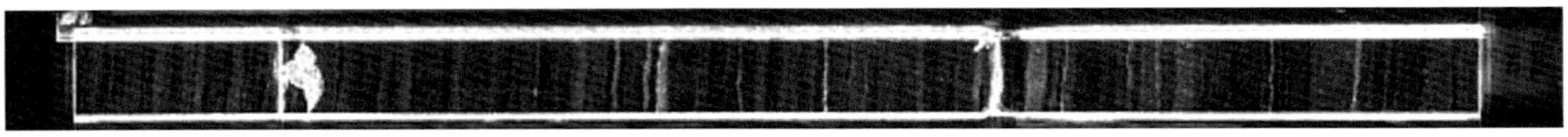

1944

1945

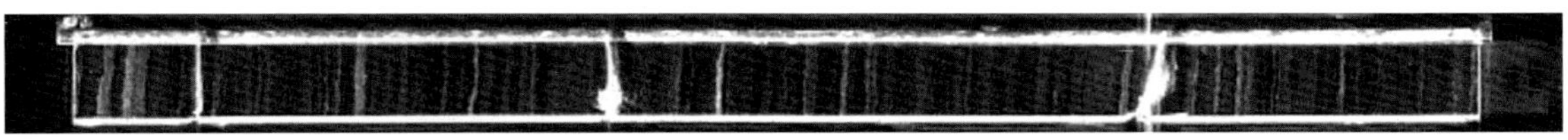

1946

1947

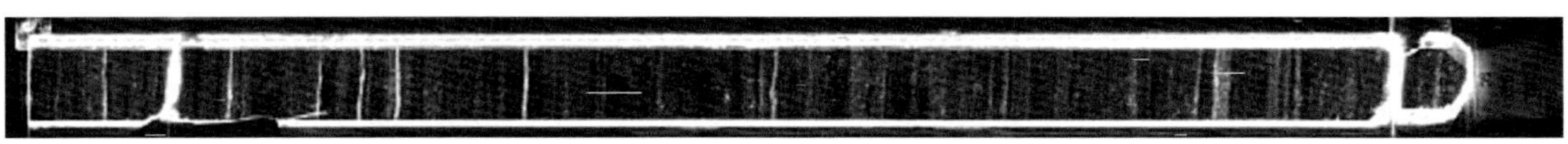

1948

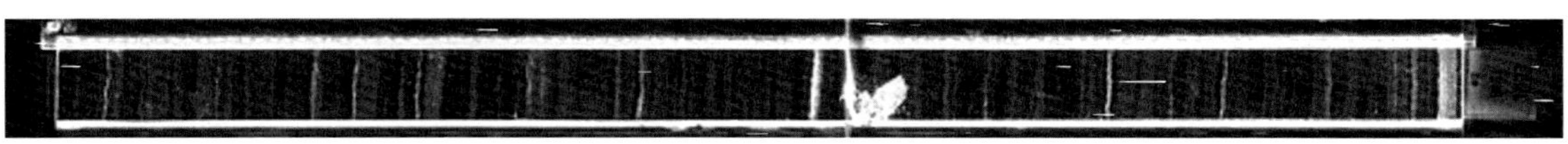

1949

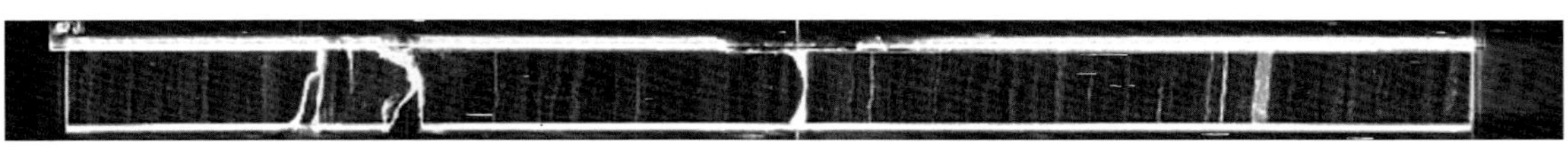

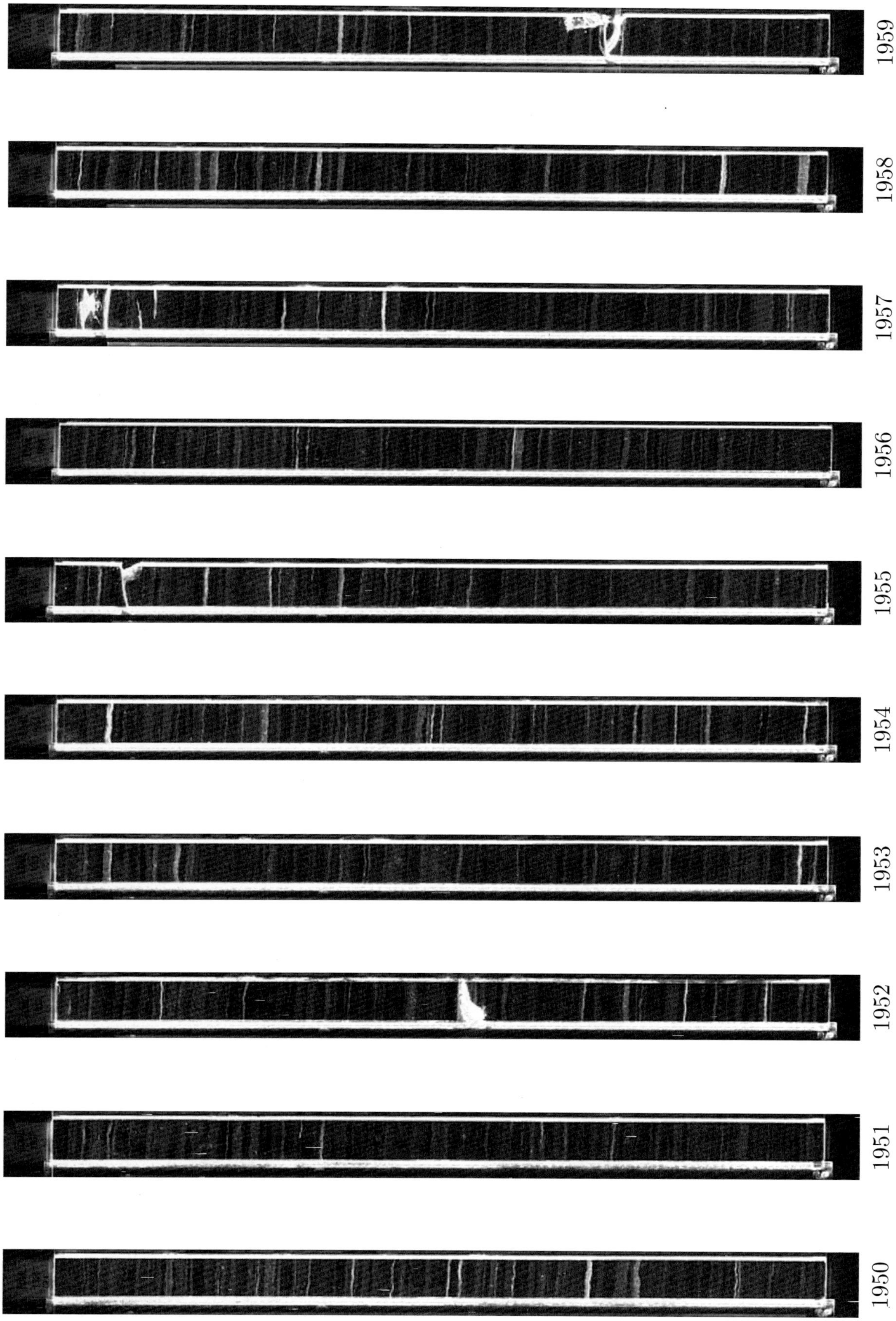
1959
1958
1957
1956
1955
1954
1953
1952
1951
1950

1960

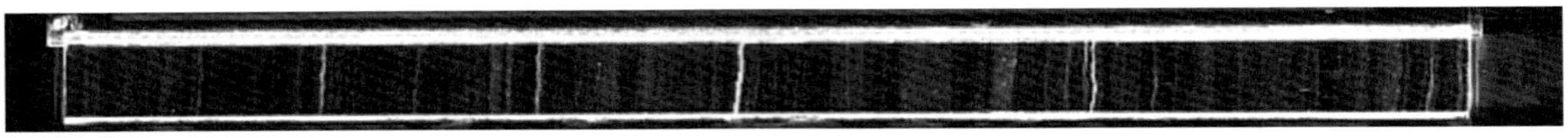

1961

1962

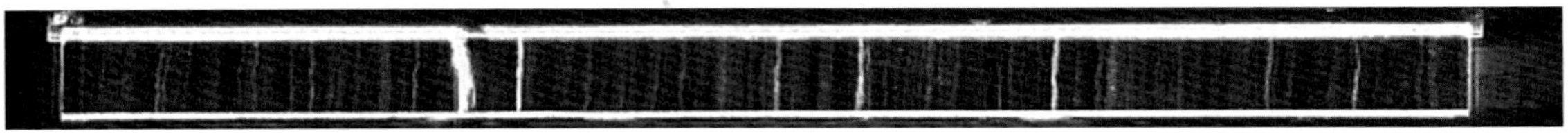

1963

1964

1965

1966

1967

1968

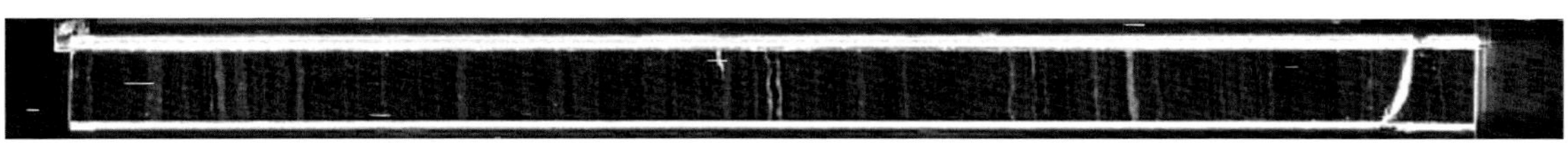

1969

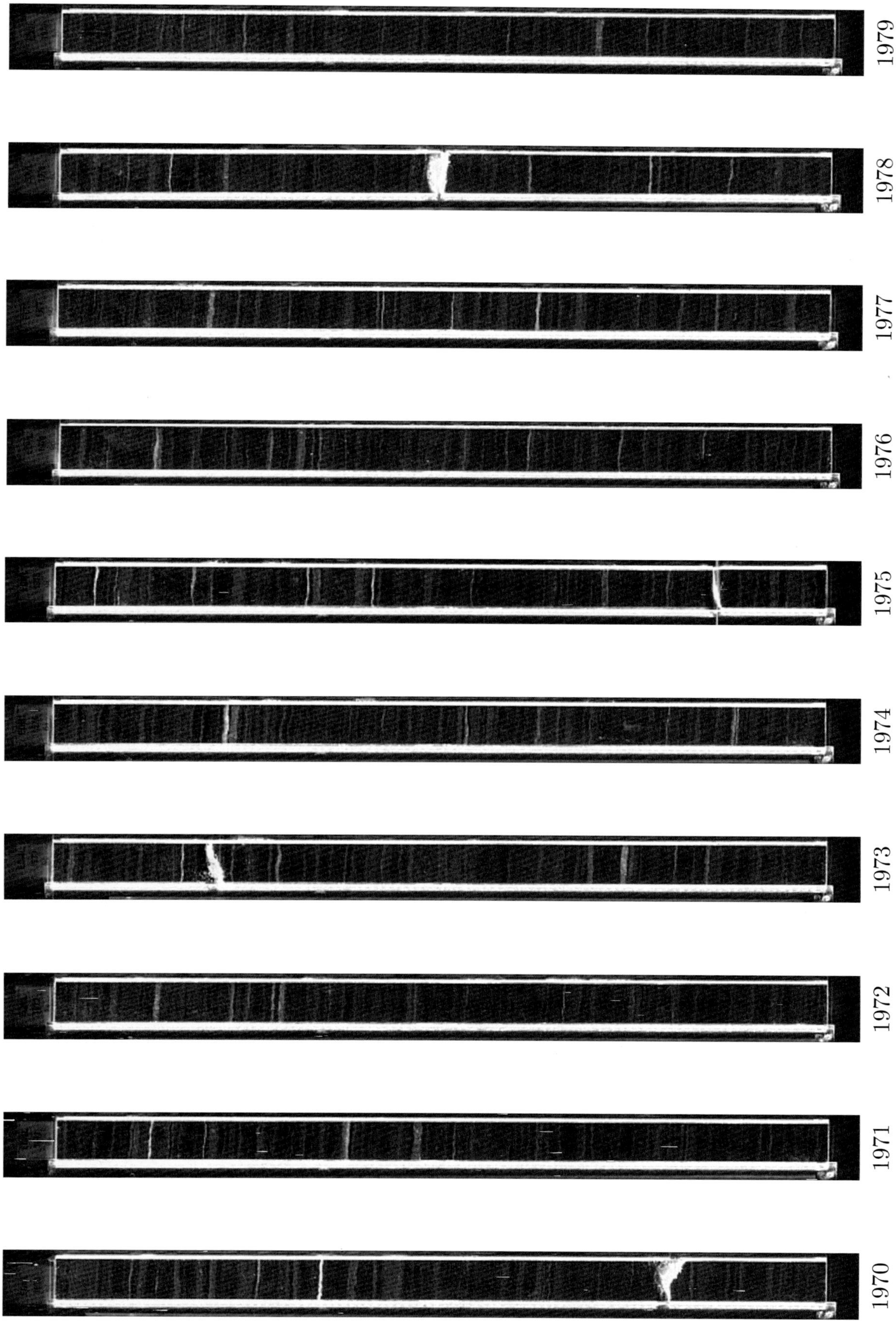
1979
1978
1977
1976
1975
1974
1973
1972
1971
1970

1980

1981

1982

1983

1984

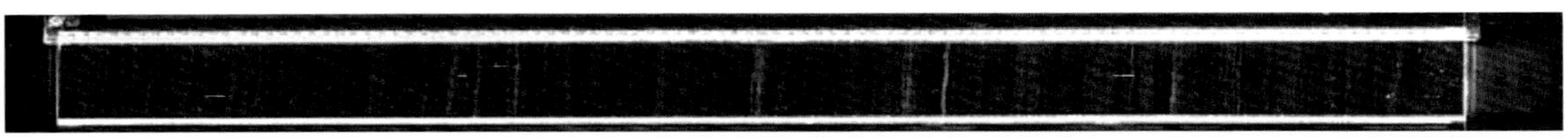

1985

1986

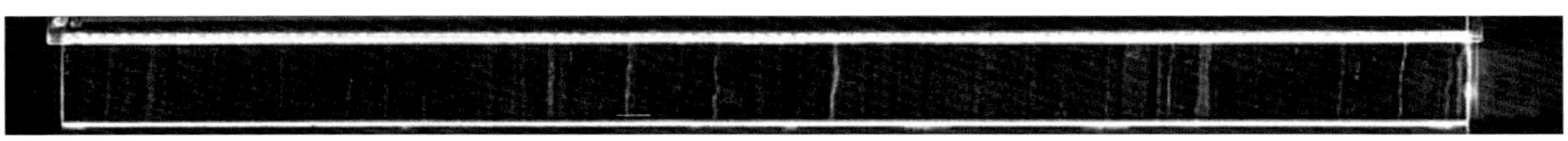

1987

1988

1989

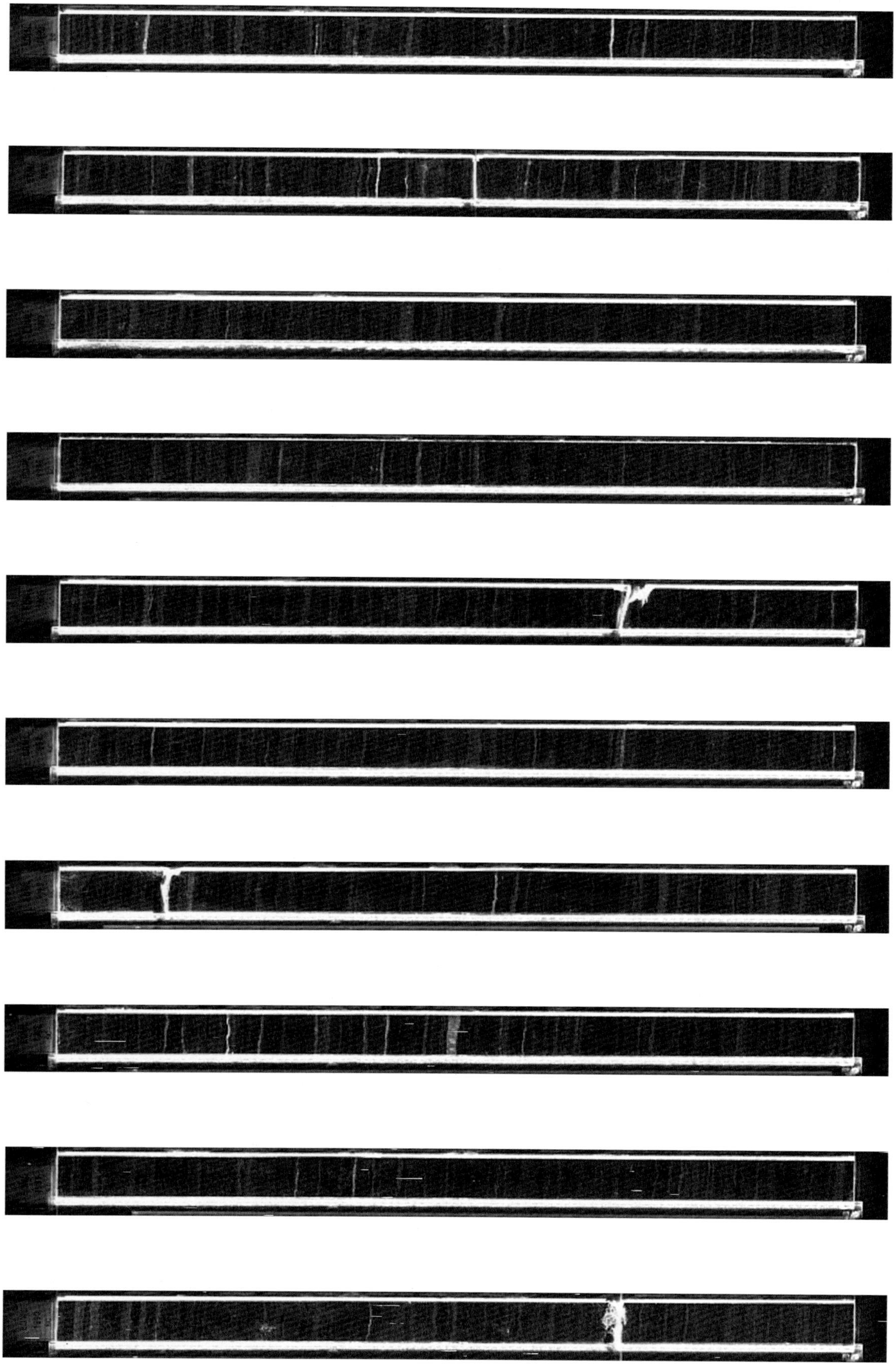
1999
1998
1997
1996
1995
1994
1993
1992
1991
1990

2000

2001

2002

2003

2004

2005

2006

2007

2008

2009

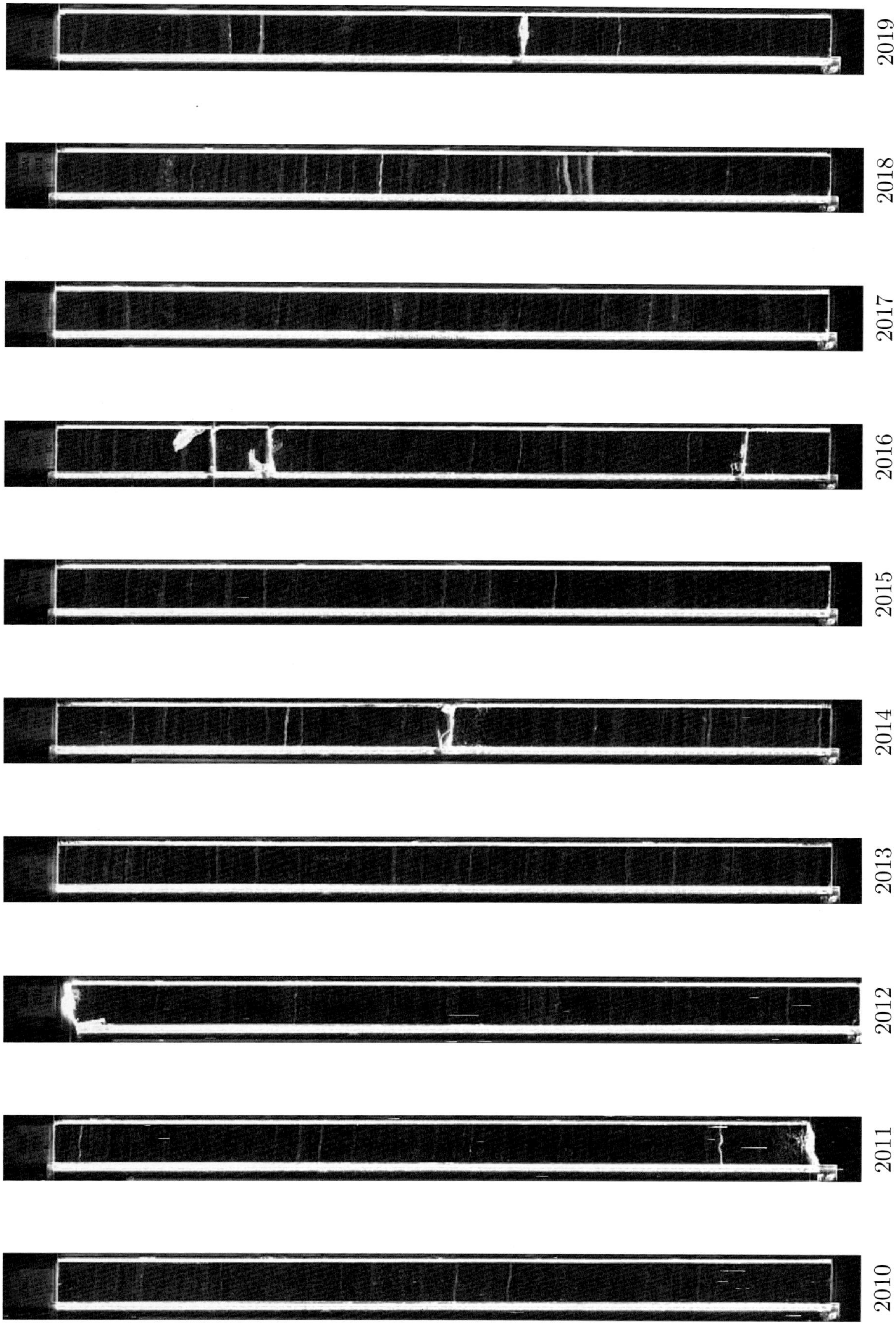
2019
2018
2017
2016
2015
2014
2013
2012
2011
2010

2020

2021

2022

2023

2024

2025

2026
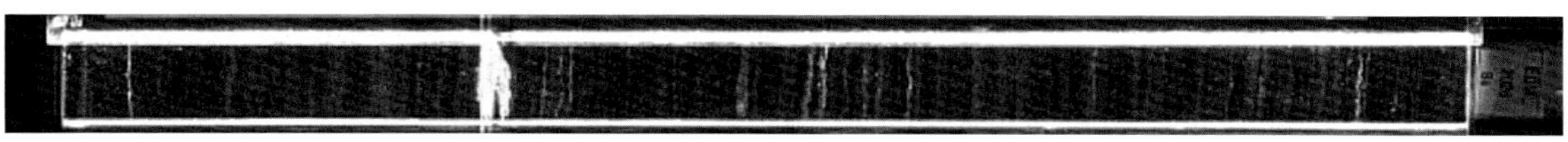

2027

2028
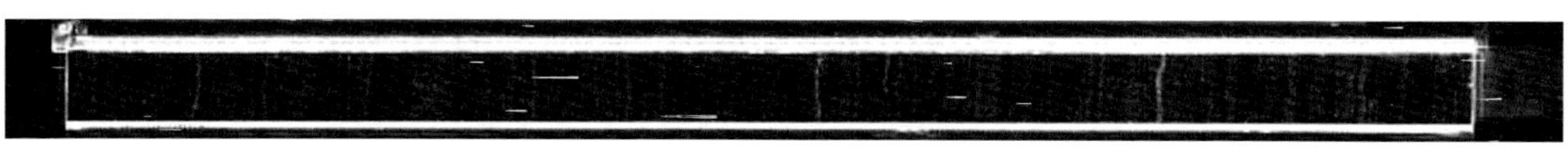

2029
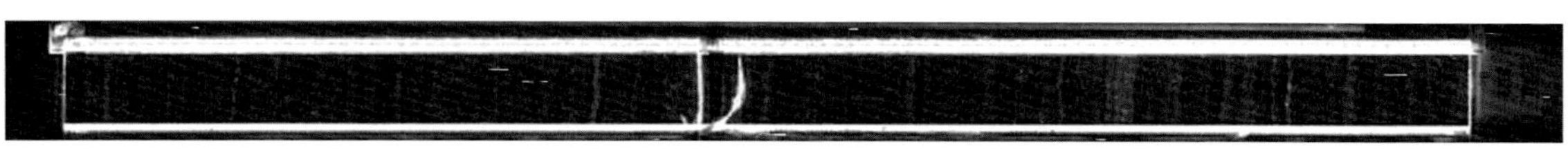

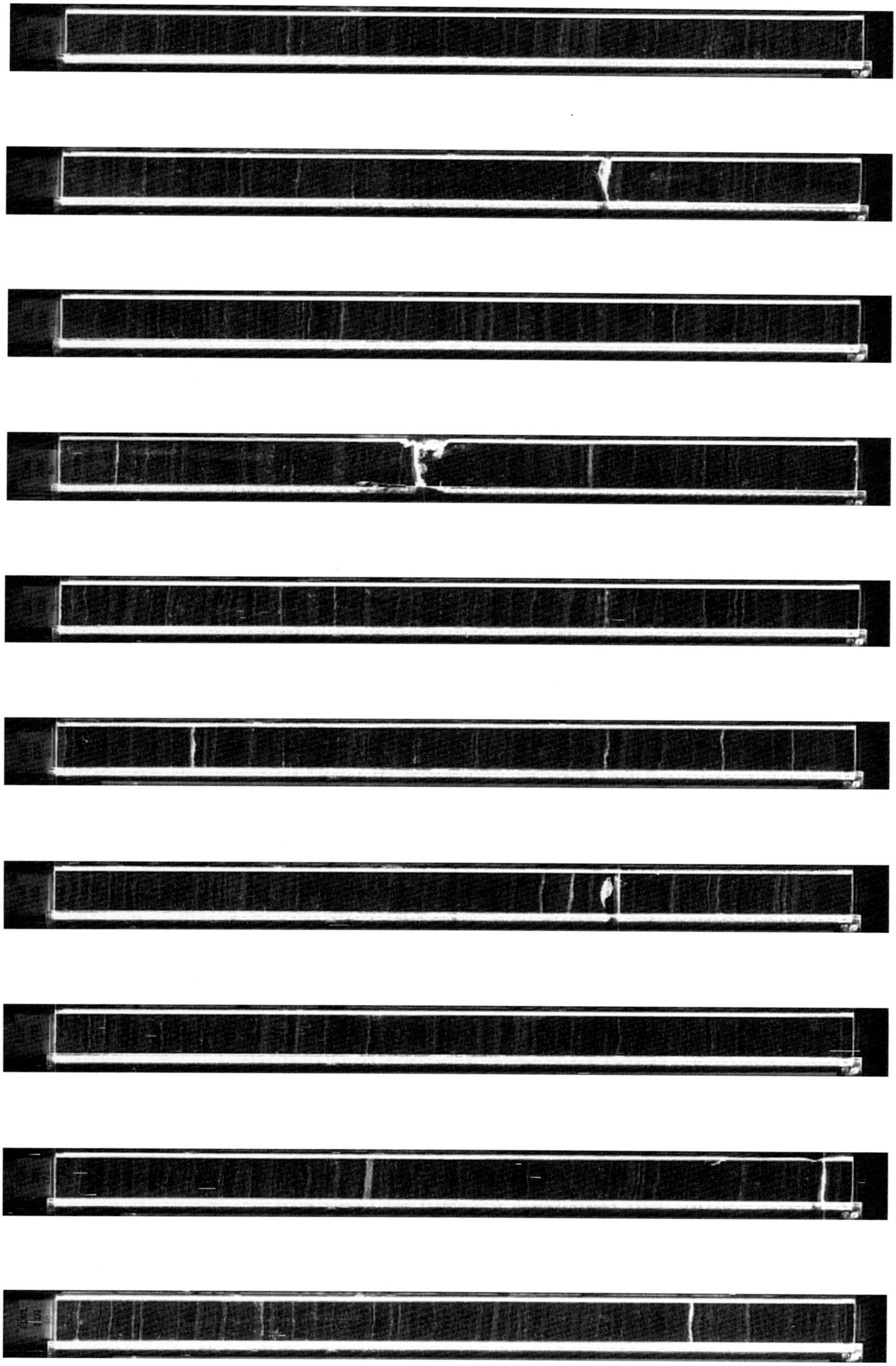

2039

2038

2037

2036

2035

2034

2033

2032

2031

2030

2040

2041

2042

2043

2044
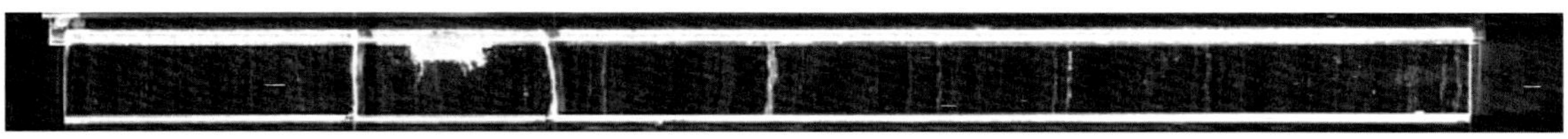

2045

2046

2047

2048
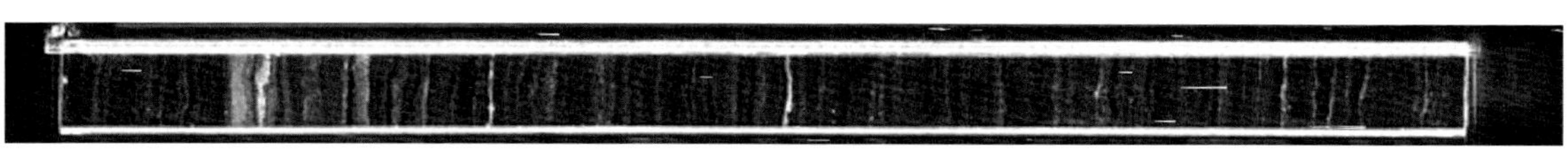

2049
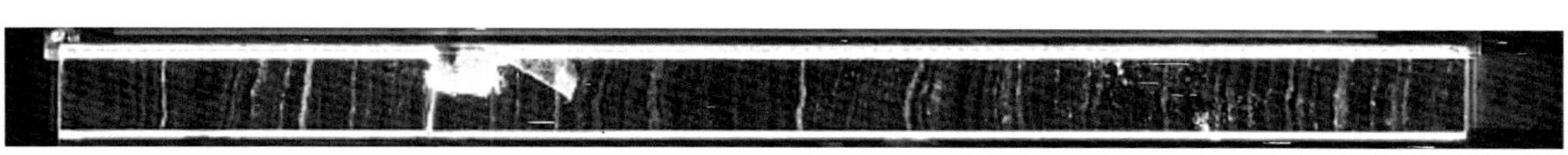

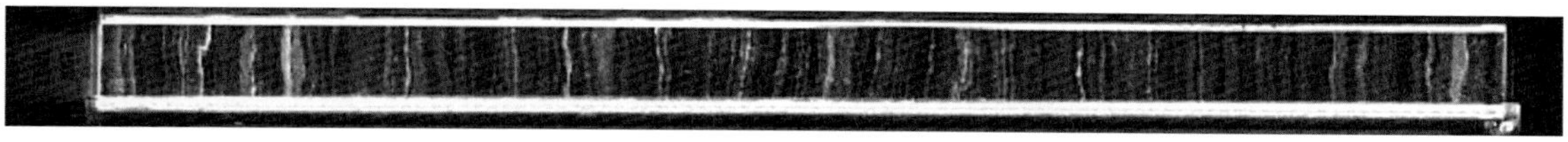
2059

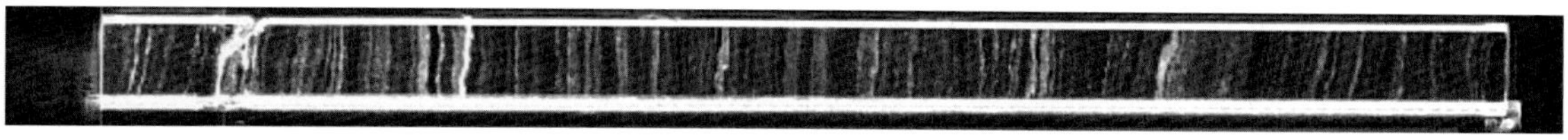
2058

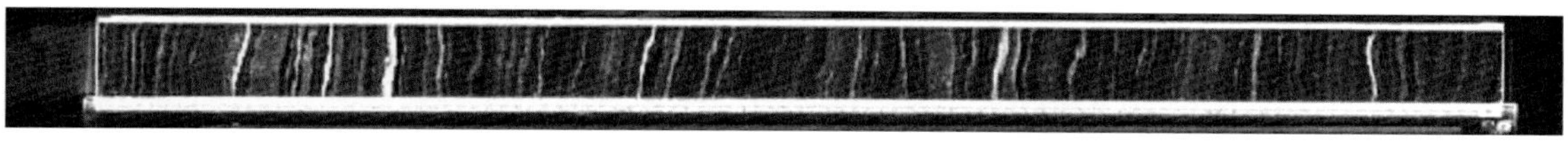
2057

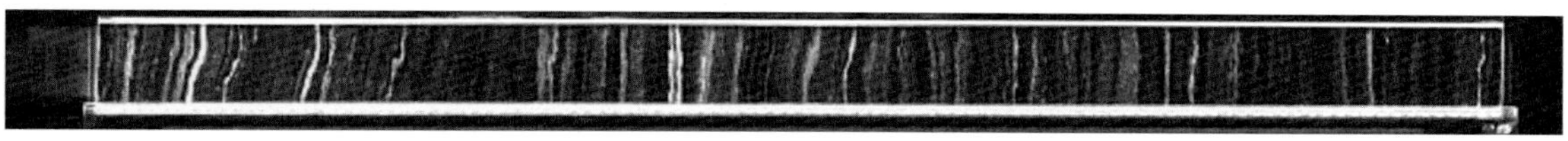
2056

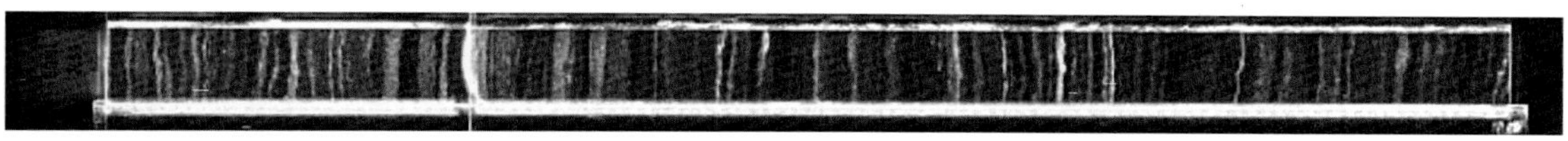
2055

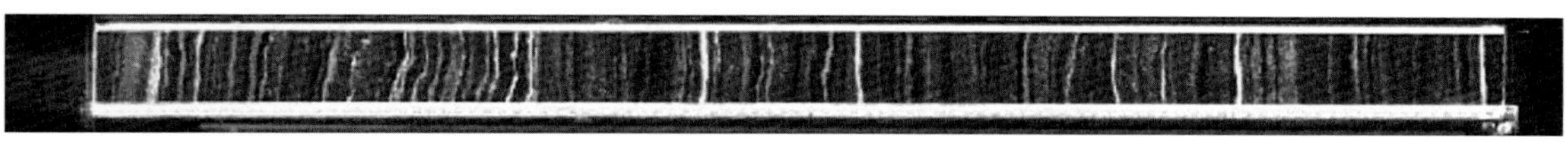
2054

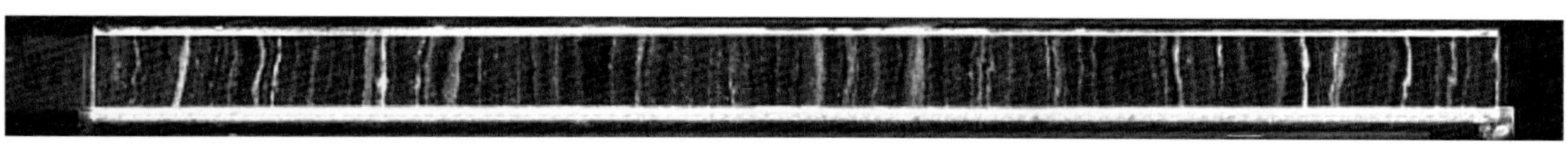
2053

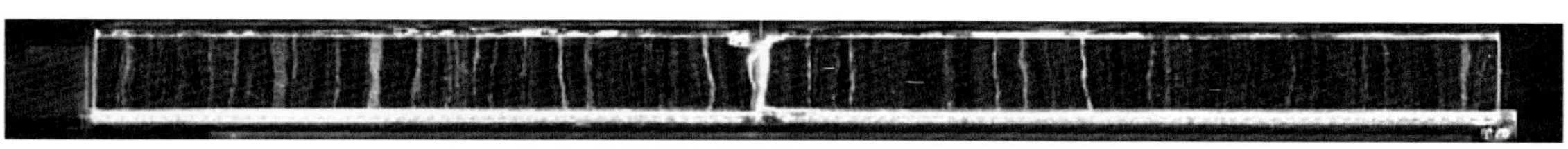
2052

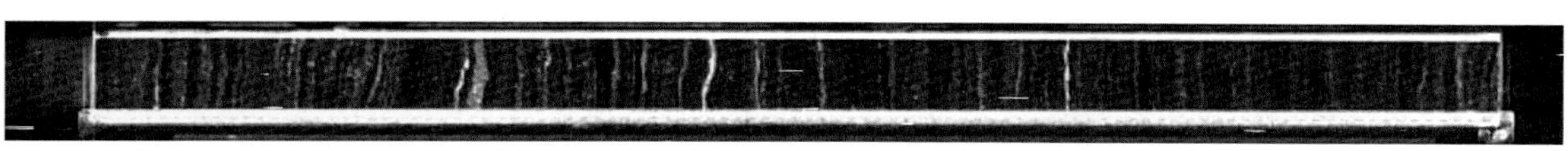
2051

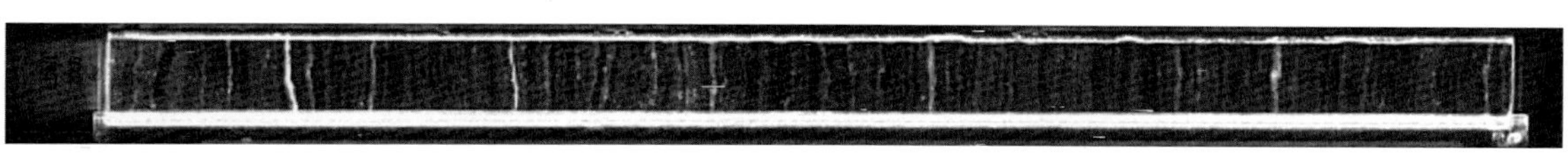
2050

2060

2061

2062
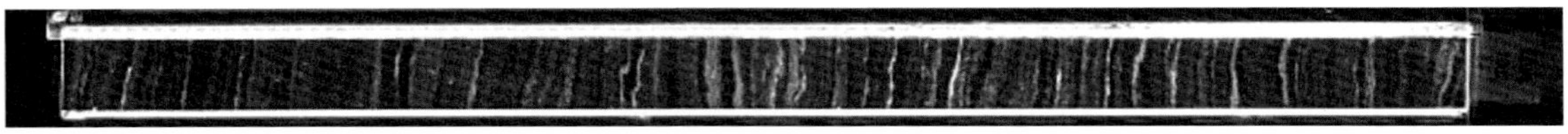

2063
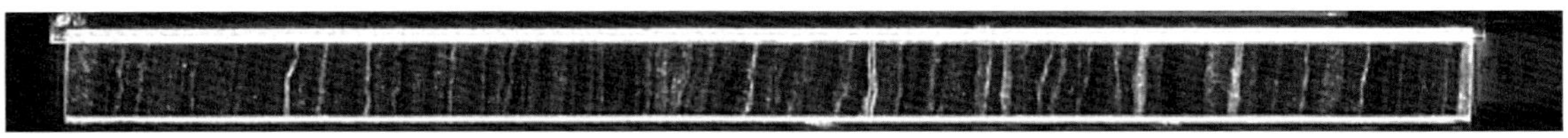

2064

2065
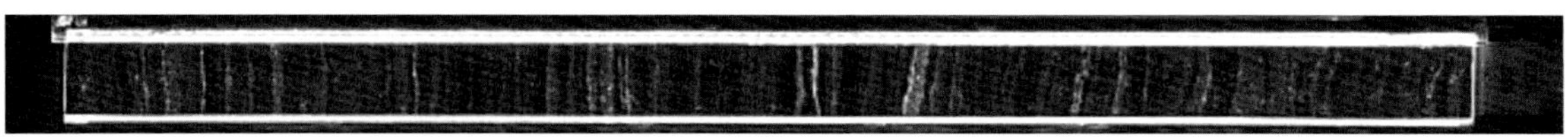

2066

2067
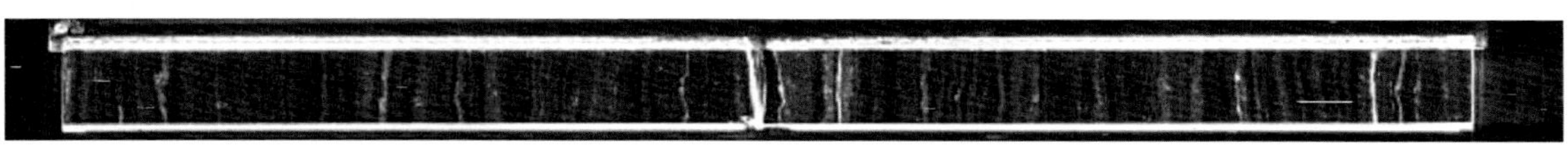

2068
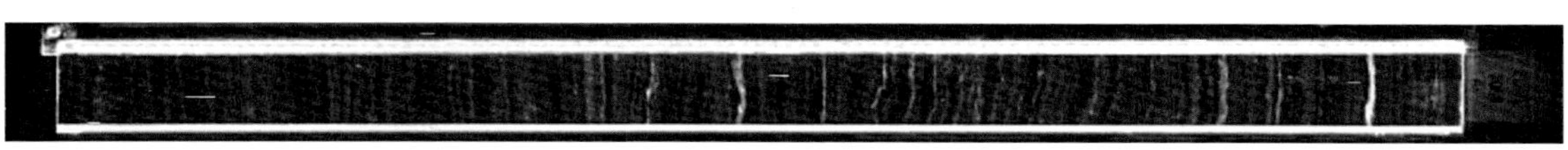

2069

2079

2078

2077

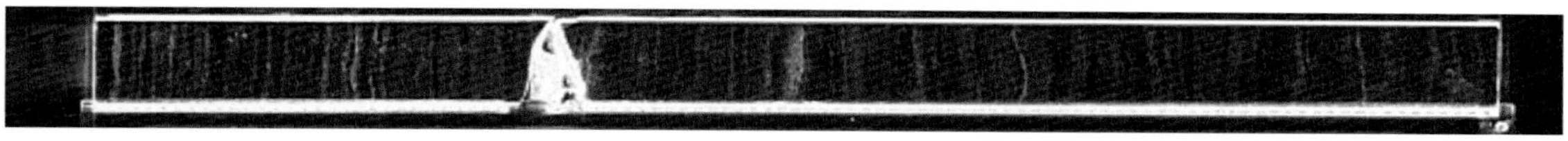
2076

2075

2074

2073

2072

2071

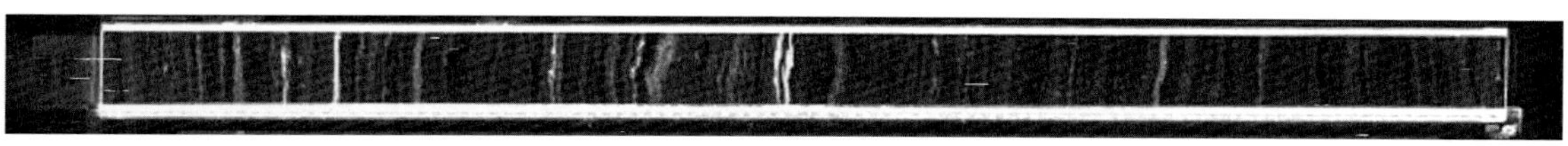
2070

2080

2081

2082

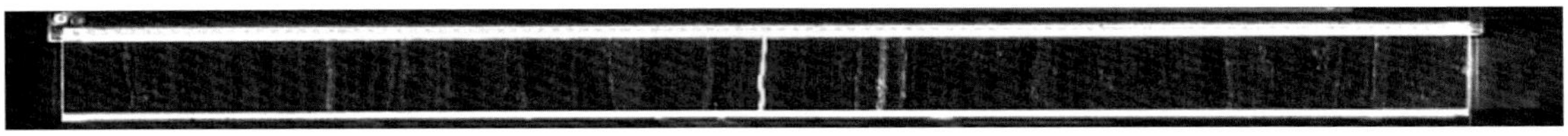

2083

2084

2085

2086

2087

2088

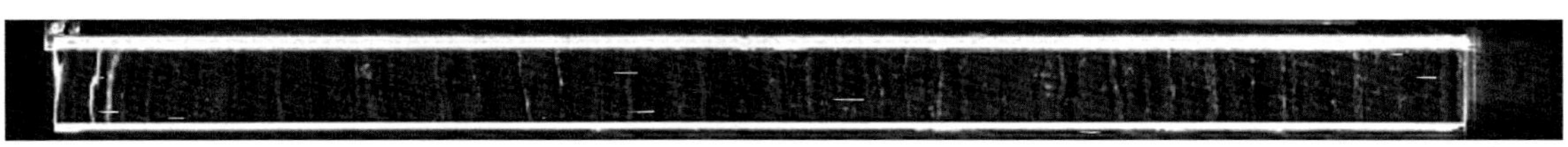

2089

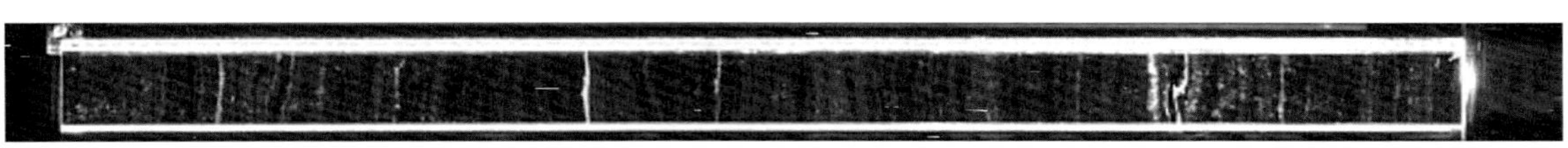

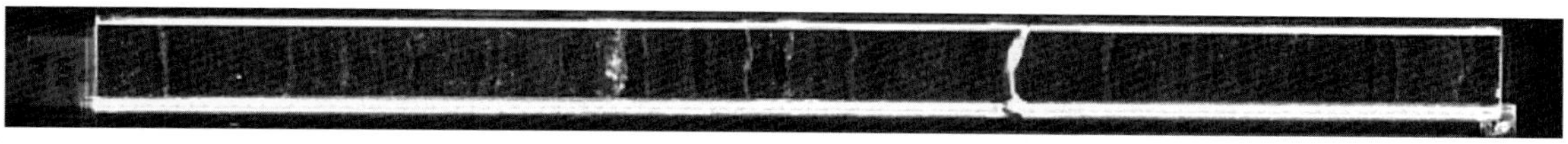

2099

2098

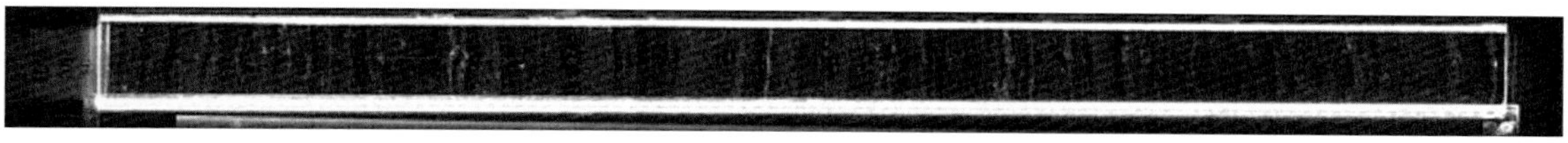

2097

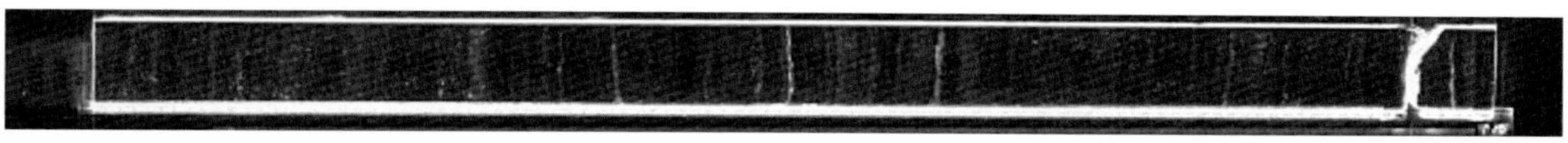

2096

2095

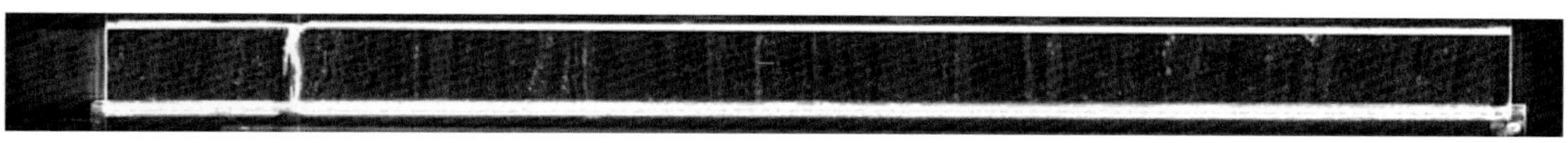

2094

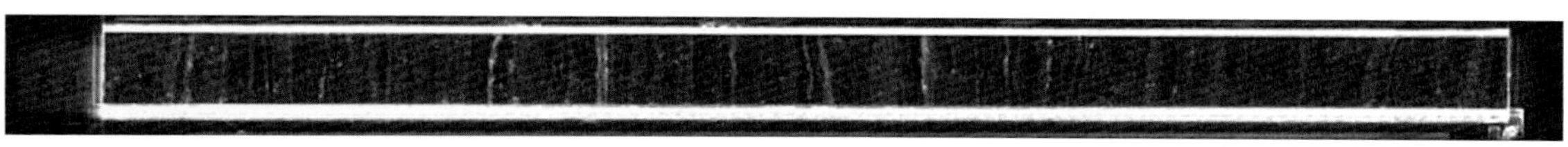

2093

2092

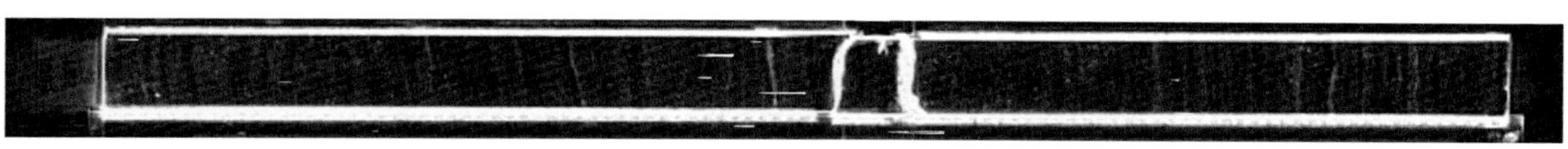

2091

2090

2100

2101

2102

2103

2104

2105

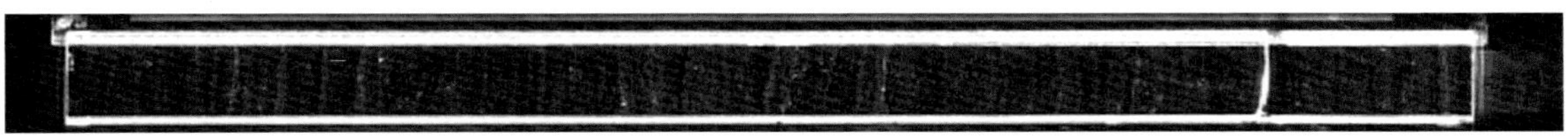

2106

2107

2108

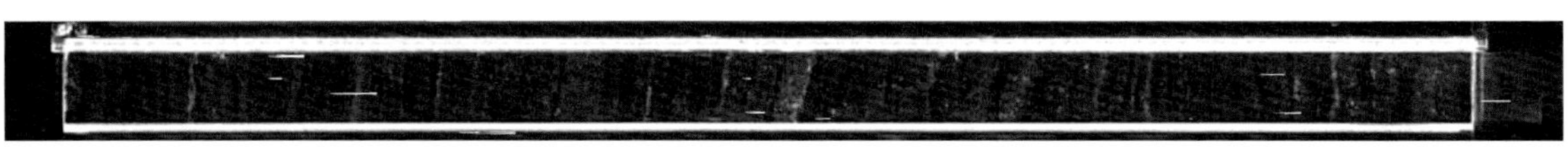

2109

2119

2118

2117

2116

2115

2114

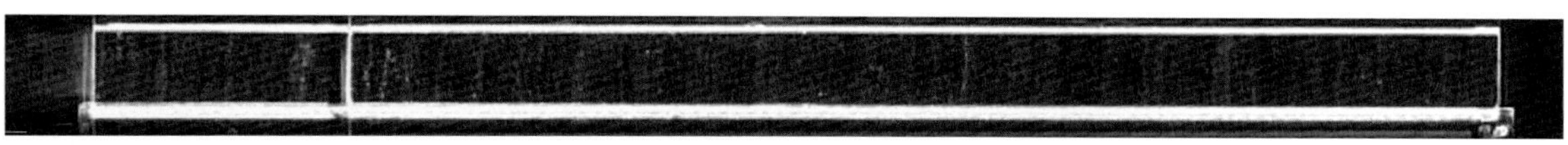

2113

2112

2111

2110

2120
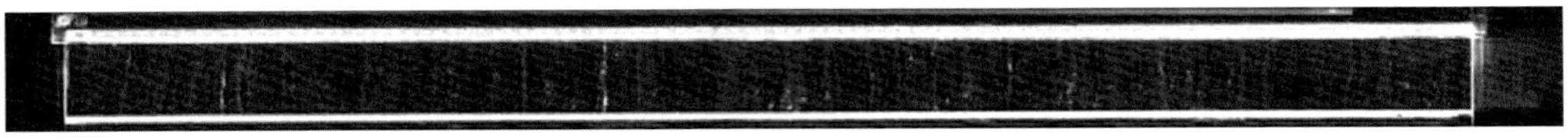

2121

2122

2123

2124

2125

2126

2127

2128

2129

2139

2138

2137

2136

2135

2134

2133

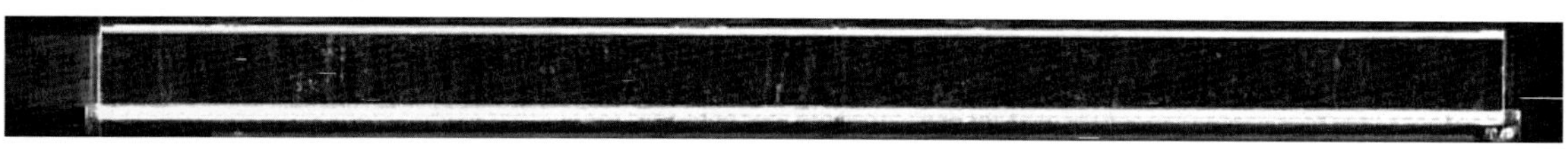
2132

2131

2130

2140

2141

2142

2143

2144

2145

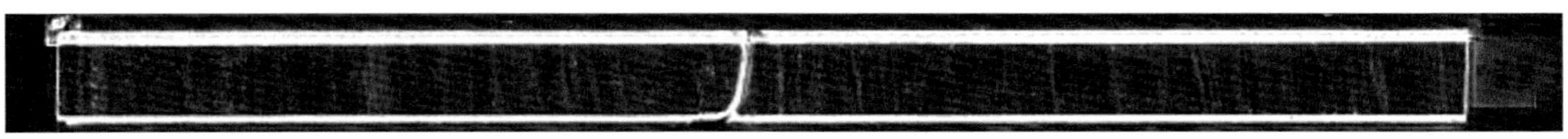

2146

2147

2148

2149

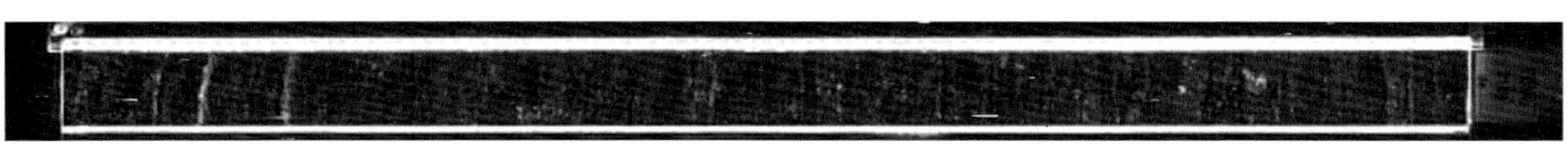

2159

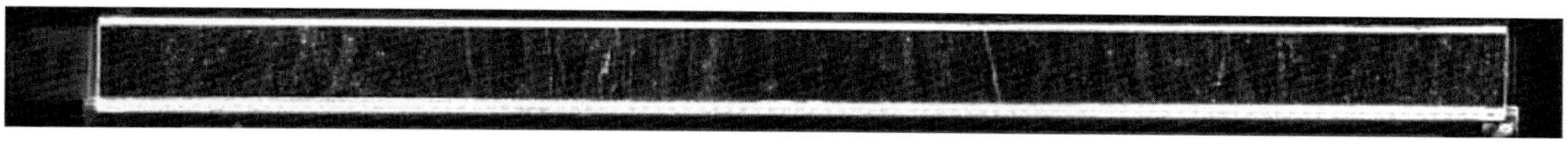
2158

2157

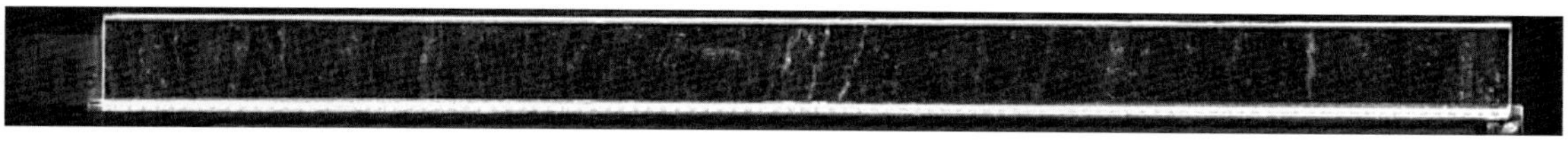
2156

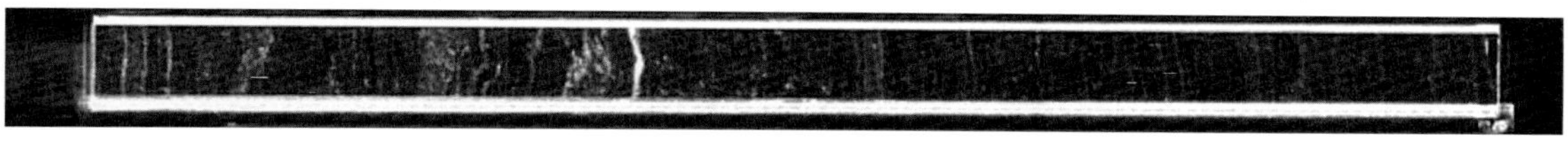
2155

2154

2153

2152

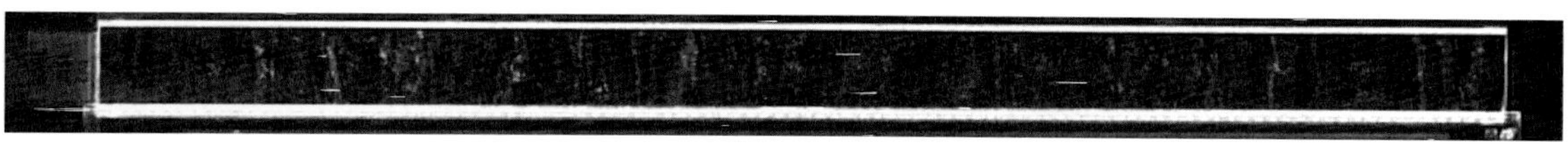
2151

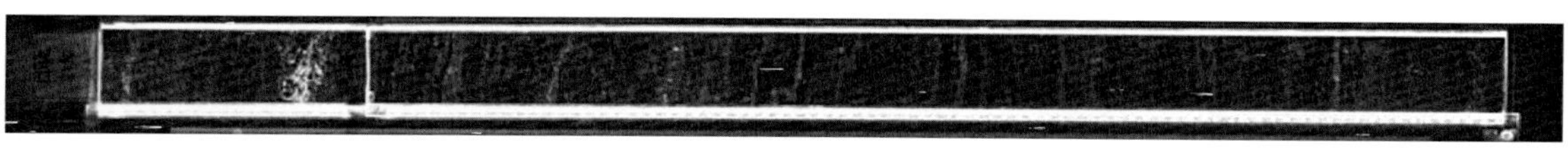
2150

2160

2161

2162

2163

2164

2165

2166
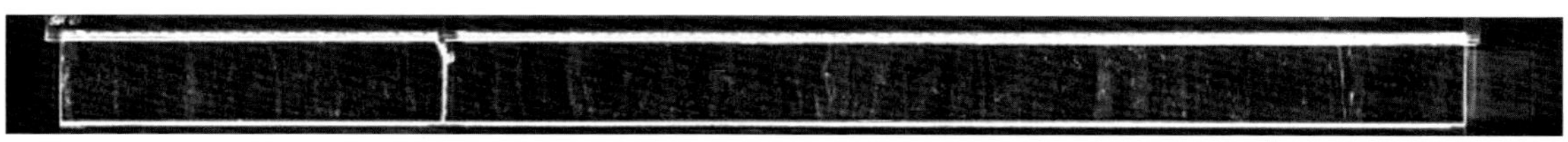

2167

2168
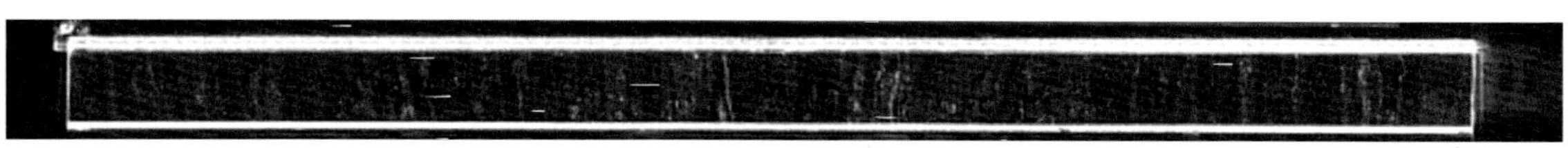

2169
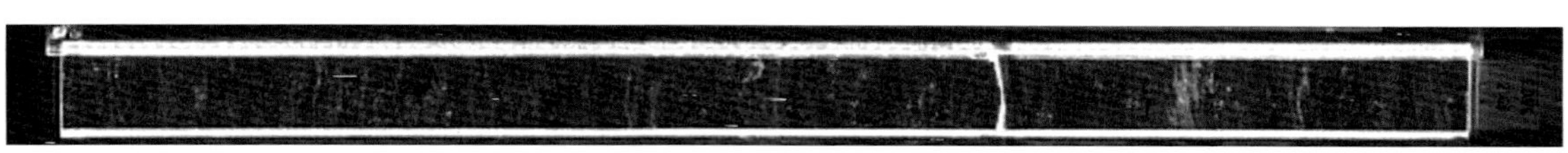

2179

2178

2177

2176

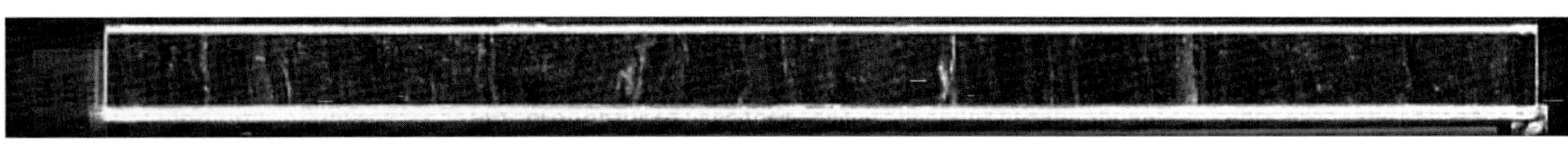

2175

2174

2173

2172

2171

2170

2180

2181

2182

2183
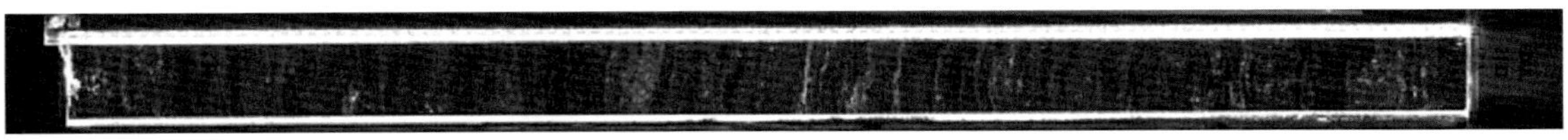

2184

2185
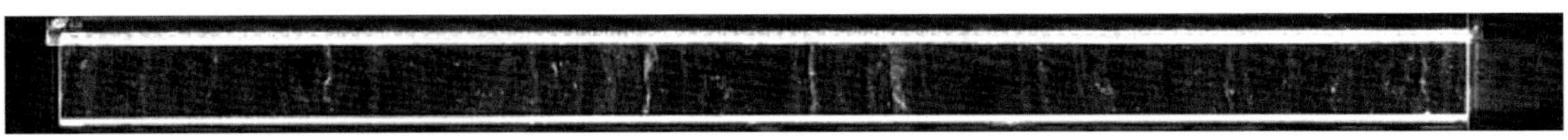

2186

2187
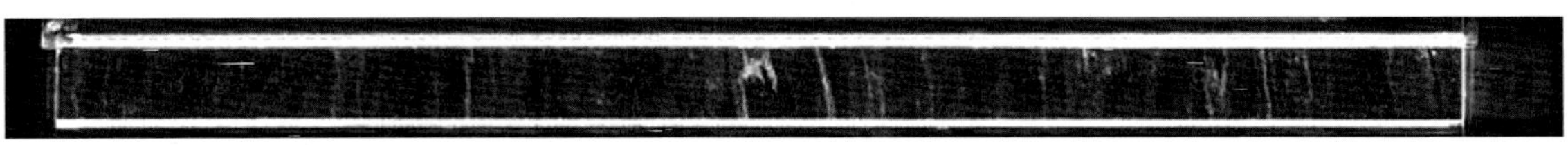

2188

2189
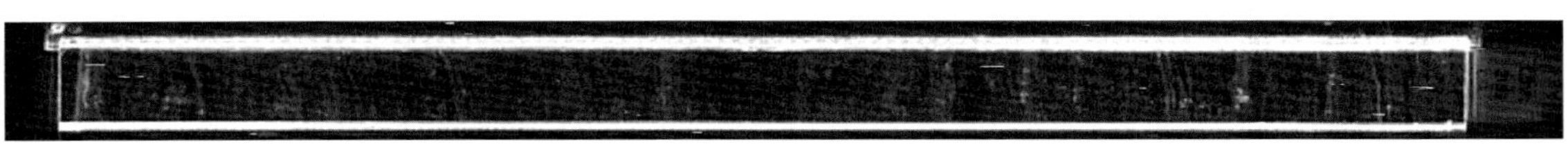

2199

2198

2197

2196

2195

2194

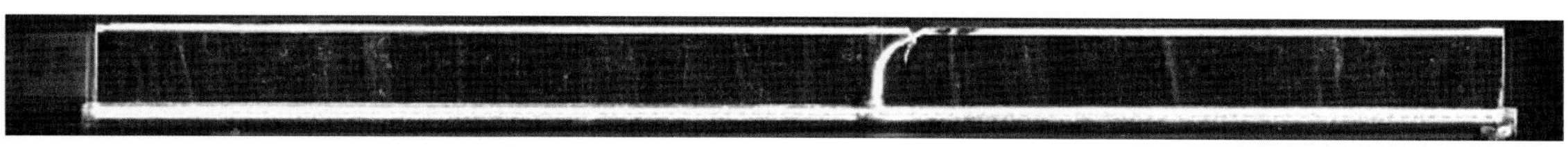

2193

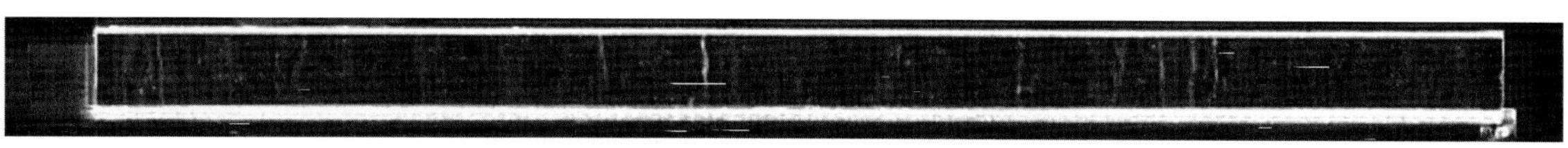

2192

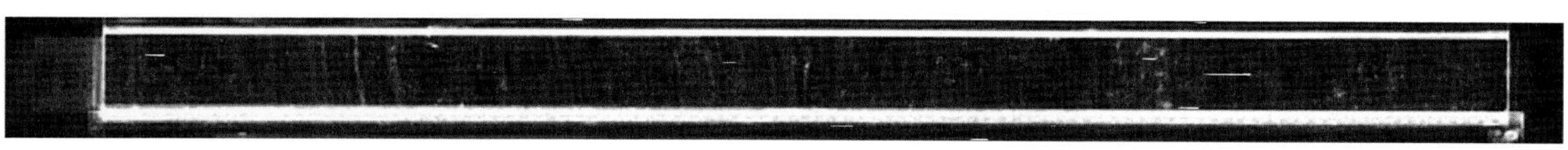

2191

2190

2200

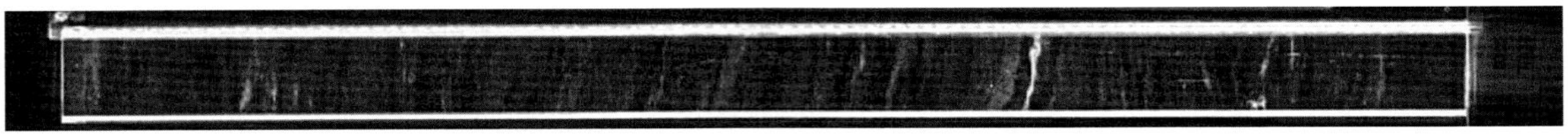

2201

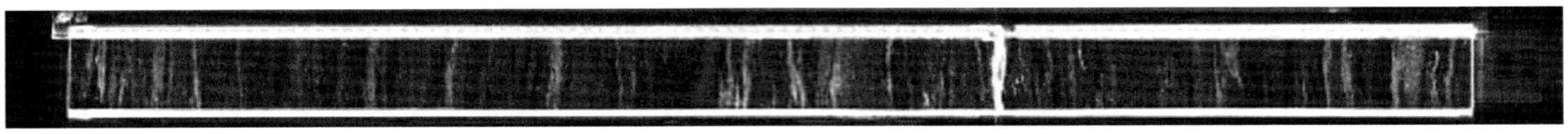

2202

2203

2204

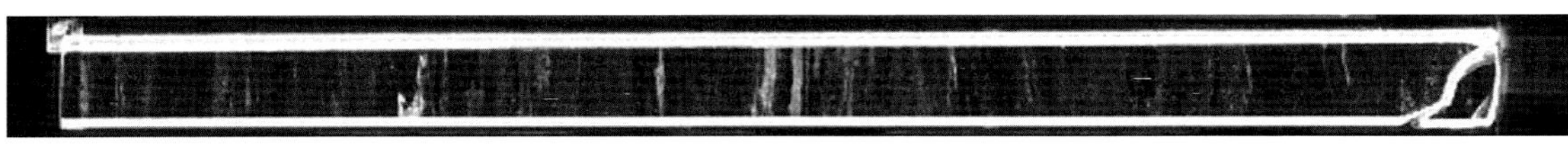

2205

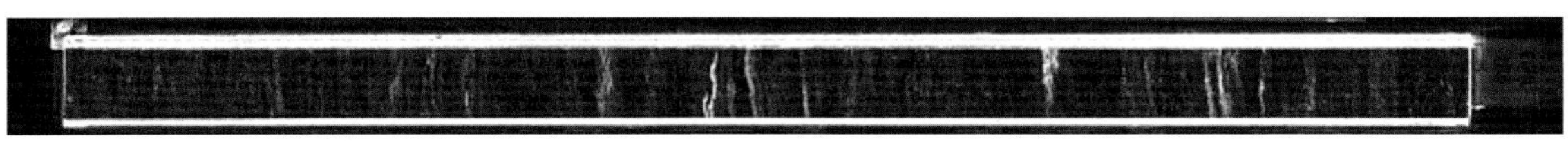

2206

2207

2208

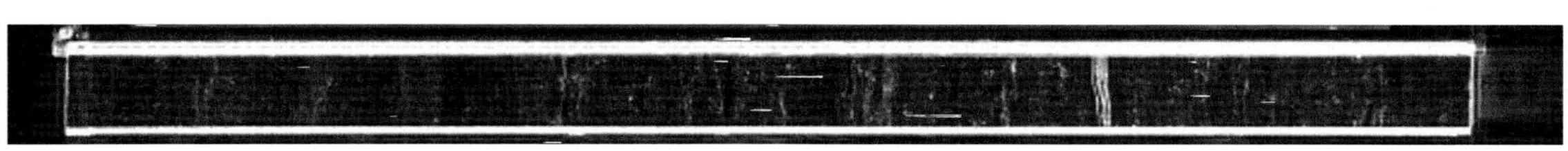

2209

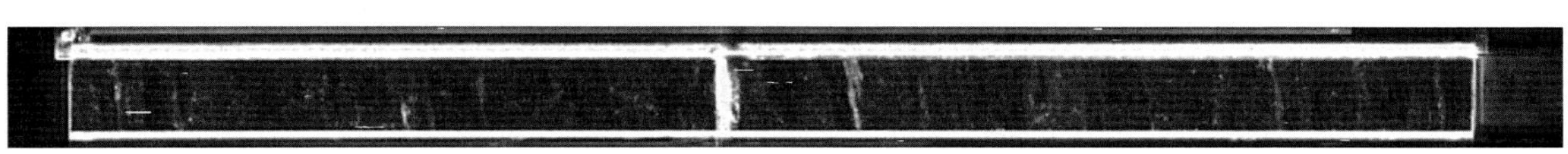

2219

2218

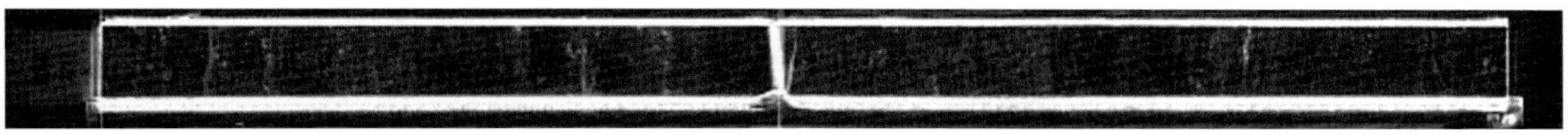
2217

2216

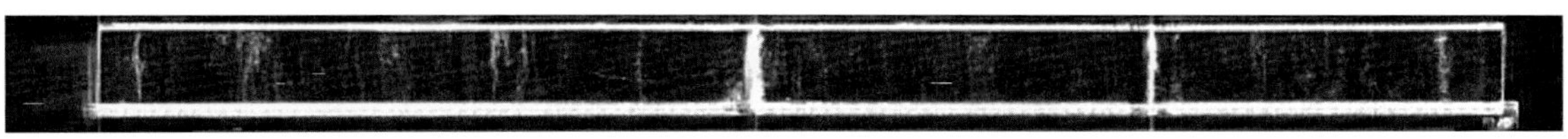
2215

2214

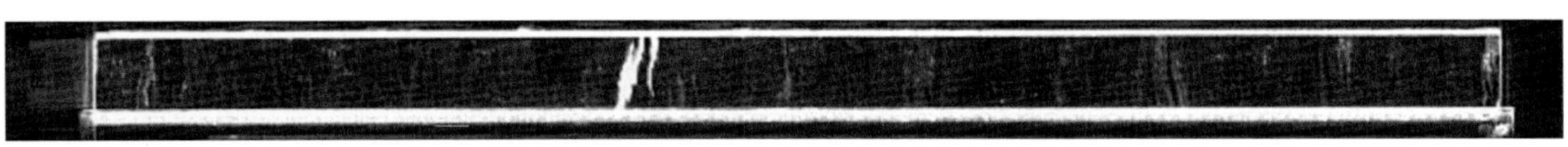
2213

2212

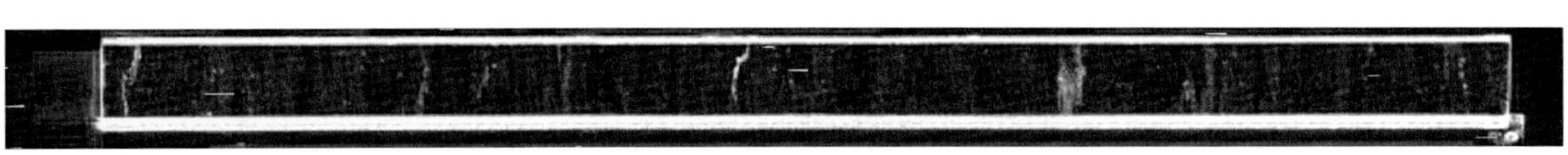
2211

2210

2220

2221

2222

2223

2224

2225

2226

2227

2228

2229

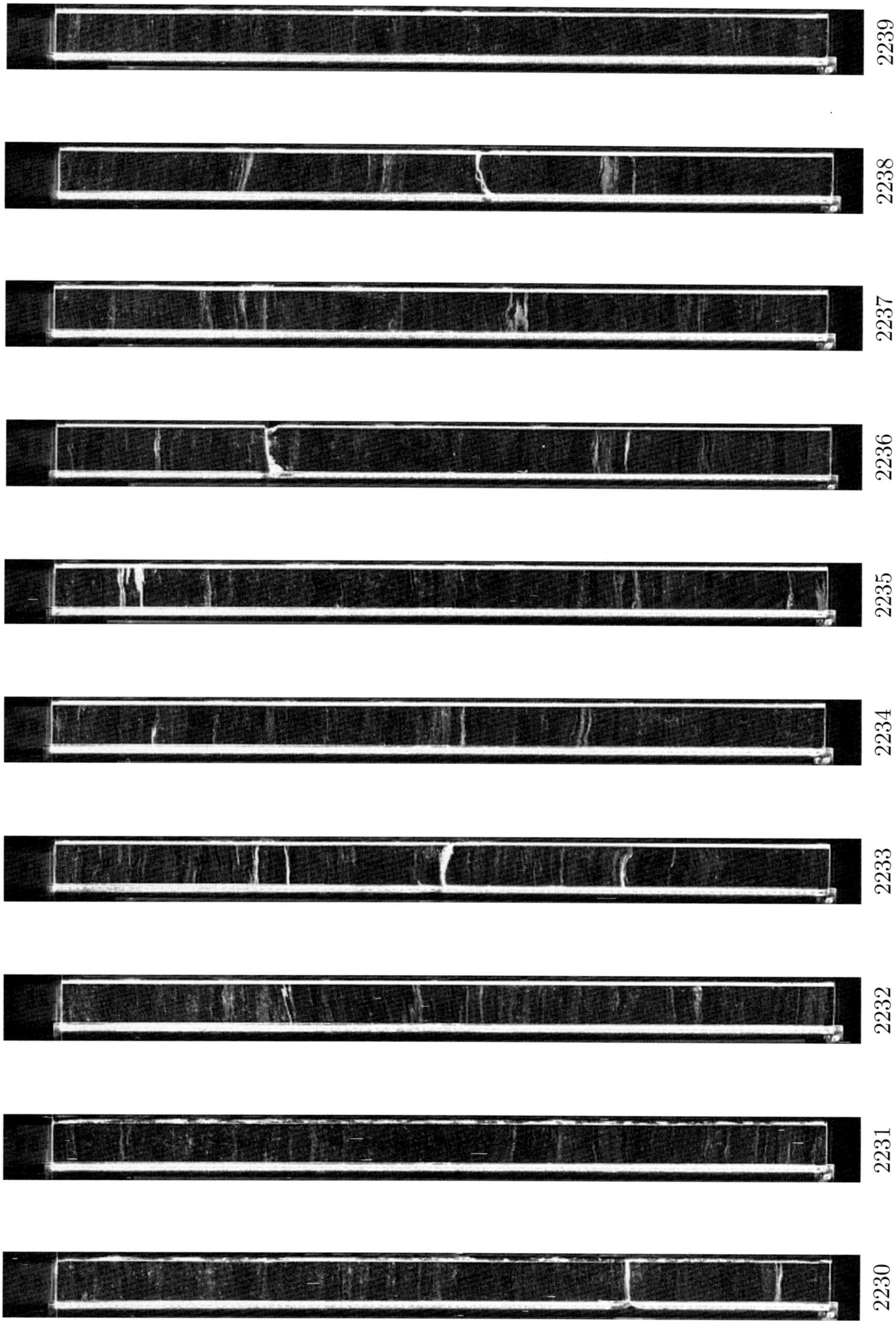
2239
2238
2237
2236
2235
2234
2233
2232
2231
2230

2240

2241

2242

2243

2244

2245

2246

2247

2248

2249

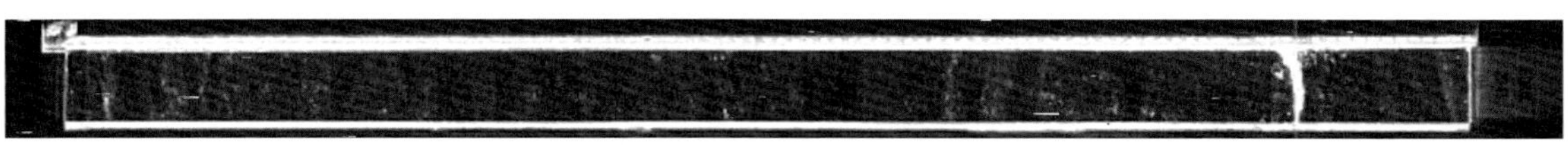

2259

2258

2257

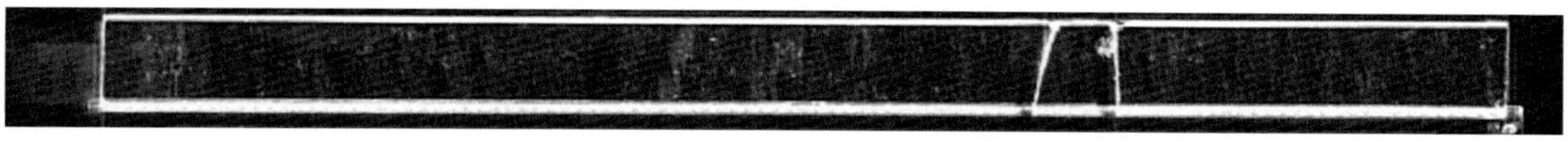
2256

2255

2254

2253

2252

2251

2250

2260

2261

2262

2263

2264

2265

2266

2267
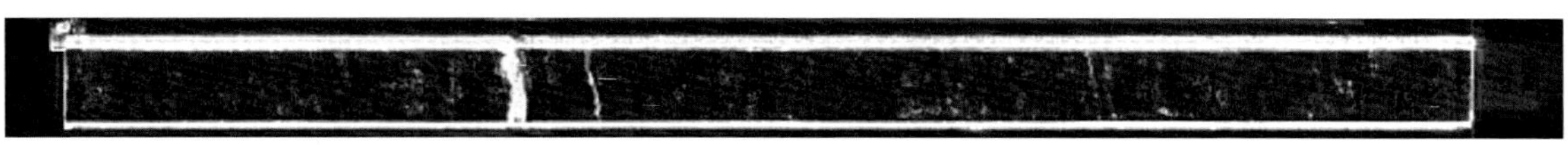

2268

2269
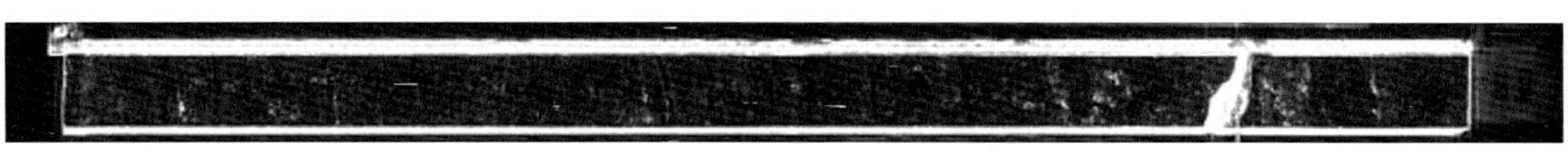

2279

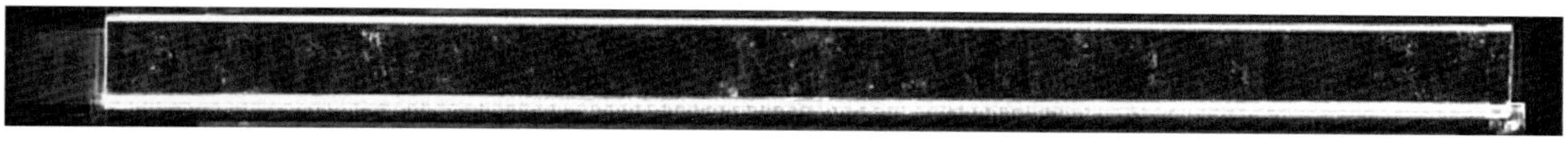
2278

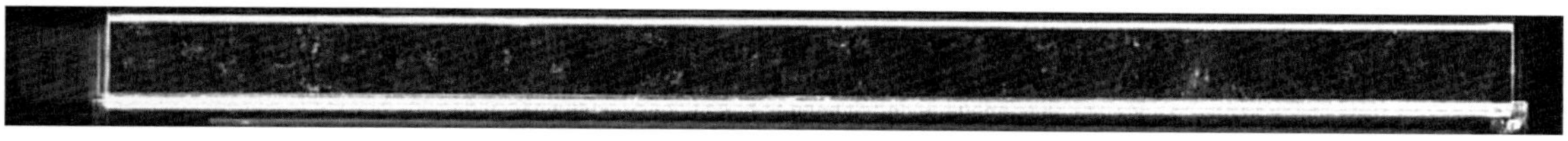
2277

2276

2275

2274

2273

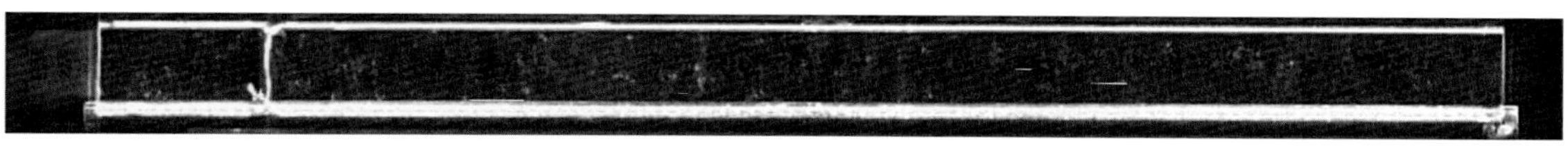
2272

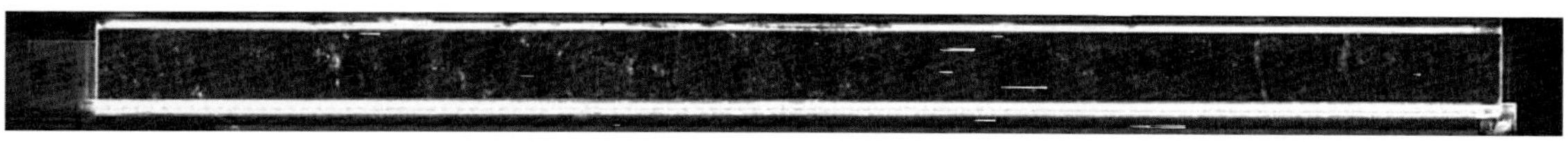
2271

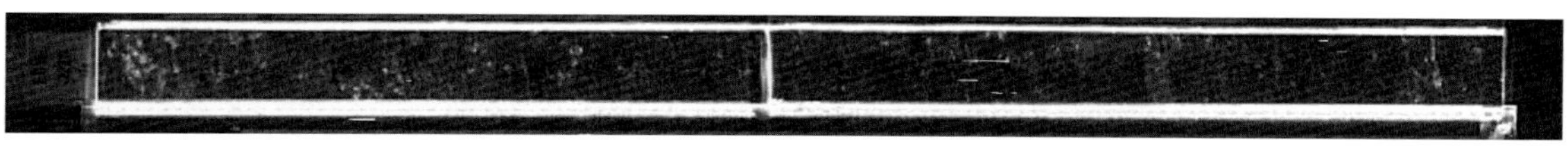
2270

2280

2281
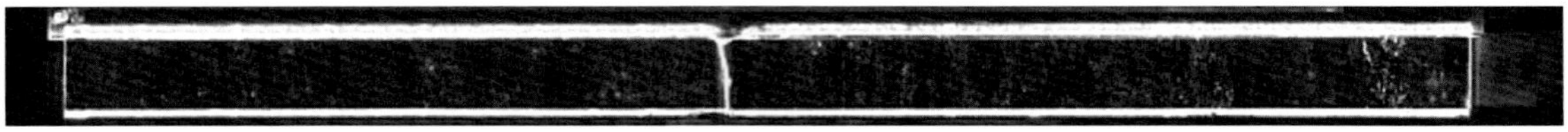

2282

2283

2284

2285

2286

2287
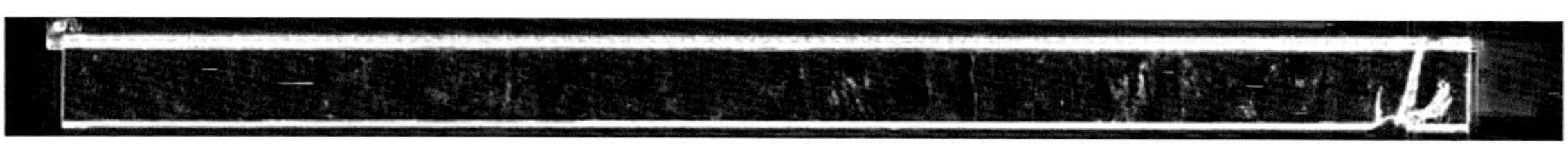

2288
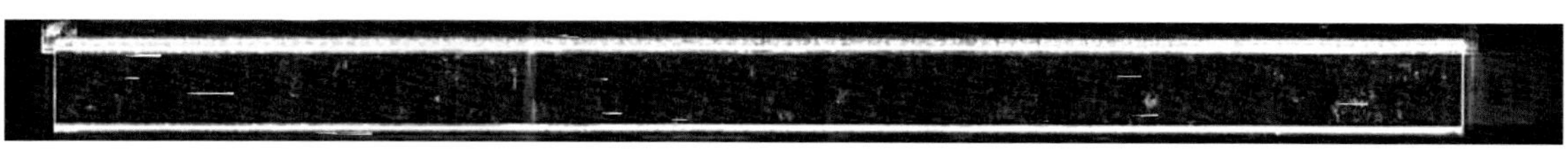

2289
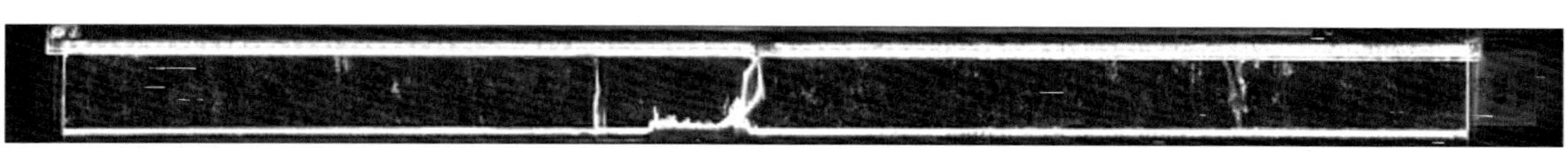

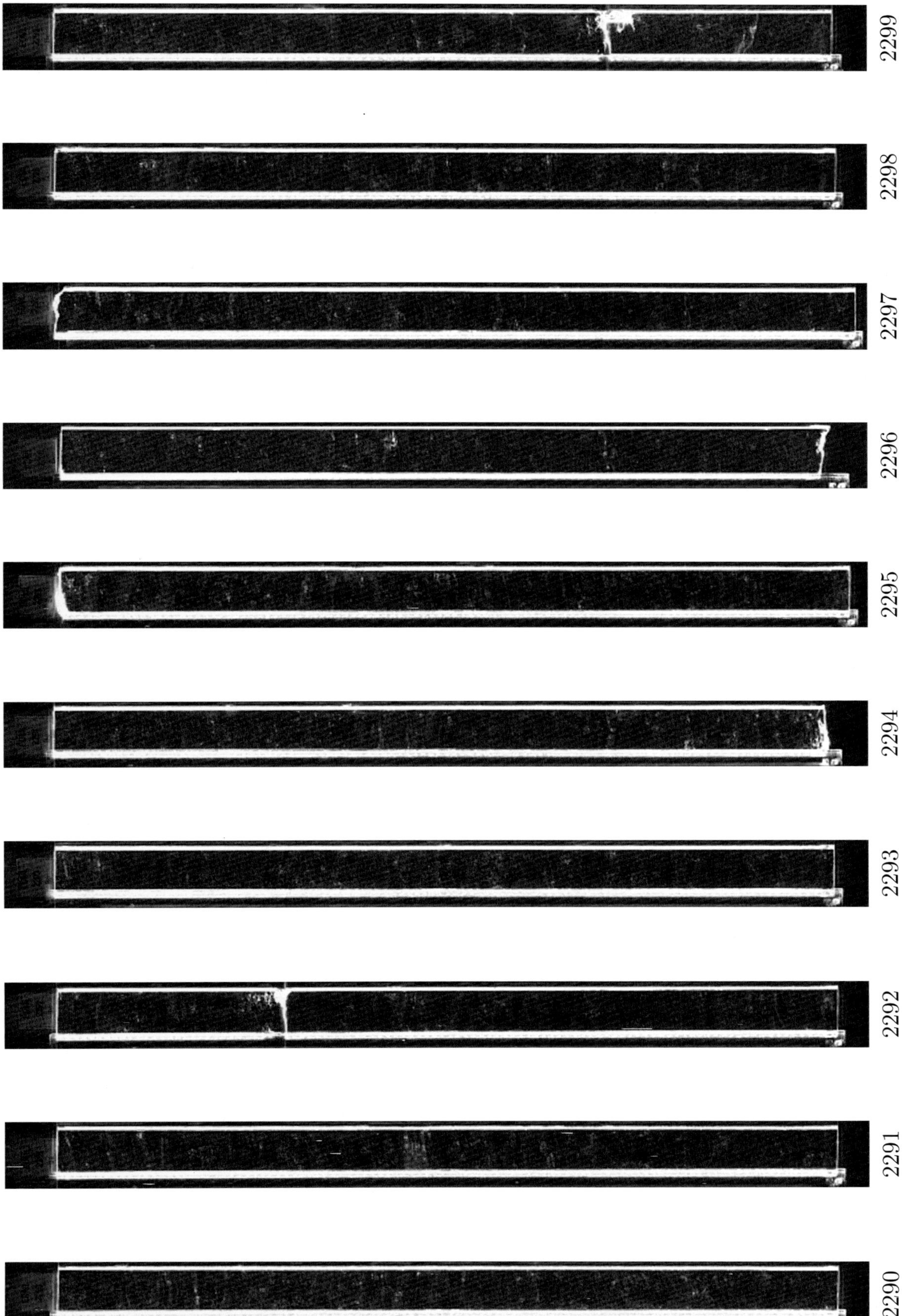
2299
2298
2297
2296
2295
2294
2293
2292
2291
2290

2300

2301

2302

2303

2304

2305

2306

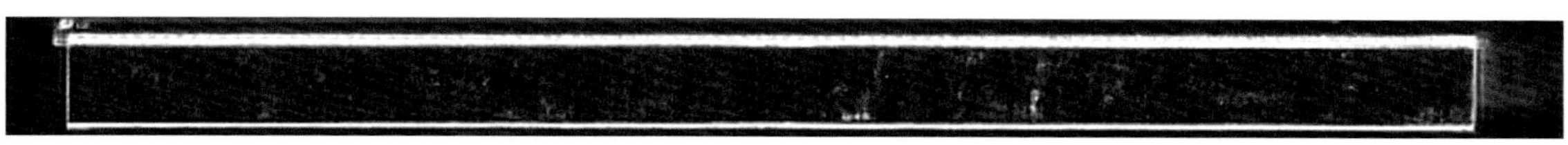

2307

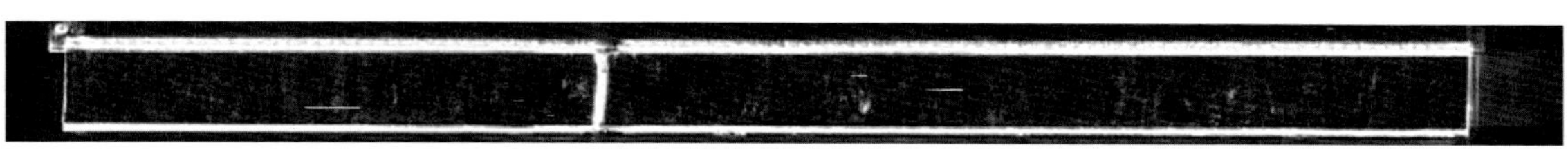

2308

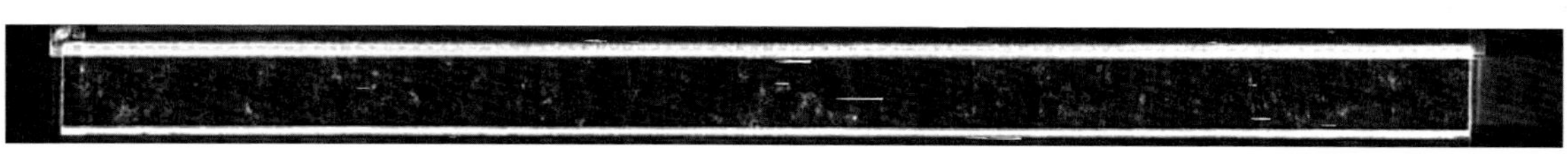

2309

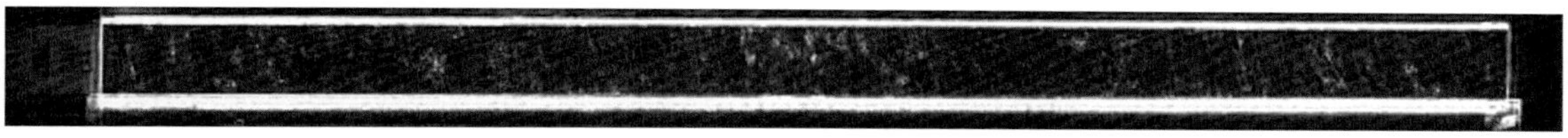
2319

2318

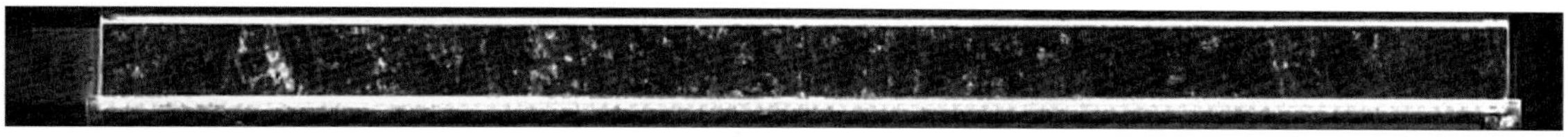
2317

2316

2315

2314

2313

2312

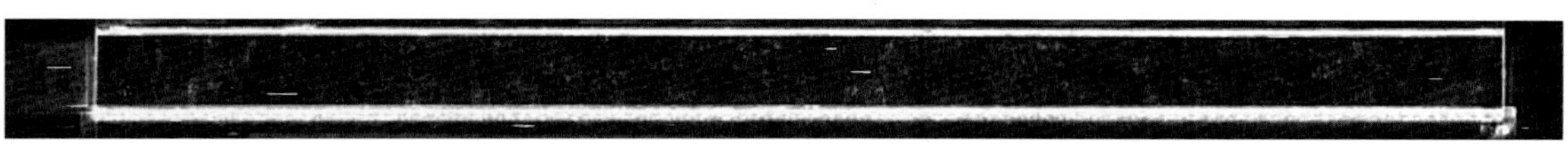
2311

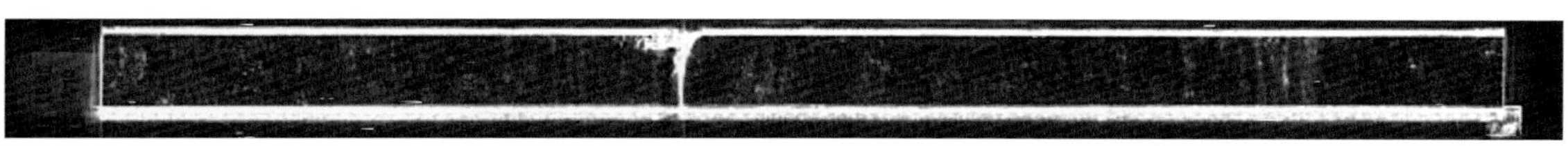
2310

2320

2321

2322
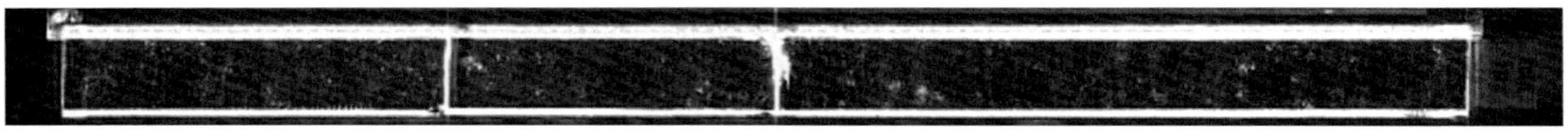

2323

2324
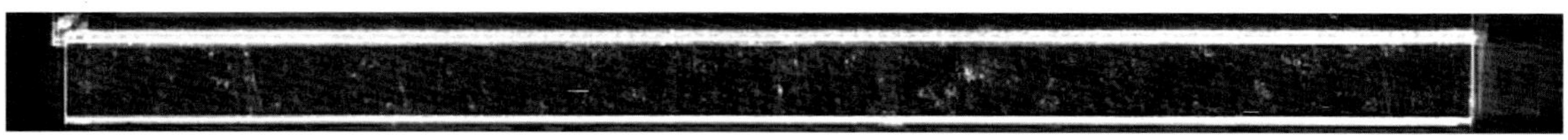

2325

2326

2327
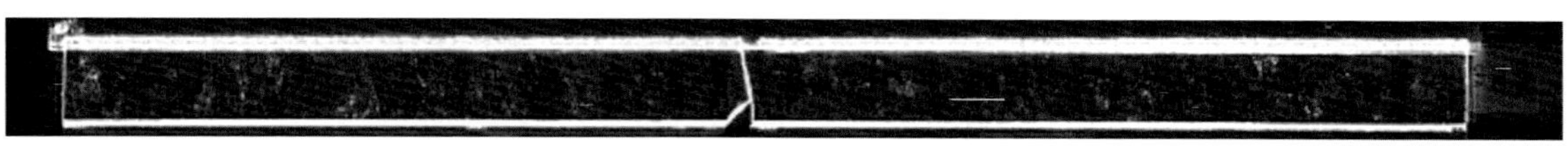

2328

2329
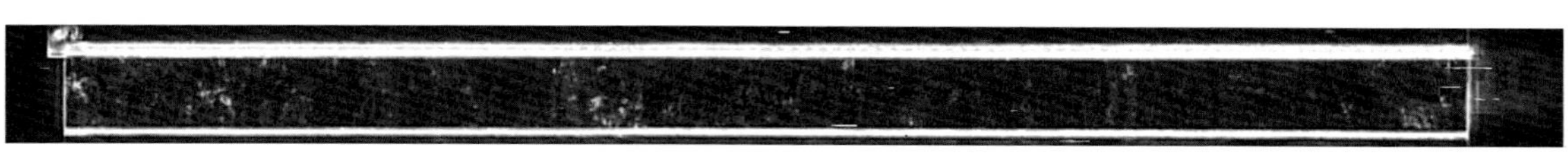

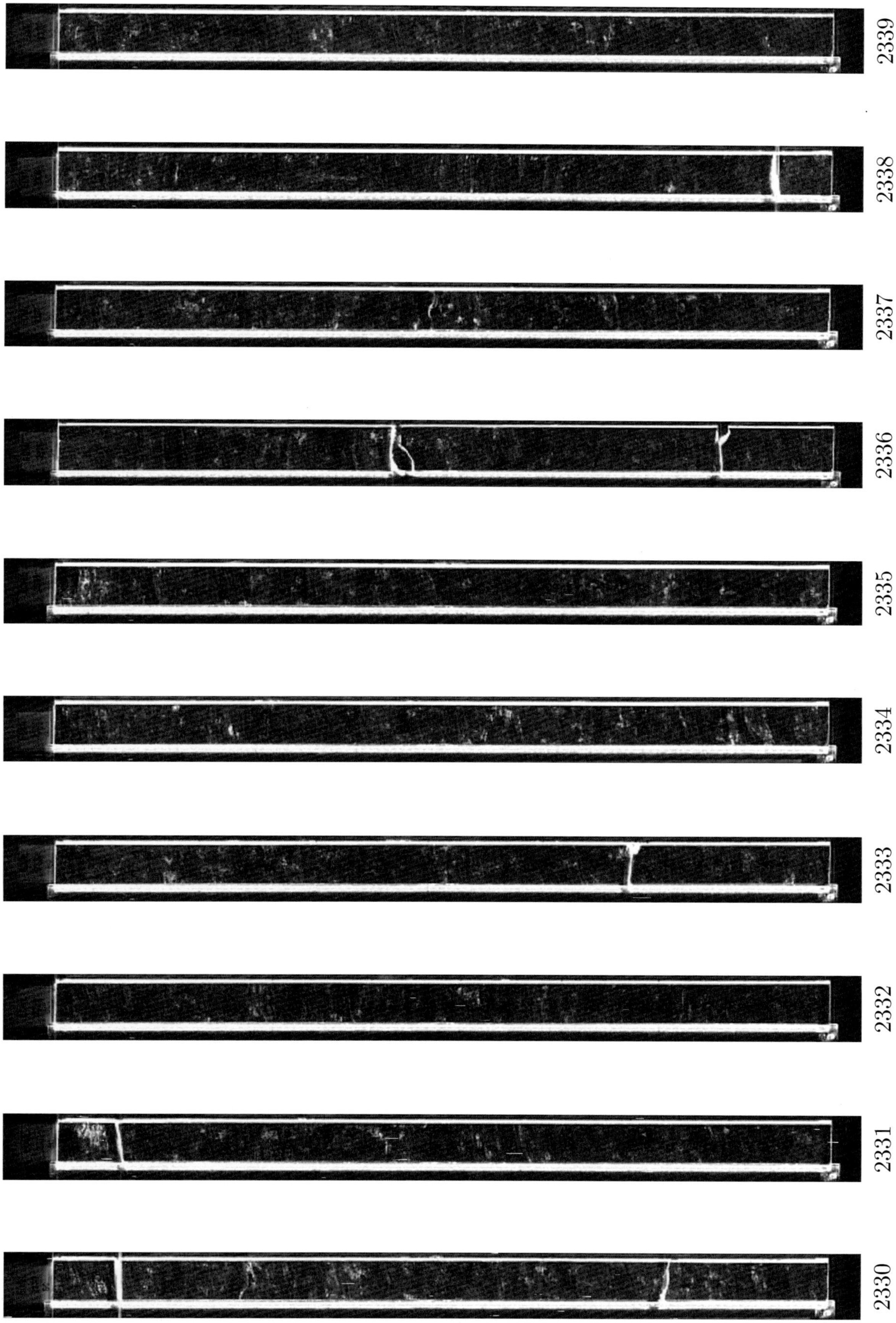
2339
2338
2337
2336
2335
2334
2333
2332
2331
2330

2340

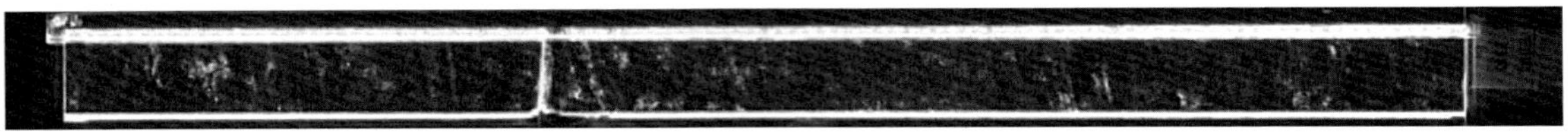

2341

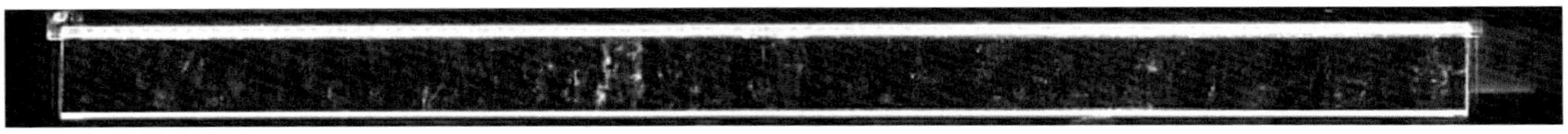

2342

2343

2344

2345

2346

2347

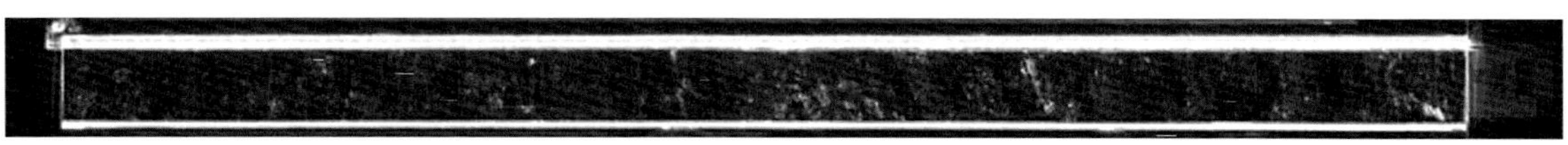

2348

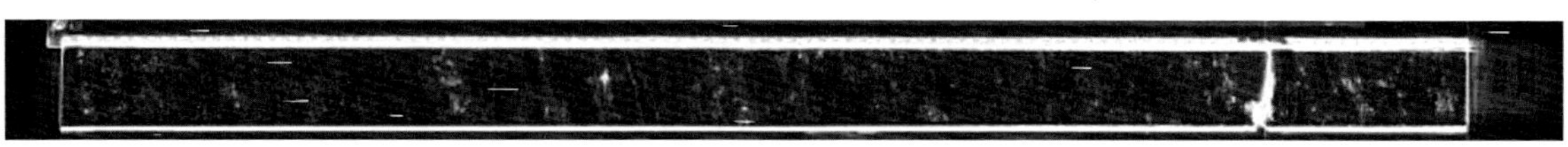

2349

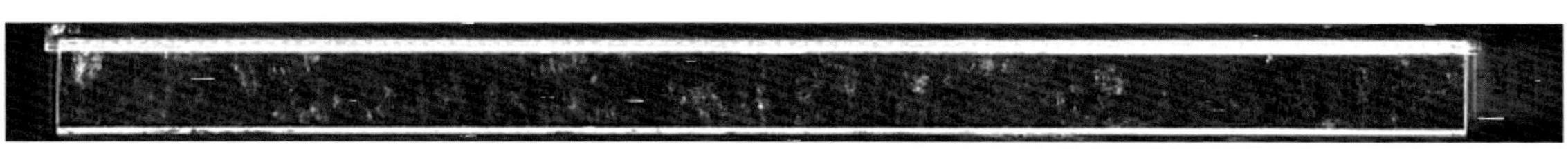

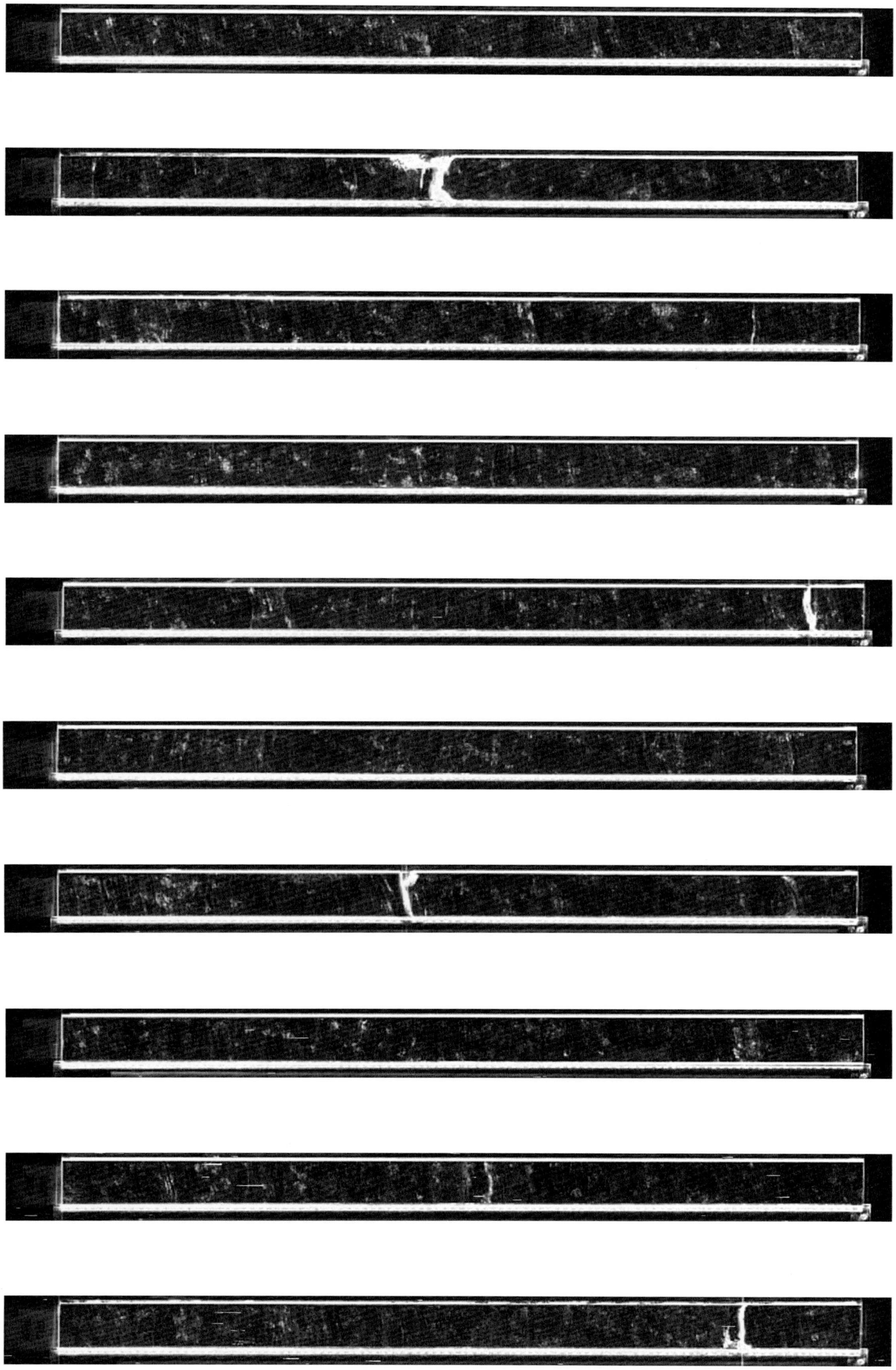

2359

2358

2357

2356

2355

2354

2353

2352

2351

2350

2360

2361

2362

2363

2364

2365
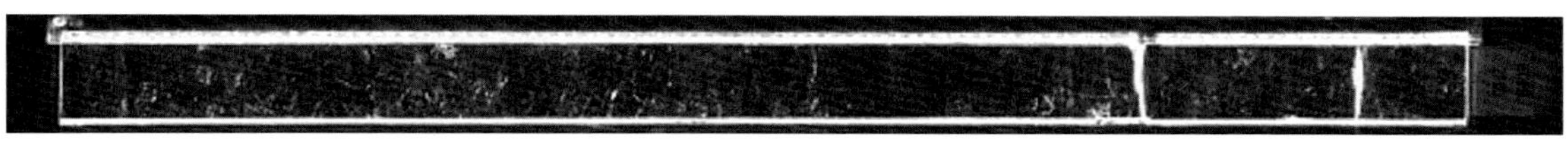

2366
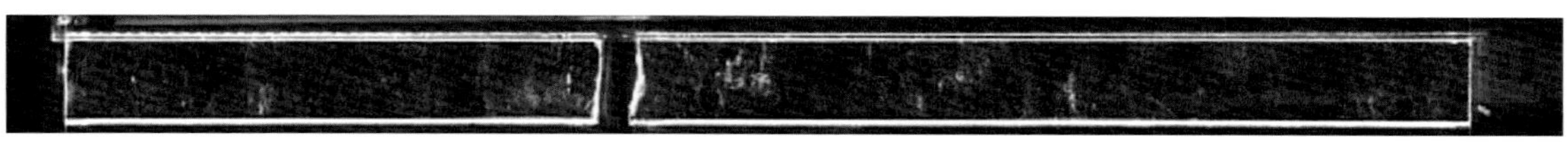

2367
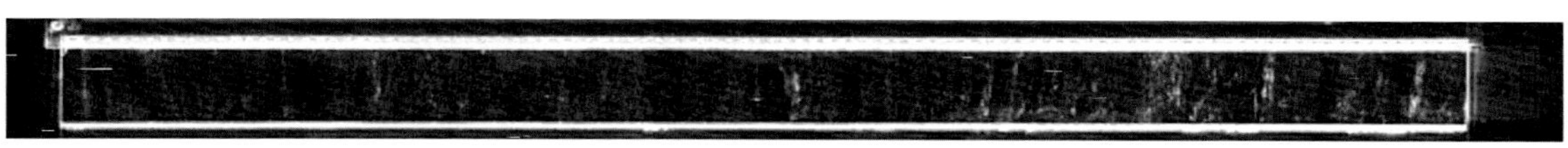

2368

2369
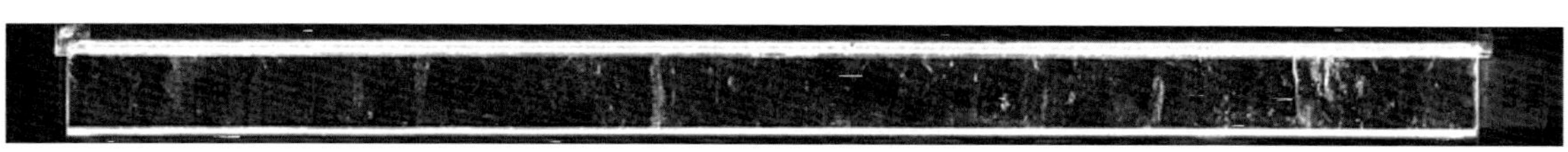

2379

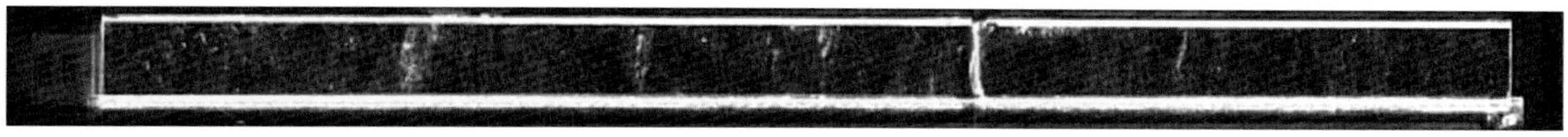
2378

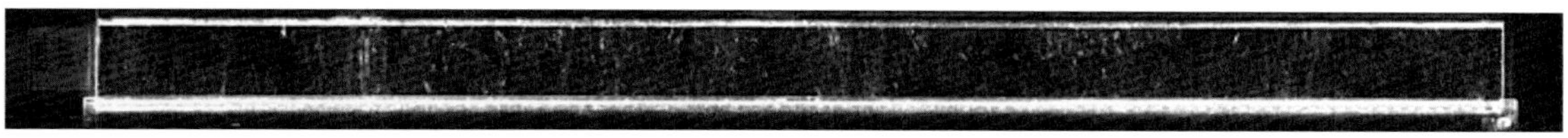
2377

2376

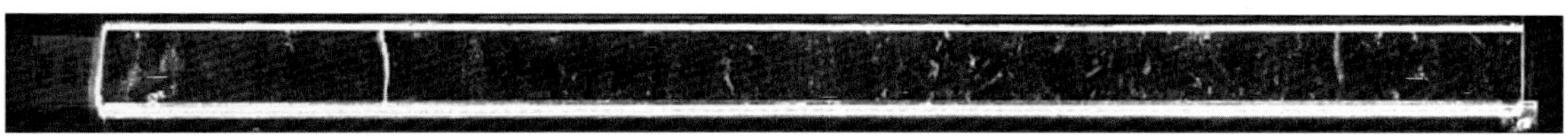
2375

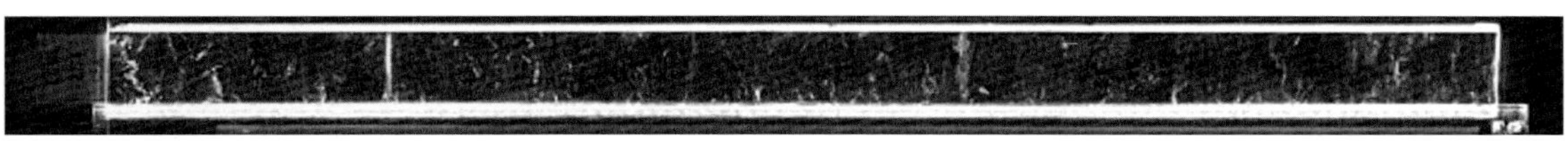
2374

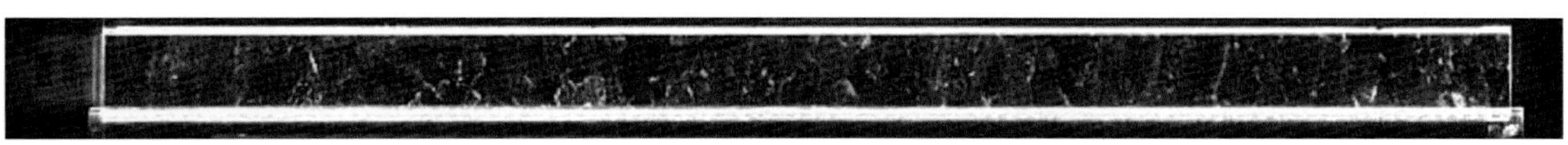
2373

2372

2371

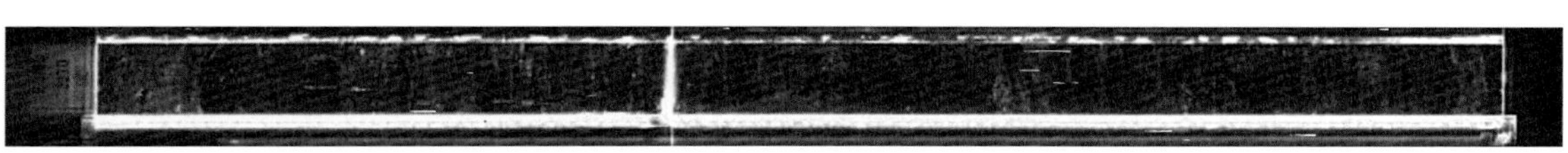
2370

2380

2381

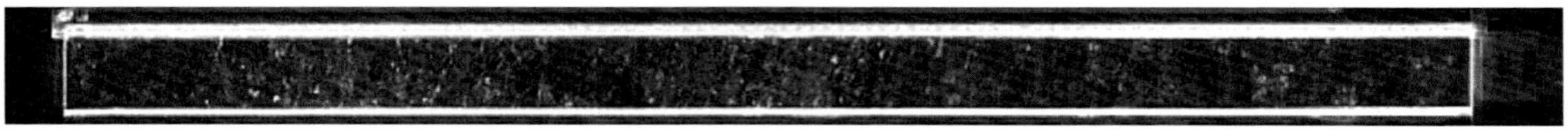

2382

2383

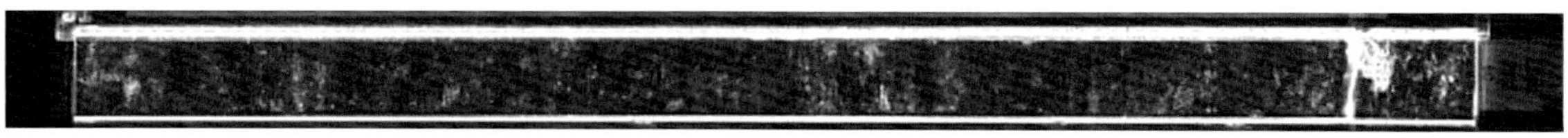

2384

2385

2386

2387

2388

2389

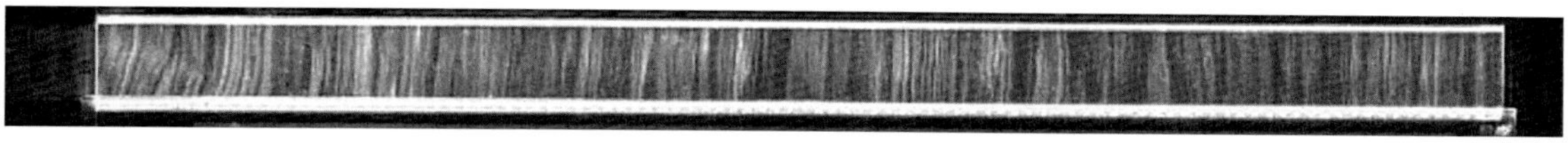
2399

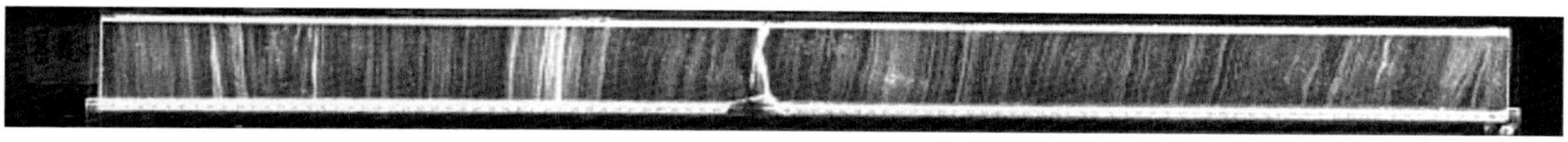
2398

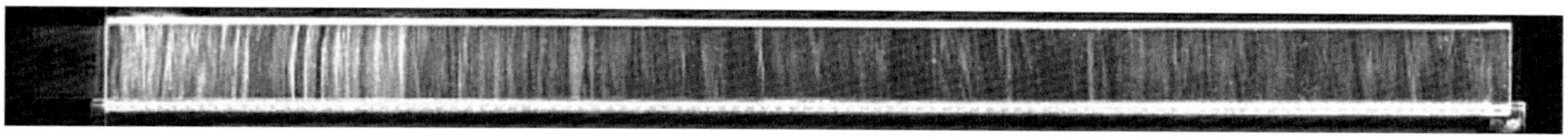
2397

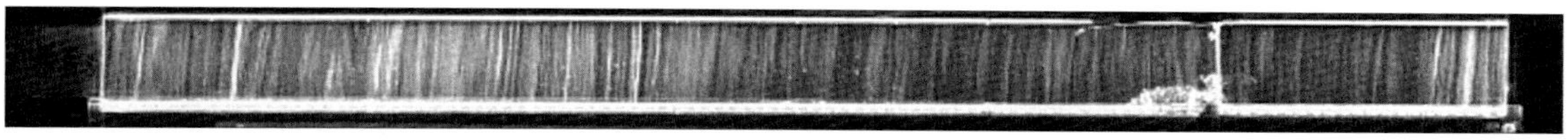
2396

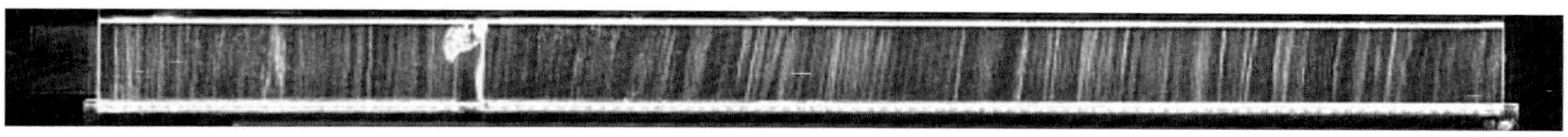
2395

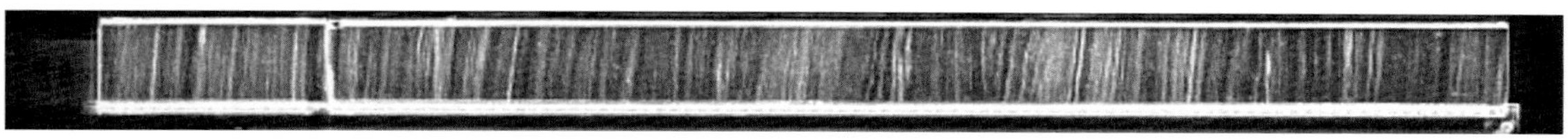
2394

2393

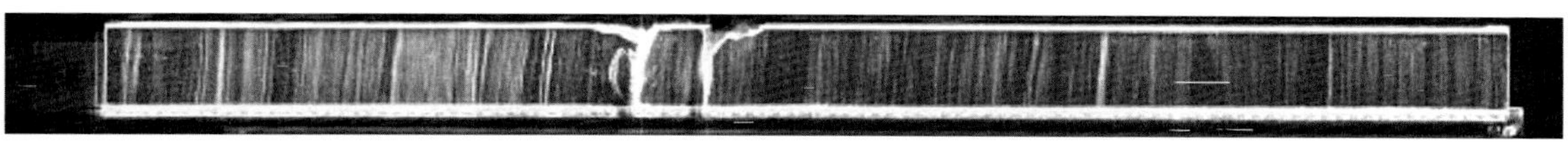
2392

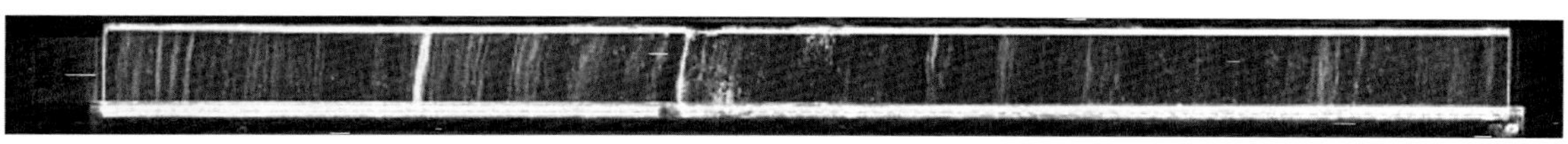
2391

2390

2400

2401

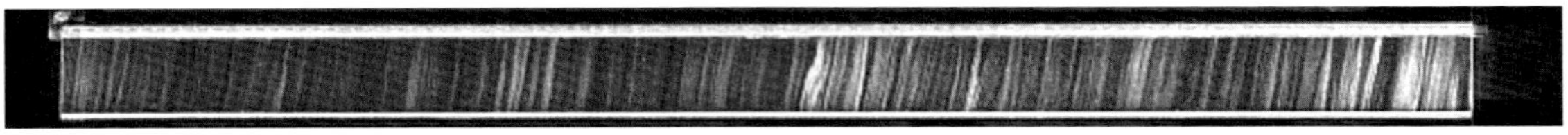

2402

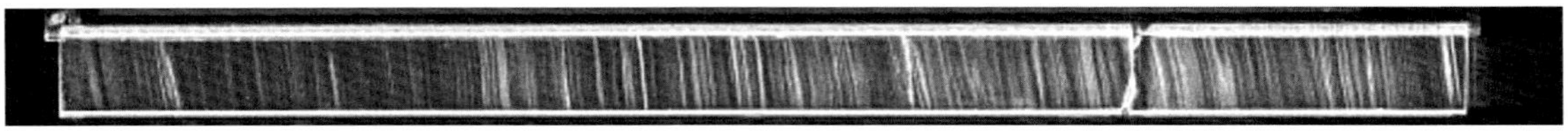

2403

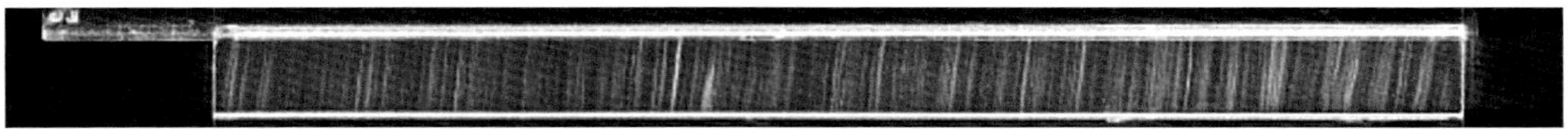

2404

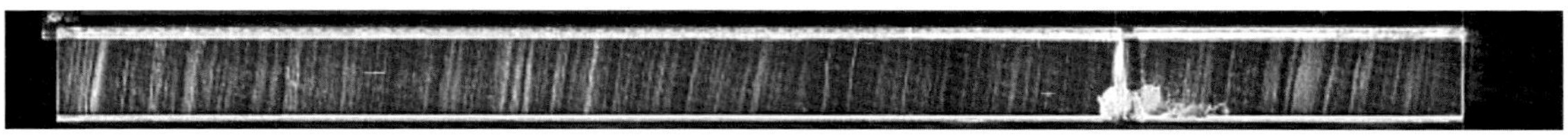

2405

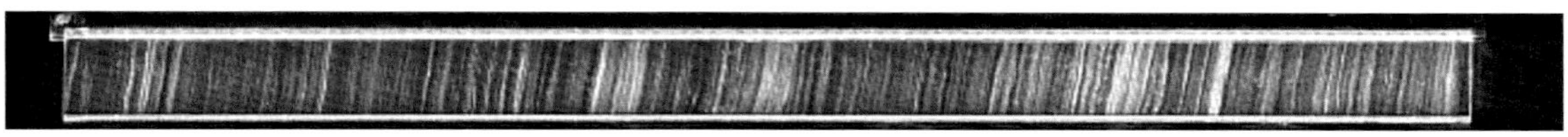

2406

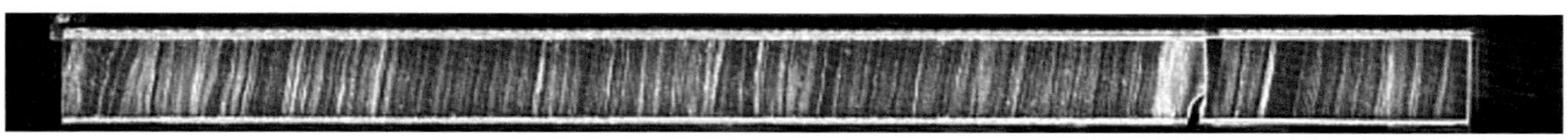

2407

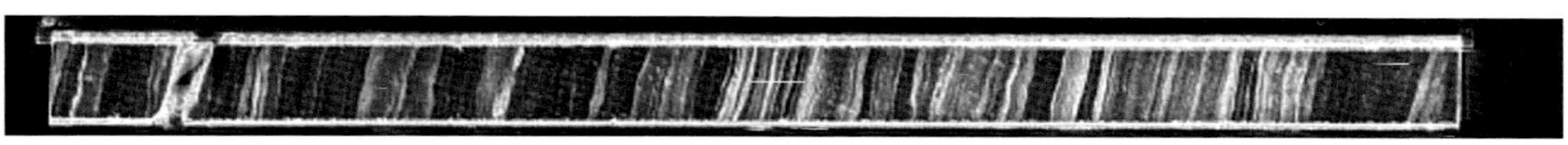

2408

2409

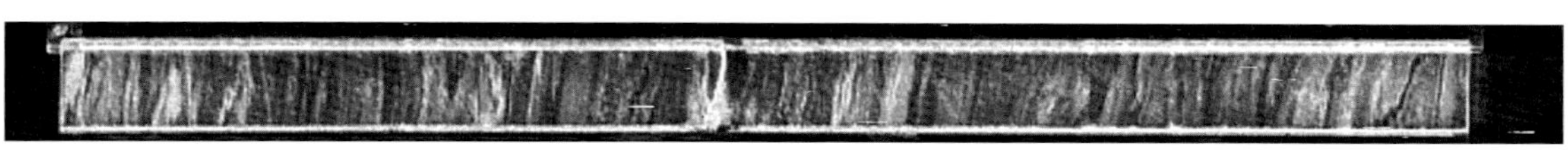

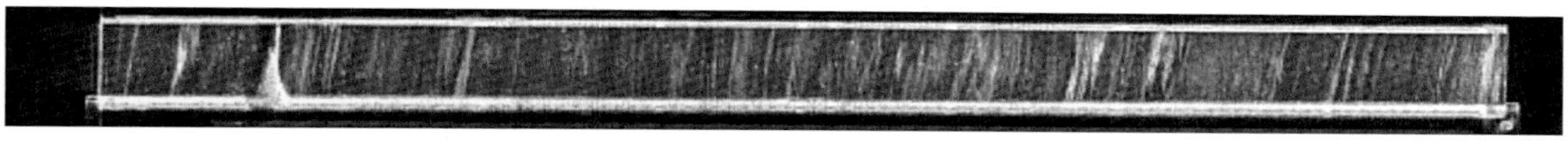

2419

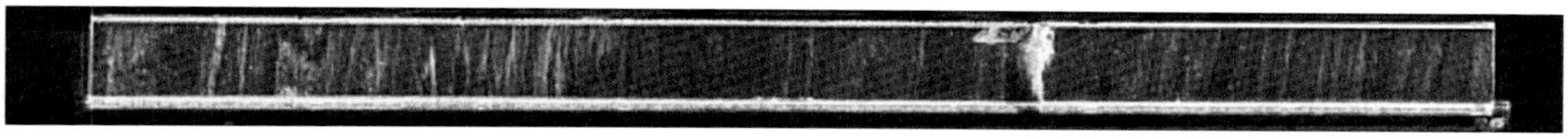

2418

2417

2416

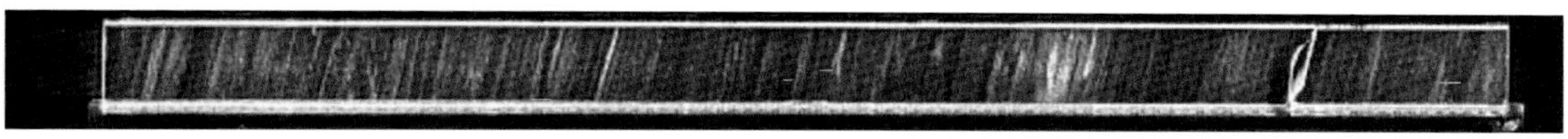

2415

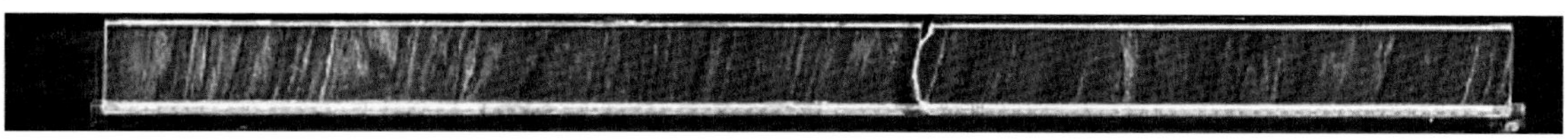

2414

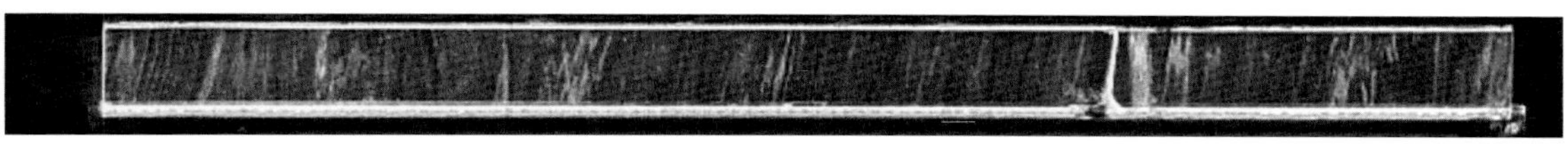

2413

2412

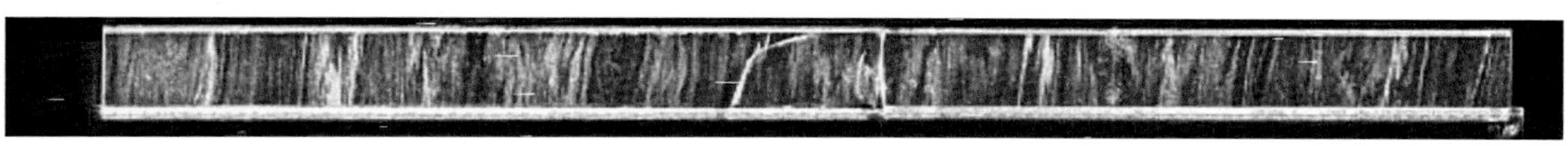

2411

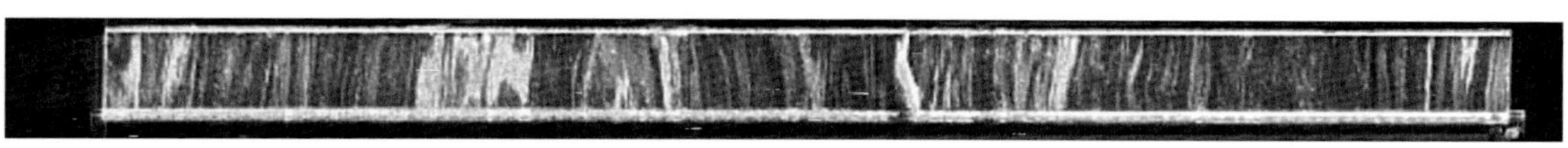

2410

2420

2421

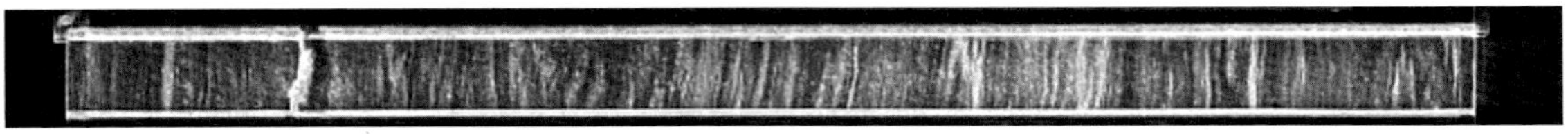

2422

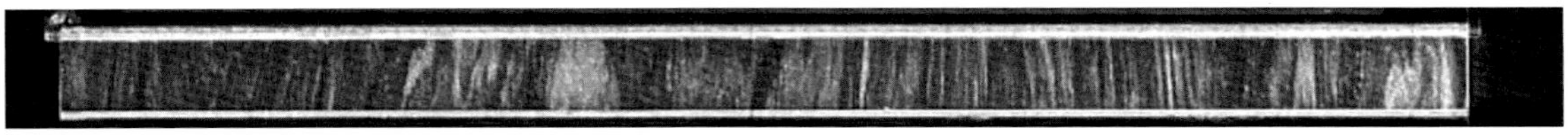

2423

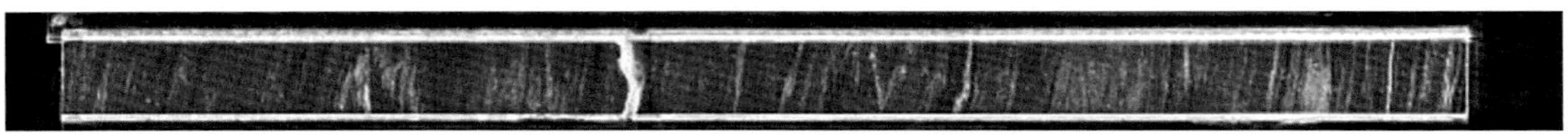

2424

2425

2426

2427

2428

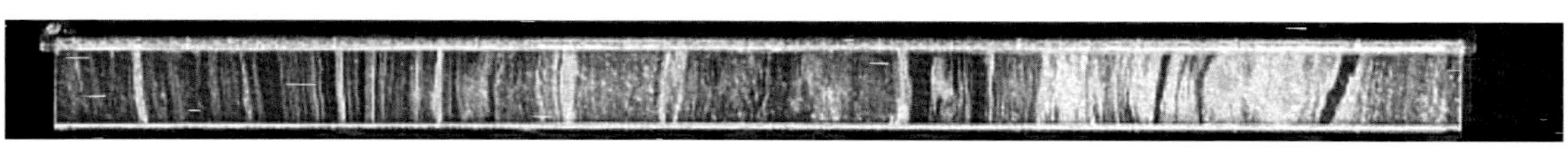

2429

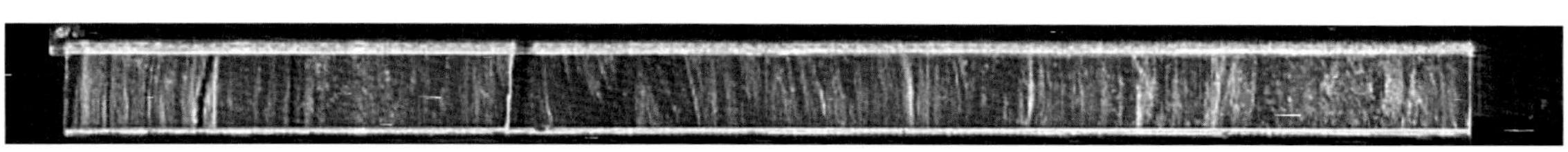

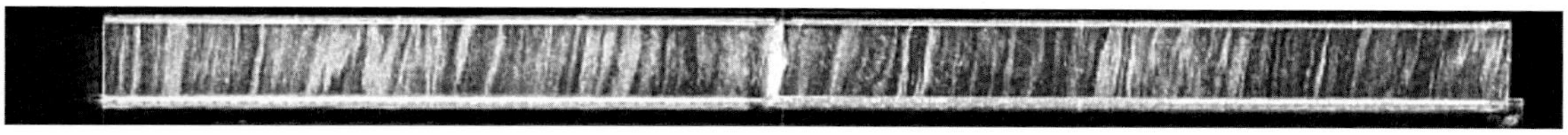
2439

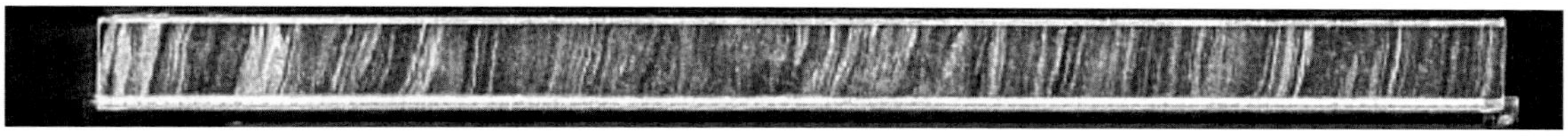
2438

2437

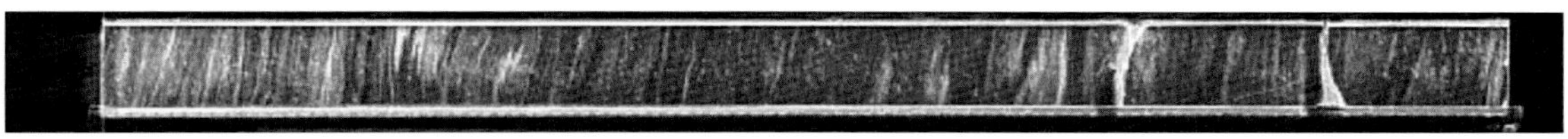
2436

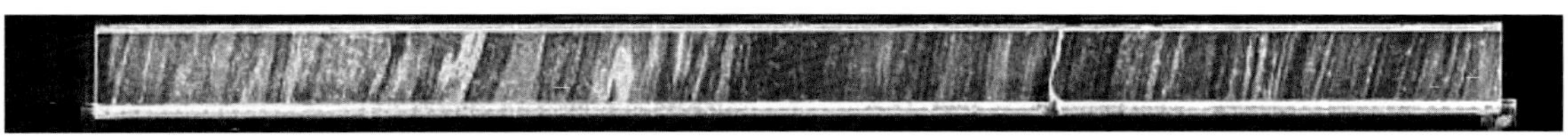
2435

2434

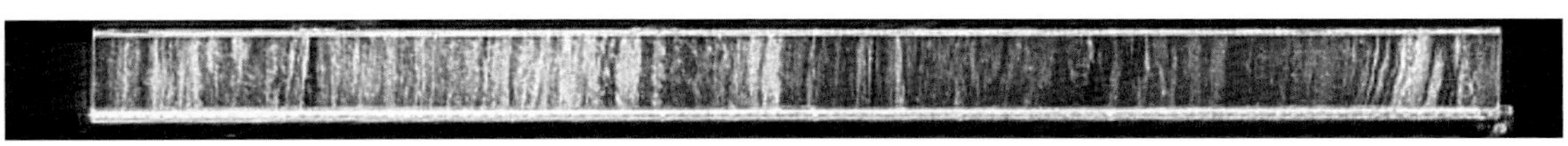
2433

2432

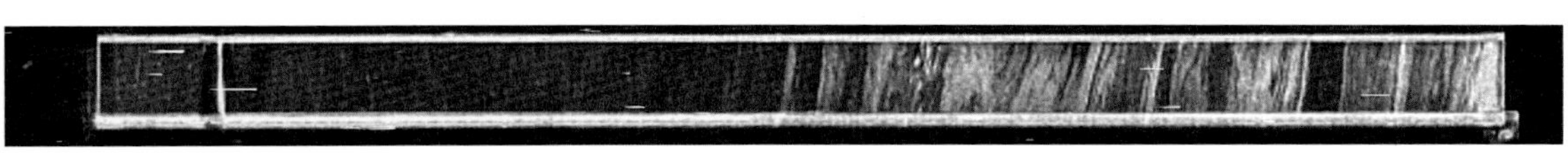
2431

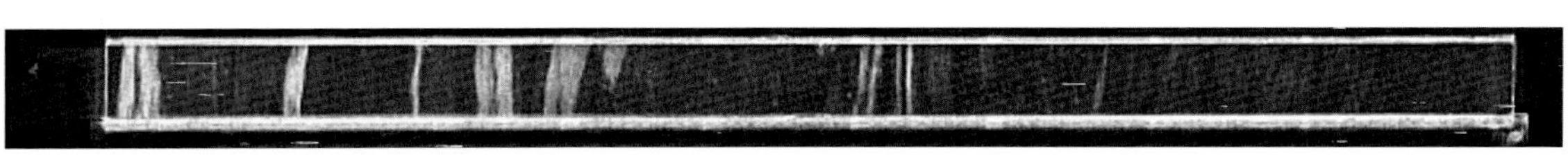
2430

2440

2441

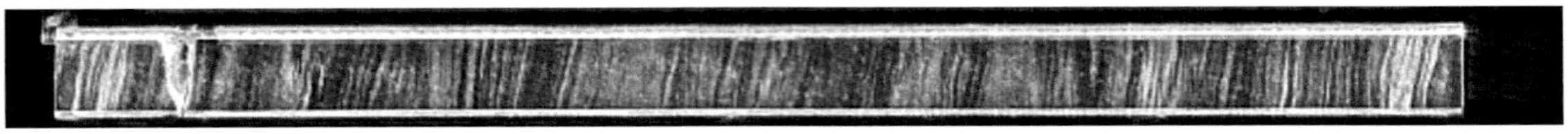

2442

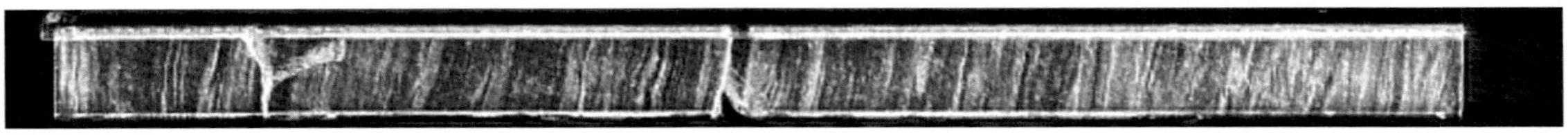

2443

2444

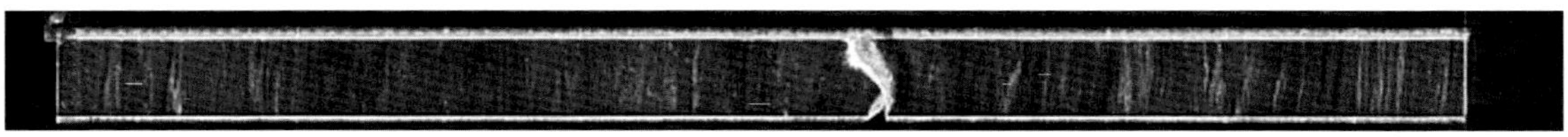

2445

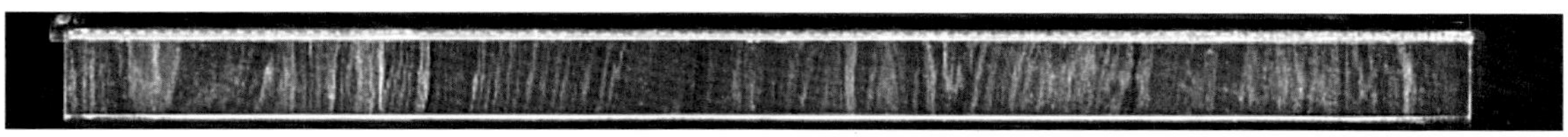

2446

2447

2448

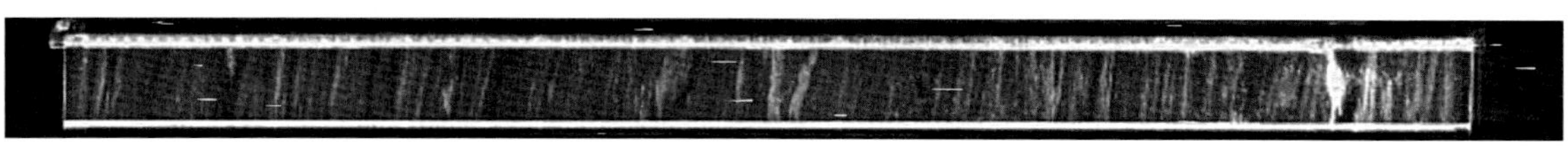

2449

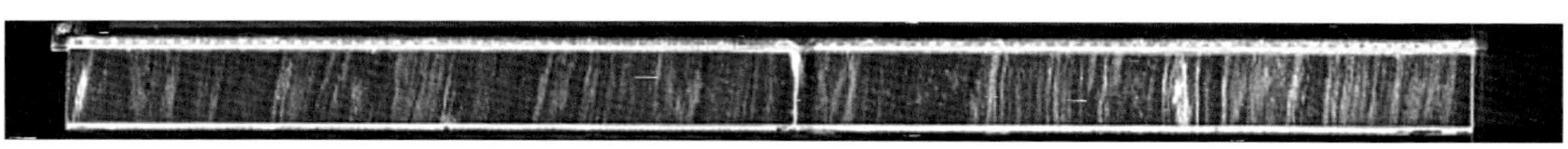

2459

2458

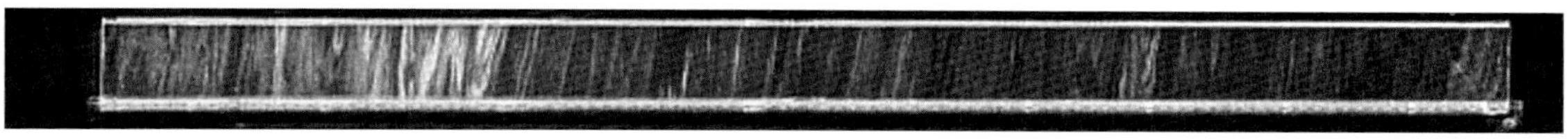
2457

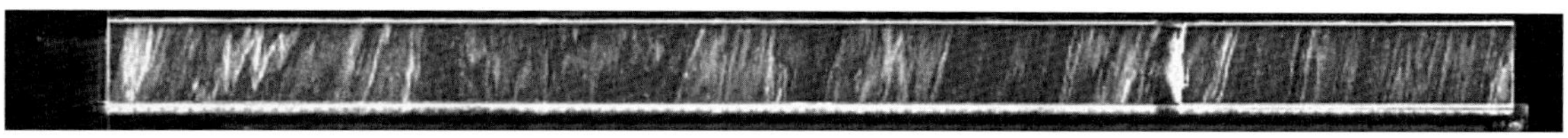
2456

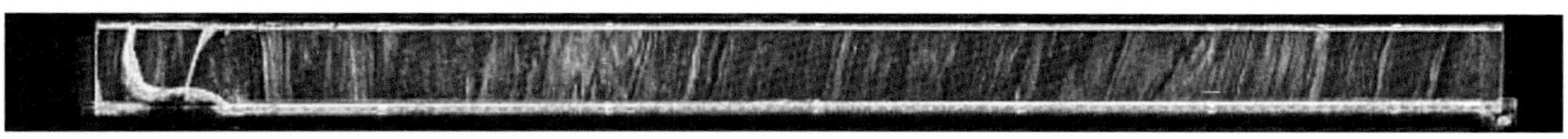
2455

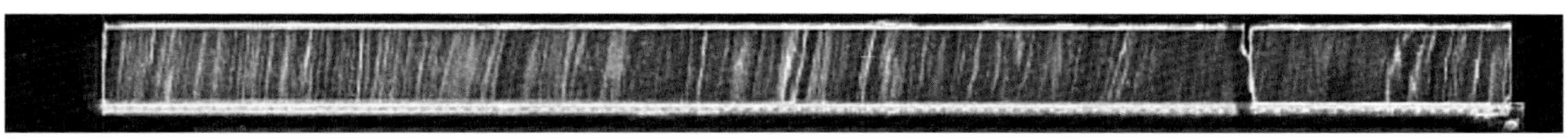
2454

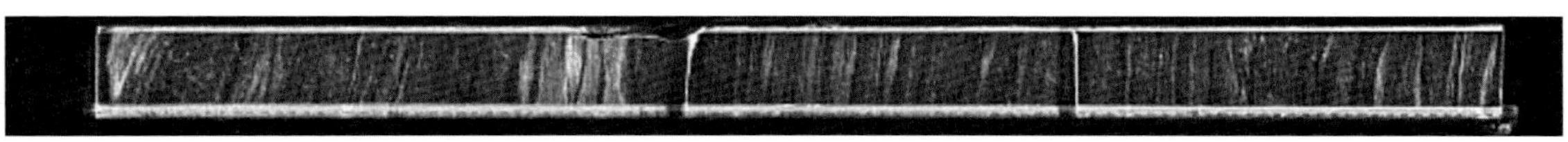
2453

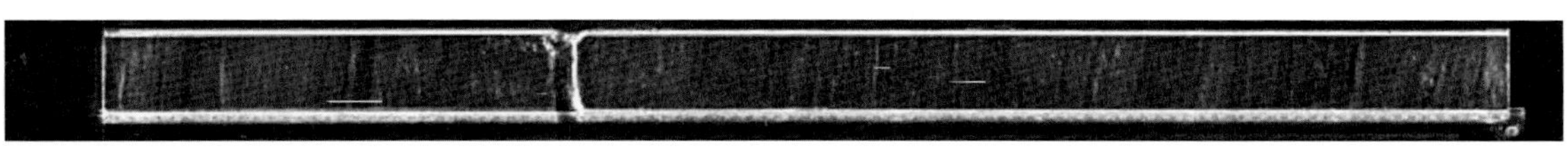
2452

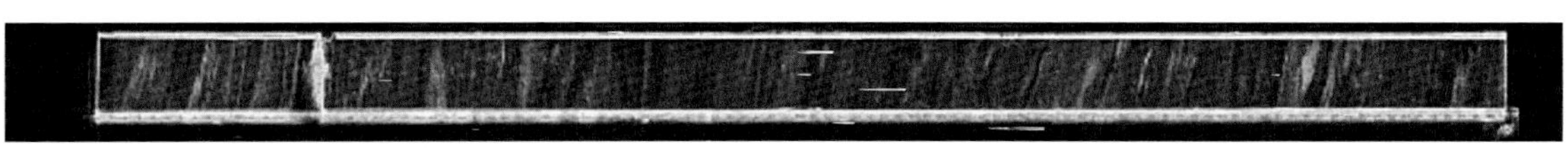
2451

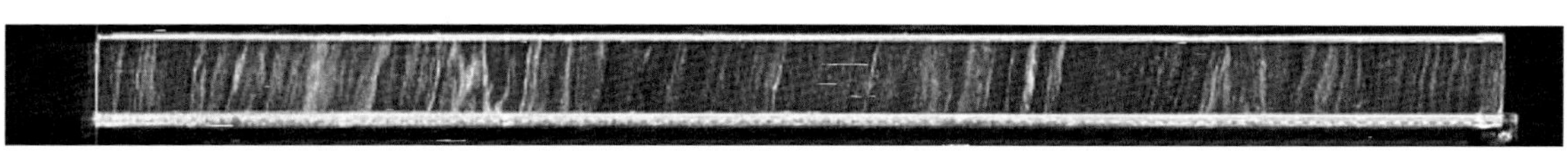
2450

2460

2461

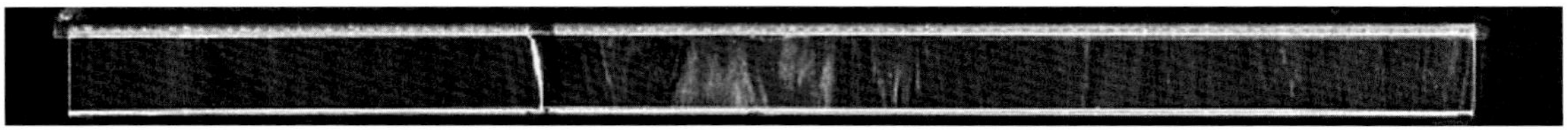

2462

2463

2464

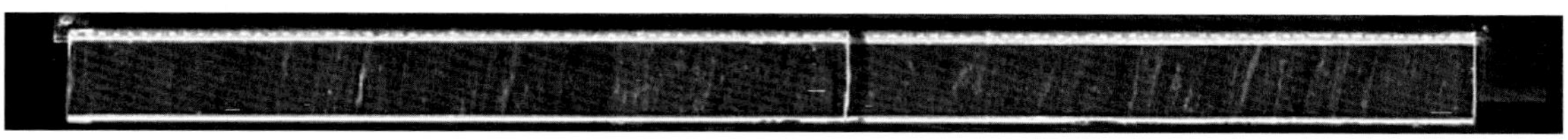

2465

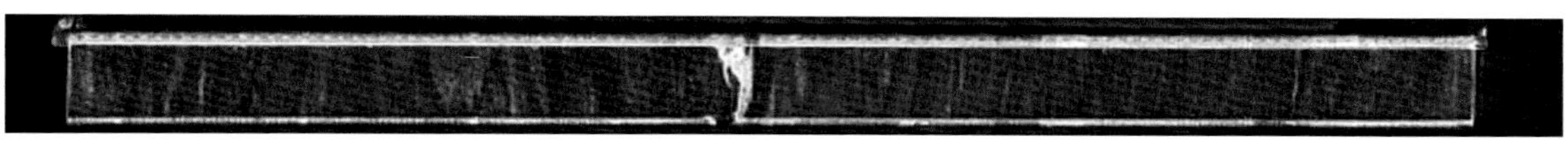

2466

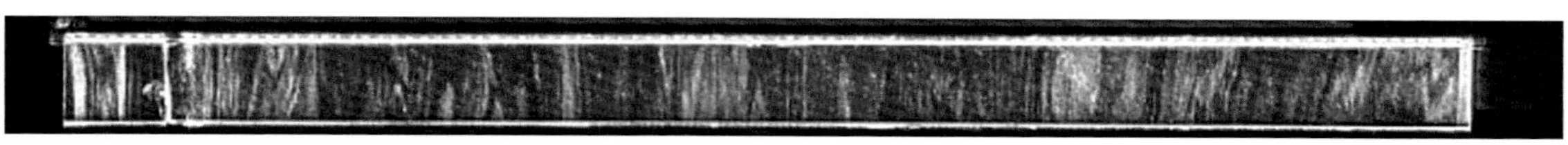

2467

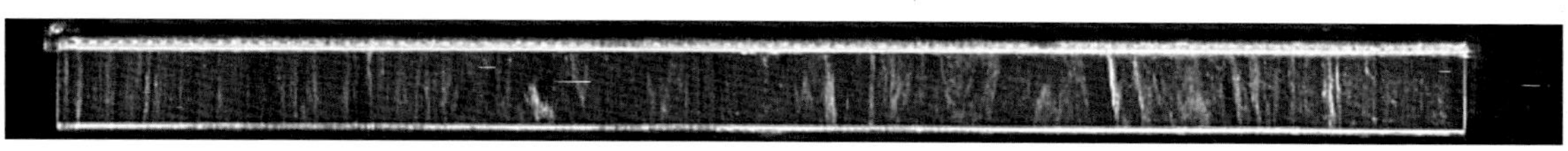

2468

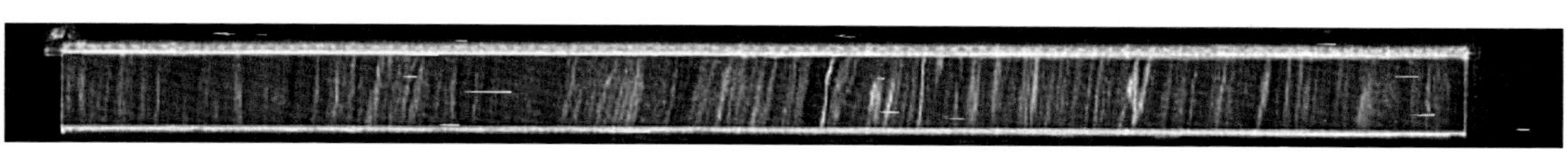

2469

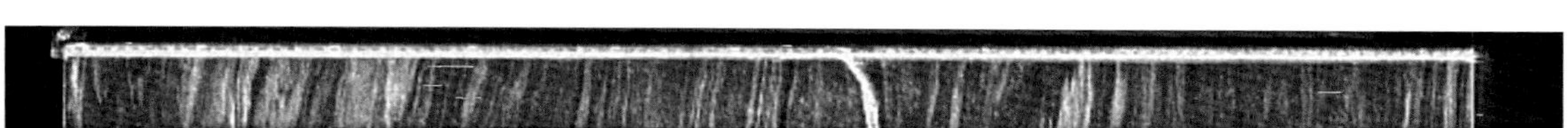

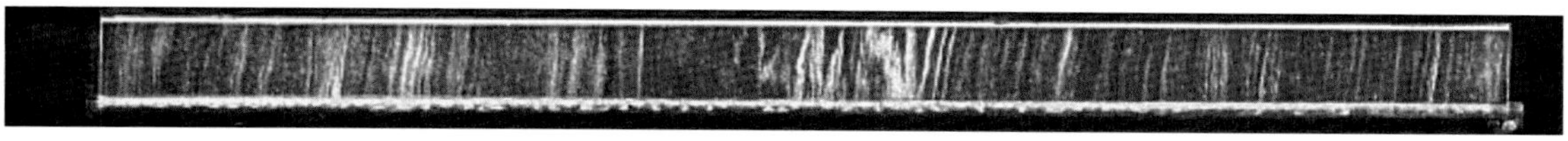
2479

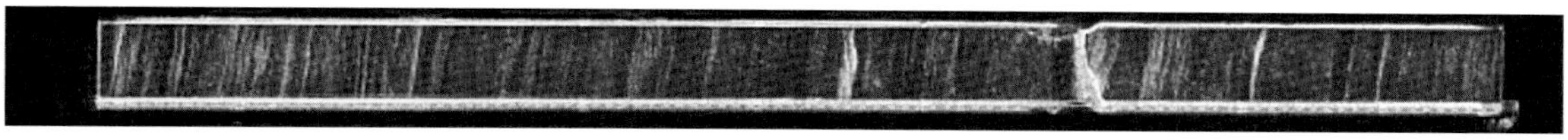
2478

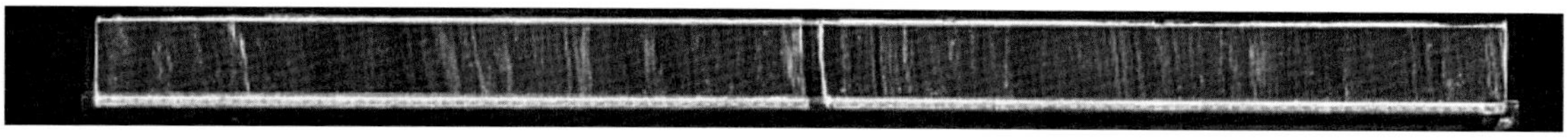
2477

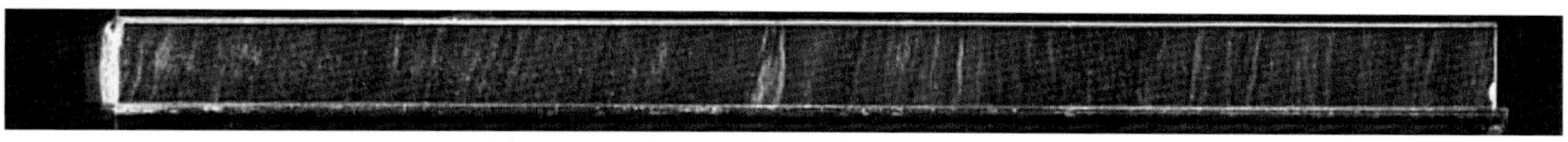
2476

2475

2474

2473

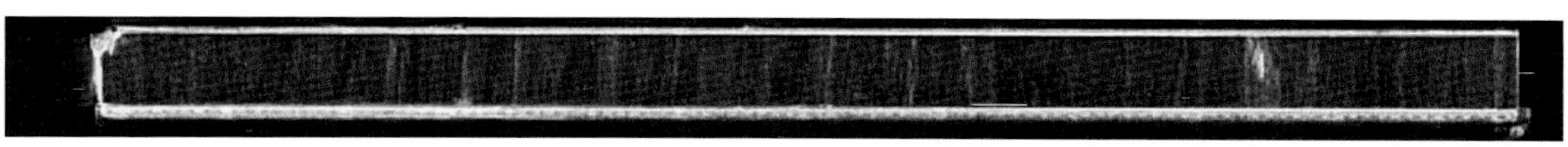
2472

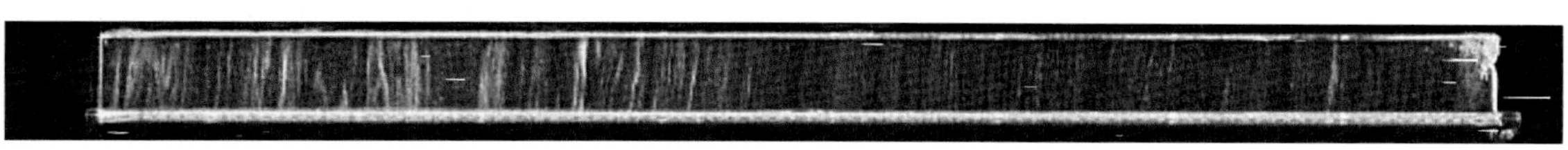
2471

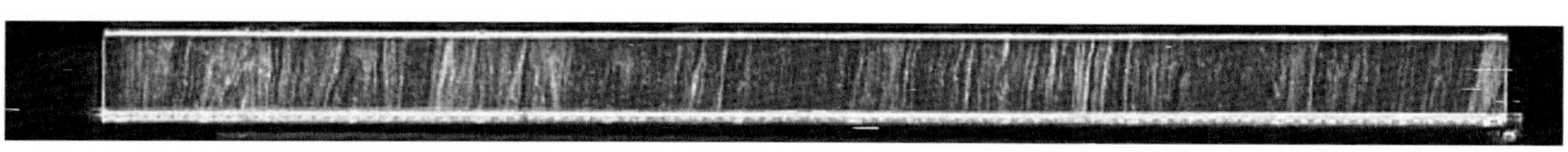
2470

2480

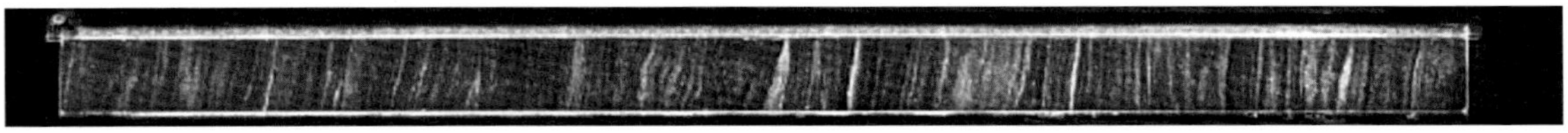

2481

2482

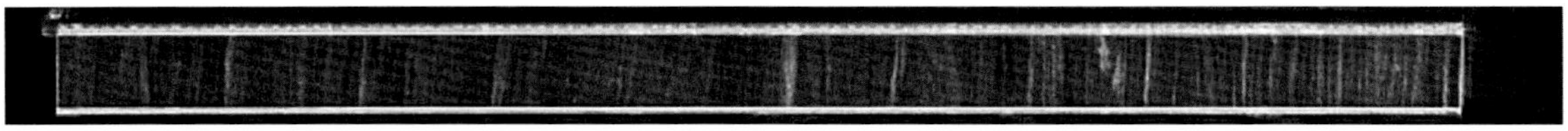

2483

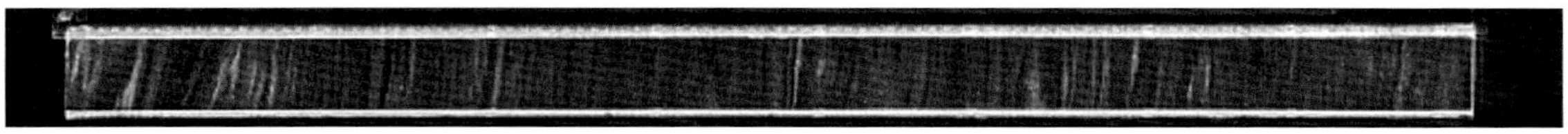

2484

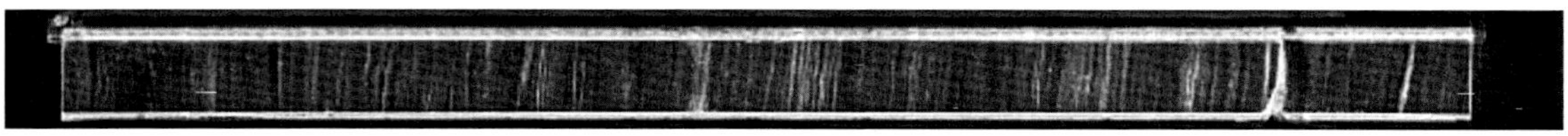

2485

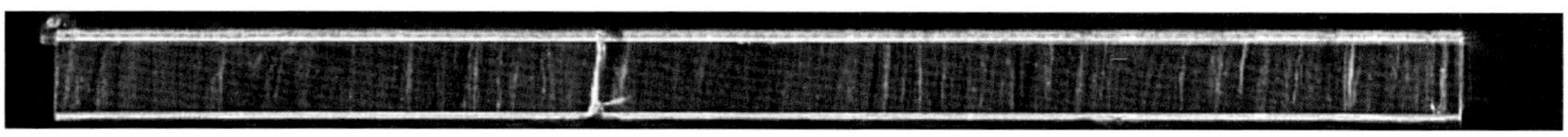

2486

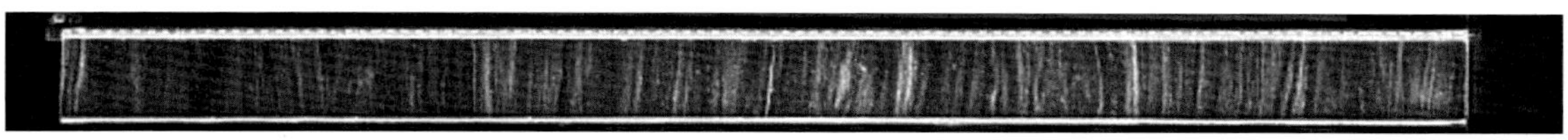

2487

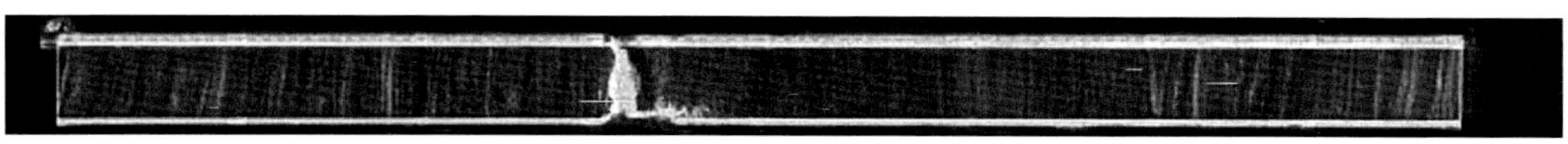

2488

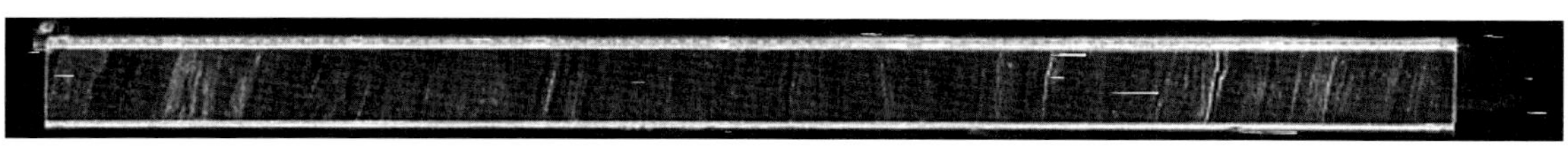

2489

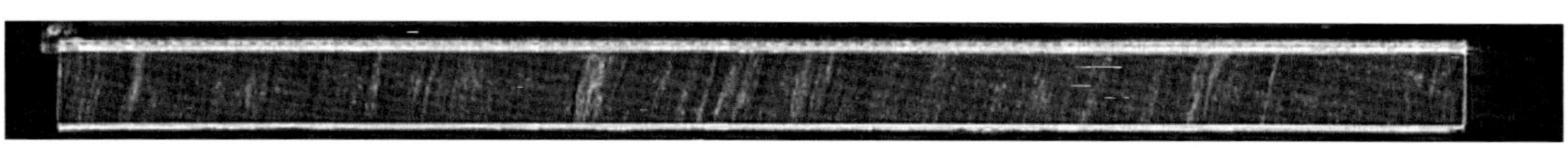

2499

2498

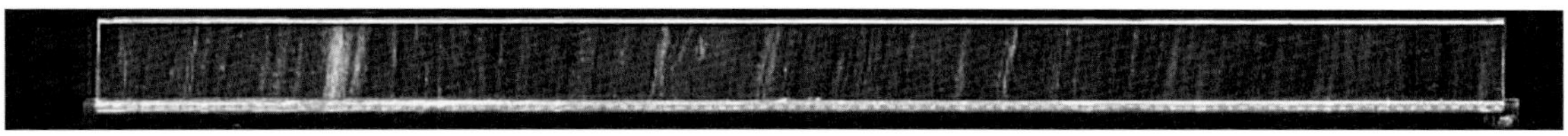

2497

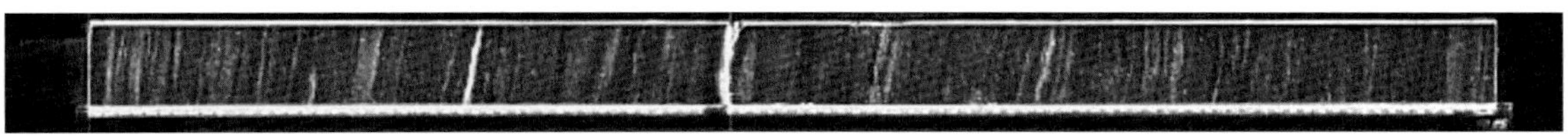

2496

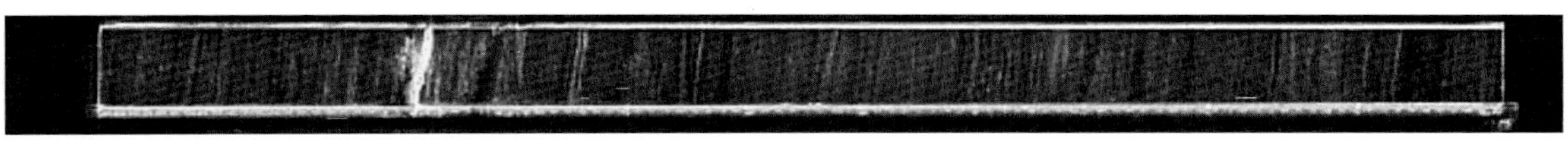

2495

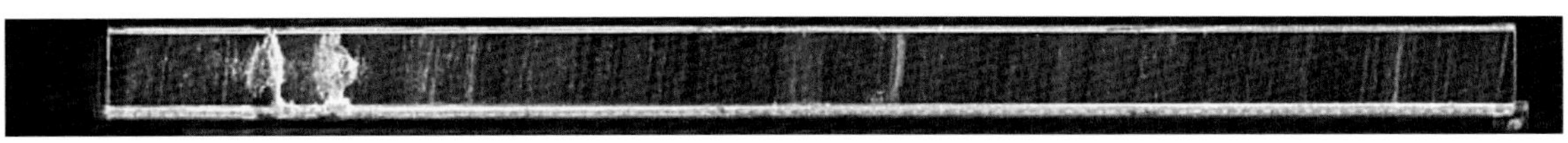

2494

2493

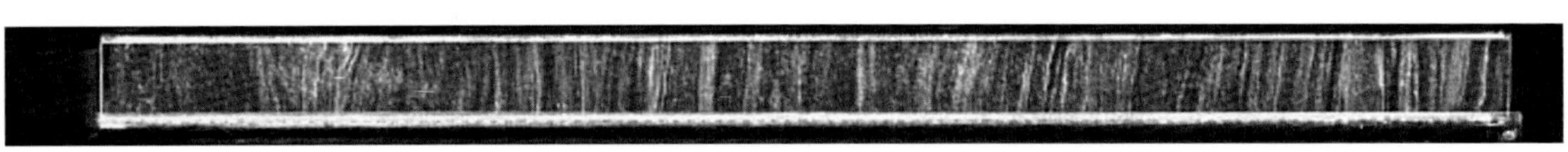

2492

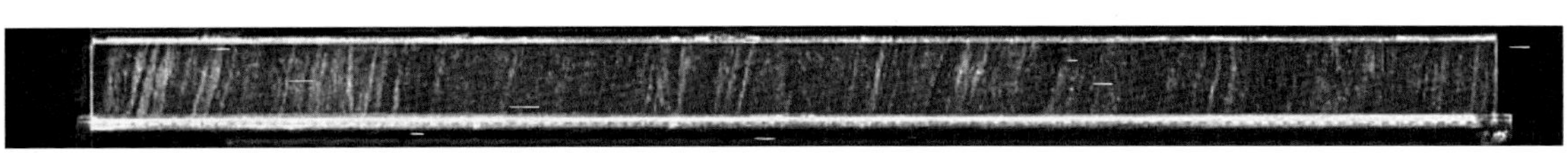

2491

2490

2500

2501

2502
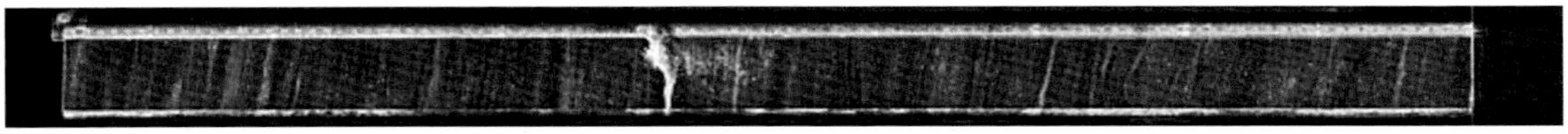

2503
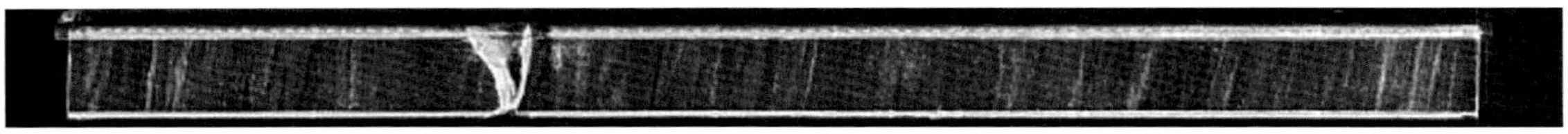

2504
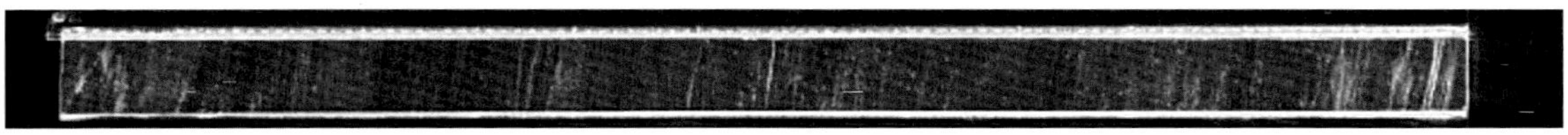

2505
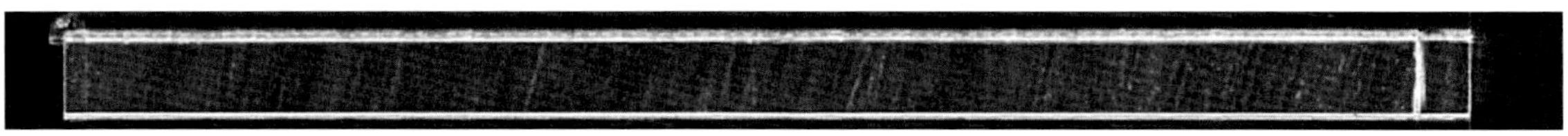

2506

2507
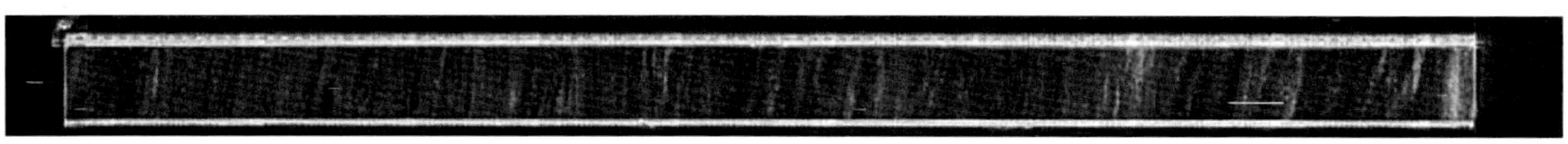

2508
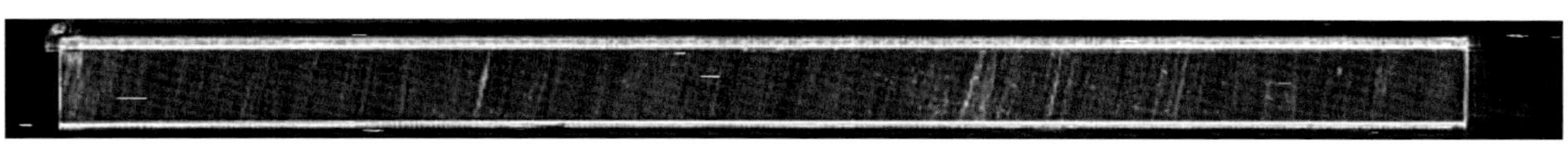

2509
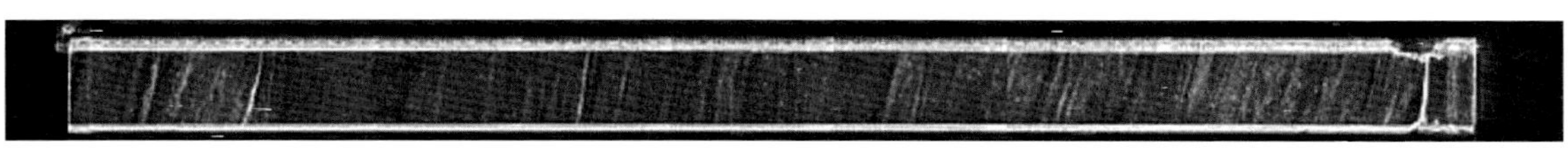

2519

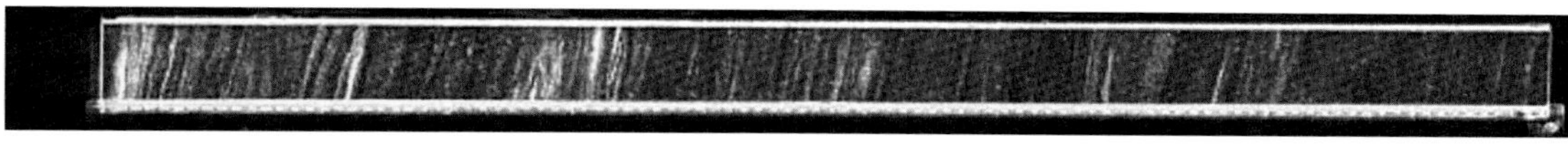

2518

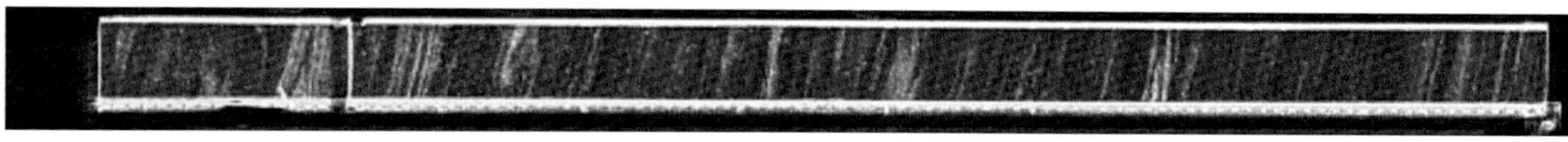

2517

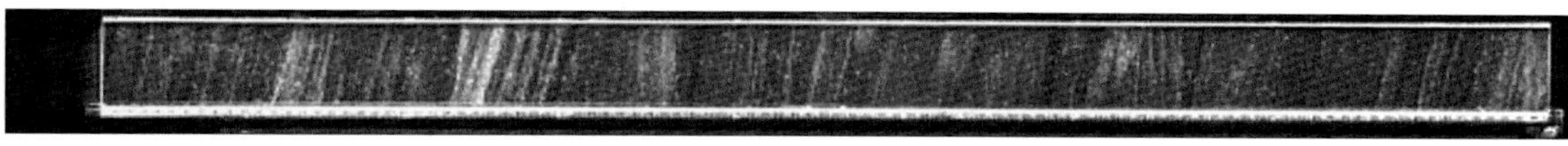

2516

2515

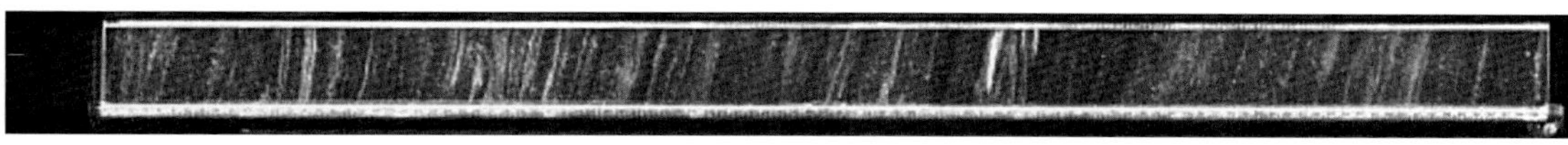

2514

2513

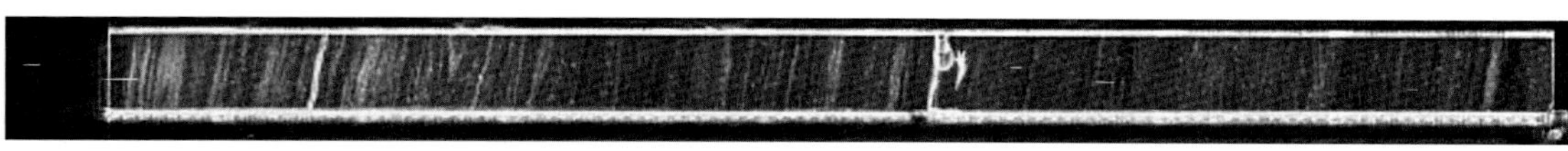

2512

2511

2510

2520

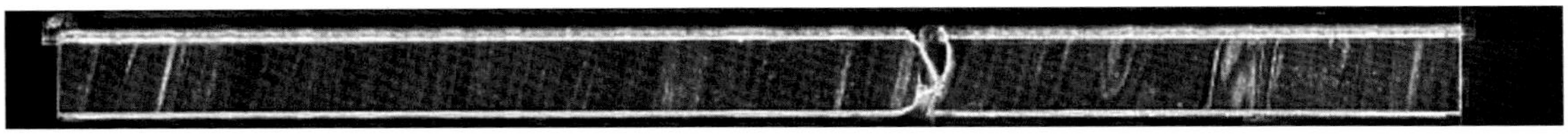

2521

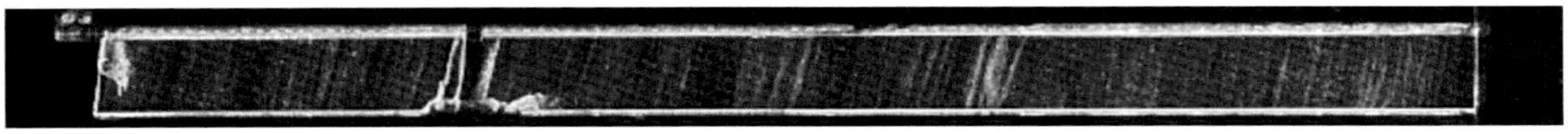

2522

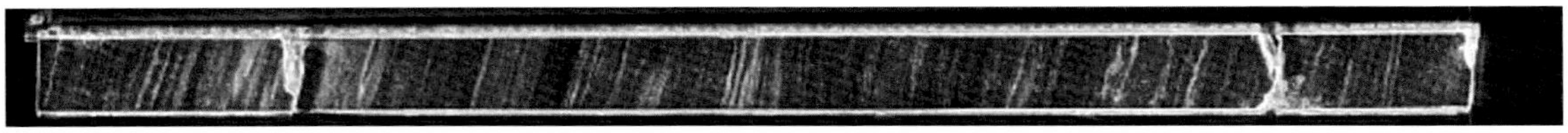

2523

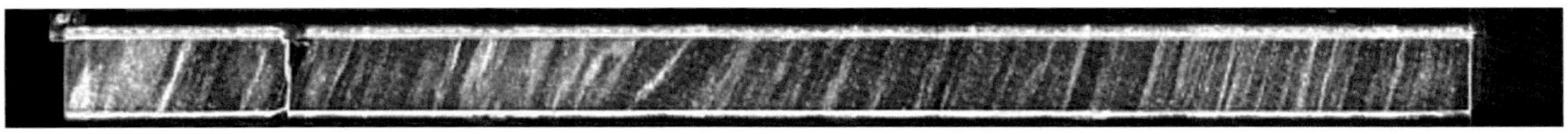

2524

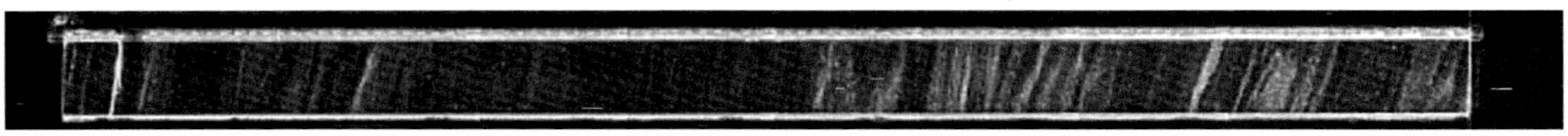

2525

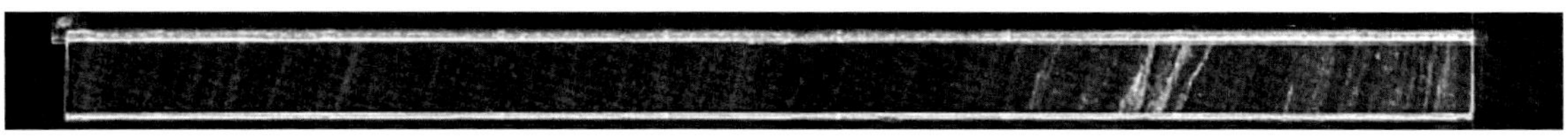

2526

2527

2528

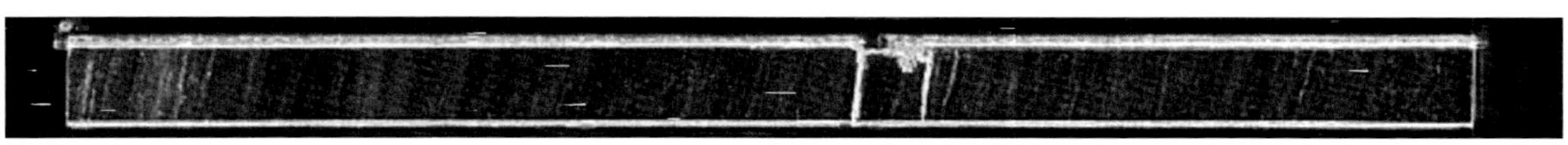

2529

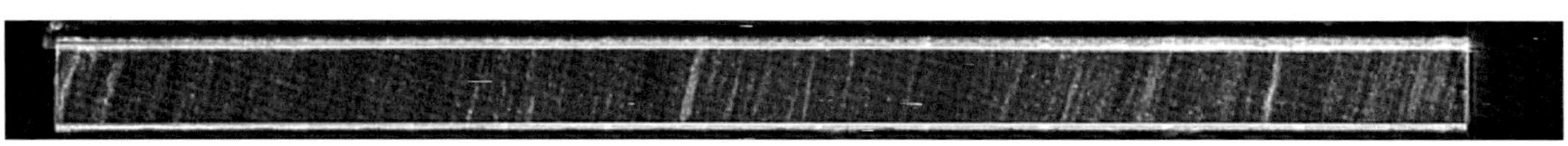

2539

2538

2537

2536

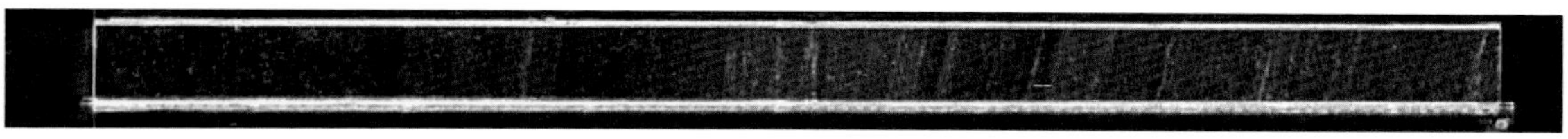
2535

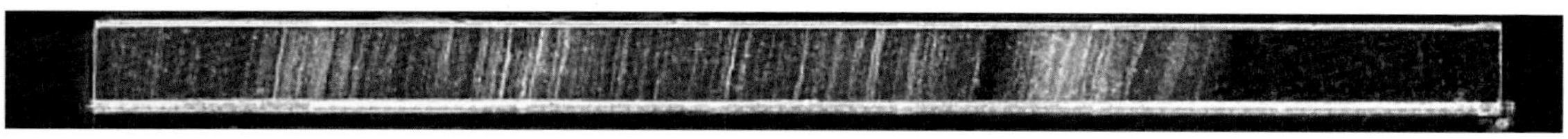
2534

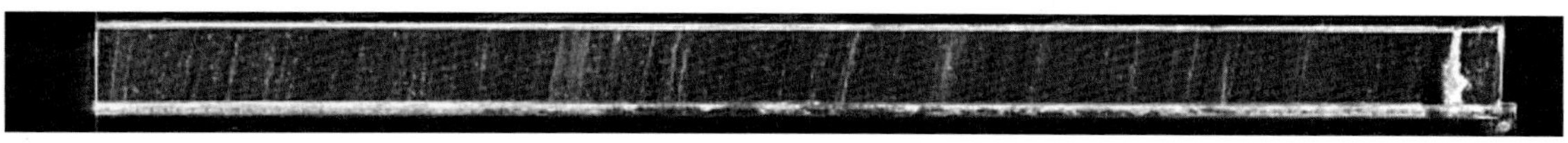
2533

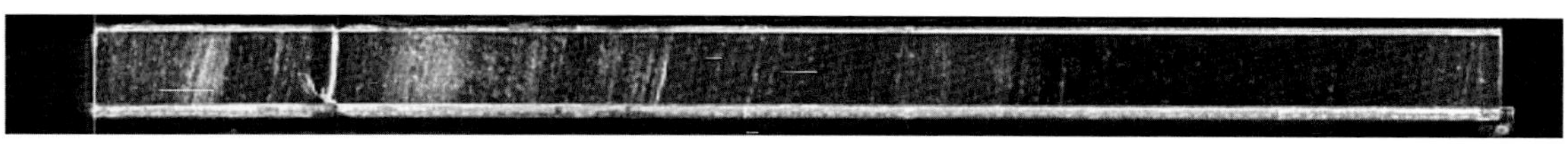
2532

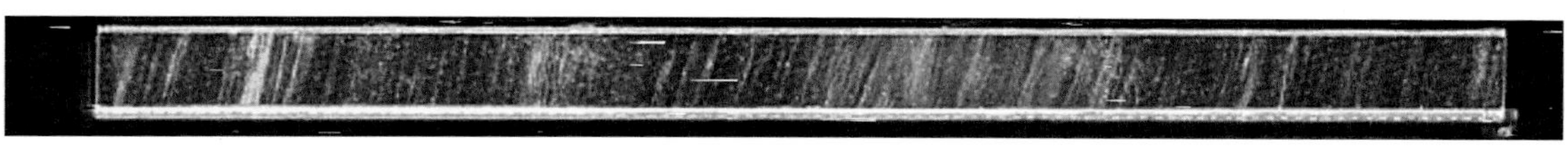
2531

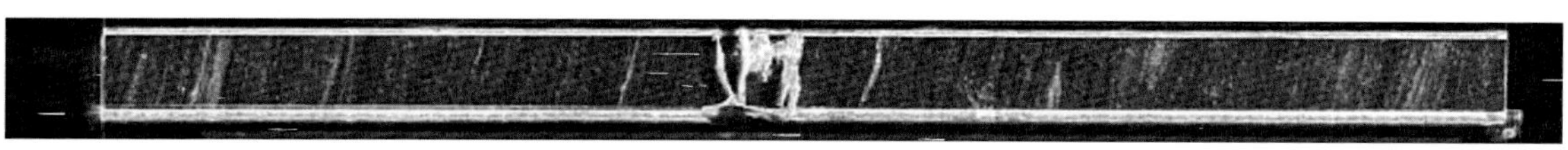
2530

2540

2541

2542

2543

2544

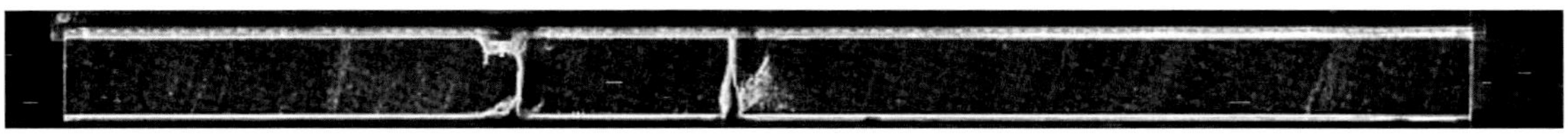

2545

2546

2547

2548

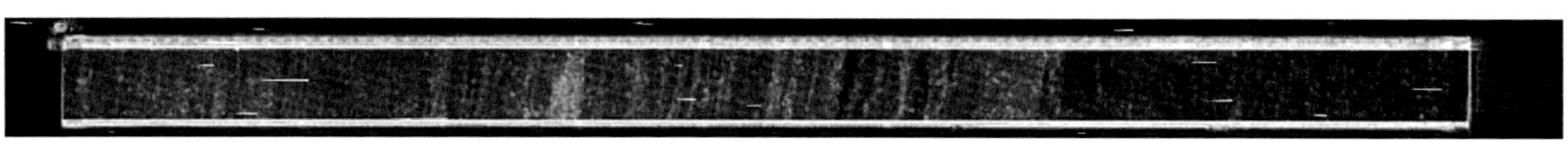

2549

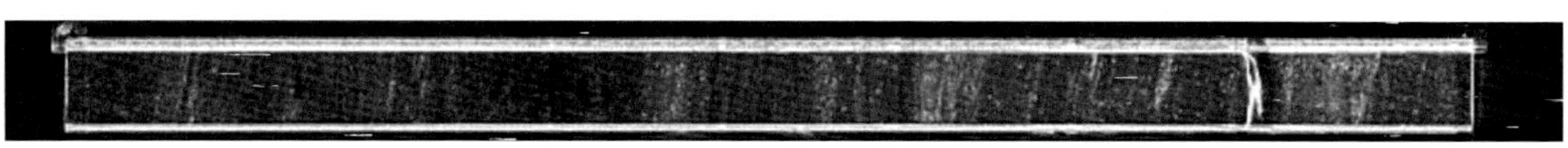

2559

2558

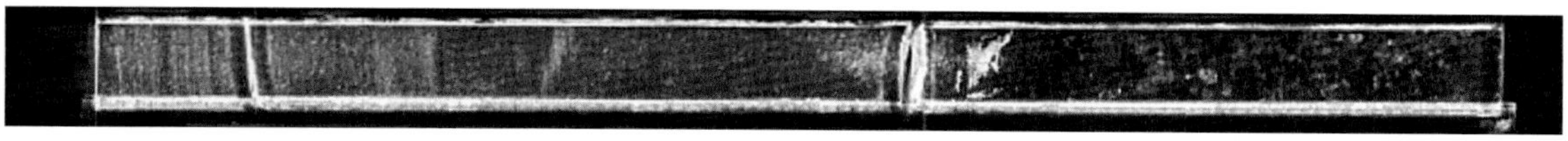

2557

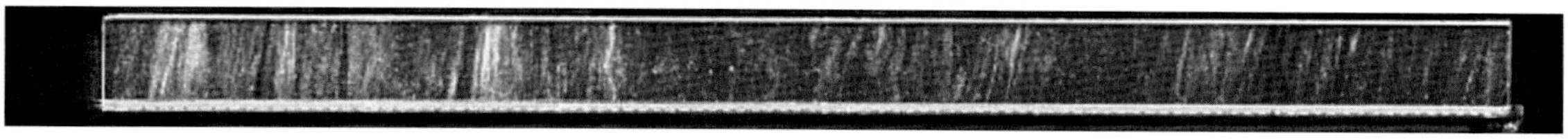

2556

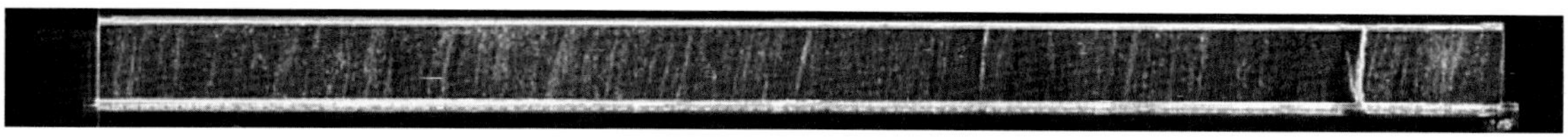

2555

2554

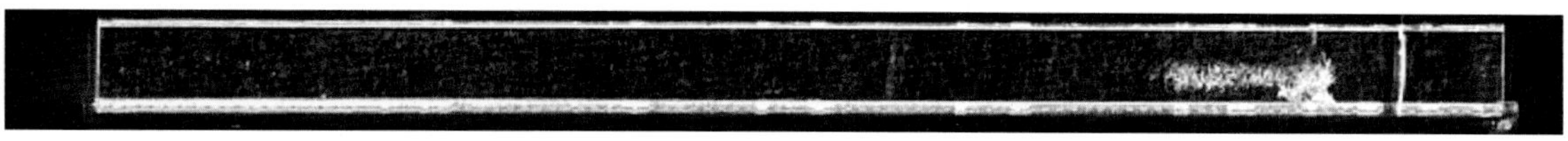

2553

2552

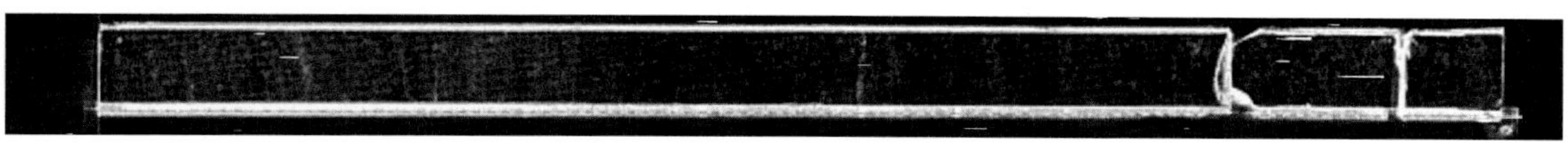

2551

2550

2560

2561

2562
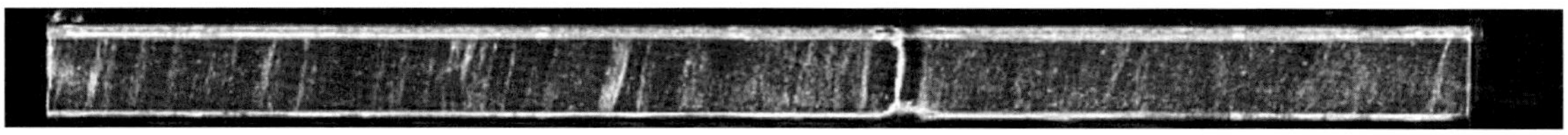

2563

2564
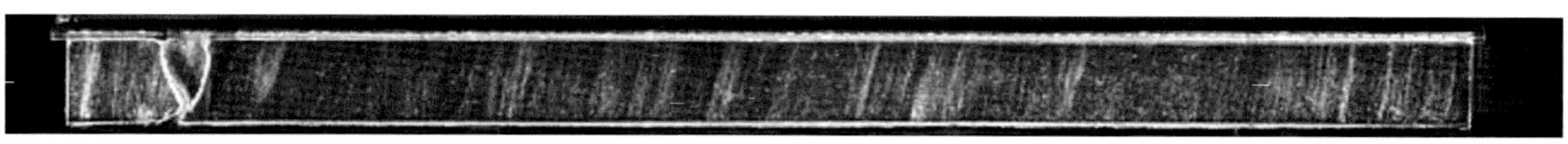

2565
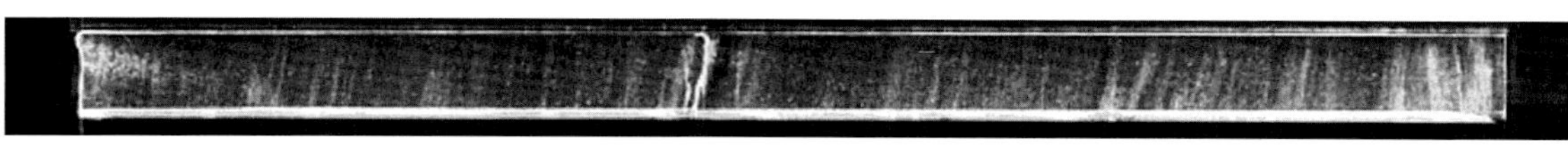

2566

2567
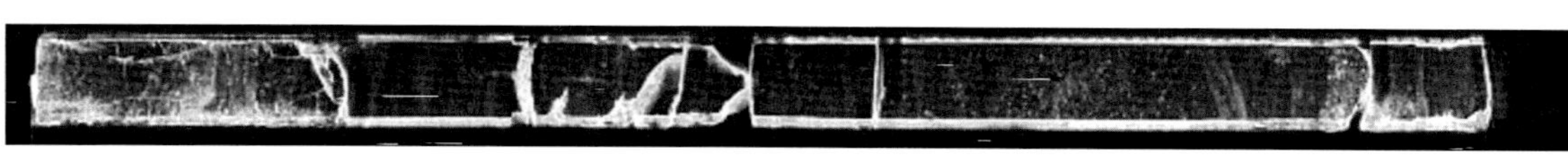

2568
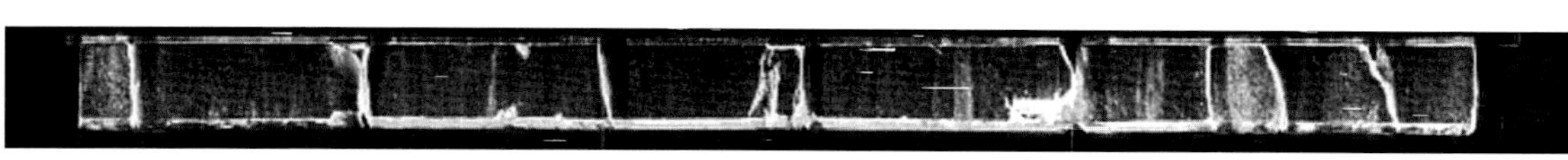

2569

2579

2578

2577

2576

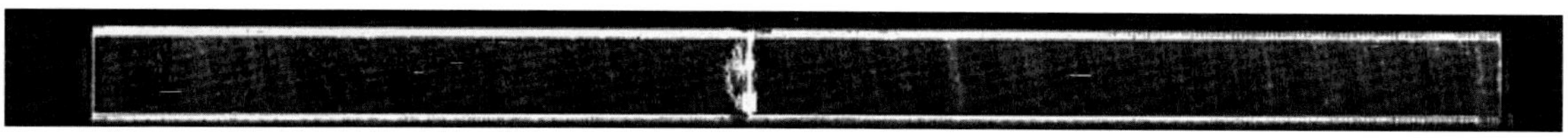
2575

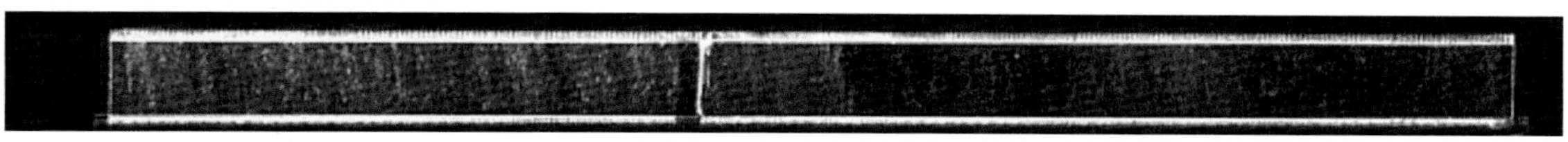
2574

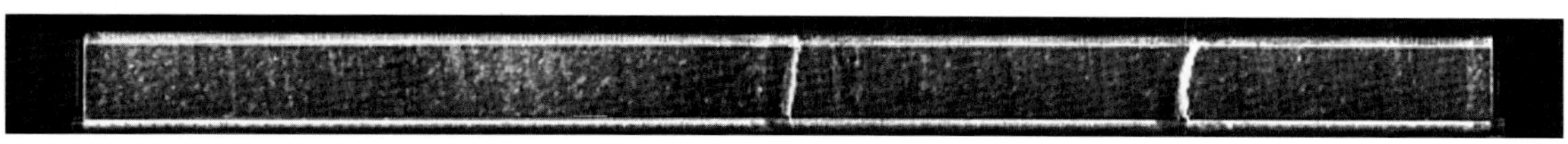
2573

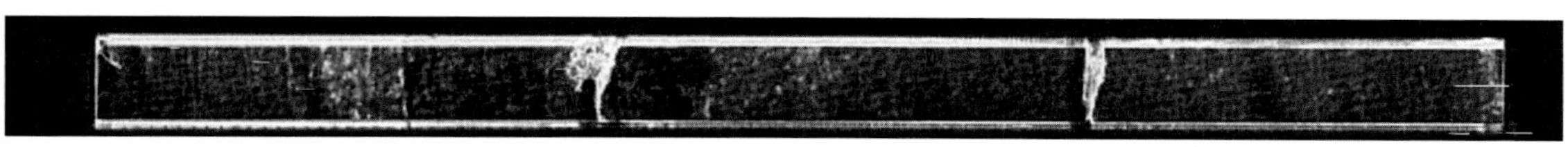
2572

2571

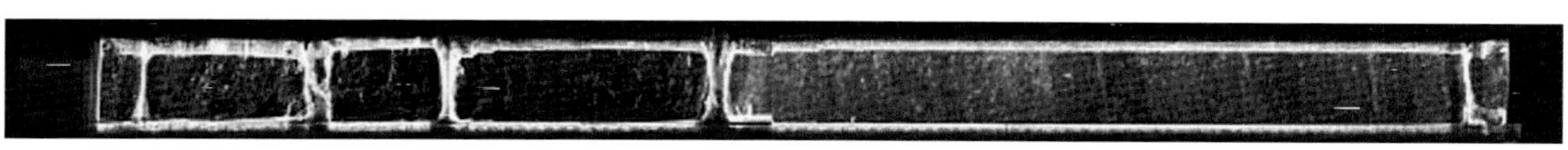
2570

2580
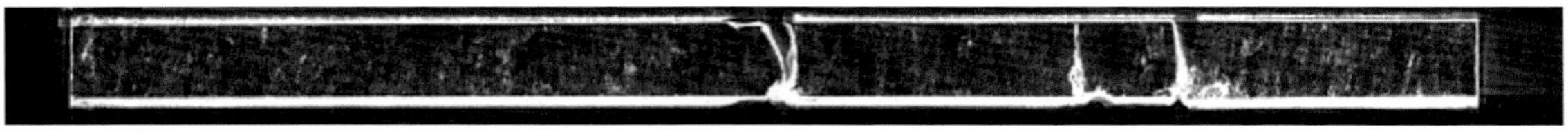

2581

2582

2583

2584

2585

2586
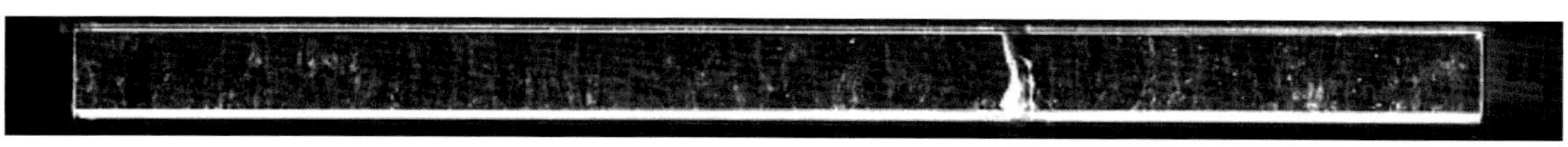

2587

2588

2589

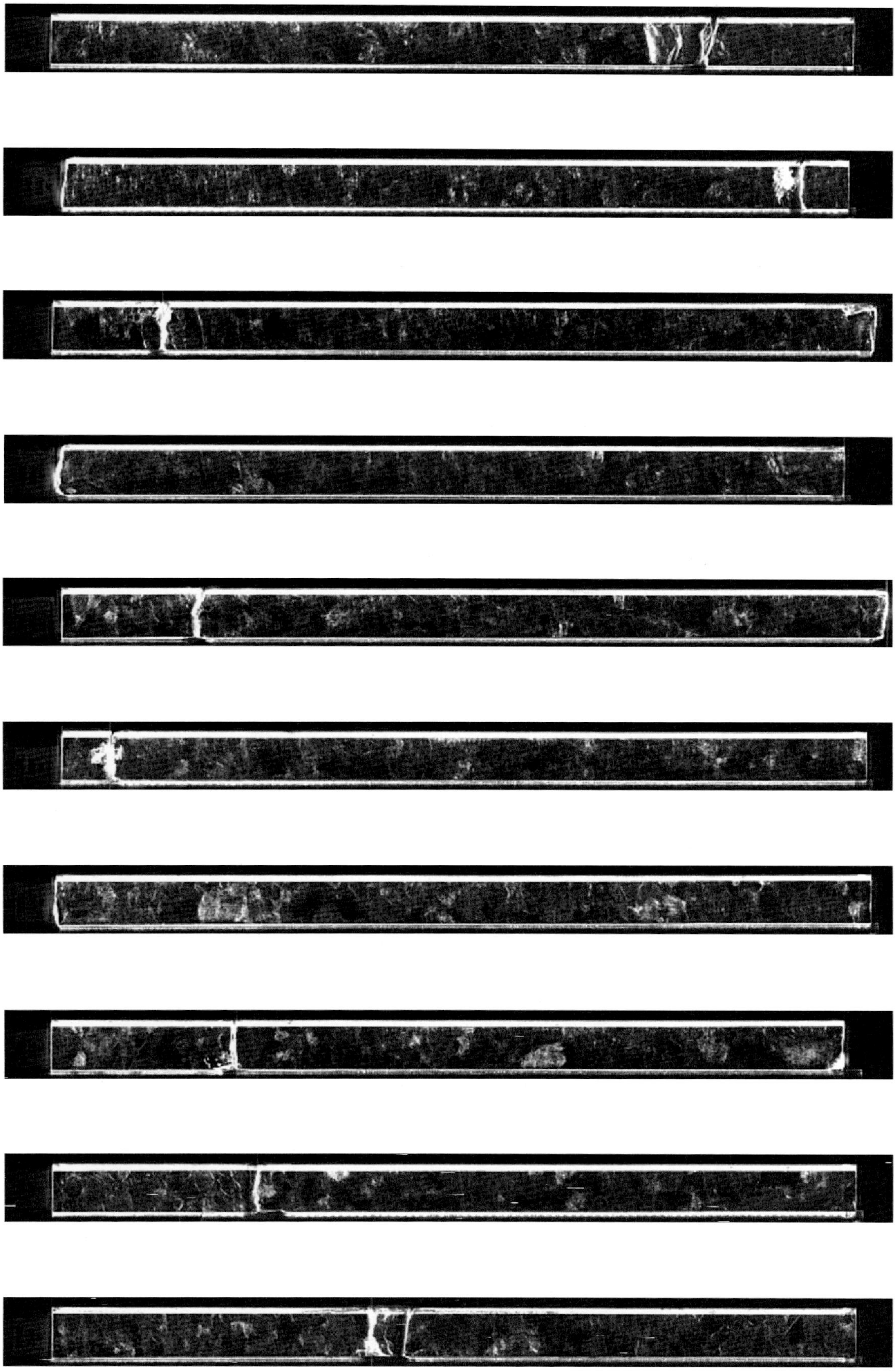

2599
2598
2597
2596
2595
2594
2593
2592
2591
2590

2600
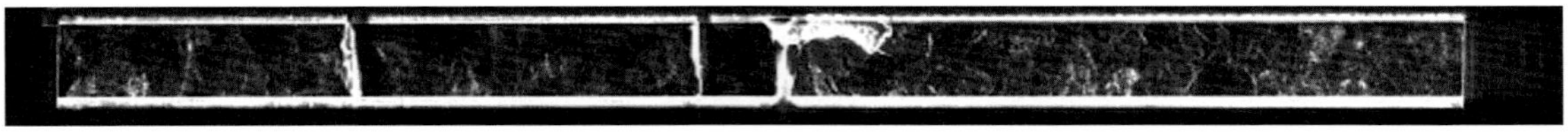

2601

2602

2603
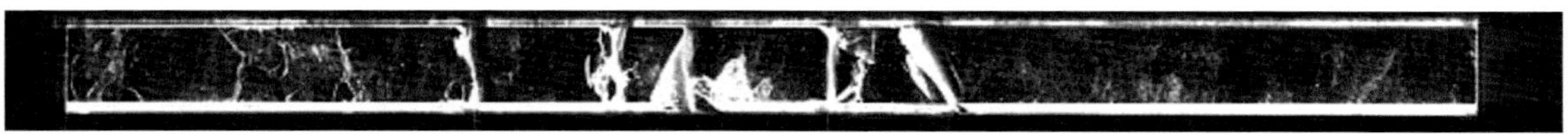

2604
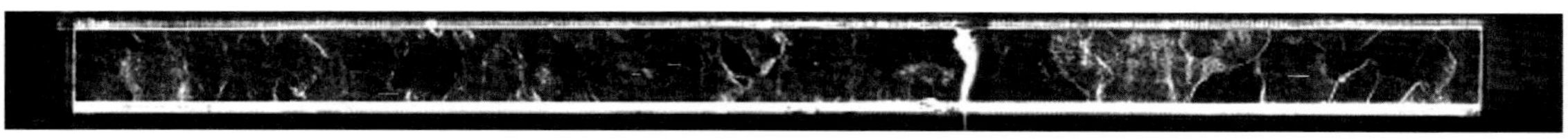

2605
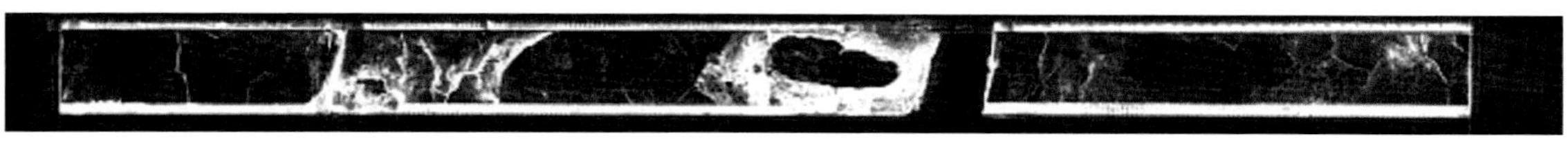

2606
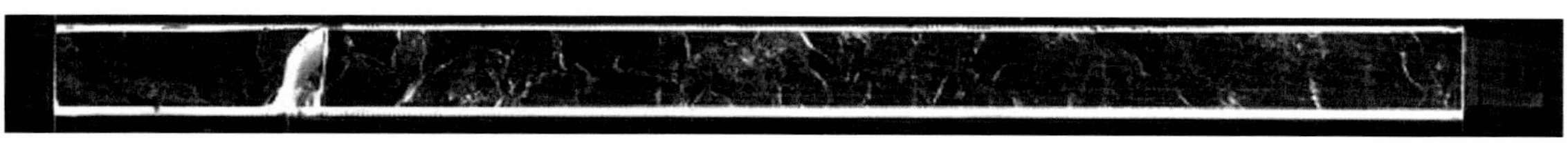

2607
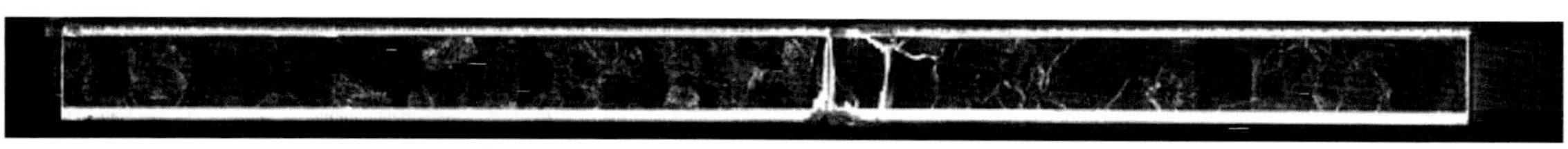

2608

2609
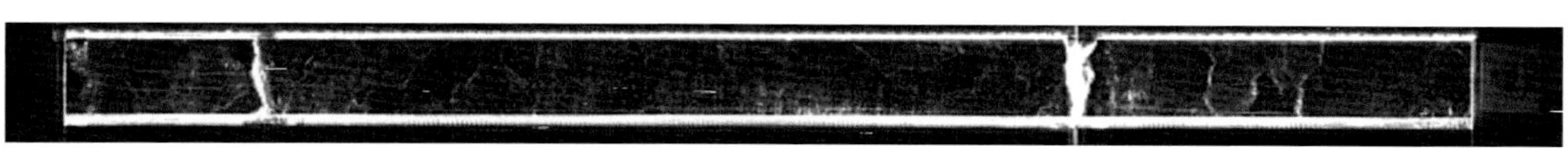

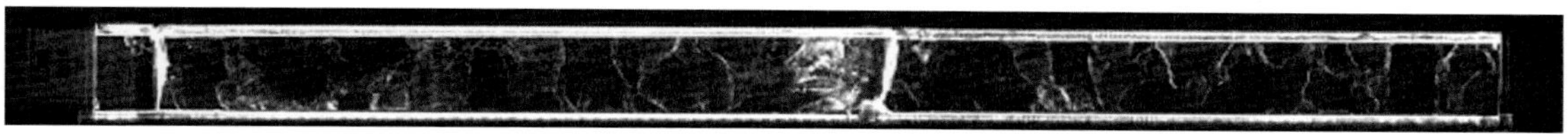
2619

2618

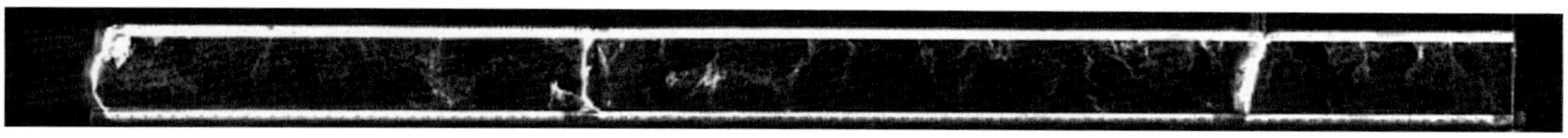
2617

2616

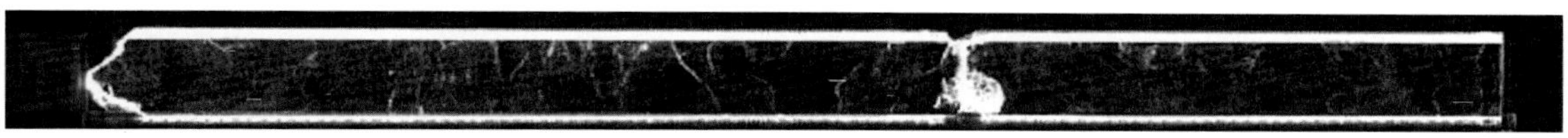
2615

2614

2613

2612

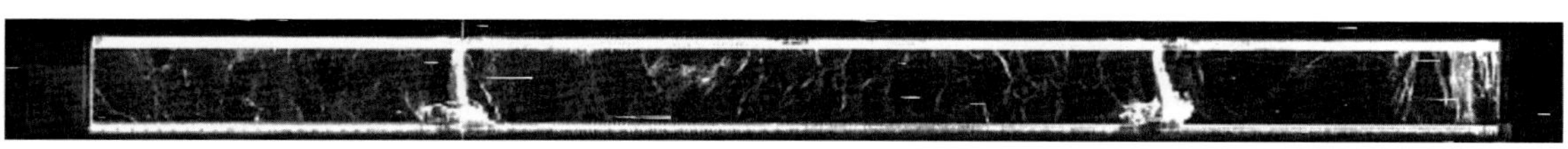
2611

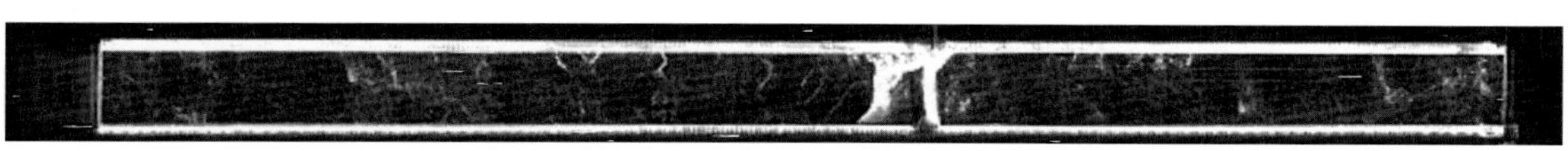
2610

2620

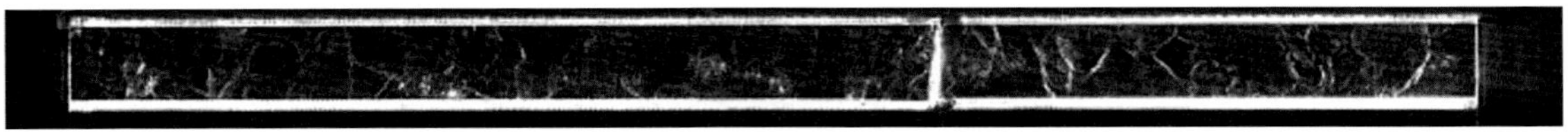

2621

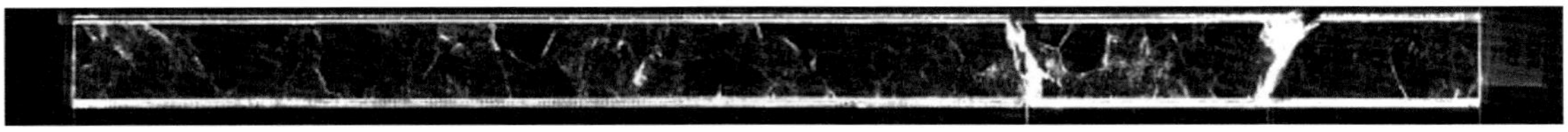

2622

2623

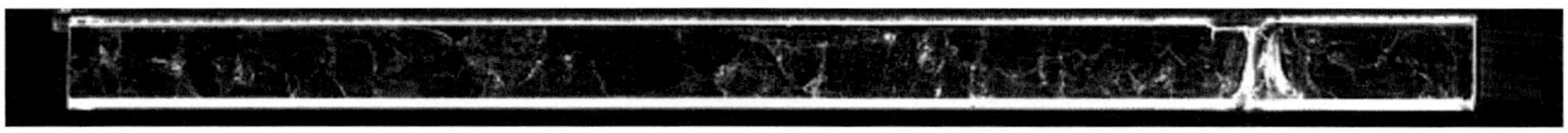

2624

2625

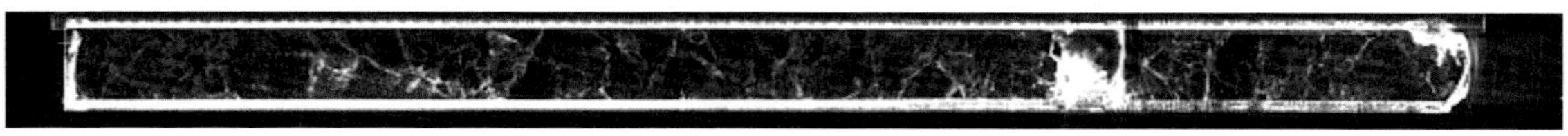

2626

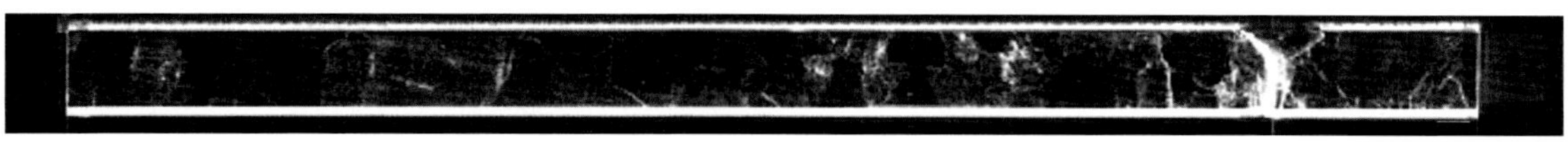

2627

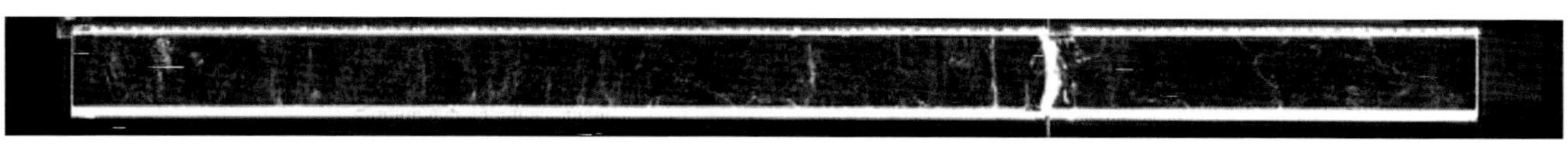

2628

2629

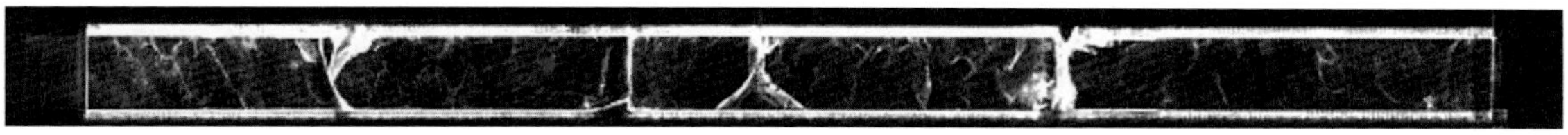

2639

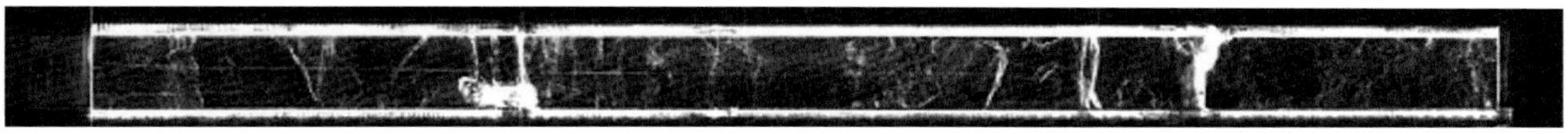

2638

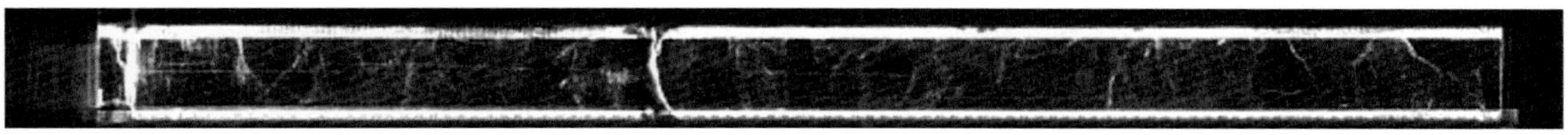

2637

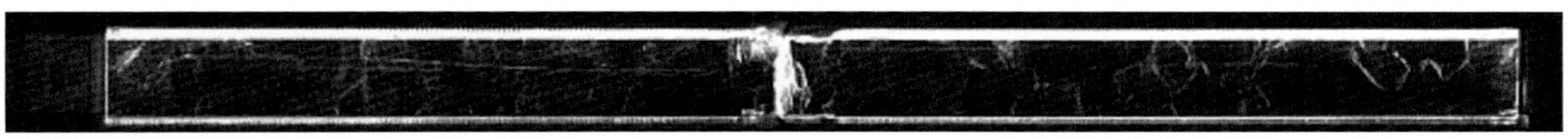

2636

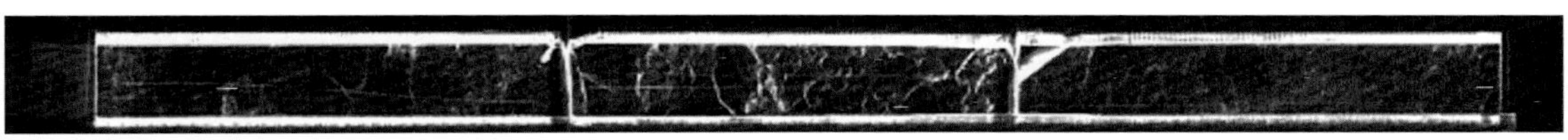

2635

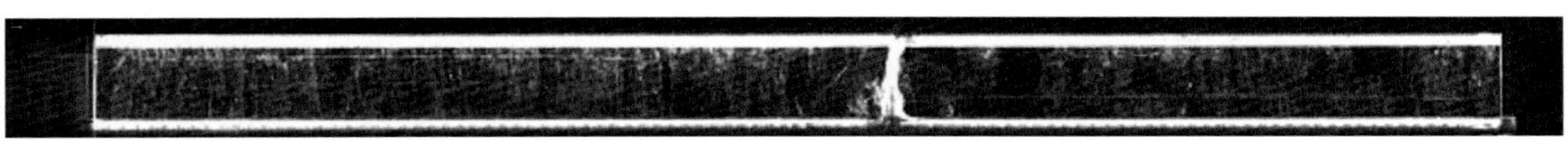

2634

2633

2632

2631

2630

2640

2641

2642

2643

2644

2645
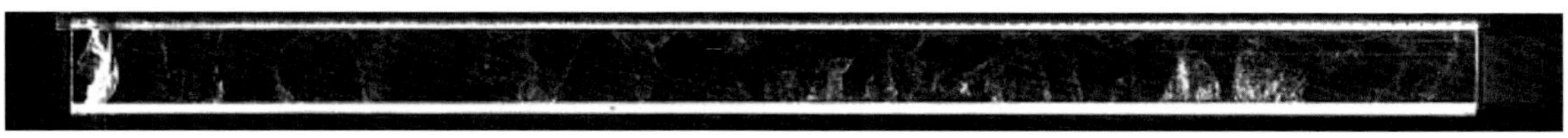

2646

2647
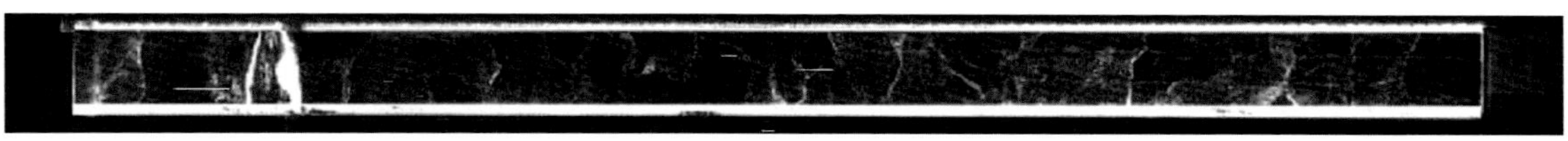

2648
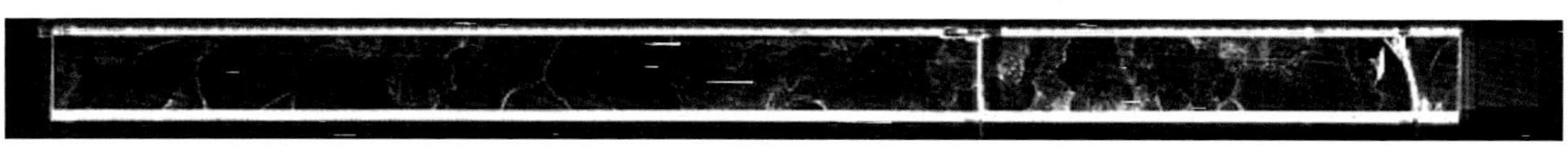

2649
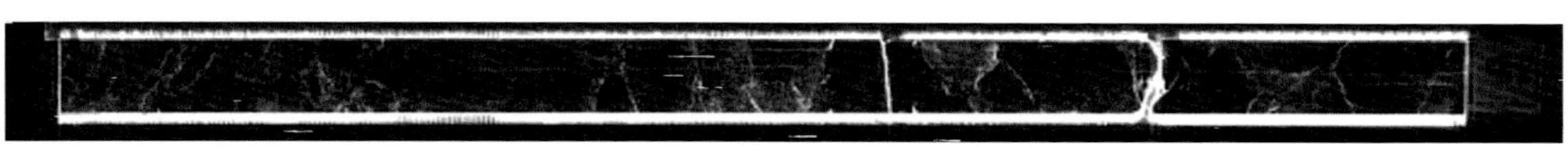

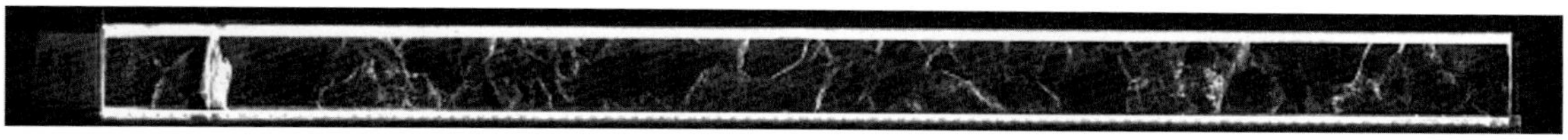

2659

2658

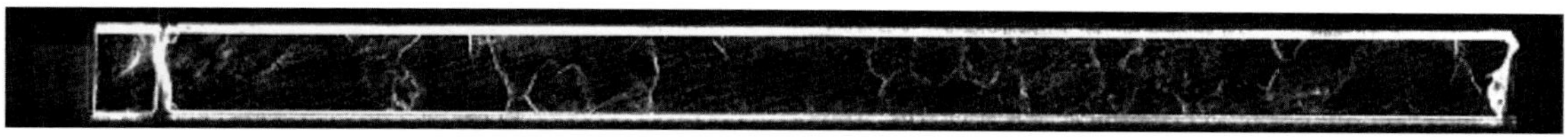

2657

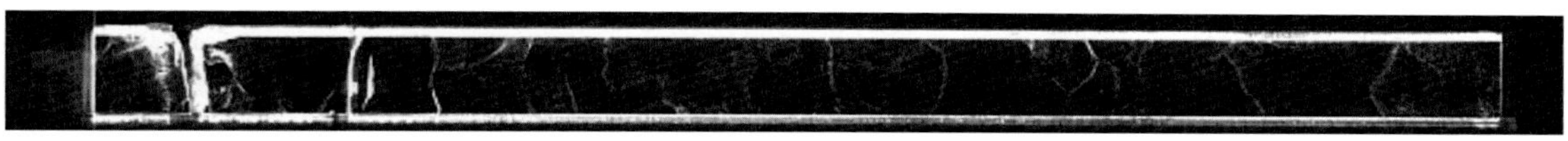

2656

2655

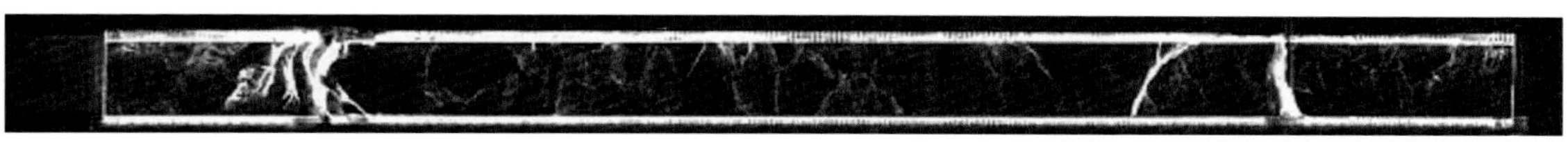

2654

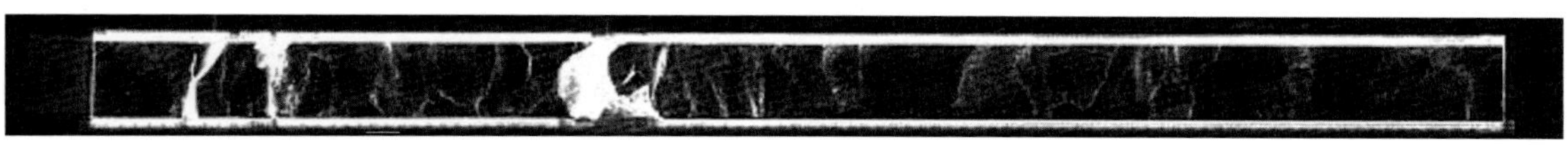

2653

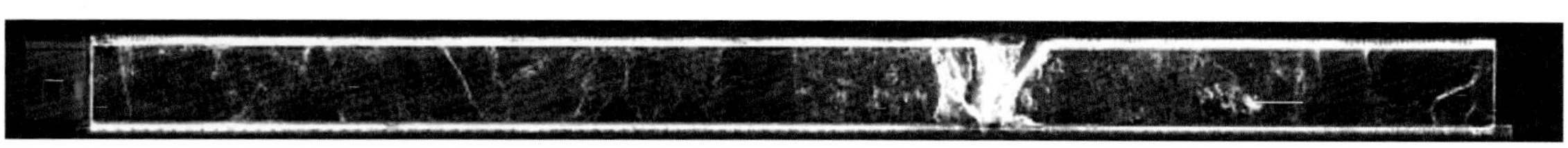

2652

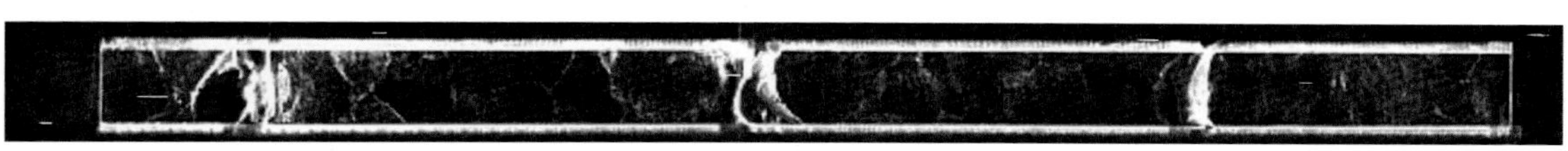

2651

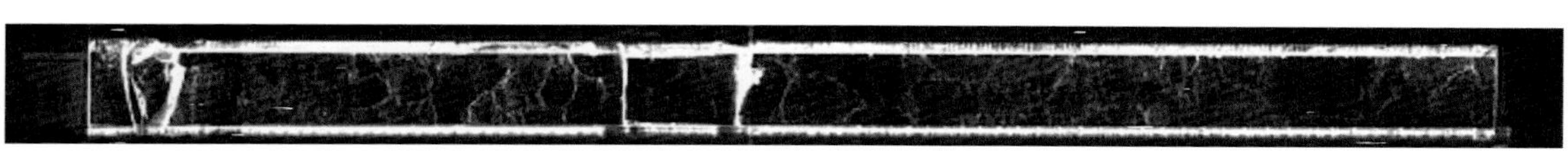

2650

2660

2661

2662

2663
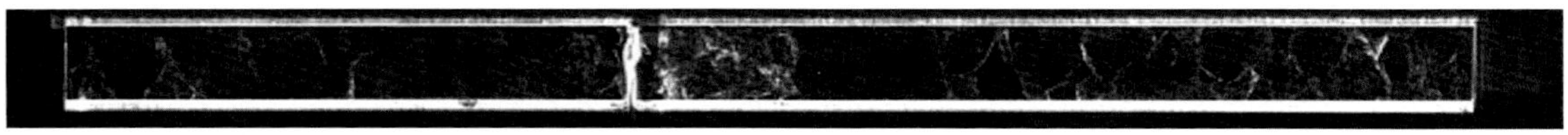

2664
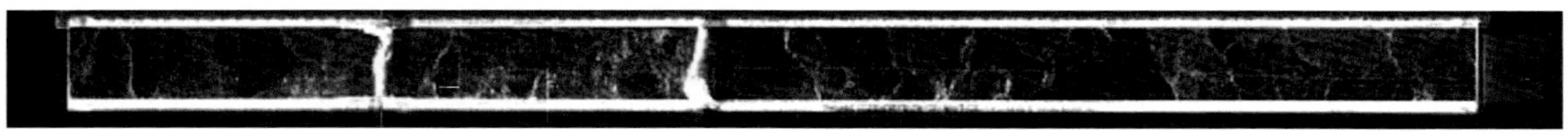

2665
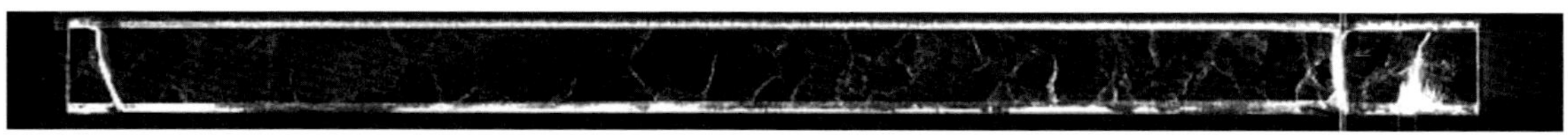

2666
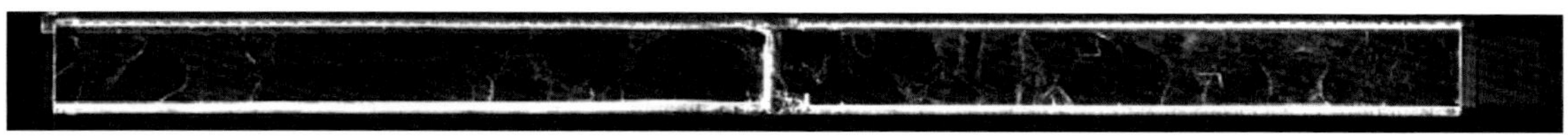

2667
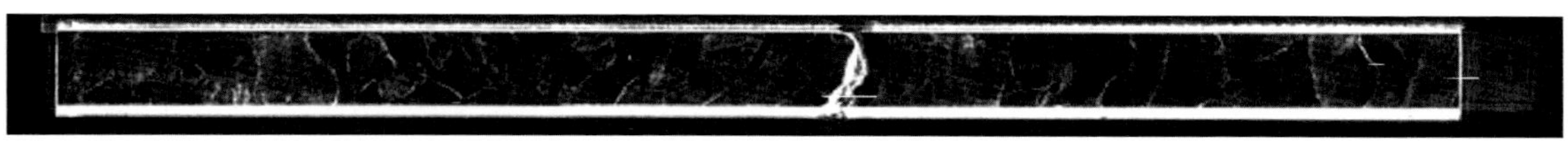

2668
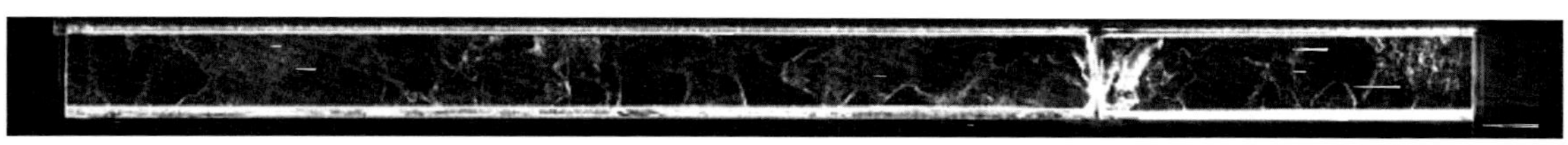

2669
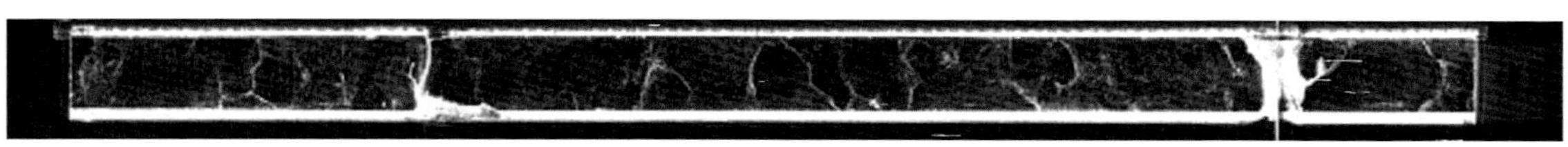

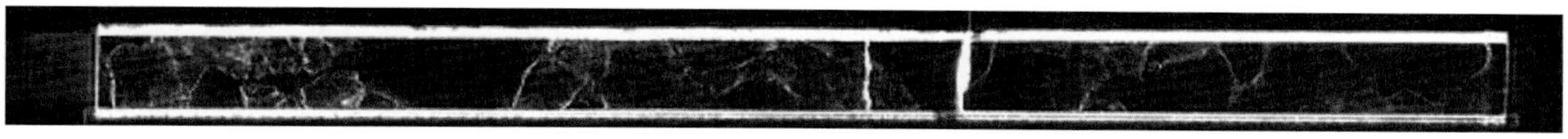

2679

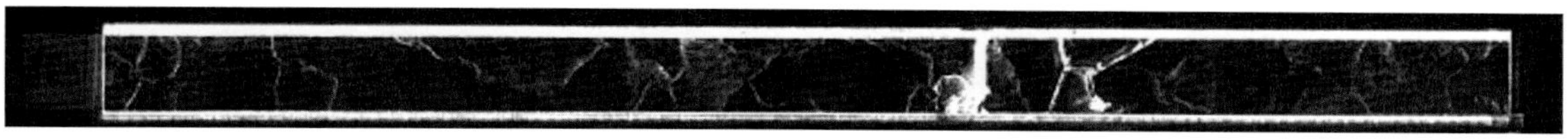

2678

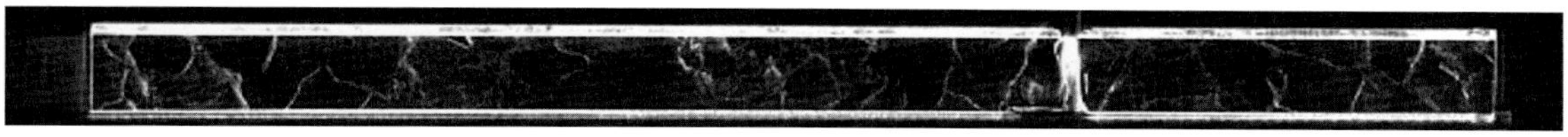

2677

2676

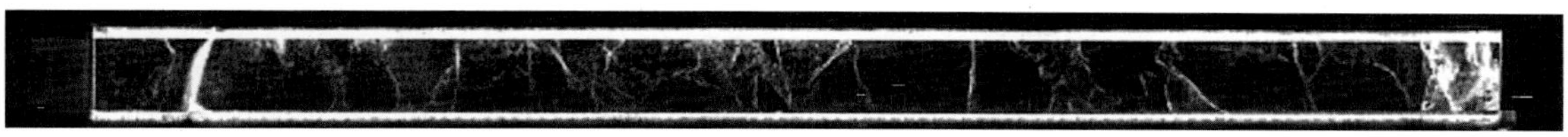

2675

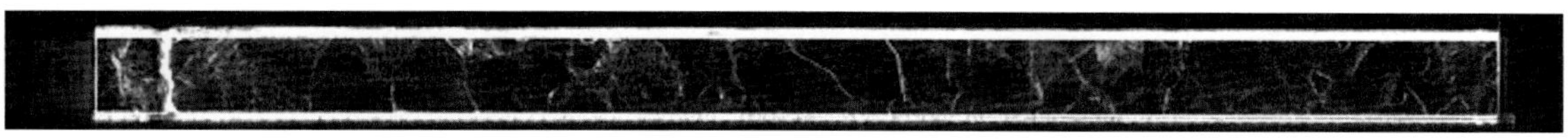

2674

2673

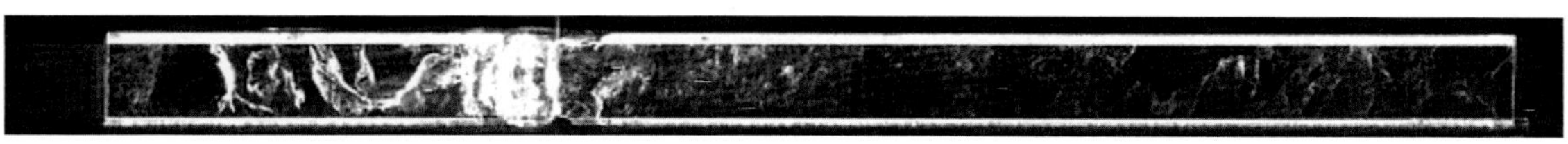

2672

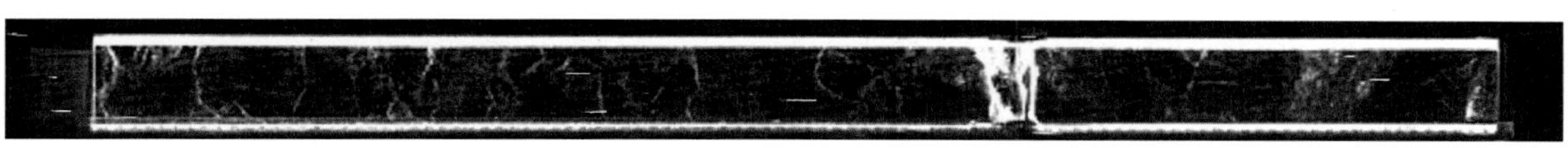

2671

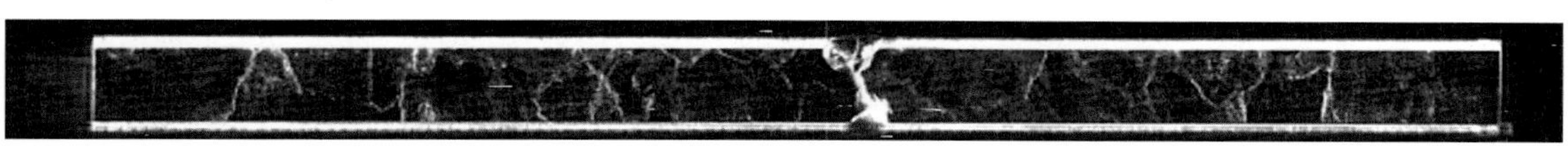

2670

2680

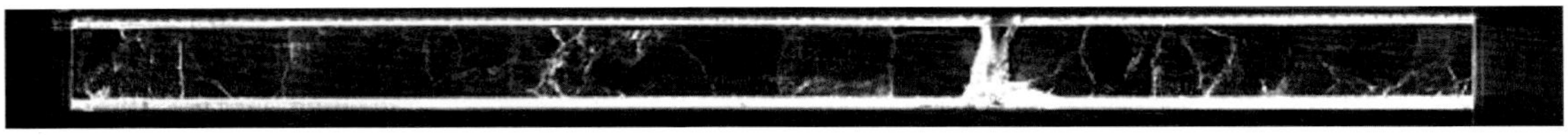

2681

2682

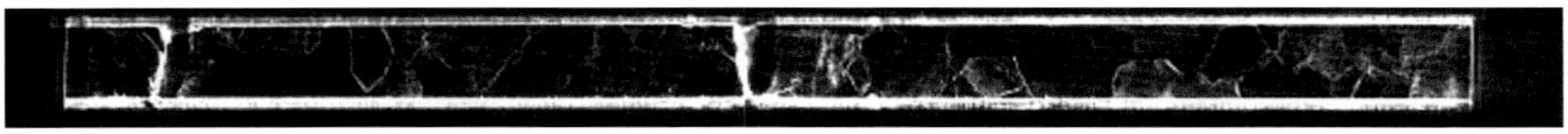

2683

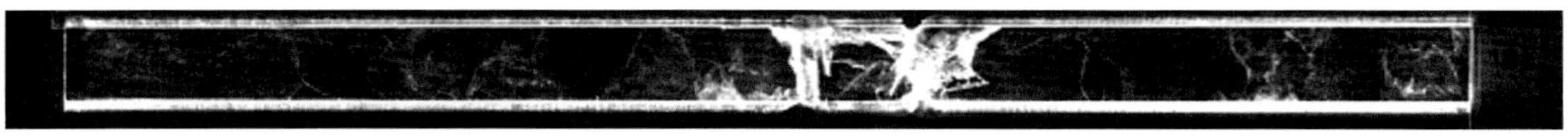

2684

2685

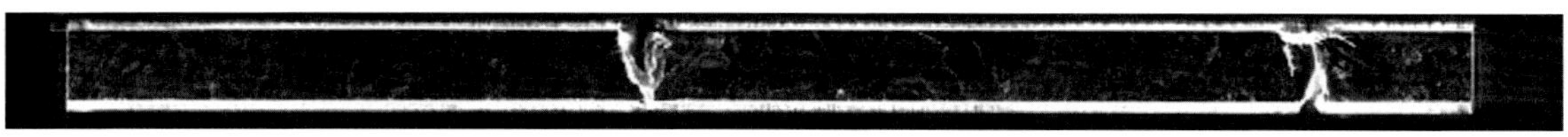

2686

2687

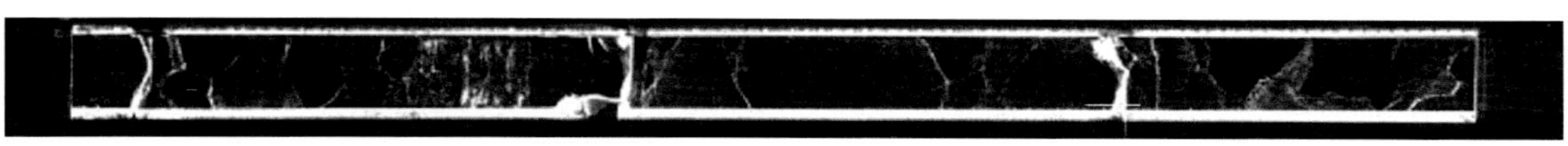

2688

2689

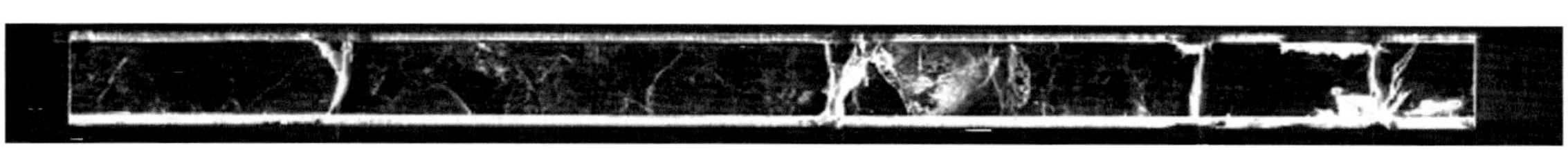

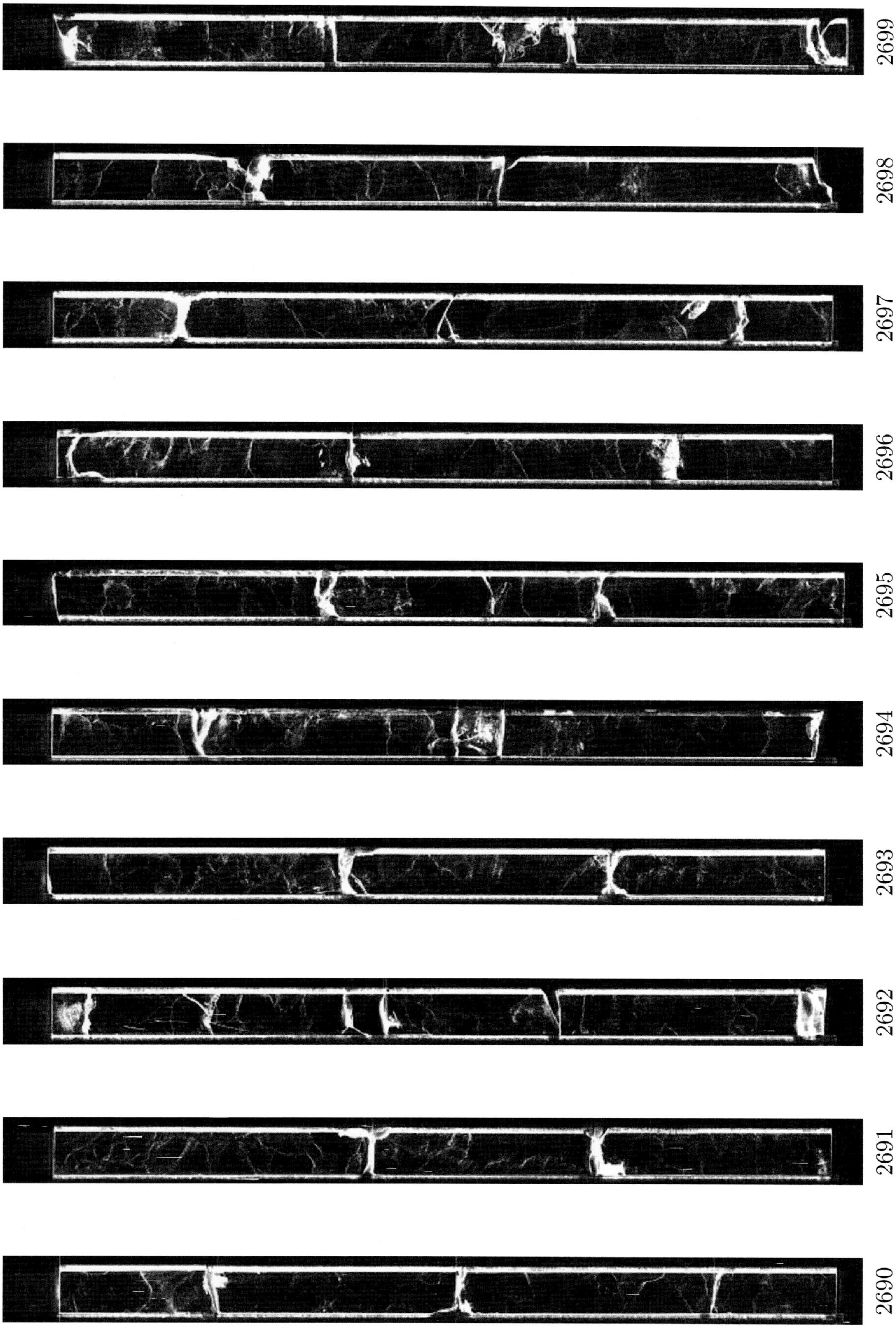
2699
2698
2697
2696
2695
2694
2693
2692
2691
2690

2700

2701

2702

2703

2704

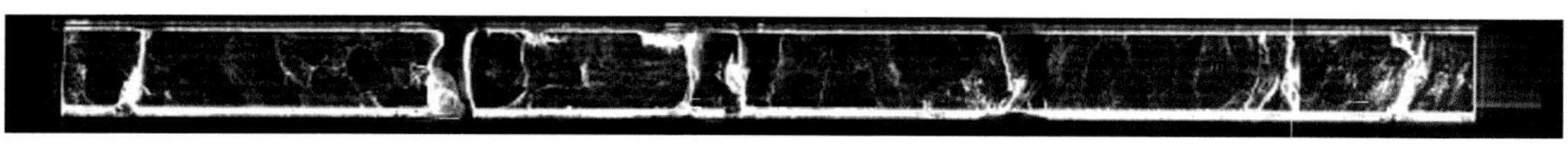

2705

2706

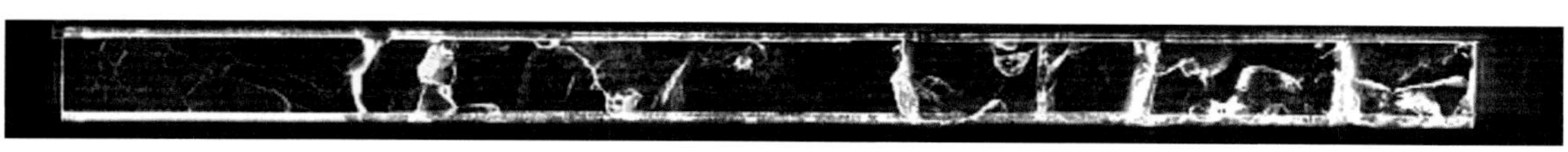

2707

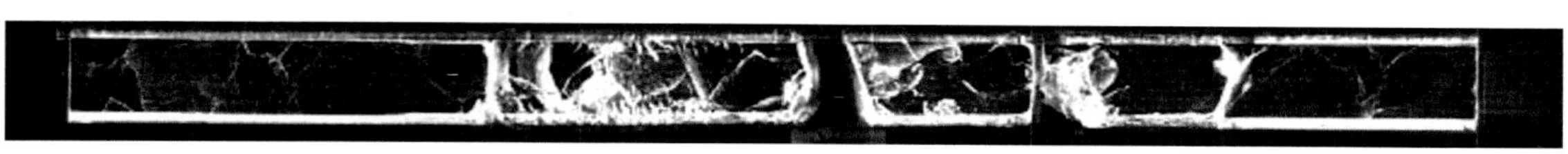

2708

2709

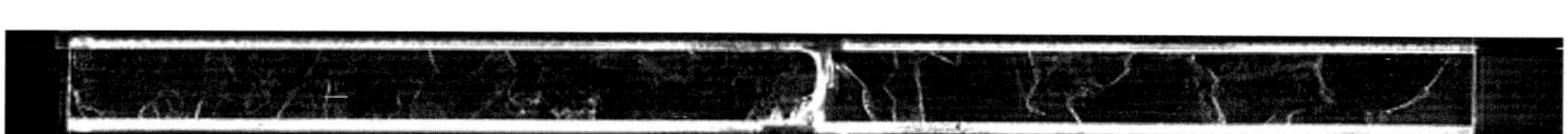

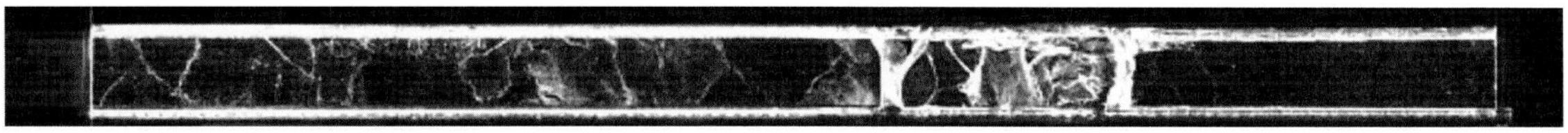
2719

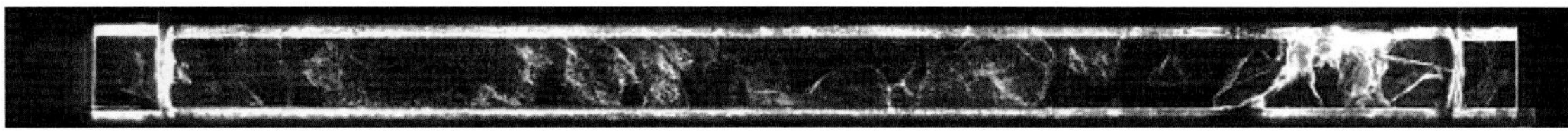
2718

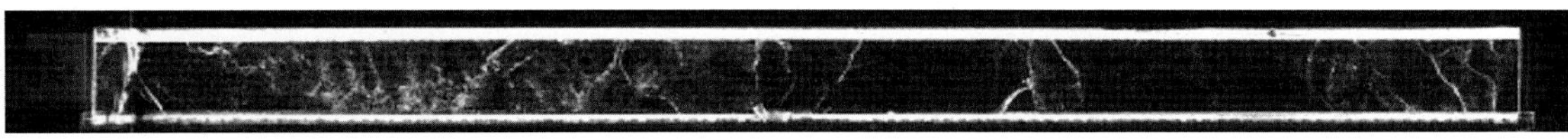
2717

2716

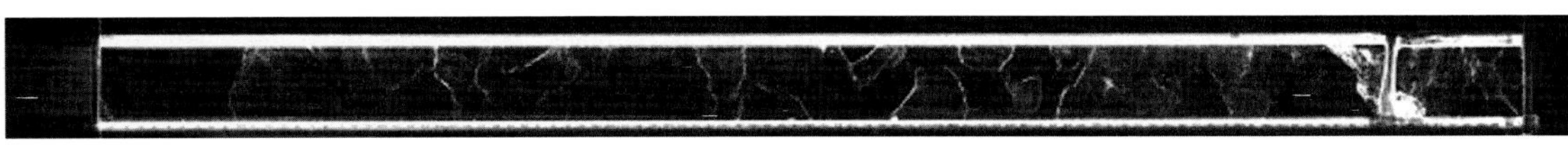
2715

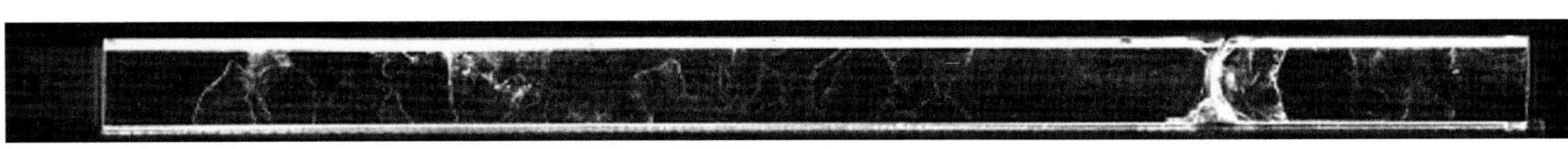
2714

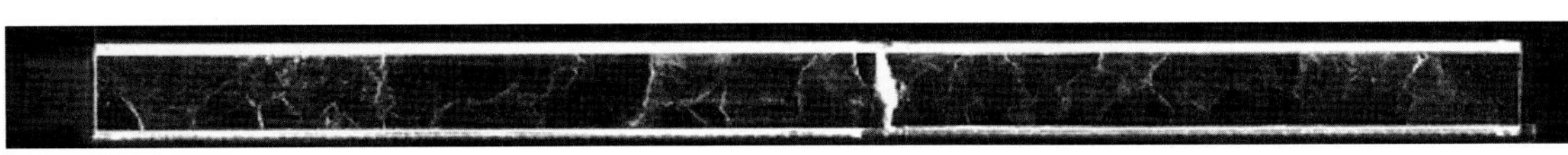
2713

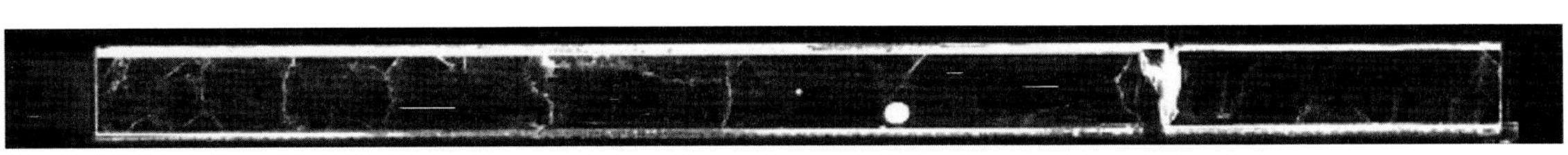
2712

2711

2710

2720

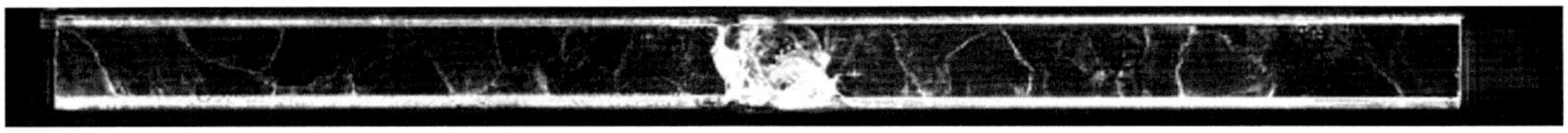

2721

2722

2723

2724

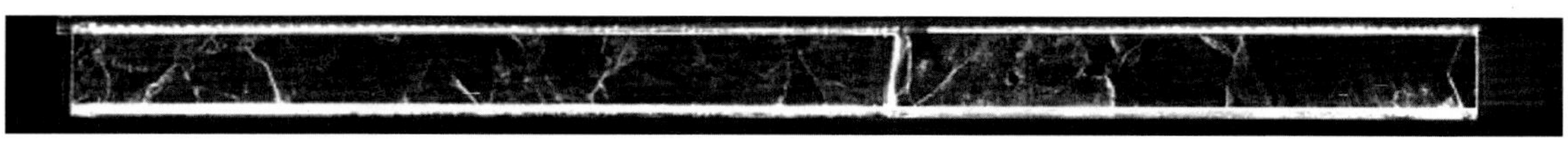

2725

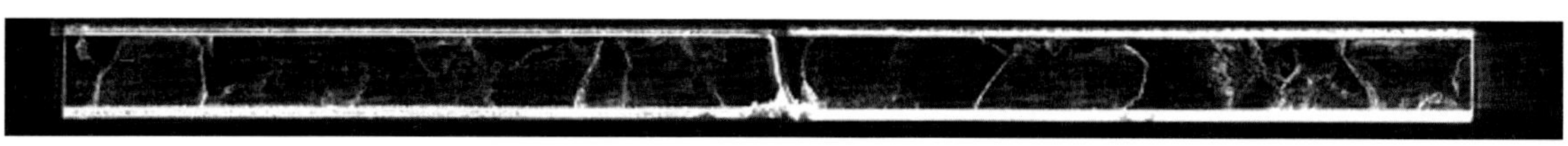

2726

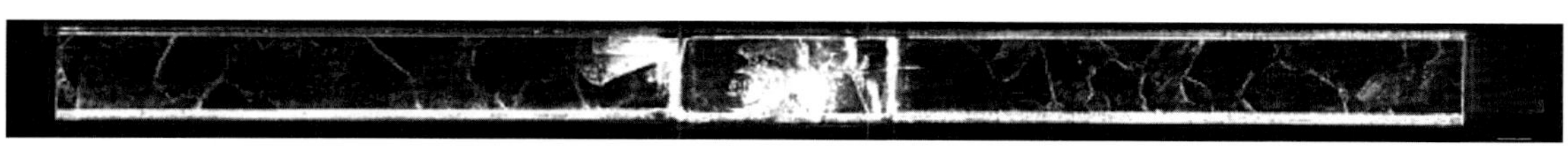

2727

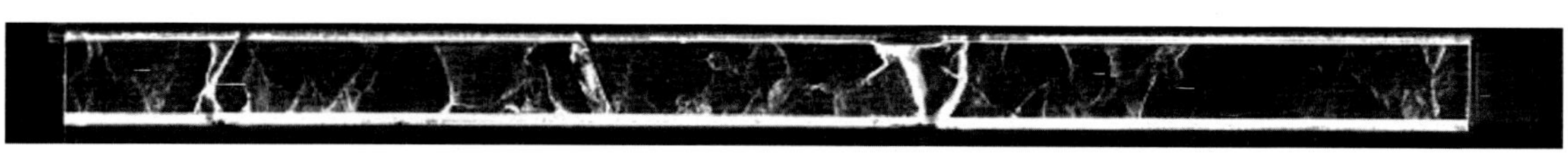

2728

2729

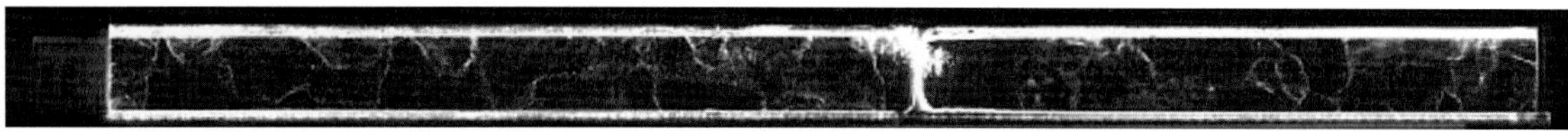
2739

2738

2737

2736

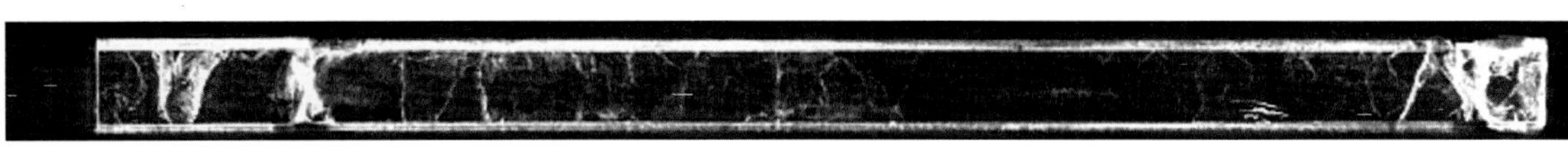
2735

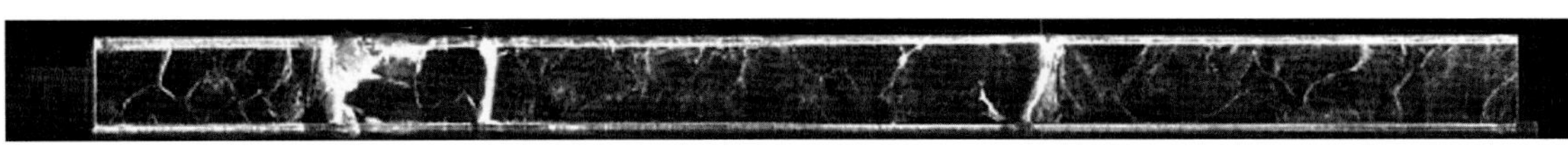
2734

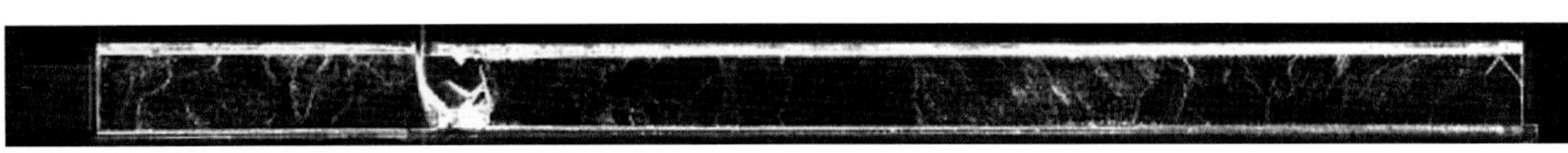
2733

2732

2731

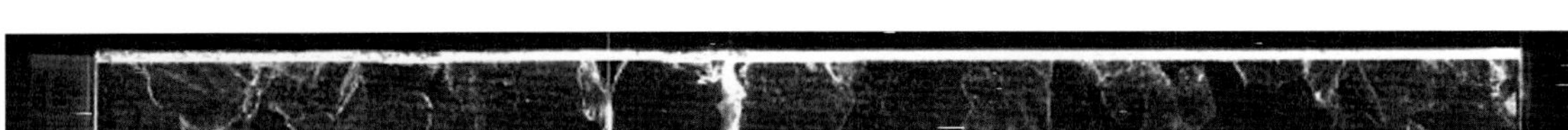
2730

2740

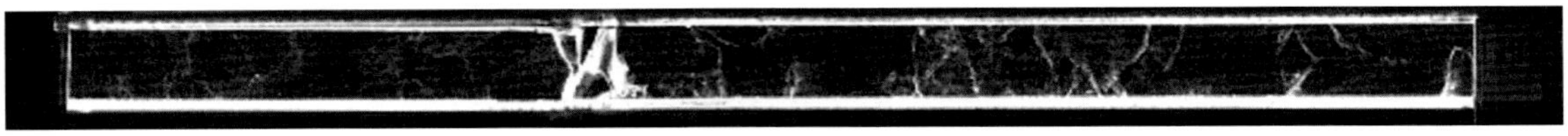

2741

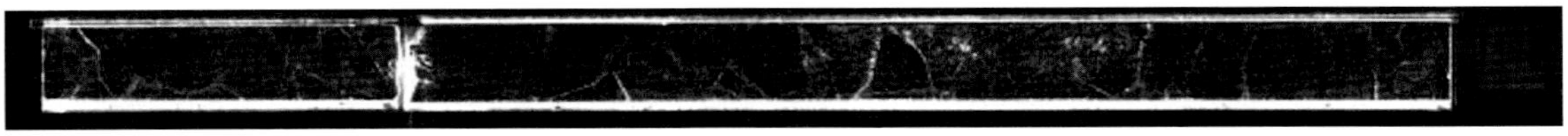

2742

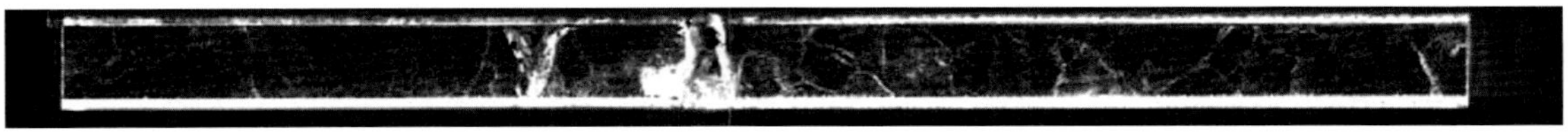

2743

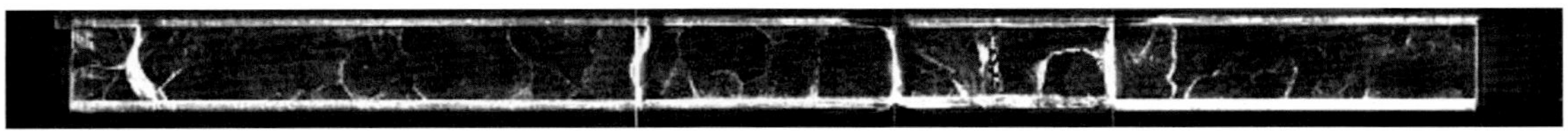

2744

2745

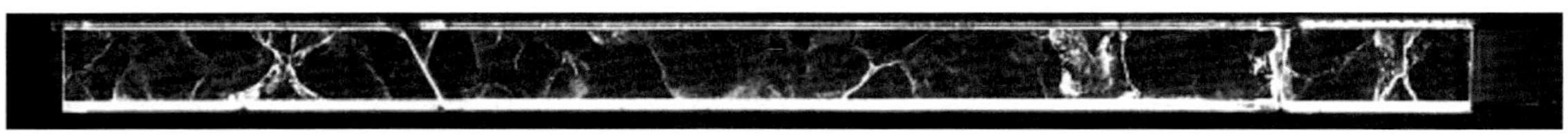

2746

2747

2748

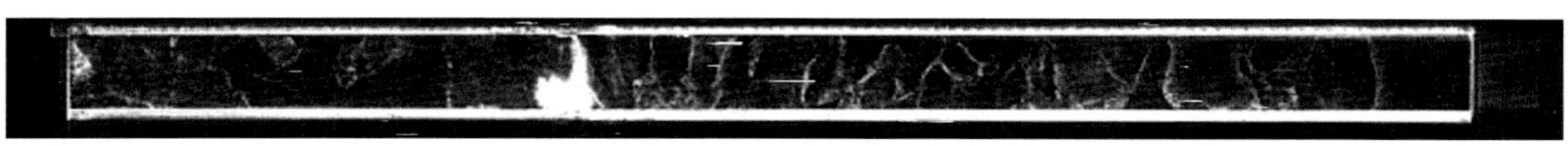

2749

2759

2758

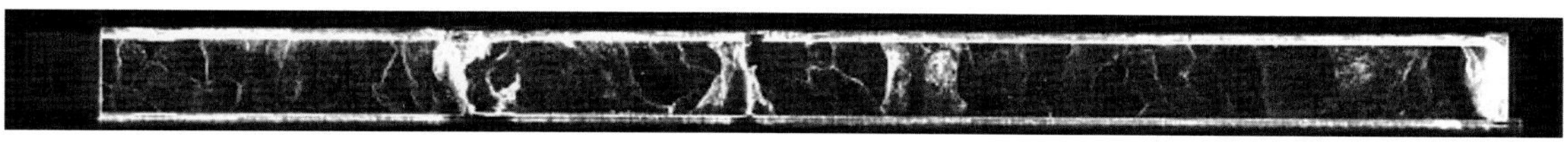

2757

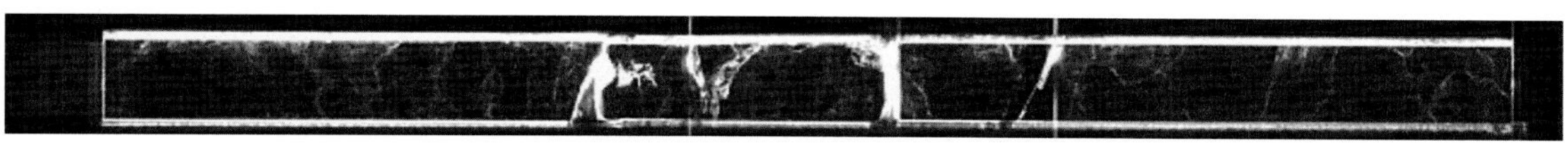

2756

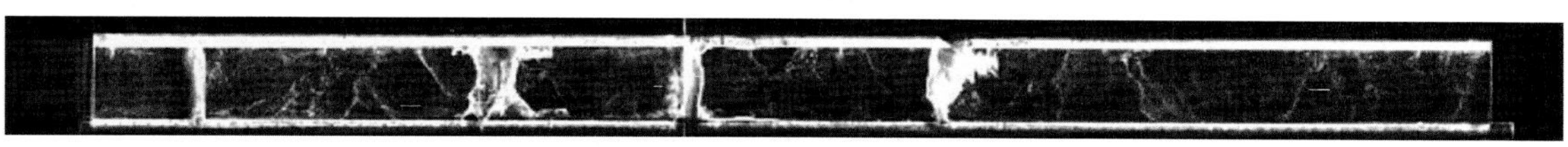

2755

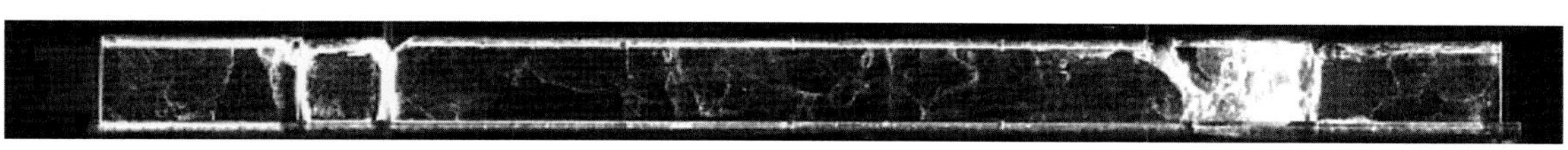

2754

2753

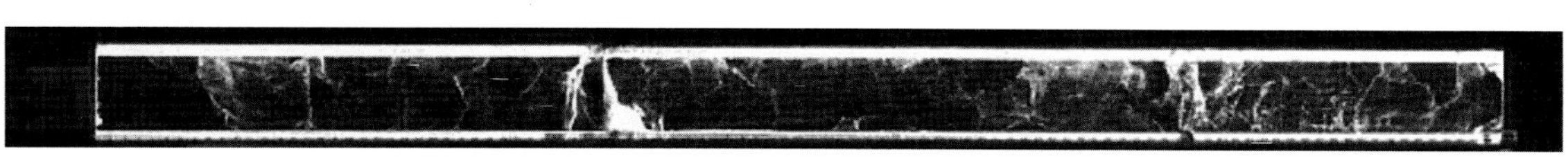

2752

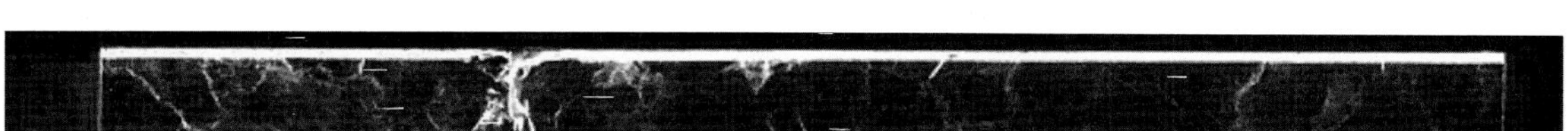

2751

2750

2760

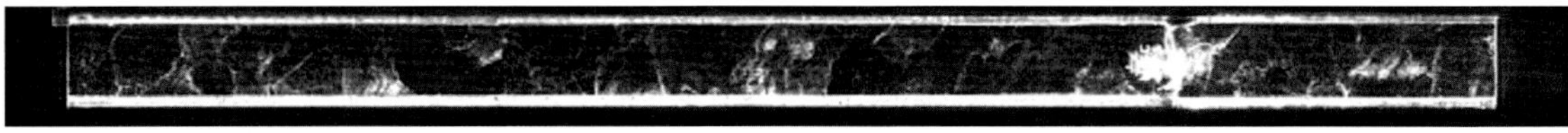

2761

2762

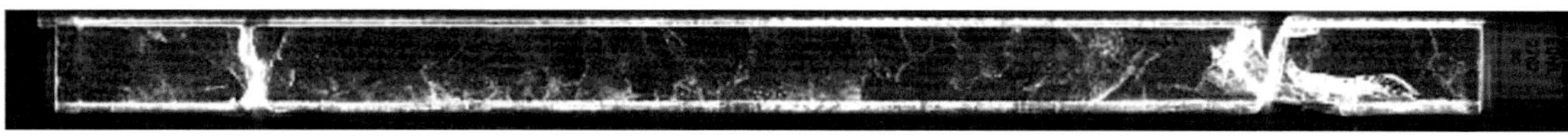

2763

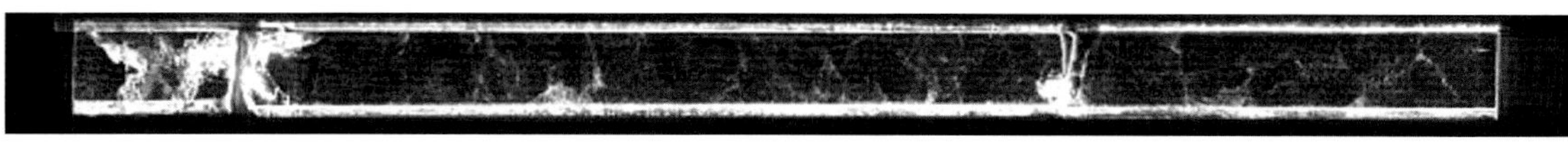

2764

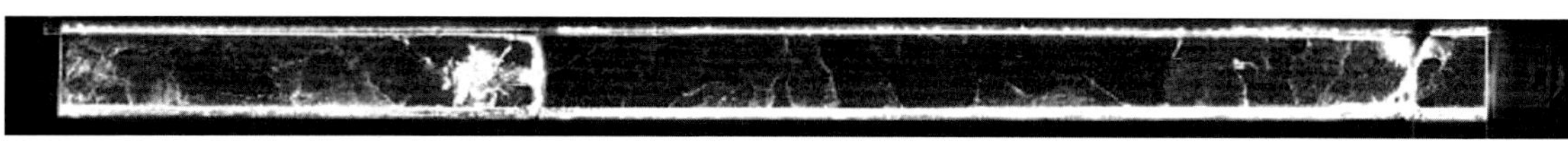

2765

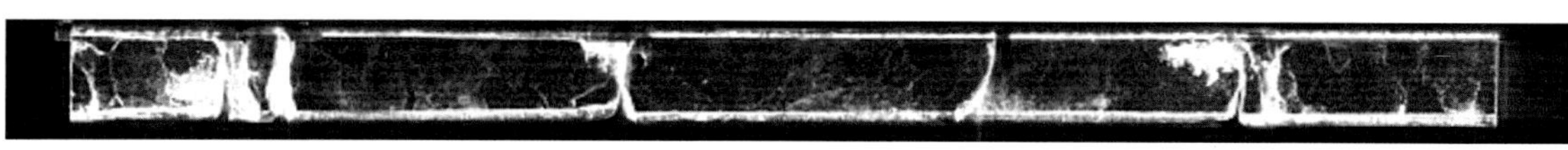

2766

2767

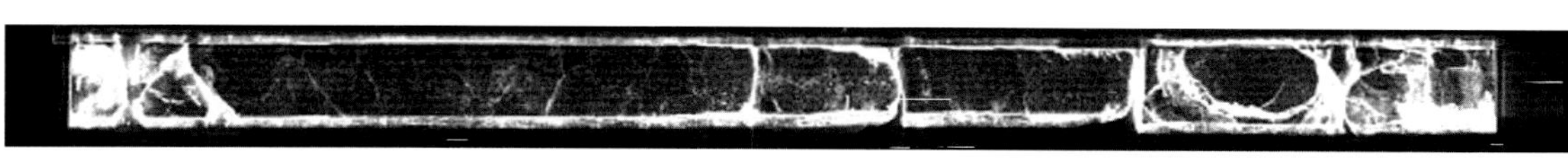

2768

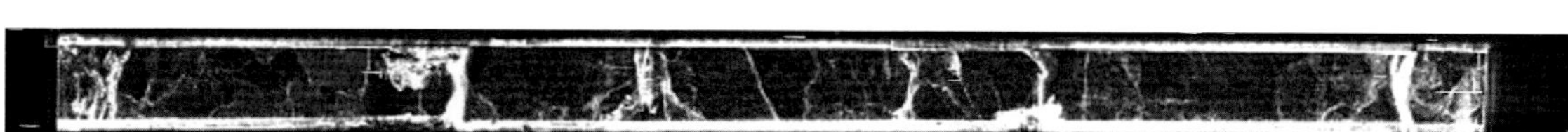

2769

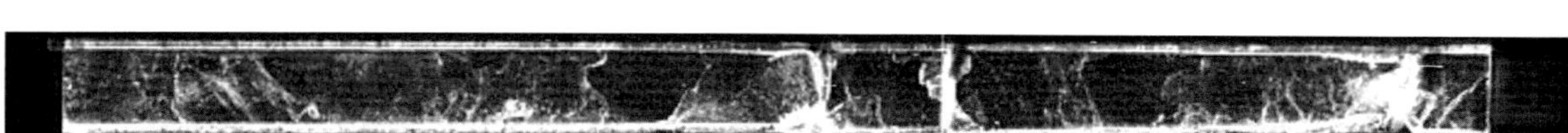

2774

2773

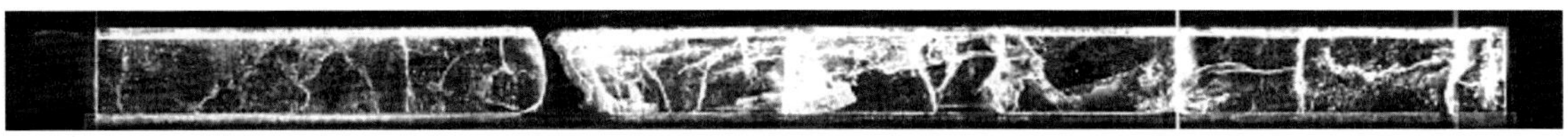

2772

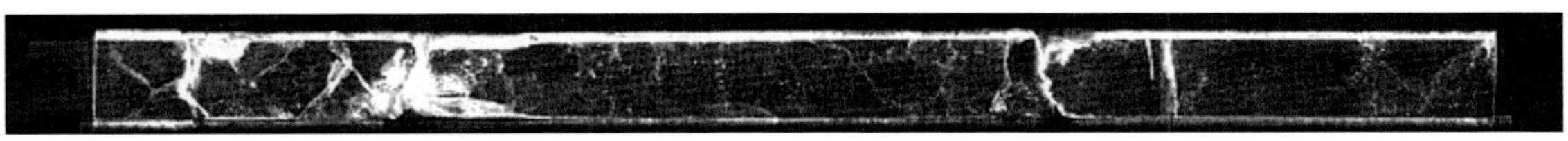

2771

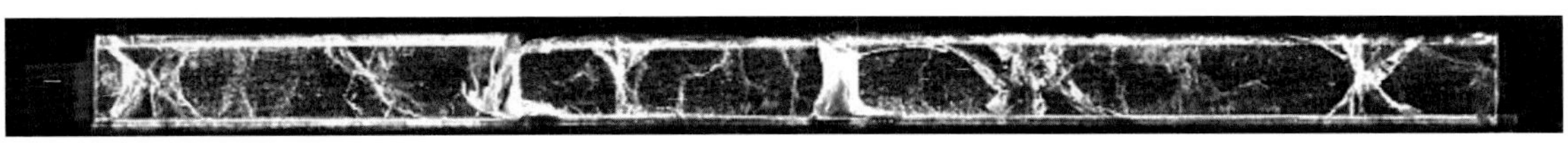

2770

References

The numbers within curly brackets at the end of each entry denote the pages in which that reference has been cited.

Alley, R. B. (1987). Firn densification by grain boundary sliding: A first model. *Journal de Physique Colloques (Paris)*, *48*(C1, Suppl. 3), 249–256. {22}

Alley, R. B. (1988). Fabrics in polar ice sheets: Development and prediction. *Science*, *240*, 493–495. {67}

Alley, R. B. (1992). Flow-law hypothesis for ice-sheet modelling. *Journal of Glaciology*, *38*, 245–256. {67, 68}

Alley, R. B., Bolzan, J. F., & Whillans, I. M. (1982). Polar firn densification and grain growth. *Annals of Glaciology*, *3*, 7–11. {21}

Alley, R. B., Gow, A. J., & Meese, D. A. (1995). Mapping c-axis fabrics to study physical processes in ice. *Journal of Glaciology*, *41*(137), 197–203. {67, 68}

Alley, R. B., Gow, A. J., Meese, D. A., Fitzpatrick, J. J., Waddington, E. D., & Bolzan, J. F. (1997). Grain-scale processes, folding, and stratigraphic disturbance in the Greenland Ice Sheet Project 2 ice core. *Journal of Geophysical Research*, *102*, 26819–26830. {17, 59}

Alley, R. B., Perepezko, J. H., & Bentley, C. R. (1986a). Grain growth in polar ice: I. Theory. *Journal of Glaciology*, *32*(112), 415–424. {68, 69}

Alley, R. B., Perepezko, J. H., & Bentley, C. R. (1986b). Grain growth in polar ice: II. Application. *Journal of Glaciology*, *32*(112), 425–433. {68}

Ams, J., Brehme, A., Janneck, J., Kaiser, W., Lensch, N., Porgarzalek, J., et al. (2001). Dismantling of the Filchner Station. In E. Fahrbach & S. E. Naggar (Eds.), The Expeditions ANTARKTIS XVI/1-2 of the Research Vessel POLARSTERN in 1998/1999. no. 380 in Reports on Polar and Marine Research (Ber. Polarforsch. Meeresforsch., pp. 126–131). Bremerhaven: Alfred Wegener Institute for Polar and Marine Research. {45}

Anderson, D. L., & Benson, C. S. (1963). The densification and diagenesis of snow. In W. D. Kingery (Ed.), *Ice and Snow* (pp. 391–411). Cambridge, MA: MIT Press. {22}

Arnaud, L., Barnola, J. M., & Duval, P. (2000). Physical modeling of the densification of snow/firn and ice in the upper part of polar ice sheets. In T. Hondoh (Ed.), *Physics of Ice Core Records* (pp. 285–305). Sapporo: Hokkaido University Press. {22}

Arnaud, L., Gay, M., Barnola, J. M., & Duval, P. (1998). Imaging of firn and bubbly ice in coaxial reflected light: A new technique for the characterization of these porous media. *Journal of Glaciology*, *44*(147), 326–332. {63}

Arthern, R. J., Vaughan, D. G., Rankin, A. M., Mulvaney, R. & Thomas, E. R. (2010). In situ measurements of Antarctic snow compaction compared with predictions of models. *Journal of Geophysical Research*, *115*(F3): F03011/1–12. {22}

Azuma, N., & Goto-Azuma, K. (1996). An anisotropic flow law for ice-sheet ice and its implications. *Annals of Glaciology*, *23*, 202–208. {59, 68}

Azuma, N., Miyakoshi, T., Yokoyama, S., & Takata, M. (2012). Impeding effect of air bubbles on normal grain growth of ice. *Journal of Structural Geology*, *42*, 184–193. {69}

Bader, H. (1962) Scope, problems, and potential value of deep core drilling in ice sheets. Special Report 58, U.S. Army Cold Regions Research and Engineering Laboratory, Hanover, NH. {30}

Bamber, J. L., Alley, R. B., & Joughin, I. (2007). Rapid response of modern day ice sheets to external forcing. *Earth and Planetary Science Letters*, *257*, 1–13. {16}

Bamber, J. L., & Bindschadler, R. A. (1997). An improved elevation dataset for climate and ice-sheet modelling: Validation with satellite imagery. *Annals of Glaciology*, *25*, 439–444. {38}

Bamber, J. L., Vaughan, D. G., & Joughin, I. (2000). Widespread complex flow in the interior of the Antarctic ice sheet. *Science*, *287*, 1248–1250. {16}

Barbante, C., Fischer, H., Masson-Delmotte, V., Waelbroeck, C., & Wolff, E. W. (2010). Climate of the last million years: New insights from EPICA and other records. *Quaternary Science Reviews*, *29*(1), 1–7. {35}

Barnes, P. R. F., & Wolff, E. W. (2004). Distribution of soluble impurities in cold glacial ice. *Journal of Glaciology*, *50*(170), 311–324. {59}

Barnes, P. R. F., Wolff, E. W., Mader, H. M., Udisti, R., Castellano, E. & Röthlisberger, R. (2003). Evolution of chemical peak shapes in the Dome C, Antarctica, ice core. *Journal of Geophysical Research*, *108*(D3): 4126/1–14. {59}

Barnola, J. M., Pimienta, P., Raynaud, D., & Korotkevich, Y. S. (1991). CO_2 climate relationship as deduced from the Vostok ice core: A re-examination based on new measurements and on a re-evaluation of the air dating. *Tellus B*, *43*(2), 83–90. {22}

Barnola, J. M., Raynaud, D., Korotkevich, Y. S., & Lorius, C. (1987). Vostok ice core provides 160,000-year record of atmospheric carbon dioxide. *Nature*, *394*, 738–743. {24}

Baschin, O. (1929). Entdeckungsflüge im Südpolargebiet. *Naturwissenschaften*, *17*(13), 215–217. {37}

Baschin, O. (1931). Die Südpolarforschung im Sommer 1929–1930. *Naturwissenschaften*, *19*(16), 333–341. {37}

Beck, P. A., & Sperry, P. R. (1950). Strain induced grain boundary migration in high purity aluminum. *Journal of Applied Physics*, *21*, 150–152. {10}

Bendel, V., Ueltzhöffer, K. J., Freitag, J., Kipfstuhl, S., Kuhs, W. F., Garbe, C. S., et al. (2013). High-resolution variations in size, number,

S.H. Faria et al., *The EPICA-DML Deep Ice Core*, Frontiers in Earth Sciences,
https://doi.org/10.1007/978-3-662-55308-4

and arrangement of air bubbles in the EPICA DML ice core. *Journal of Glaciology*, *59*(217), 972–980. {23, 25, 57, 75, 79}

Bender, M. L. (2002). Orbital tuning chronology for the Vostok climate record supported by trapped gas composition. *Earth and Planetary Science Letters*, *204*, 275–289. {24, 27}

Bender, M. L., Sowers, T., Barnola, J. M., & Chappellaz, J. (1994). Changes in the O_2-N_2 ratio of the atmosphere during recent decades reflected in the composition of air in the firn at Vostok station, Antarctica. *Geophysical Research Letters*, *21*, 189–192. {21}

Birnbaum, G., Freitag, J., Brauner, R., König-Langlo, G., Fischer, E., Kipfstuhl, S., et al. (2010). Strong-wind events and their influence on the formation of snow dunes: Observations from Kohnen Station, Dronning Maud Land. *Journal of Glaciology*, *5*(199), 891–902. {19}

Bond, G., Broecker, W., Johnsen, S., McManus, J., Labeyrie, L., Jouzel, J., et al. (1993). Correlations between climate records from North Atlantic sediment and Greenland ice. *Nature*, *365*, 143–147. {31}

Bond, G., Heinrich, H., Broecker, W., Labeyrie, L., McManus, J., Andrews, J., et al. (1992). Evidence for massive discharges of icebergs into the North Atlantic ocean during the last glacial period. *Nature*, *360*, 245–249. {16}

Budd, W. F., & Warner, R. C. (1996). A computer scheme for rapid calculations of balance-flux distributions. *Annals of Glaciology*, *23*, 21–27. {45}

Byrd, R. E. (1930). The conquest of Antarctica by air. *National Geographic Magazine*, *58*(2), 127–227. {37}

Cahn, R. W. (1970). Recovery and recrystallization. In R. W. Cahn (Ed.), *Physical Metallurgy* (pp. 1129–1197). Amsterdam: North-Holland. {7}

Calov, R., Savvin, A., Greve, R., Hansen, I., & Hutter, K. (1998). Simulation of the Antarctic ice sheet with a three-dimensional polythermal ice-sheet model, in support of the EPICA project. *Annals of Glaciology*, *27*, 201–206. {39}

Cogley, J. G., Hock, R., Rasmussen, L. A., Arendt, A. A., Bauder, A., Braithwaite, et al. (2011). Glossary of glacier mass balance and related terms. IHP-VII Technical Documents in Hydrology 86, UNESCO-IHP, Paris. IACS Contribution No. 2. {4}

Colbeck, S. C. (1983). Theory of metamorphism of dry snow. *Journal of Geophysical Research*, *88*(C9), 5475–5482. {19, 21}

Colbeck, S. C. (1989). Air movement in snow due to windpumping. *Journal of Glaciology*, *35*(120), 209–213. {21}

Cuffey, K. M. (2008). A matter of firn. *Science*, *320*, 1596–1597. {25}

Cuffey, K. M., & Steig, E. J. (1998). Isotopic diffusion in polar firn: Implications for interpretation of seasonal climate parameters in ice-core records, with emphasis on central Greenland. *Journal of Glaciology*, *44*, 273–284. {19}

Dahl-Jensen, D. & Gundestrup, N.S. (1987) Constitutive properties of ice at Dye 3, Greenland. In: *IAHS Red Book 170, The Physical Basis of Ice Sheet Modelling*, pp. 31–43. International Association of Hydrological Sciences. {9}

Dansgaard, W., Johnsen, S. J., Clausen, H. B., Dahl-Jensen, D., Gundestrup, N. S., Hammer, C. U., et al. (1993). Evidence for general instability of past climate from a 250-kyr ice-core record. *Nature*, *364*, 218–220. {31}

Davis, R., Arons, E., & Albert, M. (1996). Metamorphism of polar firn: Significance of microstructure in energy, mass and chemical species transfer. *Chemical Exchange Between the Atmosphere and Polar Snow, NATO ASI Series*, *143*, 379–401. {19}

De Angelis, H., & Kleman, J. (2007). Palaeo-ice streams in the Foxe/Baffin sector of the Laurentide Ice Sheet. *Quaternary Science Reviews*, *26*, 1313–1331. {16}

De la Chapelle, S., Castelnau, O., Lipenkov, V., & Duval, P. (1998). Dynamic recrystallization and texture development in ice as revealed by the study of deep ice cores in Antarctica and Greenland. *Journal of Geophysical Research*, *103*, 5091–5105. {67, 68, 69}

Descartes, R. (1637). Les météores. In: *Discours de la Méthode*, pp. 223–233. Ian Maire, Leyden. {3}

Descartes, R. (1644). *Principia Philosophiae*. Amsterdam: Elsevier. {3}

Dominé, F., & Shepson, P. B. (2002). Air-snow interactions and atmospheric chemistry. *Science*, *297*, 1506–1510. {19}

Drücker, C., Wilhelms, F., Oerter, H., Frenzel, A., Gernandt, H., & Miller, H. (2002). Design, transport, construction, and operation of the summer base Kohnen for ice-core drilling in Dronning Maud Land, Antarctica. *Memoirs of National Institute of Polar Research. Special issue*, *56*, 302–312. {39}

Drury, M. R., & Urai, J. L. (1990). Deformation-related recrystallization processes. *Tectonophys*, *172*, 235–253. {6}

Durand, G., Weiss, J., Lipenkov, V., Barnola, J. M., Krinner, G., Parrenin, F., et al. (2006). Effect of impurities on grain growth in cold ice sheets. *Journal of Geophysical Research*, *111*, F01015. {68, 77}

Duval, P., Ashby, M. F., & Anderman, I. (1983). Rate-controlling processes in the creep of polycrystalline ice. *Journal of Physical Chemistry*, *87*, 4066–4074. {67}

Duval, P. & Castelnau, O. (1995). Dynamic recrystallization of ice in polar ice sheets. *Journal de Physique IV (Paris), Colloque C3*, *5*: 197–205. {68}

Duval, P., & Montagnat, M. (2002). Comment on "Superplastic deformation of ice: Experimental observations" by D. L. Goldsby and D. L. Kohlstedt. *Journal of Geophysical Research*, *107*(B4), 2082. {68}

Eisen, O., Hamann, I., Kipfstuhl, S., Steinhage, D., & Wilhelms, F. (2007). Direct evidence for radar reflector originating from changes in crystal-orientation fabric. *The Cryosphere*, *1*(1), 1–10. {79}

Eliot, T. S. (1960). *The Family Reunion*. London: Faber and Faber. {57}

EPICA Community Members. (2004). Eight glacial cycles from an Antarctic ice core. *Nature*, *429*(6992), 623–628. {16, 33, 35}

EPICA Community Members. (2006). One-to-one coupling of glacial climate variability in Greenland and Antarctica. *Nature*, *444*(7116), 195–197. {16, 34, 35, 67}

Etheridge, D. M. (1989). Dynamics of the Law Dome ice cap, Antarctica, as found from bore-hole measurements. *Annals of Glaciology*, *12*, 46–50.

Faria, S. H. (2006). Creep and recrystallization of large polycrystalline masses. Part III: continuum theory of ice sheets. *Proceedings of the Royal Society of London A*, *462*(2073), 2797–2816. {60}

Faria, S. H. (2009). The multidisciplinary ice core. *Low Temperature Science*, *68*, 35–37. {*IX*}

Faria, S. H., Freitag, J., & Kipfstuhl, S. (2010). Polar ice structure and the integrity of ice-core paleoclimate records. *Quaternary Science Reviews*, *29*(1), 338–351. {16, 17, 26, 34, 59, 63, 69, 70, 75, 76, 77, 78, 79, 80, 81}

Faria, S. H., Hamann, I., Kipfstuhl, S., & Miller, H. (2006). Is Antarctica like a birthday cake? Preprint 33/2006. Max Planck Institute for Mathematics in the Sciences, Leipzig. {77, 81}

Faria, S. H., & Kipfstuhl, S. (2004). Preferred slip band orientations and bending observed in the Dome Concordia (East Antarctica) ice core. *Annals of Glaciology*, *39*, 386–390. {63}

Faria, S. H., Kipfstuhl, S., Azuma, N., Freitag, J., Hamann, I., Murshed, M. M., et al. (2009). The multiscale structure of Antarctica. Part I: inland ice. *Low Temperature Science*, *68*, 39–59. {4, 17, 26, 59, 62, 69, 70, 77, 78, 81}

Faria, S. H., Ktitarev, D., & Hutter, K. (2002). Modelling evolution of anisotropy in fabric and texture of polar ice. *Annals of Glaciology*, *35*, 545–551. {68}

Faria, S. H., Weikusat, I., & Azuma, N. (2014a). The microstructure of polar ice. Part I: highlights from ice core research. *Journal of Structural Geology*, *61*, 2–20. {16, 29, 59, 67, 76, 77}

Faria, S. H., Weikusat, I., & Azuma, N. (2014b). The microstructure of polar ice. Part II: state of the art. *Journal of Structural Geology*, *61*, 21–49. {4, 17, 24, 34, 58, 59, 60, 62, 63, 65, 67, 70, 77, 78, 81}

Fischer, H., Fundel, F., Ruth, U., Twarloh, B., Wegner, A., Udisti, R. et al. (2007a). Reconstruction of millennial changes in dust emission, transport and regional sea ice coverage using the deep EPICA ice cores from the Atlantic and Indian Ocean sector of Antarctica. *Earth and Planetary Science Letters*, *260*(1–2), 340–354. {67}

Fischer, H., Siggaard-Andersen, M. L., Ruth, U., Roethlisberger, R. & Wolff, E. (2007b). Glacial/interglacial changes in mineral dust and sea-salt records in polar ice cores: Sources, transport, and deposition. *Reviews of Geophysics*, *45*(1): RG1002/1–26. {19}

Fischer, H., Wagenbach, D., & Kipfstuhl, S. (1998). Sulfate and nitrate firn concentrations on the Greenland Ice Sheet 1: Large-scale geographical deposition changes. *Journal of Geophysical Research*, *103*(D17), 21927–21934. {19}

Freitag, J., Kipfstuhl, S., & Faria, S. H. (2008). The connectivity of crystallite agglomerates in low density firn at Kohnen station, Dronning Maud Land, Antarctica. *Annals of Glaciology*, *49*, 114–120. {59}

Freitag, J., Kipfstuhl, S., Laepple, T., & Wilhelms, F. (2013). Impurity-controlled densification: A new model for stratified polar firn. *Journal of Glaciology*, *59*(218), 1163–1169. {22, 25}

Freitag, J., Wilhelms, F., & Kipfstuhl, S. (2004). Microstructure-dependent densification of polar firn derived from X-ray microtomography. *Journal of Glaciology*, *50*, 243–250. {24}

Frezzotti, M., Gandolfi, S. & Urbini, S. (2002). Snow megadunes in Antarctica: Sedimentary structure and genesis. *Journal of Geophysical Research*, *107*(D18). Art. No. 4344. {19}

Frost, H. J., & Ashby, M. F. (1982). *Deformation-mechanism Maps*. Oxford: Pergamon. {15}

Fujita, S., Okuyama, J., Hori, A., & Hondoh, T. (2009). Metamorphism of stratified firn at Dome Fuji, Antarctica: A mechanism for local insolation modulation of gas transport conditions during bubble close off. *Journal of Geophysical Research*, *114*(F3), F03023. {27}

Garfield, D. E., & Ueda, H. T. (1976). Resurvey of the "Byrd" Station, Antarctica, drill hole. *Journal of Glaciology*, *17*(75), 29–34. {9}

Gerland, S., Oerter, H., Kipfstuhl, J., Wilhelms, F., Miller, H., & Miners, W. D. (1999). Density log of a 181m long ice core from Berkner Island, Antarctica. *Annals of Glaciology*, *29*, 215–219. {24}

Gödert, G., & Hutter, K. (1998). Induced anisotropy in large ice sheets: Theory and its homogenization. *Continuum Mechanics and Thermodynamics*, *13*, 91–120. {68}

Goujon, C., Barnola, J. M. & Ritz, C. (2003). Modeling the densification of polar firn including heat diffusion: Application to close-off characteristics and gas isotopic fractionation for Antarctica and Greenland sites. *Journal of Geophysical Research*, *108*(D24): 4792/1–18. {22}

Gow, A. J. (1969). On the rates of growth of grains and crystals in south polar firn. *Journal of Glaciology*, *8*(53), 241–252. {67, 68, 69}

Gow, A. J. (1971). Relaxation of ice in deep drill cores from Antarctica. *Journal of Glaciology*, *76*, 2533–2541. {27, 28}

Gow, A. J., & Meese, D. (2007a). The distribution and timing of tephra deposition at Siple Dome, Antarctica: Possible climatic and rheologic implications. *Journal of Glaciology*, *53*(183), 585–596. {16}

Gow, A. J., & Meese, D. (2007b). Physical properties, crystalline textures and c-axis fabrics of the Siple dome (Antarctica) ice core. *Journal of Glaciology*, *53*(183), 573–584. {17}

Gow, A. J., & Williamson, T. (1975). Gas inclusions in the Antarctic ice sheet and their glaciological significance. *Journal of Geophysical Research*, *80*(36), 5101–5108. {27}

Gow, A. J., & Williamson, T. (1976). Rheological implications of the internal structure and crystal fabrics of the West Antarctic ice sheet as revealed by deep core drilling at Byrd Station. *Geological Society of America Bulletin*, *87*, 1665–1677. {16, 76, 77}

Grannas, A. M., et al. (2007). An overview of snow photochemistry: Evidence, mechanisms and impacts. *Atmospheric Chemistry and Physics*, *7*, 4165–4283. {19}

Gundestrup, N. S., & Hansen, B. L. (1984). Bore-hole survey at Dye 3, south Greenland. *Journal of Glaciology*, *30*, 282–288. {9}

Helsen, M. M., van den Broeke, M. R., van de Wal, R. S. W., van de Berg, W. J., van Meijgaard, E., Davis, C. H., et al. (2008). Elevation changes in Antarctica mainly determined by accumulation variability. *Science*, *320*, 1626–1629. {25}

Hemming, S. R. (2004). Massive late Pleistocene detritus layers of the North Atlantic and their global climate imprints. *Reviews of Geophysics*, *42*(1): pRG1005. {16}

Herron, M. M., & Langway, C. C, Jr. (1980). Firn densification: An empirical model. *Journal of Glaciology*, *25*(93), 373–385. {22}

Hobbs, P. V. (1974). *Ice Physics*. Oxford: Clarendon. {63}

Hörhold, M. W., Kipfstuhl, S., Wilhelms, F., Freitag, J. & Frenzel, A. (2011). The densification of layered polar firn. *Journal of Geophysical Research*, *116*(F1): F01001/1–15. {24}

Hörhold, M. W., Laepple, T., Freitag, J., Bigler, M., Fischer, H., & Kipfstuhl, S. (2012). On the impact of impurities on the densification of polar firn. *Earth and Planetary Science Letters*, *325–326*, 93–99. {24, 25}

Hori, A., Tayuki, K., Narita, H., Hondoh, T., Fujita, S., Kameda, T., et al. (1999). A detailed density profile of the Dome Fuji (Antarctica) shallow ice core by X-ray transmission method. *Annals of Glaciology*, *29*, 211–214. {24}

Humphreys, F. J. & Hatherly, M. (2004). *Recrystallization and Related Annealing Phenomena* (2nd edn). Oxford: Pergamon. {7, 9, 10, 69}

Huybrechts, P., Steinhage, D., Wilhelms, F., & Bamber, J. (2000). Balance velocities and measured properties of the Antarctic ice sheet from a new compilation of gridded data for modelling. *Annals of Glaciology*, *30*, 52–60. {38, 39}

Hvidberg, C. S., Steffensen, J. P., Clausen, H. B., Shoji, H., & Kipfstuhl, J. (2002). The NorthGRIP ice-core logging procedure: description and evaluation. *Annals of Glaciology*, *35*, 5–8. {9}

Ikeda, T., Fukazawa, H., Mae, S., Pepin, L., Duval, P., Champagnon, B., et al. (1999). Extreme fractionation of gases caused by formation of clathrate hydrates in Vostok Antarctic ice. *Geophysical Research Letters*, *26*(1), 91–94. {26}

Ikeda-Fukazawa, T., Fukumizu, K., Kawamura, K., Aoki, S., Nakazawa, T., & Hondoh, T. (2005). Effects of molecular diffusion on trapped gas composition in polar ice cores. *Earth and Planetary Science Letters*, *229*, 183–192. {27}

Ikeda-Fukazawa, T., Hondoh, T., Fukumura, T., Fukazawa, H., & Mae, S. (2001). Variation in N_2/O_2 ratio of occluded air in Dome Fuji antarctic ice. *Journal of Geophysical Research*, *106*(D16), 17799–17810. {27}

Iliescu, D., Baker, I., & Chang, H. (2004). Determining the orientations of ice crystals using electron backscatter patterns. *Microscopy Research and Technique*, *63*, 183–187. {59, 67}

Jansen, D., Sandhäger, H. & Rack, W. (2005a) Evolution of tabular iceberg A-38B, observation and simulation. FRISP Report 16, Forum for Research into Ice Shelf Processes. {45}

Jansen, D., Sandhäger, H., & Rack, W. (2005b). Model experiments on large tabular iceberg evolution: ablation and strain thinning. *Journal of Glaciology*, *51*(174), 363–372. {45}

Johnsen, S., Hansen, S. B., Sheldon, S. G., Dahl-Jensen, D., Steffensen, J. P., Augustin, L., et al. (2007). The Hans Tausen drill: design, performance, further developments and some lessons learned. *Annals of Glaciology*, *47*, 89–98. {48, 53}

Johnsen, S. J., Clausen, H. B., Cuffey, K. M., Hoffmann, G., Schwander, J., & Creyts, T. (2000). Diffusion of stable isotopes in polar firn and ice: the isotope effect in firn diffusion. In T. Hondoh (Ed.), *Physics of Ice Core Records* (pp. 121–140). Sapporo: Hokkaido University Press. {59}

Johnsen, S. J., Clausen, H. B., Dansgaard, W., Fuhrer, K., Gundestrup, N. S., Hammer, C. U., et al. (1992). Irregular glacial interstadials recorded in a new greenland ice core. *Nature*, *359*, 311–313. {31}

Jouzel, J., Barkov, N. I., Barnola, J. M., Bender, M., Chappellaz, J., Genthon, C., et al. (1993). Extending the Vostok ice-core record of paleoclimate to the penultimate glacial period. *Nature*, *364*, 407–412. {31}

Jouzel, J., Masson-Delmotte, V., Cattani, O., Dreyfus, G., Falourd, S., Hoffmann, G., et al. (2007). Orbital and millennial Antarctic climate variability over the past 800,000 years. *Science*, *317*(5839), 793–797. {33, 34}

Kaufmann, P., Fundel, F., Fischer, H., Bigler, M., Ruth, U., Udisti, et al. (2010). Ammonium and non-sea salt sulfate in the EPICA ice cores as indicator of biological activity in the Southern Ocean. *Quaternary Science Reviews*, *29*(1–2), 313–323. {67}

Kepler, J. (1611). *Strena seu de nive sexangula*. Gottfried Tambach, Frankfurt. {3}

Kepler, J. (1619). *Harmonices Mundi*. Johann Planck (for Gottfried Tambach), Linz. {3}

Kipfstuhl, S., Faria, S. H., Azuma, N., Freitag, J., Hamann, I., Kaufmann, P., et al. (2009). Evidence of dynamic recrystallization in polar firn. *Journal of Geophysical Research*, *114*, B05204. {24, 58, 62, 63, 67, 69, 70, 78}

Kipfstuhl, S., Hamann, I., Lambrecht, A., Freitag, J., Faria, S. H., Grigoriev, D., et al. (2006). Microstructure mapping: A new method for imaging deformation-induced microstructural features of ice on the grain scale. *Journal of Glaciology*, *52*(178), 398–406. {62, 63, 67, 69, 70, 78}

Kipfstuhl, S., Pauer, F., Kuhs, W. F., & Shoji, H. (2001). Air bubbles and clathrate hydrates in the transition zone of the NGRIP deep ice core. *Geophysical Research Letters*, *28*, 591–594. {25, 26, 28}

Kirchner, N. P., & Faria, S. H. (2009). The multiscale structure of Antarctica. Part II: ice shelves. *Low Temperature Science*, *68*, 61–71. {8, 16}

Kristensen, M. (1983). Iceberg calving and deterioration in Antarctica. *Progress in Physical Geography*, *7*(3), 313–328. {16}

Ktitarev, D., Gödert, G. & Hutter, K. (2002). Cellular automaton model for recrystallization, fabric and texture development in polar ice. *Journal of Geophysical Research*, *107*(B8), EPM 5–1–EPM 5–9. {68}

Kuroiwa, D., & Hamilton, W. L. (1963). Studies of ice etching and dislocation etch pits. In W. D. Kingery (Ed.), *Ice and snow: Properties, processes, and applications* (pp. 34–55). Cambridge, MA: MIT Press. {63}

Langway, Jr., C. C. (2008) The history of early polar ice cores. Technical Report ERDC/CRREL TR-08-1, U.S. Army Engineer Research and Development Center, Cold Regions Research and Engineering Laboratory, Hanover, NH. {30}

Legrand, M., & Mayewski, P. A. (1997). Glaciochemistry of polar ice cores: a review. *Reviews of Geophysics*, *35*, 219–243. {25}

Lemke, P., Ren, J., Alley, R. B., Allison, I., Carrasco, J., Flato, G., et al. (2007). Observations: changes in snow, ice and frozen ground. In S. Solomon, D. Qin, M. Manning, Z. Chen, M. Marquis, K. B. Averyt, M. Tignor, & H. L. Miller (Eds.), *Climate Change 2007: The Physical Science Basis, Contribution of Working Group I to the Fourth Assessment Report of the Intergovernmental Panel on Climate Change (IPCC)*. Cambridge: Cambridge University Press. {15}

Lhomme, N., Clarke, G. K. C., & Marshall, S. J. (2005). Tracer transport in the Greenland Ice Sheet: Constraints on ice cores and glacial history. *Quaternary Science Reviews*, *24*, 173–194. {17}

Lipenkov, V. Y. (2000). Air bubbles and air-hydrate crystals in the Vostok ice core. In T. Hondoh (Ed.), *Physics of ice core records* (pp. 327–358). Sapporo: Hokkaido University Press. {75}

Lipenkov, V. Y., Salamatin, A., & Duval, P. (1997). Bubbly-ice densification in ice sheets: II. Applications. *Journal of Glaciology*, *43*, 397–407. {25}

Lüthi, D., Bereiter, B., Stauffer, B., Winkler, R., Schwander, J., Kindler, P., et al. (2010). CO_2 and O_2/N_2 variations in and just below the bubble-clathrate transformation zone of Antarctic ice cores. *Earth and Planetary Science Letters*, *297*, 226–233. {59, 75, 79}

Maeno, N., & Ebinuma, T. (1983). Pressure sintering of ice and its implication to the densification of snow at polar glaciers and ice sheets. *Journal of Physical Chemistry*, *87*(21), 4103–4110. {22, 24, 69}

McConnell, J. R., Maselli, O. J., Sigl, M., Vallelonga, P., Neumann, T., Anschütz, H., et al. (2014). Antarctic-wide array of high-resolution ice core records reveals pervasive lead pollution began in 1889 and persists today. *Scientific Reports*, *4*, 5848/1–5. {19}

McGwire, K. C., Hargreaves, G. M., Alley, R. B., Popp, T. J., Reusch, D. B., Spencer, M. K., et al. (2008). An integrated system for optical imaging of ice cores. *Cold Regions Science and Technology*, *53*, 216–228. {72}

Miller, S. L. (1969). Clathrate hydrates of air in Antarctic ice. *Science*, *165*, 489–490. {25}

Miyamoto, A., Saito, T., & Hondoh, T. (2009). Visual observation of volume relaxation under different storage temperatures in the Dome Fuji ice core, Antarctica. *Low Temperature Science*, *68*, 73–79. {27}

Montagnat, M., Azuma, N., Dahl-Jensen, D., Eichler, J., Fujita, S., Gillet-Chaulet, F., et al. (2014). Fabric along the NEEM ice core, Greenland, and its comparison with GRIP and NGRIP ice cores. *The Cryosphere*, *8*(4), 1129–1138. {61}

Montagnat, M., & Duval, P. (2000). Rate controlling processes in the creep of polar ice, influence of grain boundary migration associated with recrystallization. *Earth and Planetary Science Letters*, *183*, 179–186. {68}

Mullins, W. W. (1957). Theory of thermal grooving. *Journal of Applied Physics*, *28*(3), 333–339. {63}

Nakahara, J., Shigesato, Y., Higashi, A., Hondoh, T., & Langway, C. C. (1988). Raman spectra of natural clathrates in deep ice cores. *Philosophical Magazine B*, *57*(3), 421–430. {26}

Narita, H., Azuma, N., Hondoh, T., Fujii, M., Kawaguchi, M., Mae, S., et al. (1999). Characteristics of air bubbles and hydrates in the Dome Fuji ice core, Antarctica. *Journal of Glaciology*, *29*(1), 207–210. {26}

Nedelcu, A. F., Faria, S. H., & Kuhs, W. F. (2009). Raman spectra of plate-like inclusions in the EPICA-DML (Antarctica) ice core. *Journal of Glaciology*, *55*(189), 183–184. {27, 28}

NEEM Community Members. (2013). Eemian interglacial reconstructed from a Greenland folded ice core. *Nature*, *493*(7433), 489–494. {17, 59}

Nishida, K., & Narita, H. (1996). Three-dimensional observations of ice crystal characteristics in polar ice sheets. *Journal of Geophysical Research*, *101*(D16), 21311–21317. {63}

Nordenskjöld, A. E. (1861). Beitrag zur Kenntnis der Krystallformen einiger Oxide. *Annual Review of Physical Chemistry*, *4*(24): 612–627. Poggendorff's Ann. 114. {60}

Obbard, R., & Baker, I. (2007). The microstructure of meteoric ice from Vostok, Antarctica. *Journal of Glaciology*, *53*(180), 41–62. {59}

Oerlemans, J. (2001). *Glaciers and climate change*. Lisse: A. A. Balkema. {16}

Oerter, H., Drücker, C., Kipfstuhl, S., & Wilhelms, F. (2009). Kohnen Station, the drilling camp for the EPICA deep ice core in Dronning Maud Land. *Polarforschung*, *78*(1/2), 1–23. {32, 33, 34, 38, 39, 47, 49}

Ohno, H., Igarashi, M. & Hondoh, T. (2006). Characteristics of salt inclusions in polar ice from Dome Fuji, East Antarctica. *Geophysical Research Letters*, *33*: L08501/1–5. {59}

Ohno, H., Lipenkov, V. Y., & Hondoh, T. (2010). Formation of air clathrate hydrates in polar ice sheets: Heterogeneous nucleation induced by micro-inclusions. *Journal of Glaciology*, *56*(199), 917–921. {75}

Parrenin, F., Barnola, J. M., Beer, J., Blunier, T., Castellano, E., Chappellaz, J., et al. (2007). The EDC3 chronology for the EPICA Dome C ice core. *Climate of the Past*, *3*, 485–497. {34}

Paterson, W. S. B. (1991). Why ice-age ice is sometimes "soft". *Cold Regions Science and Technology*, *20*(1), 75–98. {17, 77}

Paterson, W. S. B. (1994). *The physics of glaciers*, 3rd edn. Oxford: Pergamon. {9, 17, 31, 69, 77}

Pauer, F., Kipfstuhl, S., & Kuhs, W. (1996). Raman spectroscopic study on the spatial distribution of nitrogen and oxygen in natural ice clathrates and their decomposition to air bubbles. *Geophysical Research Letters, 2*, 177–180. {26}

Pauer, F., Kipfstuhl, S., Kuhs, W. F., & Shoji, H. (1999). Air clathrate crystals from the GRIP deep ice core, Greenland: A number, size, and shape distribution study. *Journal of Glaciology, 45*, 22–30. {28}

Petit, J. R., Jouzel, J., Raynaud, D., Barkov, N. I., Barnola, J. M., Basile, I., et al. (1999). Climate and atmospheric history of the past 420,000 years from the Vostok ice core, Antarctica. *Nature, 399*, 429–436. {16, 31}

Phillips, R. (2001). *Crystals, defects and microstructures*. Cambridge: Cambridge University Press. {15}

Piazolo, S., Montagnat, M., & Blackford, J. R. (2008). Sub-structure characterization of experimentally and naturally deformed ice using cryo-EBSD. *Journal of Microscopy, 230*(3), 509–519. {67}

Pimienta, P. & Duval, P. (1987) Rate controlling processes in the creep of polar glacier ice. *Journal de Physique (Paris), 48*(C1, Suppl. 3): 243–248. {68}

Poirier, J. P. (1985). *Creep of crystals*. Cambridge: Cambridge University Press. {10}

Popper, K.R. (1994) *The Myth of the Framework*. Routledge, London. Edited by M.A. Notturno. {3}

Prior, D. J., Diebold, S., Obbard, R., Daghlian, C., Goldsby, D. L., Durham, W. B., et al. (2012). Insight into the phase transformations between ice Ih and ice II from electron backscatter diffraction data. *Scripta Materialia, 66*(2), 69–72. {67}

Raynaud, D., Lipenkov, V., Lemieux, B., Duval, P., Loutre, M. F., & Lhomme, N. (2007). The local insolation signature of air content in Antarctic ice: a new step toward an absolute dating of ice records. *Earth and Planetary Science Letters, 261*, 337–349. {24}

Rignot, E., Casassa, G., Gogieni, P., Krabill, W., Rivera, A., & Thomas, R. (2004). Accelerated ice discharge from the Antarctic Peninsula following the collapse of Larsen B ice shelf. *Geophysical Research Letters, 31*, L18401. {16}

Roessiger, J., Bons, P. D., & Faria, S. H. (2014). Influence of bubbles on grain growth in ice. *Journal of Structural Geology, 61*, 123–132. {69}

Roessiger, J., Bons, P. D., Griera, A., Jessell, M. W., Evans, L., Montagnat, M., et al. (2011). Competition between grain growth and grain-size reduction in polar ice. *Journal of Glaciology, 57*(205), 942–948. {69}

Ruth, U., Barnola, J. M., Beer, J., Bigler, M., Blunier, T., Castellano, E., et al. (2007). "EDML1": A chronology for the EPICA deep ice core from Dronning Maud Land, Antarctica, over the last 150 000 years. *Climate of the Past, 3*, 475–484. {34, 59, 78, 81}

Ryser, C., Lüthi, M. P., Andrews, L. C., Hoffman, M. J., Catania, G. A., Hawley, R. L., et al. (2014). Sustained high basal motion of the Greenland ice sheet revealed by borehole deformation. *Journal of Glaciology, 60*(222), 647–660. {9}

Sagan, C. (1997). *Billions and billions*. London: Headline. {15}

Salamatin, A., Lipenkov, V. Y., & Duval, P. (1997). Bubbly-ice densification in ice sheets: I. Theory. *Journal of Glaciology, 43*, 387–396. {25}

Salamatin, A. N., Lipenkov, V. Y., Barnola, J. M., Hori, A., Duval, P., & Hondoh, T. (2009). Snow/firn densification in polar ice sheets. *Low Temperature Science, 68*, 195–222. {22, 25}

Salamatin, A. N., Lipenkov, V. Y., Ikeda-Fukazawa, T., & Hondoh, T. (2001). Kinetics of air-hydrate nucleation in polar ice sheets. *J. Cryst. Growth, 223*, 285–305. {25}

Saylor, D. M., & Rohrer, G. S. (1999). Measuring the influence of grain-boundary misorientation on thermal groove geometry in ceramic polycrystals. *Journal of the American Ceramic Society, 82*, 1529–1536. {65}

Schwander, J., & Stauffer, B. (1984). Age difference between polar ice and the air trapped in its bubbles. *Nature, 311*, 45–47. {24}

Schytt, V. (1958). *Snow and ice studies in Antarctica*. Ph.D. thesis, University of Stockholm, Oslo. {29}

Schytt, V. (1974). Obituary Hans W:son Ahlmann - 1889–1974. *Journal of Glaciology, 13*(69), 541–542. {29}

Severinghaus, J. P., & Battle, M. O. (2006). Fractionation of gases in polar ice during bubble close-off: new constraints from firn air, Ne, Kr and Xe observations. *Earth and Planetary Science Letters, 244*(1–2), 474–500. {27}

Shimada, W., & Hondoh, T. (2004). In situ observation of the transformation from air bubbles to air clathrate hydrate crystals using a Mizuho ice core. *Journal of Crystal Growth, 265*(1–2), 309–317. {75}

Shoji, H., & Langway, C. C, Jr. (1982). Air hydrate inclusions in fresh ice core. *Nature, 298*, 548–550. {25, 26, 28}

Shoji, H. & Langway, Jr., C.C. (1987) Microscopic observations of the air hydrate-bubble transformation process in glacier ice. *Journal de Physique (Paris), 48*(C1, Suppl. 3): 141–148. {26}

Siegenthaler, U., Monnin, E., Kawamura, K., Spahni, R., Schwander, J., Stauffer, B., et al. (2005). Supporting evidence from the EPICA Dronning Maud Land ice core for atmospheric CO_2 changes during the past millennium. *Tellus B, 57*(1), 51–57. {67}

Siegert, M. J. (2001). *Ice sheets and late quaternary environmental change*. Chichester: Wiley. {16}

Sloan, Jr., E. D. (1998).*Clathrate Hydrates of Natural Gases*, 2nd edn. New York: Marcel Dekker. {25}

Smith, C. S. (1952). Grain shapes and other metallurgical applications of topology. In: *Metal Interfaces*, pp. 65–108. Cleveland, OH: American Society for Metals (ASM). {11}

Smith, C. S. (1954). The microstructure of polycrystalline materials. *Transactions of Chalmers University of Technology*. Gothenburg: Chalmers University of Technology. {58}

Staroszczyk, R., & Morland, L. W. (2001). Strengthening and weakening of induced anisotropy in polar ice. *Proceedings of the Royal Society of London. Series A, 451*(2014), 2419–2440. {68}

Stauffer, B., Schwander, J., & Oeschger, H. (1985). Enclosure of air during metamorphosis of dry firn to ice. *Annals of Glaciology, 6*, 108–112. {24}

Stauffer, B., & Tschumi, J. (2000). Reconstruction of past atmospheric CO_2 concentrations by ice core analyses. In T. Hondoh (Ed.), *Physics of ice core records* (pp. 217–241). Sapporo: Hokkaido University Press. {27}

Steinhage, D. (2001) Contributions of geophysical measurements in Dronning Maud Land, Antarctica, locating an optimal drill site for a deep ice core drilling. Reports on polar and marine research (ber. polarforsch. meeresforsch.), Alfred Wegener Institute for Polar and Marine Research, Bremerhaven. {32, 34, 38, 39, 40, 41, 42, 43, 44, 45}

Steinhage, D., Nixdorf, U., Meyer, U., & Miller, H. (1999). New maps of the ice thickness and subglacial topography in Dronning Maud Land, Antarctica, determined by means of airborne radio echo sounding. *Annals of Glaciology, 29*, 267–272. {38}

Steinhage, D., Nixdorf, U., Meyer, U., & Miller, H. (2001). Subglacial topography and internal structure of central and western Dronning Maud Land, Antarctica, determined from airborne radio echo sounding. *Journal of Applied Physics, 47*, 183–189. {38}

Stenni, B., Masson-Delmotte, V., Selmo, E., Oerter, H., Meyer, H., Roethlisberger, R., et al. (2010). The deuterium excess records of EPICA Dome C and Dronning Maud Land ice cores (East Antarctica). *Quaternary Science Reviews, 29*(1–2), 146–159. {67}

Stephenson, P. J. (1967). Some considerations of snow metamorphism in the Antarctic ice sheet in the light of ice crystal studies. In: H. Oura (Ed.), *Physics of Snow and Ice*, vol. 1, Part 2, pp. 725–740. Sapporo: Hokkaido University Press. Proceedings of the International Conference on Low Temperature Science, 1966, Sapporo, Japan. {67, 68, 69}

Suwa, M., & Bender, M. L. (2008). O_2/N_2 ratios of occluded air in the GISP2 ice core. *Journal of Geophysical Research, 113*(D11), D11119. {27}

Svensson, A., Durand, G., Mathiesen, J., Persson, A., & Dahl-Jensen, D. (2009). Texture of the upper 1000 m in the GRIP and NorthGRIP ice cores. *Low Temperature Science*, *68*, 107–113. {61}

Svensson, A., Nielsen, S. W., Kipfstuhl, S., Johnsen, S. J., Steffensen, J. P., Bigler, M., et al. (2005). Visual stratigraphy of the North Greenland Ice Core Project (NorthGRIP) ice core during the last glacial period. *Journal of Geophysical Research*, *110*, D02108. {72, 73}

Takata, M., Iizuka, Y., Hondoh, T., Fujita, S., Fuji, Y., & Shoji, H. (2004). Stratigraphic analysis of Dome Fuji Antarctic ice core using an optical scanner. *Annals of Glaciology*, *39*, 467–472. {72, 73}

The Holy Bible. (2003). *Authorized (King James) Version*. Iowa Falls: World Bible Publishers. {37}

Thomson, D. W. (1917). *On growth and form*. Cambridge: Cambridge University Press. {58}

Thorsteinsson, T. (2002). Fabric development with nearest-neighbor interaction and dynamic recrystallization. *Journal of Geophysical Research*, *107*(B1): ECV3–1–ECV3–13. {68}

Thorsteinsson, T., Kipfstuhl, J., & Miller, H. (1997). Textures and fabrics in the GRIP ice core. *Journal of Geophysical Research*, *102*, 26583–26599. {68}

Thorsteinsson, T., & Waddington, E. D. (2002). Folding in strongly anisotropic layers near ice-sheet centers. *Annals of Glaciology*, *35*, 480–486. {59}

Uchida, T., Hondoh, T., Mae, S., Lipenkov, V. Y., & Duval, P. (1994a). Air-hydrate crystals in deep ice-core samples from Vostok Station, Antarctica. *Journal of Glaciology*, *40*, 79–86. {26, 28}

Uchida, T., Hondoh, T., Mae, S., Shoji, H., & Azuma, N. (1994b). Optimized storage condition of deep ice core samples from the viewpoint of air-hydrate analysis. *Memoirs of National Institute of Polar Research. Special issue*, *49*, 306–313. {27}

Uchida, T., Yasuda, K., Oto, Y., Shen, R., & Ohmura, R. (2014). Natural supersaturation conditions needed for nucleation of air-clathrate hydrates in deep ice sheets. *Journal of Glaciology*, *60*(224), 1111–1116. {26}

Ueda, H. T. & Garfield, D. E. (1968). Drilling through the Greenland ice sheet. Special Report 126, U.S. Army Cold Regions Research and Engineering Laboratory, Hanover, NH. {30}

Ueda, H. T. & Garfield, D. E. (1969) Core drilling through the Antarctic ice sheet. Technical Report 231, U.S. Army Cold Regions Research and Engineering Laboratory, Hanover, NH. {30}

Ueltzhöffer, K. J., Bendel, V., Freitag, J., Kipfstuhl, S., Wagenbach, D., Faria, S. H., et al. (2010). Distribution of air bubbles in the EDML and EDC ice cores from a new method of automatic image analysis. *Journal of Glaciology*, *56*(196), 339–348. {23, 25, 57, 79}

Van der Veen, C. J., & Whillans, I. M. (1994). Development of fabric in ice. *Cold Regions Science and Technology*, *22*, 171–195. {68}

Vaughan, D. G., Comiso, J. C., Allison, I., Carrasco, J., Kaser, G., Kwok, R., et al. (2013). Observations: Cryosphere. In T. F. Stocker, D. Qin, G. K. Plattner, M. Tignor, S. Allen, J. Boschung, A. Nauels, Y. Xia, V. Bex, & P. Midgley (Eds.), *Climate Change 2013: The Physical Science Basis, Contribution of Working Group I to the Fifth Assessment Report of the Intergovernmental Panel on Climate Change (IPCC)*. Cambridge: Cambridge University Press. {15}

Wang, Y., Kipfstuhl, S., Azuma, N., Thorsteinsson, T., & Miller, H. (2003). Ice-fabrics study in the upper 1500 m of the Dome C (East Antarctica) deep ice core. *Annals of Glaciology*, *37*, 97–104. {59, 63}

Watanabe, O., Jouzel, J., Johnsen, S., Parrenin, F., Shoji, H., & Yoshida, N. (2003). Homogeneous climate variability across East Antarctica over the past three glacial cycles. *Nature*, *422*, 509–512. {16}

Weikusat, C., Freitag, J., & Kipfstuhl, S. (2012). Raman spectroscopy of gaseous inclusions in EDML ice core: First results - microbubbles. *Journal of Glaciology*, *58*(210), 761–766. {28}

Weikusat, I., de Winter, D. A. M., Pennock, G. M., Hayles, M., Schneijdenberg, C. T. W. M., & Drury, M. R. (2011a). Cryogenic EBSD on ice: Preserving a stable surface in a low pressure SEM. *Journal of Microscopy*, *242*(3), 295–310. {59, 67}

Weikusat, I., Jansen, D., Binder, T., Eichler, J., Faria, S. H., Wilhelms, F., et al. (2017). Physical analysis of an Antarctic ice core–towards an integration of micro- and macrodynamics of polar ice. *Philosophical Transactions of the Royal Society A*, *375*, 20150347. {9, 45, 61, 78}

Weikusat, I., Kipfstuhl, S., Azuma, N., Faria, S. H., & Miyamoto, A. (2009a). Deformation microstructures in an Antarctic ice core (EDML) and in experimentally deformed artificial ice. *Low Temperature Science*, *68*, 115–123. {69}

Weikusat, I., Kipfstuhl, S., Faria, S. H., Azuma, N., & Miyamoto, A. (2009b). Subgrain boundaries and related microstructural features in EDML (Antarctica) deep ice core. *Journal of Glaciology*, *55*(191), 461–472. {63, 65, 69, 70, 80}

Weikusat, I., Miyamoto, A., Faria, S. H., Kipfstuhl, S., Azuma, N., & Hondoh, T. (2011b). Subgrain boundaries in Antarctic ice quantified by X-ray Laue diffraction. *Journal of Glaciology*, *57*(201), 85–94. {59, 62, 63, 65, 69, 70}

Wesche, C., Eisen, O., Oerter, H., Schulte, D., & Steinhage, D. (2007). Surface topography and ice flow in the vicinity of the EDML deep-drilling site. *Journal of Glaciology*, *53*(182), 442–448. {39}

West, R. D., Winebrenner, D. P., Tsang, L., & Rott, H. (1996). Microwave emission from density-stratified Antarctic firn at 6 cm wavelength. *Journal of Glaciology*, *42*, 63–76. {19}

Wilen, L. A., DiPrinzio, C. L., Alley, R. B., & Azuma, N. (2003). Development, principles, and applications of automated ice fabric analyzers. *Microscopy Research and Technique*, *62*, 2–18. {59}

Wilhelms, F., Miller, H., Gerasimoff, M. D., Drücker, C., Frenzel, A., Fritzsche, D., et al. (2014). The EPICA Dronning Maud Land deep drilling operation. *Journal of Glaciology*, *55*(68), 355–366. {33}

Wilson, C. J. L., & Peternell, M. (2011). Evaluating ice fabrics using fabric analyser techniques in Sørsdal Glacier, East Antarctica. *Journal of Glaciology*, *57*(205), 881–894. {61}

Wilson, C. J. L., Russell-Head, D. S., & Sim, H. M. (2003). The application of an automated fabric analyzer system to the textural evolution of folded ice layers in shear zones. *Annals of Glaciology*, *37*, 7–17. {59}

Wittgenstein, L. (1983). *Remarks on the Foundations of Mathematics*. Cambridge, MA: MIT Press. Edited by G. H. Wright R. Rhees and G. E. M. Anscombe. Translated to the English by G. E. M. Anscombe. {83}

Zachos, J., Pagani, M., Sloan, L., Thomas, E., & Billups, K. (2001). Trends, rythms and aberrations in global climate 65Ma to present. *Science*, *292*(5517), 686–693. {16}

Subject Index

S.H. Faria et al., *The EPICA-DML Deep Ice Core*, Frontiers in Earth Sciences,
https://doi.org/10.1007/978-3-662-55308-4

Author Index

S.H. Faria et al., *The EPICA-DML Deep Ice Core*, Frontiers in Earth Sciences,
https://doi.org/10.1007/978-3-662-55308-4